Open Channel Hydraulics

To My Family

Open Channel Hydraulics

A. Osman Akan

AMSTERDAM • BOSTON • HEIDELBERG • LONDON • NEW YORK • OXFORD
PARIS • SAN DIEGO • SAN FRANCISCO • SINGAPORE • SYDNEY • TOKYO
Butterworth-Heinemann is an imprint of Elsevier

Butterworth-Heinemann is an imprint of Elsevier
The Boulevard, Langford Lane, Kidlington, Oxford, OX5 1GB
30 Corporate Drive, Suite 400, Burlington, MA 01803, USA

First edition 2006
Reprinted 2008 (twice)

British Library Cataloguing in Publication Data
A catalogue record for this book is available from the British Library

Library of Congress Cataloging-in-Publication Data
A catalog record for this book is available from the Library of Congress

ISBN: 978-0-7506-6857-6

Transferred to Digital Printing in 2010

Contents

Preface

This book was conceived as a textbook for undergraduate seniors and first-year graduate students in civil and environmental engineering. However, I am confident the book will also appeal to practising engineers. As a registered professional engineer, and having taught a number of graduate courses over the years attended by full-time engineers, I am familiar with what is needed in the engineering practice.

The students are expected to have had a fluid mechanics course before studying this book. Chapter 1 presents a review of fluid mechanics as applied to open-channel flow. The conservation laws are revisited, and the equations of continuity, momentum, and energy are derived. In Chapter 2, the applications of the energy and momentum principles are discussed along with the problem of choking in steady flow. It is also demonstrated that the hydraulic behavior of open-channel flow can be very different under the subcritical and supercritical conditions. Also, the phenomenon of hydraulic jump is introduced. Chapter 3 is devoted to normal flow. A brief description of flow resistance formulas is first provided in relation to the boundary layer theory, and then the normal flow calculations for uniform, grass-lined, riprap, composite, and compound channels are presented. Chapter 4 deals with water surface profile calculations for gradually-varied flow. I realize that this can be a difficult subject at first, since the boundary conditions needed to calculate a water surface profile depend on the type of the profile itself. Therefore, in Chapter 4, I have attempted to emphasize how to identify the flow controls, predict the profile, and formulate a solution accordingly. Once the solution is correctly formulated, the numerical calculations are easily performed. Chapter 5 is devoted to the hydraulic design of different types of open channels. Several charts are provided to facilitate the lengthy trial-and-error procedures we often need. Chapter 6 discusses various flow-measurement structures, culverts, spillways, stilling basins, and channel transitions. Chapter 7 is devoted to bridge hydraulics. First the flow calculations are discussed in the vicinity of bridge sections, then the contraction and local scour phenomena are described, and finally empirical equations are given to estimate the total bridge scour. The subject of unsteady open-channel flow, by itself, could be an advanced-level graduate course. Therefore, no attempt is made in this book to cover this subject thoroughly. However, while Chapter 8 is only an introduction to unsteady flow, it includes enough information to help a student to develop an implicit finite difference model. Simpler channel routing schemes are also discussed.

I mean to give the students a solid background on the fundamental principles and laws of open-channel flow in this book. However, the book also includes numerous detailed, worked-out examples. Where applicable, these examples are enriched with underlying arguments derived from the basic laws and principles discussed in earlier sections.

I believe that the first five chapters provide adequate material for an undergraduate open-channel hydraulics course for civil and environmental engineering students. Selected sections from Chapter 6 can also be included instead of Chapter 5. It is suggested that all eight chapters be covered if the book is used for a graduate course. However, in that event, less time should be spent on the first three chapters.

Most of the equations adopted in the book are dimensionally homogeneous, and can be used in conjunction with any consistent unit system. The unit-specific equations are clearly identified.

Various design procedures are included in the book. These procedures heavily rely upon the available experimental and field data, such as the allowable shear stress for earthen channels or various coefficients for bridge scour equations. The reader should understand that all this empirical information is subject to change as more effort is devoted to open-channel studies. Also, for real-life design problems, the reader is urged to review the references cited since it is impossible to include all the details, assumptions, and limitations of the procedures that can be found only in the design manuals. Moreover, obviously, local manuals and ordinances should be followed for designing hydraulic structures where available.

Acknowledgments

I am thankful to Professor Cahit Çıray, who introduced me to the fascinating subject of open-channel hydraulics when I was an undergraduate student at Middle East Technical University. I attended the University of Illinois for my graduate studies, and received my MS and PhD degrees under the supervision of Ben C. Yen, from whom I learned so much. Dr Yen, a gentleman and scholar, remained my friend, teacher, and mentor until he passed away in 2001. He always has a warm place in my heart. I only hope that he would be proud if he saw this book published. I am indebted to Ven Te Chow and F. M. Henderson for their earlier books on open-channel hydraulics, which I studied as a student. I still use these books frequently for reference. I have learned from the work of many other authors and colleagues that I cannot enumerate here, and I am grateful to all. I would like to thank John Paine for reviewing parts of chapter 5 and for his suggestions. I would like to thank my students for pointing out some errors when the draft manuscript was used as a course-pack. I also would like to thank Old Dominion University for the institutional support I received during the preparation of this book. Old Dominion University is a wonderful institution for students to learn and for faculty to teach and conduct research.

I am most indebted to my wife, Güzin, and my son, Doruk, for all the happiness, love, inspiration, and support they have given me throughout this project and always.

Acknowledgements

1 Fundamentals of open-channel flow

Open channels are natural or manmade conveyance structures that normally have an open top, and they include rivers, streams and estuaries. An important characteristic of open-channel flow is that it has a free surface at atmospheric pressure. Open-channel flow can occur also in conduits with a closed top, such as pipes and culverts, provided that the conduit is flowing partially full. For example, the flow in most sanitary and storm sewers has a free surface, and is therefore classified as open-channel flow.

1.1 Geometric Elements of Open Channels

A channel section is defined as the cross-section taken perpendicular to the main flow direction. Referring to Figure 1.1, the geometric elements of an open channel are defined as follows:

Flow depth, y	Vertical distance from the channel bottom to the free surface.
Depth of flow section, d	Flow depth measured perpendicular to the channel bottom. The relationship between d and y is $d = y \cos\theta$. For most manmade and natural channels $\cos\theta \approx 1.0$, and therefore $y \approx d$. The two terms are used interchangeably.
Top width, T	Width of the channel section at free surface.
Wetted perimeter, P	Length of the interface between the water and the channel boundary.
Flow area, A	Cross-sectional area of the flow.
Hydraulic depth, D	Flow area divided by top width, $D = A/T$.
Hydraulic radius, R	Flow area divided by wetted perimeter, $R = A/P$.
Bottom slope, S_0	Longitudinal slope of the channel bottom, $S_0 = \tan\theta \approx \sin\theta$.

Table 1.1 presents the relationship between various section elements. A similar, more detailed table was previously presented by Chow (1959).

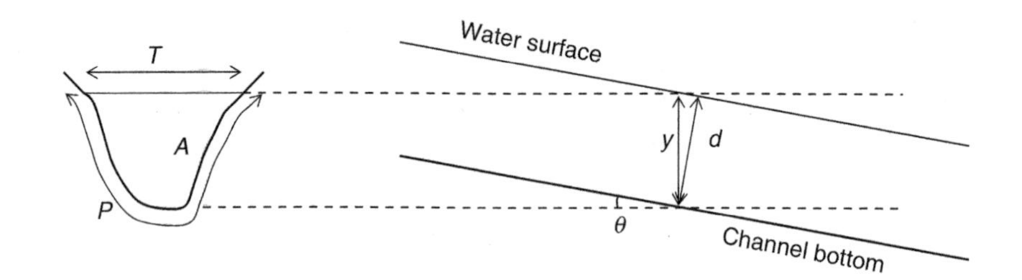

FIGURE 1.1
Definition sketch for section elements

1.2 VELOCITY AND DISCHARGE

At any point in an open channel, the flow may have velocity components in all three directions. For the most part, however, open-channel flow is assumed to be one-dimensional, and the flow equations are written in the main flow direction. Therefore, by *velocity* we usually refer to the velocity component in the main flow direction. The velocity varies in a channel section due to the friction forces on the boundaries and the presence of the free-surface. We use the term *point velocity* to refer to the velocity at different points in a channel section. Figure 1.2 shows a typical distribution of point velocity, v, in a trapezoidal channel.

The volume of water passing through a channel section per unit time is called the *flow rate* or *discharge*. Referring to Figure 1.3, the incremental discharge, dQ, through an incremental area, dA, is

$$dQ = vdA \tag{1.1}$$

where v = point velocity.

Then by definition,

$$Q = \int_A dQ = \int_A vdA \tag{1.2}$$

where Q = discharge.

In most open-channel flow applications we use the *cross-sectional average velocity*, V, defined as

$$V = \frac{Q}{A} = \frac{1}{A}\int_A vdA \tag{1.3}$$

1.3 HYDROSTATIC PRESSURE

Pressure represents the force the water molecules push against other molecules or any surface submerged in water. The molecules making up the water are in

TABLE 1.1 Geometric elements of channel sections

Section type	Area A	Wetted perimeter P	Hydraulic radius R	Top width T	Hydraulic depth D
Rectangular	by	$b+2y$	$\dfrac{by}{b+2y}$	b	y
Trapezoidal	$(b+my)y$	$b+2y\sqrt{1+m^2}$	$\dfrac{(b+my)y}{b+2y\sqrt{1+m^2}}$	$b+2my$	$\dfrac{(b+my)y}{b+2my}$
Triangular	my^2	$2y\sqrt{1+m^2}$	$\dfrac{my}{2\sqrt{1+m^2}}$	$2my$	$\dfrac{y}{2}$
Circular	$\frac{1}{8}(2\theta - \sin 2\theta)d_0^2$ $\theta = \pi - \text{arcCos}$ $\left[\left(y - \frac{d_0}{2}\right)/(d_0/2)\right]$	θd_0	$\dfrac{1}{4}\left(1 - \dfrac{\sin 2\theta}{2\theta}\right)d_0$	$(\sin\theta)d_0$ or $2\sqrt{y(d_0-y)}$	$\dfrac{1}{8}\left(\dfrac{2\theta - \sin 2\theta}{\sin\theta}\right)d_0$

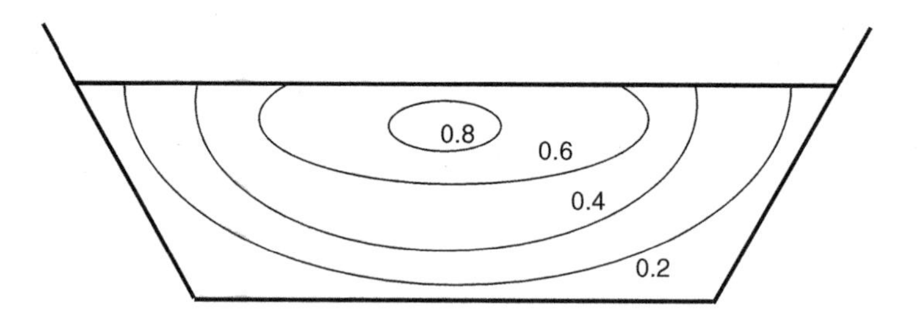

FIGURE 1.2 Velocity distribution in a trapezoidal channel section

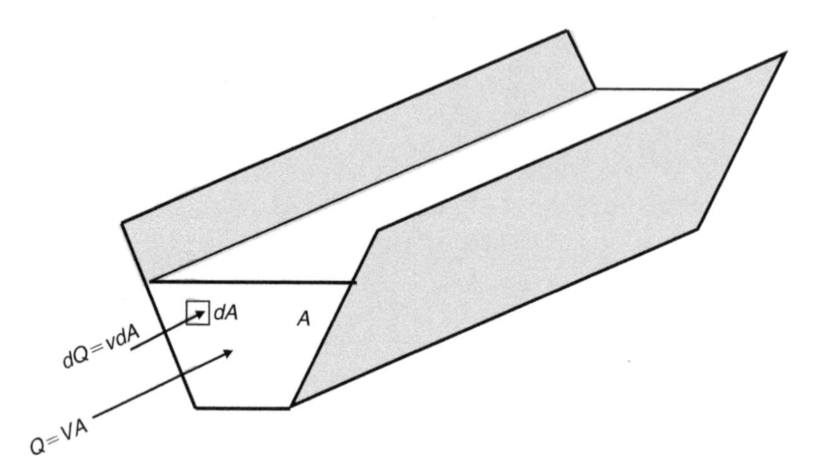

FIGURE 1.3 Definition of discharge

constant motion even when a body of water is at rest in the macroscopic sense. The pressure results from the collisions of these molecules with one another and with any submerged surface like the walls of a container holding a water body. Because, the molecular motion is random, the resulting pressure is the same in every direction at any point in water.

The water surface in an open channel is exposed to the atmosphere. Millions of collisions take place every second between the molecules making up the atmosphere and the water surface. As a result, the atmosphere exerts some pressure on the water surface. This pressure is called *atmospheric pressure*, and it is denoted by p_{atm}.

The pressure occurring in a body of water at rest is called *hydrostatic pressure.* In Figure 1.4, consider a column of water extending from the water surface to point B at depth of Y_B. Let the horizontal cross-sectional area of the column be A_0. This column of water is pushed downward at the surface by a force equal to $p_{atm}A_0$ due to the atmospheric pressure and upward at the bottom by a force $(p_{abs})_B A_0$ due to the absolute water pressure, $(p_{abs})_B$ at point B. In addition, the weight of the water column, a downward force, is $W = \gamma Y_B A_0$ where $\gamma =$ specific weight of water. Because the water column is in equilibrium,

$$(p_{abs})_B A_0 = p_{atm} A_0 + \gamma Y_B A_0$$

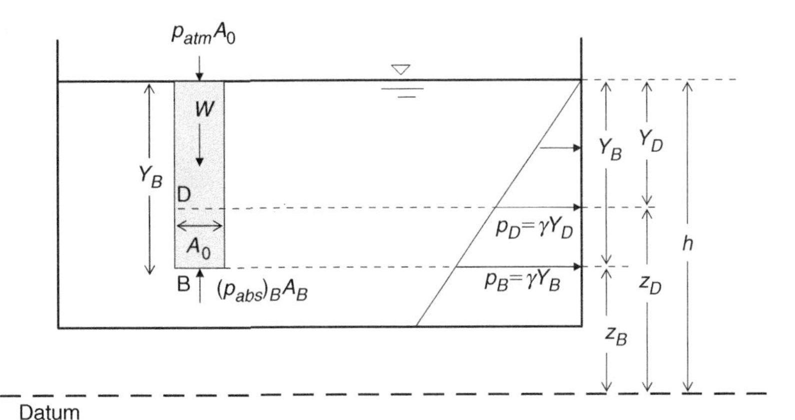

FIGURE 1.4
Hydrostatic pressure distribution

or

$$(p_{\text{abs}})_B - p_{\text{atm}} = \gamma Y_B$$

Pressure is usually measured using atmospheric pressure as base. Therefore, the difference between the absolute pressure and the atmospheric pressure is usually referred to as *gage pressure*. In this text we will use the term *pressure* interchangeably with gage pressure. Denoting the gage pressure or pressure by p,

$$p_B = (p_{\text{abs}})_B - p_{\text{atm}} = \gamma Y_B \tag{1.4}$$

In other words, the hydrostatic pressure at any point in the water is equal to the product of the specific weight of water and the vertical distance between the point and the water surface. Therefore, the hydrostatic pressure distribution over the depth of water is triangular as shown in Figure 1.4.

Let the elevation of point B be z_B above a horizontal datum as shown in Figure 1.4. Let us now consider another point D, which is a distance z_D above the datum and Y_D below the water surface. The pressure at this point is $p_D = \gamma Y_D$. Thus, $Y_D = p_D/\gamma$. An inspection of Figure 1.4 reveals that

$$z_B + \frac{p_B}{\gamma} = z_D + \frac{p_D}{\gamma} = h \tag{1.5}$$

where h is the elevation of the water surface above the datum. As we will see later, $(z + p/\gamma)$ is referred to as *piezometric head*. Equation 1.5 indicates that the piezometric head is the same at any point in a vertical section if the pressure distribution is hydrostatic.

The hydrostatic pressure distribution is valid even if there is flow as long as the flow lines are horizontal. Without any vertical acceleration, the sum of the vertical forces acting on a water column should be zero. Then, the derivation given above for the hydrostatic case is valid for horizontal flow as well. If the flow lines are inclined but parallel to the channel bottom, we can show that

$$p_B = \gamma Y_B \cos^2 \theta \tag{1.6}$$

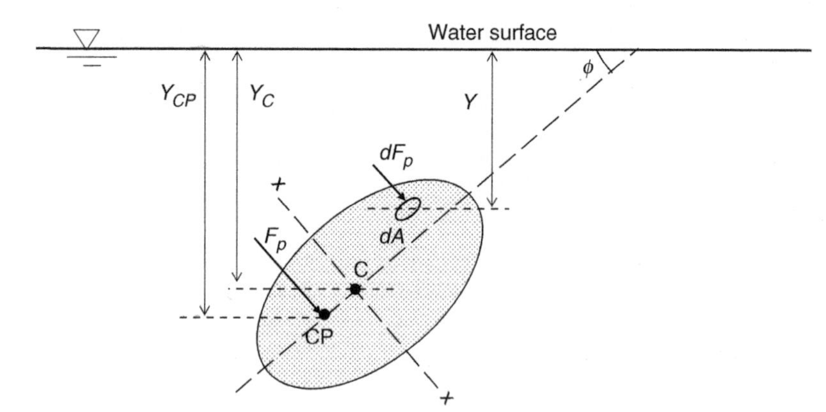

FIGURE 1.5 Hydrostatic pressure force

where θ = angle between the horizontal and the bottom of the channel. Therefore, strictly speaking, the pressure distribution is not hydrostatic when the flow lines are inclined. However, for most manmade and natural open channels θ is small and $\cos\theta \approx 1$. We can assume that the pressure distribution is hydrostatic as long as θ is small and the flow lines are parallel.

The hydrostatic forces resulting from the hydrostatic pressure act in a direction normal to a submerged surface. Consider a submerged, inclined surface as shown in Figure 1.5. Let C denote the centroid of the surface. The pressure force acting on the infinitesimal area dA is $dF_p = pdA$ or $dF_p = \gamma YdA$. To find the total hydrostatic force, we integrate dF_p over the total area A of the surface. Thus

$$F_p = \int_A \gamma YdA \tag{1.7}$$

Noting that γ is constant, and recalling the definition of the centroid (point C in Figure 1.5) as

$$Y_C = \frac{\int_A YdA}{A} \tag{1.8}$$

we obtain

$$F_p = \gamma Y_C A \tag{1.9}$$

In other words, the hydrostatic pressure force acting on a submerged surface, vertical, horizontal, or inclined, is equal to the product of the specific weight of water, area of the surface, and the vertical distance from the free surface to the centroid of the submerged surface. Again, the direction of the hydrostatic force is normal to the submerged surface. The point of application of the resultant hydrostatic force is called the *center of pressure* (point CP in Figure 1.5). The location of the center of pressure can be found by equating the moment of the resultant F_p around the centroidal horizontal axis (axis xx in Figure 1.5) to that of dF_p integrated over the area. This will result in the relationship

$$Y_{CP} = Y_C + \frac{I_x(\sin\phi)^2}{AY_C} \tag{1.10}$$

where ϕ = angle between the water surface and the plane of the submerged surface, and I_x = moment of inertia of the surface with respect to the centroidal horizontal axis.

1.4 Mass, Momentum and Energy Transfer in Open-Channel Flow

1.4.1 MASS TRANSFER

The *mass* of an object is the quantity of matter contained in the object. The *volume* of an object is the space it occupies. The *density*, ρ, is the mass per unit volume. Water is generally assumed to be incompressible in open-channel hydraulics, and the density is constant for incompressible fluids. The *mass transfer rate* or *mass flux* in open-channel flow is the rate with which the mass is transferred through a channel section. Recalling that Q = *discharge* is the volume transfer rate, we can write

$$\text{Rate of } \textit{mass transfer} = \rho Q \tag{1.11}$$

1.4.2 MOMENTUM TRANSFER

Momentum or *linear momentum* is a property only moving objects have. An object of mass M moving with velocity V_M has a momentum equal to MV_M. In the absence of any external forces acting on the object in (or opposite to) the direction of the motion, the object will continue to move with the same velocity. From everyday life, we know that it is more difficult to stop objects that are moving faster or that are heavier (that is objects with higher momentum). Thus we can loosely define the *momentum* as a numerical measure of the tendency of a moving object to keep moving in the same manner.

The rate of mass transfer at any point in a channel section through an incremental area dA (as in Figure 1.3) is $\rho dQ = \rho v dA$, and therefore the momentum transfer rate is $\rho v^2 dA$. Integrating this over the area A, we obtain the momentum transfer rate through the section as

$$\text{Rate of momentum transfer} = \rho \int_A v^2 dA \tag{1.12}$$

We often express the momentum transfer rate in terms of the average cross-sectional velocity, V, as

$$\text{Rate of momentum transfer} = \beta \rho V^2 A = \beta \rho Q V \tag{1.13}$$

where β = *momentum coefficient* (or *momentum correction coefficient*) introduced to account for the non-uniform velocity distribution within the channel section.

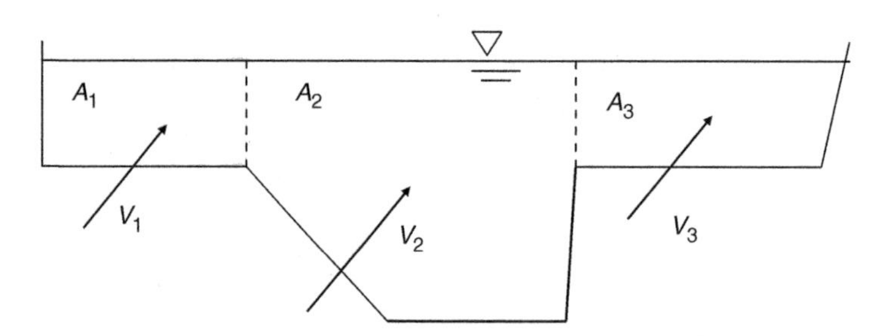

FIGURE 1.6
Compound channel

Then, from Equations 1.12 and 1.13, we obtain

$$\beta = \frac{\int_A v^2 dA}{V^2 A} \tag{1.14}$$

For regular channels β is often set equal to 1.0 for simplicity. For compound channels, as in Figure 1.6, it can be substantially higher. For a compound channel as in Figure 1.6, we can evaluate β by using

$$\beta = \frac{V_1^2 A_1 + V_2^2 A_2 + V_3^2 A_3}{V^2 A} \tag{1.15}$$

in which $A = A_1 + A_2 + A_3$ and V is obtained as

$$V = \frac{V_1 A_1 + V_2 A_2 + V_3 A_3}{A_1 + A_2 + A_3}.2 \tag{1.16}$$

Note that if $V_1 = V_2 = V_3$, Equation 1.15 yields $\beta = 1.0$.

1.4.3 ENERGY TRANSFER

Energy is generally defined as a measure of an object's capability to perform work. It can be in different forms. For open-channel flow problems, potential energy, kinetic energy, and internal energy are of interest. We will define the *total energy* as the sum of these three forms.

In the earth's gravitational field, every object has *potential energy*, or capability to perform work due to its position (elevation). The potential energy cannot be defined as an absolute quantity; it is defined as a relative quantity. For example, with respect to a horizontal datum (a reference elevation), the potential energy of an object of mass M is Mgz_C where g = gravitational acceleration and z_C = elevation of the center of mass of the object above the datum. In open channel flow, Q = rate of volume transfer, and ρQ = rate of mass transfer. Therefore, we can define the rate of potential energy transfer through a channel section as

$$\text{Rate of potential energy transfer} = \rho Q g z_C \tag{1.17}$$

where z_C = the elevation of the center of gravity or center of mass (the same as the centroid, since ρ is constant) of the channel section above the datum.

A moving object has the capability of performing work because of its motion. *Kinetic energy* is a measure of this capability. The kinetic energy of a mass M traveling with velocity V_M is defined as $M(V_M)^2/2$. In open-channel flow, we are concerned with the rate of kinetic energy transfer or the kinetic energy transfer through a channel section per unit time. The mass rate at any point in a channel section through an incremental area dA (as in Figure 1.3) is $\rho dQ = \rho v dA$. Therefore, the kinetic energy transfer per unit time through the incremental area is $\rho v^3 dA/2$. Integrating over the section area, and assuming ρ is constant for an incompressible fluid like water, we obtain

$$\text{Rate of kinetic energy transfer} = \frac{\rho}{2}\int_A v^3 dA \tag{1.18}$$

Note that in the above equation v stands for the point velocity, which varies over the channel section. In practice, we work with the average cross-sectional velocity, V. We define the rate of kinetic energy transfer in terms of the average cross-sectional velocity as

$$\text{Rate of kinetic energy transfer} = \alpha\frac{\rho}{2}V^3 A = \alpha\frac{\rho}{2}QV^2 \tag{1.19}$$

where α = *energy coefficient* (or *kinetic energy correction coefficient*) to account for the non-uniform point velocity distribution within a section. From Equations 1.18 and 1.19 we obtain

$$\alpha = \frac{\int_A v^3 dA}{V^3 A} \tag{1.20}$$

For regular channels, α is usually set equal to 1.0. However, in compound channels, like an overflooded river with a main channel and two overbank channels, α can be substantially higher. For the casc for Figure 1.6, Equation 1.20 can be approximated using

$$\alpha = \frac{V_1^3 A_1 + V_2^3 A_2 + V_3^3 A_3}{V^3 A} \tag{1.21}$$

where $A = A_1 + A_2 + A_3$ and V is as defined by Equation 1.16. As expected, Equation 1.21 yields $\alpha = 1.0$ if $V_1 = V_2 = V_3$.

Internal energy results from the random motion of the molecules making up an object and the mutual attraction between these molecules. Denoting the internal energy per unit mass of water by e, the rate of internal energy transfer through an incremental area dA (as in Figure 1.3) is $\rho e v dA$. Integrating this over the area, and assuming e is distributed uniformly,

$$\text{Rate of internal energy transfer} = \rho e V A = \rho e Q \tag{1.22}$$

1.5 Open-Channel Flow Classification

Open-channel flow is classified in various ways. If time is used as the criterion, open-channel flow is classified into steady and unsteady flows. If, at a given flow section, the flow characteristics remain constant with respects to time, the flow is said to be *steady*. If flow characteristics change with time, the flow is said to be *unsteady*. If space is used as a criterion, flow is said to be *uniform* if flow characteristics remain constant along the channel. Otherwise the flow is said to be *non-uniform*. A non-uniform flow can be classified further into *gradually-varied* and *rapidly-varied* flows, depending on whether the variations along the channel are gradual or rapid. For example, the flow is gradually varied between Sections 1 and 2 and 2 and 3 in Figure 1.7. It is rapidly varied between 3 and 4 and uniform between 4 and 5. Usually, the pressure distribution can be assumed to be hydrostatic for uniform and gradually-varied flows.

Various types of forces acting on open-channel flow affect the hydraulic behavior of the flow. The Reynolds Number, R_e, defined as

$$R_e = \frac{4VR}{\nu} \tag{1.23}$$

where ν = kinematic viscosity of water, represents the ratio of inertial to viscous forces acting on the flow. At low Reynolds numbers, say $R_e < 500$, the flow region appears to consist of an orderly series of fluid laminae or layers conforming generally to the boundary configuration. This type of flow is called laminar flow. If we inject dye into a uniform laminar flow, the dye will flow along a straight line. Any disturbance introduced to laminar flow, due to irregular boundaries for instance, is eventually dampened by viscous forces. For $R_e > 12\,500$, the viscous forces are not sufficient to dampen the disturbances introduced to the flow. Minor disturbances are always present in moving water, and at high Reynolds numbers such disturbances will grow and spread throughout the entire zone of motion. Such flow is called turbulent, and water particles in turbulent flow follow irregular paths that are not continuous. A transitional state exists between the laminar and turbulent states. We should point out that the limits for the different states are by no means precise. Under laboratory conditions, for instance, laminar flow can be maintained for Reynolds numbers much higher than 500.

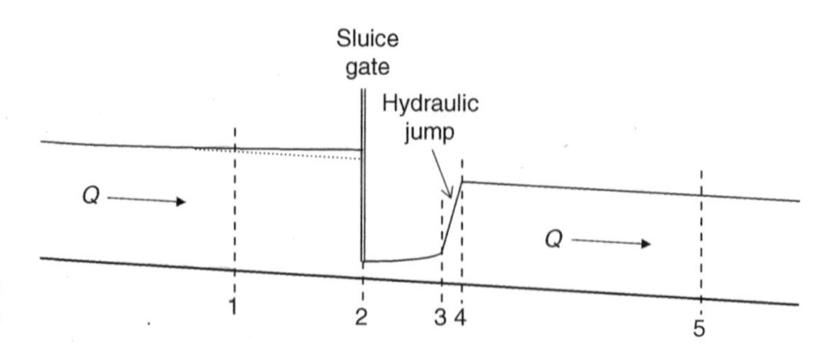

Figure 1.7 Various flow types

However, under most natural and practical open-channel flow conditions, the flow is turbulent.

The ratio of the inertial to gravitational forces acting on the flow is represented by the dimensionless *Froude number*, F_r, defined as

$$F_r = \frac{V}{\sqrt{gD}} \tag{1.24}$$

where g = gravitational acceleration. The flow is said to be at the *critical* state when $F_r = 1.0$. The flow is *subcritical* when $F_r < 1.0$, and it is *supercritical* when $F_r > 1.0$. The hydraulic behavior of open-channel flow varies significantly depending on whether the flow is critical, subcritical, or supercritical.

1.6 Conservation Laws

The laws of conservation of mass, momentum, and energy are the basic laws of physics, and they apply to open-channel flow. Rigorous treatment of the conservation laws for open-channel flow can be found in the literature (e.g. Yen, 1973). A simplified approach is presented herein.

1.6.1 CONSERVATION OF MASS

Consider a volume element of an open channel between an upstream section U and a downstream section D, as shown in Figure 1.8. The length of the element along the flow direction is Δx, and the average cross-sectional area is A. The mass of water present in the volume element is then $\rho A \Delta x$. Suppose water enters the volume element at section U at a mass transfer rate of ρQ_U (see Equation 1.11) and leaves the element at section D at a rate ρQ_D. Over a finite time increment, Δt, we can write that

$$\text{Rate of change of mass of water in the element} = \frac{\Delta(\rho A \Delta x)}{\Delta t}$$

$$\text{Net rate of mass transfer into element} = \rho Q_U - \rho Q_D$$

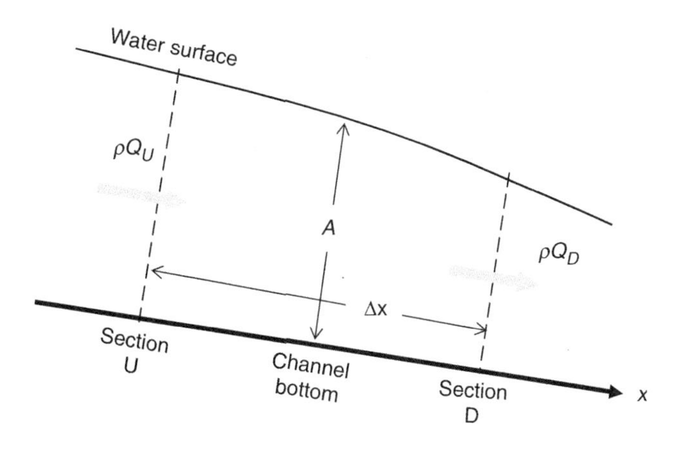

Figure 1.8 Definition sketch for conservation of mass principle

The principle of conservation of mass requires that

(Rate of change of mass of water in the element)
= (Net rate of mass transfer into elelment)

therefore

$$\frac{\Delta(\rho A \Delta x)}{\Delta t} = \rho Q_U - \rho Q_D \qquad (1.25)$$

Water is considered to be an incompressible fluid, and therefore ρ is constant. Equation 1.25 can then be written as

$$\frac{\Delta A}{\Delta t} + \frac{Q_D - Q_U}{\Delta x} = 0 \qquad (1.26)$$

For gradually-varied flow A and Q are continuous in space and time, and as Δx and Δt approach zero Equation 1.26 becomes

$$\frac{\partial A}{\partial t} + \frac{\partial Q}{\partial x} = 0 \qquad (1.27)$$

where t = time, and x = displacement in the main flow direction. We usually refer to Equation 1.27 as the *continuity equation*.

1.6.2 CONSERVATION OF MOMENTUM

Momentum is a vector quantity, and separate equations are needed if there are flow components in more than one direction. However, open-channel flow is usually treated as being one-dimensional, and the momentum equation is written in the main flow direction. Consider a volume element of an open channel between an upstream section U and a downstream section D as shown in Figure 1.9. Let the element have an average cross-sectional area of A, flow

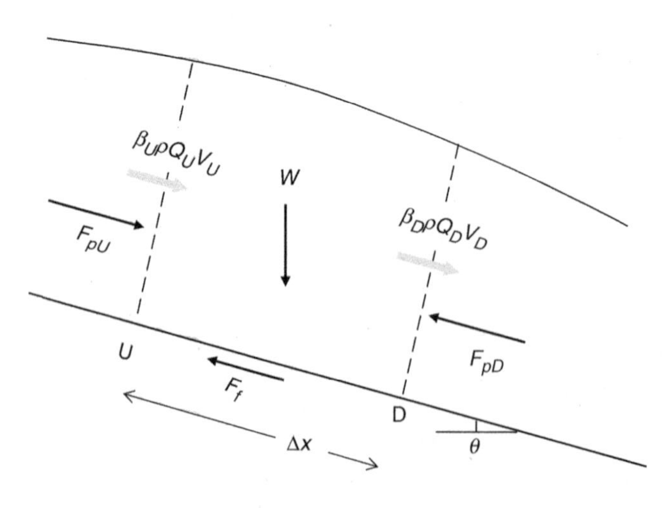

FIGURE 1.9 Definition sketch for conservation of momentum principle

velocity V, and length Δx. The momentum within this element is $\rho A \Delta x V$. The momentum is transferred into the element at section U at a rate $\beta_U \rho Q_U V_U$ (see Equation 1.13) and out of the element at section D at rate $\beta_D \rho Q_D V_D$. The external forces acting on this element in same direction as the flow are the pressure force at section U, $F_{pU} = \gamma Y_{CU} A_U$ (see Equation 1.9) and the weight component $W \sin\theta = \gamma A \Delta x \sin\theta$. The external forces acting opposite to the flow direction are the pressure force at section D, $F_{pD} = \gamma Y_{CD} A_D$, friction force on the channel bed, F_f, and any other external force, F_e, opposite to the flow direction (like a force exerted by the channel walls at a contracted section).

Therefore, we can write that

Time rate of change of the momentum accumulated within the element

$$= \Delta(\rho A \Delta x V)/\Delta t = \rho \Delta x (\Delta Q/\Delta t)$$

Net rate of momentum transfer into the element

$$= (\beta_U \rho Q_U V_U - \beta_D \rho Q_D V_D)$$

Sum of the external forces in the flow direction

$$= \gamma Y_{CU} A_U + \gamma A \Delta x \sin\theta - \gamma Y_{CD} A_D - F_f - F_e$$

The law of conservation of momentum requires that

(Time rate of change of the momentum accumulated within the element)
= (Net rate of momentum transfer into the element)
+ (Sum of the external forces in the flow direction)

Thus

$$\rho \Delta x (\Delta Q/\Delta t) = (\beta_U \rho Q_U V_U - \beta_D \rho Q_D V_D) + (\gamma Y_{CU} A_U - \gamma Y_{CD} A_D) + \gamma A \Delta x \sin\theta - F_f - F_e \tag{1.28}$$

Dividing both sides of the equation by $\rho \Delta x$, assuming $F_e = 0$, noting S_0 = longitudinal channel bottom slope = $\sin\theta$, and introducing S_f = friction slope = boundary friction force per unit weight of water as

$$S_f = \frac{F_f}{\gamma A \Delta x} \tag{1.29}$$

we obtain

$$\frac{\Delta Q}{\Delta t} + \frac{(\beta_D Q_D V_D - \beta_U Q_U V_U)}{\Delta x} + \frac{g(Y_{CD} A_D - Y_{CU} A_U)}{\Delta x} + gAS_f - gAS_0 = 0 \tag{1.30}$$

For gradually-varied flow, all the flow variables are continuous in time and space. Therefore, as Δx and Δt approach zero, Equation 1.30 becomes

$$\frac{\partial Q}{\partial t} + \frac{\partial}{\partial x}(\beta Q V) + gA\frac{\partial y}{\partial x} + gAS_f - gAS_0 = 0 \tag{1.31}$$

Note that in arriving at Equation 1.31 from Equation 1.30 we have used

$$\frac{g(Y_{CD}A_D - Y_{CU}A_U)}{\Delta x} = g\frac{\partial(AY_C)}{\partial x} = gA\frac{\partial y}{\partial x} \tag{1.32}$$

as Δx approaches zero. This equality is not obvious. However, it can be proven mathematically using the Leibnitz rule if the changes in the channel width are negligible (see Problem P.1.15). A more rigorous analysis presented by Chow *et al.* (1988) demonstrates that Equation 1.32 is valid even if the changes in channel width are not negligible.

Noting that $Q = AV$, we can expand Equation 1.31 as

$$V\frac{\partial A}{\partial t} + A\frac{\partial V}{\partial t} + \beta Q\frac{\partial V}{\partial x} + \beta V\frac{\partial Q}{\partial x} + QV\frac{\partial \beta}{\partial x} + gA\frac{\partial y}{\partial x} + gAS_f - gAS_0 = 0 \tag{1.33}$$

or

$$V\left(\frac{\partial A}{\partial t} + \beta\frac{\partial Q}{\partial x}\right) + A\frac{\partial V}{\partial t} + \beta Q\frac{\partial V}{\partial x} + QV\frac{\partial \beta}{\partial x} + gA\frac{\partial y}{\partial x} + gAS_f - gAS_0 = 0 \tag{1.34}$$

For $\beta \approx 1$ and $\partial\beta/\partial x \approx 0$, substituting Equation 1.27 into 1.34, and dividing both sides by gA, we obtain

$$\frac{1}{g}\frac{\partial V}{\partial t} + \frac{V}{g}\frac{\partial V}{\partial x} + \frac{\partial y}{\partial x} + S_f - S_0 = 0 \tag{1.35}$$

1.6.3 CONSERVATION OF ENERGY

Consider a volume element of an open channel between an upstream section U and a downstream section D as shown in Figure 1.10. Let the element have an average cross-sectional area of A, flow velocity V, and length Δx. Suppose the elevation of the center of gravity of the element above a reference datum is z_C.

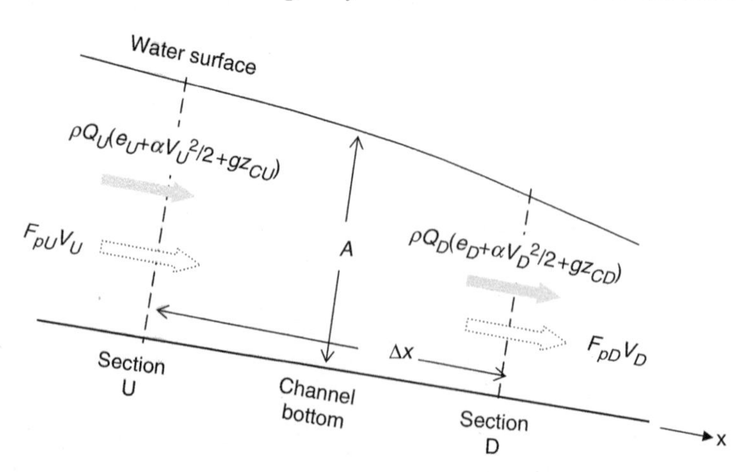

FIGURE 1.10 Definition sketch for conservation of energy principle

The total energy stored within this element is $[gz_C + (V^2/2) + e]\rho A\,\Delta x$. The energy is transferred into the element at section U at a rate $\rho Q_U[gz_{CU} + \alpha_U(V_U{}^2/2) + e_U]$ (see Equations 1.17, 1.19, and 1.22) and out of the element at section D at rate $\rho Q_D[gz_{CD} + \alpha_D(V_D{}^2/2) + e_D]$. The rate of work (or power) the surroundings perform on the volume element due to the hydrostatic pressure force at section U is $F_{pU}V_U$. The rate of work (or power) the volume element performs on the surroundings due the hydrostatic pressure force, which is opposing the flow at section D, is $F_{pD}V_D$. Referring to Equation 1.9, and noting $\gamma = \rho g$, we have $F_{pU}V_U = \rho g Y_{CU} A_U V_U$ and $F_{pD}V_D = \rho g Y_{CD} A_D V_D$.

Therefore, over a time increment Δt, we can write that

Time rate of change of total energy stored in the volume element

$$= \Delta\{(gz_C + (V^2/2) + e)\rho A\,\Delta x\}/\Delta t$$

Net rate of energy transfer into the element

$$= \rho Q_U\{gz_{CU} + \alpha_U(V_U^2/2) + e_U\} - \rho Q_D\{gz_{CD} + \alpha_D(V_D^2/2) + e_D\}$$

Net rate of energy added due to the work performed by the surroundings on the element $= \rho g Y_{CU} A_U V_U - \rho g Y_{CD} A_D V_D$

In the absence of energy added to the system due to external sources, the conservation of energy principle requires that

Time rate of change of total energy stored in the volume element
= Net rate of energy transfer into the element
+ Net rate of energy added due to the work performed by the surroundings on the element

Therefore

$$\frac{\Delta}{\Delta t}\left[\left(gz_C + \frac{V^2}{2} + e\right)\rho A\,\Delta x\right] = \rho\left[Q_U\left(gz_{CU} + \alpha_U\frac{V_U^2}{2} + e_U\right) - Q_D\left(gz_{CD} + \alpha_D\frac{V_D^2}{2} + e_D\right)\right] + \rho g(Q_U Y_{CU} - Q_D Y_{CD}) \tag{1.36}$$

Dividing both sides by Δx and rearranging gives

$$\frac{\Delta}{\Delta t}\left[\left(gz_C + \frac{V^2}{2} + e\right)\rho A\right] = \frac{\rho\left[Q_U(gz_{CU} + gY_{CU} + \alpha_U(V_U^2/2) + e_U) - Q_D(gz_{CD} + gY_{CD} + \alpha_D(V_D^2/2) + e_D)\right]}{\Delta x} \tag{1.37}$$

Let us define z_b = elevation of the channel bottom above the datum and recall that y = flow depth. Therefore, at any flow section $z_C + Y_C = z_b + y$. Then

$$\frac{\Delta}{\Delta t}\left[\left(e+\frac{V^2}{2}+gz_C\right)\rho A\right] = \frac{\rho[Q_U(gz_{bU}+gy_U+\alpha_U(V_U^2/2)+e_U)-Q_D(gz_{bD}+gy_D+\alpha_D(V_D^2/2)+e_D)]}{\Delta x} \tag{1.38}$$

As Δt and Δx approach zero Equation 1.38 becomes

$$\frac{\partial}{\partial t}\left[\left(e+\frac{V^2}{2}+gz_C\right)\rho A\right]+\frac{\partial}{\partial x}\left[\rho Q\left(gz_b+gy+\alpha\frac{V^2}{2}+e\right)\right]=0 \tag{1.39}$$

Now, substituting $z_C = z_b + y - Y_C$ we can write the first group of terms on the left side of Equation 1.39 as

$$\frac{\partial}{\partial t}\left[\left(e+\frac{V^2}{2}+gz_C\right)\rho A\right] = \rho\left[e\frac{\partial A}{\partial t}+A\frac{\partial e}{\partial t}\right]+\rho\left[\frac{V^2}{2}\frac{\partial A}{\partial t}+Q\frac{\partial V}{\partial t}\right]+\rho g\left[\frac{\partial(Az_b)}{\partial t}+\frac{\partial(Ay)}{\partial t}-\frac{\partial(AY_C)}{\partial t}\right] \tag{1.40}$$

By analogy to Equation 1.32,

$$-\frac{\partial(AY_C)}{\partial t} = -A\frac{\partial y}{\partial t} \tag{1.41}$$

Substituting Equation 1.41 into 1.40, noting that $\partial z_b/\partial t = 0$, and regrouping the terms:

$$\frac{\partial}{\partial t}\left[\left(e+\frac{V^2}{2}+gz_C\right)\rho A\right] = \rho\left(gz_b+gy+\frac{V^2}{2}+e\right)\frac{\partial A}{\partial t}+\rho A\frac{\partial e}{\partial t}+\rho Q\frac{\partial V}{\partial t} \tag{1.42}$$

Likewise,

$$\frac{\partial}{\partial x}\left[\rho Q\left(gz_b+gy+\alpha\frac{V^2}{2}+e\right)\right] = \rho\left(gz_b+gy+\alpha\frac{V^2}{2}+e\right)\frac{\partial Q}{\partial x}+\rho Q\frac{\partial}{\partial x}\left(gz_b+gy+\alpha\frac{V^2}{2}+e\right) \tag{1.43}$$

Substituting Equations 1.42 and 1.43 into 1.39 and assuming $\alpha = 1$,

$$\rho\left(gz_b+gy+\frac{V^2}{2}+e\right)\left(\frac{\partial A}{\partial t}+\frac{\partial Q}{\partial x}\right) + \rho A\frac{\partial e}{\partial t}+\rho Q\frac{\partial V}{\partial t}+\rho Q\frac{\partial}{\partial x}\left(gz_b+gy+\frac{V^2}{2}+e\right)=0 \tag{1.44}$$

Substituting Equation 1.27 into 1.44 and dividing by ρQg, we obtain

$$\frac{1}{g}\frac{\partial V}{\partial t}+\frac{\partial}{\partial x}\left(z_b+y+\frac{V^2}{2g}\right)+\frac{1}{g}\left(\frac{1}{V}\frac{\partial e}{\partial t}+\frac{\partial e}{\partial x}\right)=0 \tag{1.45}$$

We will now define $S_e =$ energy slope as

$$S_e = \frac{1}{g}\left(\frac{1}{V}\frac{\partial e}{\partial t} + \frac{\partial e}{\partial x}\right) = \frac{1}{g}\frac{de}{dx} \tag{1.46}$$

Substituting Equation 1.46 into 1.45 and noting that $\partial z_b/\partial x = -S_0$,

$$\frac{1}{g}\frac{\partial V}{\partial t} + \frac{V}{g}\frac{\partial V}{\partial x} + \frac{\partial y}{\partial x} + S_e - S_0 = 0 \tag{1.47}$$

If we recall that $e =$ internal energy per unit mass of water, Equation 1.46 indicates that positive values of S_e represent an increase in the internal energy per unit weight of water per unit distance. However, because the total energy is conserved, this increase in the internal energy is accompanied by a decrease in the mechanical (potential and kinetic) energy. Because the mechanical energy is usable energy, any conversion of mechanical energy to internal energy is commonly viewed as 'energy loss', and the energy slope is defined as the energy loss per unit weight of water per unit distance. The procedure we adopted in this text to derive the energy equation does not explain how the mechanical energy is converted to internal energy. Another approach, based on the integration of the Navier-Stokes equations presented by Strelkoff (1969) and Yen (1973), clearly demonstrates that the losses in the mechanical energy are due to the work done by the internal stresses to overcome the velocity gradients. Turbulent exchange of molecules between different velocity zones sets up an internal friction force between adjacent layers since slow-moving molecules entering a higher-velocity layer will drag the faster-moving molecules. The energy dissipated to overcome these internal friction forces in the form of heat will increase the internal energy while causing a reduction in the mechanical energy.

Although Equation 1.47 appears very similar to Equation 1.35, the two equations are fundamentally different. Momentum is a vector quantity and energy is a scalar quantity. The two equations look similar because they are both for one-dimensional flow. If we had flow components in, say, three directions, we would have three different momentum equations, while the energy approach would still yield a single equation. We assumed that $\beta = 1$ when we derived Equation 1.35 and $\alpha = 1$ for Equation 1.47. These two correction factors are actually different. The friction slope, S_f, appearing in Equation 1.35 corresponds to the (external) boundary friction forces, while the energy slope, S_e, in Equation 1.47 is related to the work done by the internal friction forces. Nevertheless, in most applications we do not differentiate between S_f and S_e and use the term *friction slope* for either.

1.6.4 STEADY FLOW EQUATIONS

The flow is said to be *steady* if the flow conditions do not vary in time. Therefore, the partial derivative terms with respect to time can be dropped from the continuity, momentum, and energy equations. As a result, we obtain

$$\frac{dQ}{dx} = 0 \tag{1.48}$$

$$\frac{V}{g}\frac{dV}{dx} + \frac{dy}{dx} + S_f - S_0 = 0 \tag{1.49}$$

and

$$\frac{V}{g}\frac{dV}{dx} + \frac{dy}{dx} + S_e - S_0 = 0 \tag{1.50}$$

Equation 1.48 shows that, under steady state conditions, the discharge is the same at any channel section. Also, Equations 1.49 and 1.50 can be rearranged to obtain

$$\frac{dy}{dx} = \frac{S_0 - S_f}{1 - F_r^2} \tag{1.51}$$

$$\frac{dy}{dx} = \frac{S_0 - S_e}{1 - F_r^2} \tag{1.52}$$

For the volume element shown in Figure 1.9, Equation 1.28 can be written for steady state conditions as

$$\left(\beta_D \frac{Q_D^2}{gA_D} + Y_{CD}A_D\right) = \left(\beta_U \frac{Q_U^2}{gA_U} + Y_{CU}A_U\right) - \frac{F_f}{\gamma} - \frac{F_e}{\gamma} + \Delta x S_0 \frac{A_D + A_U}{2} \tag{1.53}$$

Equation 1.53 is valid regardless of whether the flow between the sections U and D is gradually or rapidly varied, as long as the pressure distribution is hydrostatic at sections U and D.

Likewise, we can obtain the steady state energy equation by discretizing Equation 1.50, reintroducing the energy coefficient α, defining h_f = head loss = energy loss per unit weight = $(\Delta x)S_e$, and rearranging the terms

$$\left(z_{bU} + y_U + \alpha_U \frac{V_U^2}{2g}\right) = \left(z_{bD} + y_D + \alpha_D \frac{V_D^2}{2g}\right) + \Delta x S_e \tag{1.54}$$

1.6.5 STEADY SPATIALLY-VARIED FLOW EQUATIONS

Flow in an open channel is said to be *spatially varied* if there is lateral flow into (or out of) the channel, as shown schematically in Figure 1.11. For steady spatially-varied flow, the continuity equation becomes

$$\frac{dQ}{dx} = q_L \tag{1.55}$$

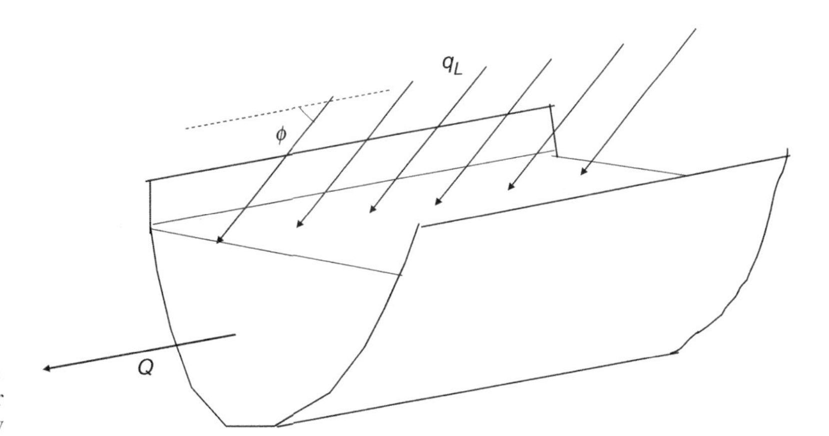

FIGURE 1.11
Definition sketch for spatially-varied flow

where q_L = lateral inflow rate per unit length of the channel. Note that the dimension of q_L is {length}^{2}/{time}.

As demonstrated by Yen and Wenzel (1970), for $\beta = 1$, the momentum equation for steady spatially-varied flow can be written as

$$\frac{dy}{dx} = \frac{S_0 - S_f - (q_L/gA)(2V - U_L \cos\phi)}{1 - F_r^2} \tag{1.56}$$

where U_L = velocity of lateral flow, and ϕ = angle between the lateral flow and channel flow directions. If lateral flow joins (or leaves) the channel in a direction perpendicular to the main flow direction, the equation becomes

$$\frac{dy}{dx} = \frac{S_0 - S_f - (2q_L V/gA)}{1 - F_r^2} \tag{1.57}$$

Yen and Wenzel (1970) also demonstrated that, for $\alpha = 1$, the energy equation can be written as

$$\frac{dy}{dx} = \frac{S_0 - S_e + (q_L/VA)\left((U_L^2/2g) - (3V^2/2g) + h_{LAT} - h\right)}{1 - F_r^2} \tag{1.58}$$

where $h = z_b + y$ = piezometric head of the main channel flow, and h_{LAT} = piezometric head of the lateral inflow. If $h = h_{LAT}$, and $V = U_L$, Equation 1.58 is simplified to obtain

$$\frac{dy}{dx} = \frac{S_0 - S_e - (q_L V/gA)}{1 - F_r^2} \tag{1.59}$$

Note that the third term in the numerator of the right side of Equation 1.59 is different from that of Equation 1.57 by a factor of 2.0. This discrepancy is due to the different assumptions involved in the two equations.

1.6.6 COMPARISON AND USE OF MOMENTUM AND ENERGY EQUATIONS

It should be clear to the reader by now that the momentum and the energy equations are obtained by using different laws of physics. Also, the friction slope, S_f, and the energy slope, S_e, appearing in these equations are fundamentally different. However, it is not practical to evaluate either S_f or S_e on the basis of their strict definitions. In practice, we employ the same empirical equations to evaluate S_f and S_e. Therefore, S_e in the energy equation is often replaced by S_f. If we also assume that $\alpha = 1$ and $\beta = 1$, then, for gradually-varied flow, the momentum and energy equation become identical (Equations 1.35 and 1.47 for unsteady flow and 1.49 and 1.50 for steady flow).

For spatially-varied flow, however, the momentum and the energy equations are different even if we assume $S_e = S_f$ and $\alpha = \beta = 1.0$. We can use the momentum equation, Equation 1.56, only if we know the direction of the lateral flow. If the lateral inflow joins a channel at an angle close to 90°, as in most natural and manmade systems, the use of Equation 1.57 is appropriate. The direction of the lateral flow is irrelevant in the energy equation, since energy is a scalar quantity. However, where lateral flow joins a main channel, some energy loss occurs due to the local mixing. This loss is not accounted for in Equation 1.59, so Equation 1.59 should not be used for lateral inflow situations. In cases involving lateral outflow, on the other hand, the assumptions of Equation 1.59 are satisfied for the most part, and the use of Equation 1.59 is allowed.

The open-channel flow is not always gradually varied. Rapid changes in the flow variables can occur near channel transitions or hydraulic structures. The momentum and energy equations given for a volume element, Equations 1.53 and 1.54, are still valid as long as the pressure distribution at sections U and D is hydrostatic. However, the term $S_e \Delta x$ in Equation 1.54 needs to be replaced by h_L = head loss, which would account for all the energy losses between the two sections. Considering Equations 1.53 and 1.54 only the former includes an external force term. Therefore, if the problem at hand involves calculation of an external force (like the force exerted by a sluice gate on the flow), the momentum equation is the only choice. The energy equation is particularly useful in situations where the energy loss between the upstream and downstream sections is negligible.

PROBLEMS

P.1.1 Derive expressions for the flow area, A, wetted perimeter, P, top width, T, hydraulic radius, R, and hydraulic depth, D, in terms of the flow depth, y, for the channel sections shown in Figure P.1.1.

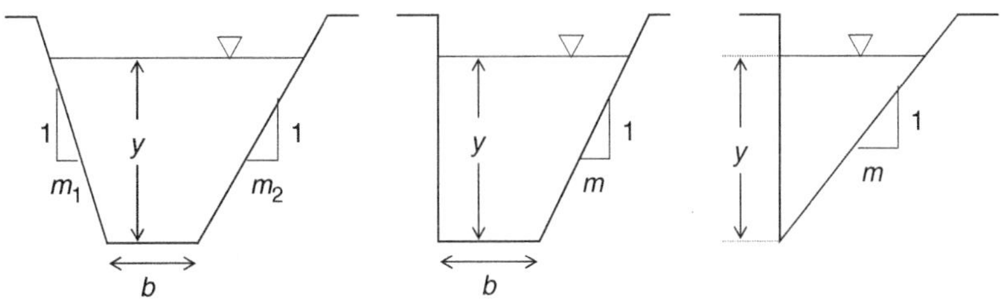

FIGURE P.1.1
Problem P.1.1

P.1.2 A nearly horizontal channel has a bottom width of 3 ft, and it carries a discharge of 60 cfs at a depth of 4 ft. Determine the magnitude and direction of the hydrostatic pressure force exerted on each of the sidewalls per unit length of the channel if

(a) the channel is rectangular with vertical sidewalls
(b) the channel is trapezoidal with each sidewall sloping outward at a slope 2 horizontal over 1 vertical, that is $m = 2$.

P.1.3 Let the point velocity in a wide rectangular channel be expressed as

$$v = 2.5 v_* \ln\left(\frac{30z}{k_s}\right)$$

where v = point velocity, $v_* = (\tau_o/\rho)^{1/2}$ = shear velocity, τ_o = average shear stress on channel bed, ρ = density, z = distance measured from channel bed, and k_s = length measure of bed roughness. The flow depth in the channel is y. Treating v_* and k_s as constants, derive an expression for the discharge per unit of the width of the channel.

Hint 1: $v = 0$ at $z = k_s/30$

Hint 2: $\int (\ln x)^n dx = x(\ln x)^n - n \int (\ln x)^{n-1} dx$.

P.1.4 Derive an expression for the average cross-sectional velocity, V, for the velocity distribution given in problem P.1.3.

Hint: $y \gg k_s$

P.1.5 At what z in Problem P.1.3 is the velocity maximum? Derive an expression for v_{max}.

P.1.6 For the channel of Problem P.1.3, show that

$$\frac{v_{max}}{V} - 1 = \frac{1}{\ln(30y/k_s - 1)}$$

P.1.7 For the velocity distribution given in Problem P.1.3, determine at what z the point velocity is equal to the average cross-sectional velocity. Often, a single velocity measurement taken at distance 0.6y from the free surface is used as an approximation to the cross-sectional velocity at a stream section. Is this a valid approximation?

P.1.8 Using the velocity distribution and the hints given in Problem P.1.3, show that

$$\beta = 1 + \left(\frac{V_{max}}{V} - 1\right)^2$$

P.1.9 Considering a unit width of the channel described in Problem P.1.3, determine the discharge, rate of momentum transfer, and rate of kinetic energy transfer if $y = 0.94$ m, $k_s = 0.001$ m, $\tau_o = 3.7\ \text{N/m}^2$ and $\rho = 1000\ \text{kg/m}^3$.

P.1.10 Determine the average cross-sectional velocity V and the discharge Q for the compound channel shown in Figure P.1.2.

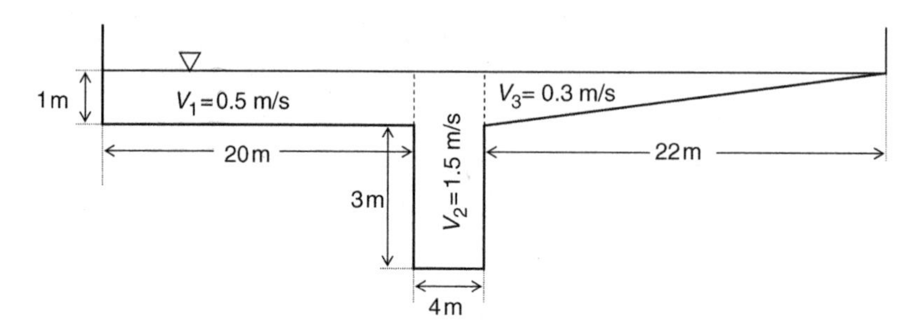

FIGURE P.1.2 Problems P.1.10 and P.1.11

P.1.11 Determine the rate of momentum transfer and the rate of kinetic energy transfer for the compound channel shown in Figure 1.P.2.

P.1.12 A trapezoidal channel with bottom width $b = 5$ ft and side slopes $m = 2$ (that is 2.0 horizontal over 1.0 vertical) carries $Q = 100$ cfs at depth $y = 3.15$. The water temperature is 60°F, and the kinematic viscosity at this temperature is $\nu = 1.217 \times 10^{-5}\ \text{ft}^2/\text{s}$.

(a) Determine if the flow is turbulent or laminar.
(b) Determine if the flow is subcritical or supercritical.

P.1.13 Is the flow likely to be uniform or non-uniform:

(a) at a natural stream section partially blocked by a fallen tree?
(b) at a drainage channel just upstream of an undersized culvert?
(c) at a section of a long prismatic, delivery channel a far distance from upstream and downstream ends?
(d) in a tidal river during high tide?

P.1.14 Is the flow likely to be steady or unsteady:

(a) in a street gutter during a short storm event?
(b) in a laboratory flume fed constant discharge at upstream end?
(c) in a drainage ditch after a long dry period?

P.1.15 Using the Leibnitz rule given below, verify Equation 1.32.

$$\frac{\partial}{\partial x}\int_{a(x)}^{b(x)} f(\eta,x)d\eta = \int_{a(x)}^{b(x)} \frac{\partial f(\eta,x)}{\partial x} d\eta + f[b(x),x]\frac{\partial b}{\partial x} - f[a(x),x]\frac{\partial a}{\partial x}$$

REFERENCES

Chow, V. T. (1959). *Open-Channel Hydraulics*. McGraw-Hill Book Co., New York, NY.

Chow, V. T, Maidment, R. M. and Mays, L. W. (1988). *Applied Hydrology*. McGraw-Hill Book Co., New York, NY.

Strelkoff, T. (1969). One-dimensional equations of open-channel flow. *Journal of the Hydraulics Division, ASCE*, **95(HY3)**, 861–876.

Yen, B. C. (1973). Open-channel flow equations revisited. *Journal of the Engineering Mechanics Division, ASCE*, **99(EM5)**, 979–1009.

Yen, B. C. and Wenzel, H. G. Jr. (1970). Dynamic equations for steady spatially varied flow. *Journal of the Hydraulics Division, ASCE*, **96(HY3)**, 801–814.

2 Energy and momentum principles

2.1 Critical Flow

Critical flow, a special type of open-channel flow, occurs under certain conditions. It is a cross-sectional flow type. In other words, critical flow is not maintained along a length of a channel. It may occur at the entrance of a steep channel, at the exit of a mild channel, and at sections where channel characteristics change.

Various concepts to be discussed in the subsequent sections will help in understanding the significance of critical flow. While deferring the definitions of certain terms to later sections, here we will provide a list of conditions associated with critical flow. Reference will be made to these conditions later where appropriate. At the critical sate of flow:

- the Froude number is equal to unity
- the specific energy is minimum for a given discharge
- the discharge is maximum for a given specific energy
- the specific momentum is minimum for a given discharge, and
- the discharge is maximum for a given specific momentum.

2.1.1 Froude Number

The Froude number, a dimensionless number, is a cross-sectional flow characteristic defined as

$$F_r = \frac{V}{\sqrt{gD}} = \frac{V}{\sqrt{g(A/T)}} = \frac{Q}{\sqrt{g(A^3/T)}} \tag{2.1}$$

where F_r = Froude number, V = velocity, Q = discharge, g = gravitational acceleration, D = hydraulic depth, A = flow area, and T = top width. The denominator, $\sqrt{gD}$, represents the speed with which gravity waves propagate in open channels. Sometimes we refer to this as *wave celerity*.

The flow is said to be *subcritical* if $F_r < 1.0$, critical if $F_r = 1.0$, and *supercritical* if $F_r > 1.0$. As will soon become clear, the hydraulic behavior of open-channel flow depends on whether the flow is subcritical or supercritical.

2.1.2 CALCULATION OF CRITICAL DEPTH

The *critical depth*, denoted by y_c, is the flow depth at a section where the flow is critical. In a given open channel, the critical flow may not occur at all. However, the critical depth is still calculated as a first step in dealing with most open-channel flow problems. As we will see in Chapter 4, the critical depth will help us to classify a channel as mild or steep in longitudinal water surface profile calculations. Also, like the Froude number, the critical depth itself can be used to identify if the flow at a section is subcritical or supercritical. The flow is *subcritical* if the flow depth is greater than the critical depth, that is if $y > y_c$. The flow is *supercritical* if $y < y_c$.

We can calculate the critical depth for a given discharge, Q, at a given channel section by expressing A and T in Equation 2.1 in terms of y (see Table 1.1), setting $F_r = 1.0$ and solving for the flow depth.

For a *rectangular* channel section of bottom width b, the Froude number is expressed as

$$F_r = \frac{Q}{\sqrt{g(A^3/T)}} = \frac{Q}{\sqrt{g(y^3b^3/b)}} = \frac{q}{\sqrt{gy^3}} \tag{2.2}$$

where, $q = Q/b$ = discharge per unit width and is defined for rectangular channels only. Then the expression for the critical depth becomes

$$y_c = \sqrt[3]{\frac{Q^2}{gb^2}} = \sqrt[3]{\frac{q^2}{g}} \tag{2.3}$$

For a triangular channel with side slopes of m (that is 1 vertical over m horizontal), we can show that

$$y_c = \sqrt[5]{\frac{2Q^2}{gm^2}} \tag{2.4}$$

For a trapezoidal channel with side slopes of m (that is 1 vertical over m horizontal), $F_r = F_r^2 = 1.0$ will lead to

$$\frac{(b + 2my_c)Q^2}{g(b + my_c)^3 y_c^3} = 1 \tag{2.5}$$

Equation 2.5 cannot be solved for y_c explicitly. Therefore, a trial-and-error procedure is needed to determine the critical depth for trapezoidal channels.

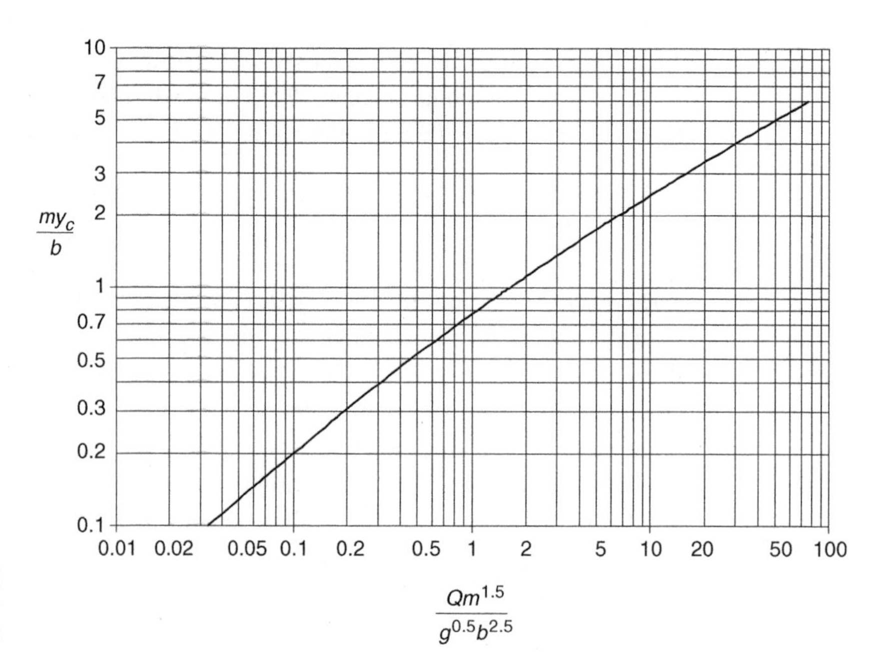

FIGURE 2.1 Critical depth chart for trapezoidal channels

Alternatively, we can use Figure 2.1, which presents a graphical solution for Equation 2.5 in dimensionless form.

Likewise, for a circular channel of diameter d_0, we can show that (Henderson, 1966)

$$\frac{Q}{g^{0.5}d_0^{2.5}} = \frac{1}{8}\sqrt{\frac{[\theta - (\sin\theta)(\cos\theta)]^3}{\sin\theta}} \tag{2.6}$$

where

$$\theta = \pi - \arccos[(y_c - d_0/2)/(d_0/2)] \tag{2.7}$$

Table 1.1 displays a graphical representation of θ. An explicit solution to Equation 2.6 for y_c is not available; we need to solve this equation by trial and error. However, a graphical solution is provided in Figure 2.2 to facilitate the calculation of the critical depth in circular channels.

EXAMPLE 2.1 A trapezoidal channel has a bottom width of $b = 6$ ft and side slopes of $m = 2$ (1V:2H). Determine the critical depth in this channel if the discharge is $Q = 290$ cfs.

We will solve Equation 2.5 by trial-and-error. Let us first guess that the critical depth is equal to 1.0 ft. Then the left-hand side of Equation 2.5 is calculated as

$$\frac{[6.0 + (2)(2.0)(1.0)](290)^2}{(32.2)[6.0 + (2.0)(1.0)]^3(1.0)^3} = 51.0$$

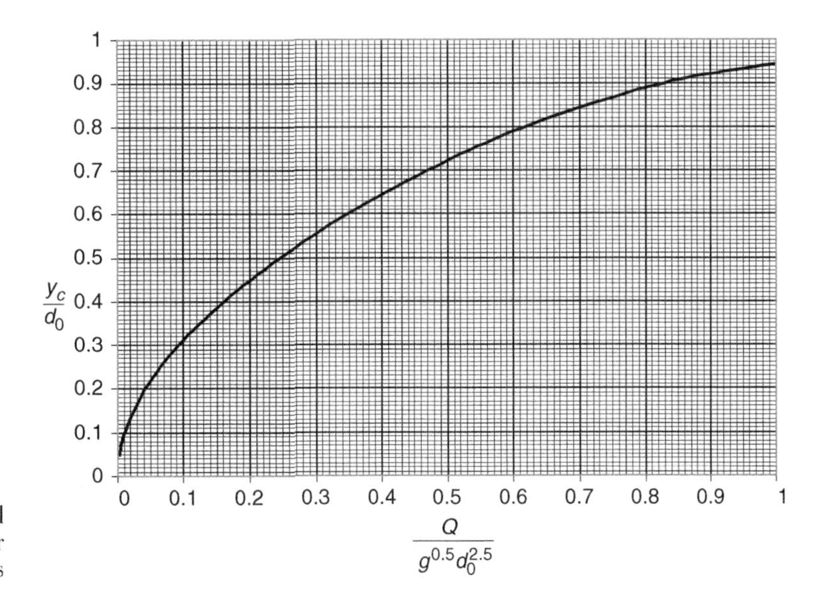

FIGURE 2.2 Critical depth chart for circular channels

Obviously we guessed the critical depth incorrectly, since the correct value should make the left-hand side of the equation 1.0.

We now try other values for the critical depth and calculate the left-hand side of the equation in the same manner. The results are tabulated as follows.

TABLE 2.1a

Trial value for y_c (ft)	Left-hand side of Equation 2.5
1.0	51.0
2.0	4.6
4.0	0.3
3.0	1.0

Therefore, for this channel, $y_c = 3.0$ ft.

Alternatively, we can use Figure 2.1 to solve this problem. Let us first evaluate

$$\frac{Qm^{1.5}}{g^{0.5}b^{2.5}} = \frac{(290)(2.0)^{1.5}}{(32.2)^{0.5}(6.0)^{2.5}} = 1.64$$

The corresponding value of my_c/b is obtained as being 1.0 from Figure 2.1. Therefore, $y_c = (1.0)\,b/m = (1.0)(6.0)/2 = 3.0$ ft.

EXAMPLE 2.2 A 36-inch storm sewer ($d_0 = 3$ ft) carries a discharge of $Q = 30$ cfs. Determine the critical depth.

We can calculate the critical depth mathematically by using a trial-and-error method. Let us evaluate the left-hand side of Equation 2.6 as

$$\frac{Q}{g^{0.5}d_0^{2.5}} = \frac{30.0}{(32.2)^{0.5}(3.0)^{2.5}} = 0.34$$

Then the correct value of y_c is the one for which the right-hand side of Equation 2.6 is equal to 0.34. Let us try $y_c = 0.6$ ft. By using Equation 2.7

$$\theta = 3.1416 - \arccos\left(\frac{0.6 - 3.0/2}{3.0/2}\right) = 0.927\,\text{rad} = 53.1^\circ$$

and, therefore, $\cos\theta = 0.60$ and $\sin\theta = 0.80$. Substituting these into the right-hand side of Equation 2.6, we obtain

$$\frac{1}{8}\sqrt{\frac{[0.927 - (0.80)(0.60)]^3}{0.8}} = 0.042$$

The guessed value is incorrect since $0.042 \neq 0.34$. Different trial values for the critical depth produces the results tabulated below.

TABLE 2.1b

Trial value for y_c (ft)	θ (rad)	Right-hand side of Equation 2.6
0.6	0.927	0.04
1.2	1.369	0.16
1.8	1.772	0.35
1.77	1.752	0.34

Therefore, for this sewer, $y_c = 1.77$ ft.

Alternatively, we can use Figure 2.2 to solve this problem. For $Q/(g^{0.5}d_0^{2.5}) = 0.34$, we obtain $y_c/d_0 = 0.59$ from the figure. Then, $y_c = (0.59)d_0 = (0.59)(3.0) = 1.77$ ft.

2.2 APPLICATIONS OF ENERGY PRINCIPLE FOR STEADY FLOW

2.2.1 ENERGY EQUATION

We have derived the conservation of energy equation in Chapter 1 (see Equation 1.54). Assuming α = energy coefficient = 1.0, for simplicity, we can rewrite the equation between an upstream channel section U and a downstream section D as

$$\left(z_{bU} + y_U + \frac{V_U^2}{2g}\right) = \left(z_{bD} + y_D + \frac{V_D^2}{2g}\right) + h_L \quad (2.8)$$

where z_b = elevation of channel bottom above a horizontal datum, y = flow depth, V = average cross-sectional velocity, g = gravitational acceleration, and h_L = energy loss per unit weight between the two sections. Figure 2.3 displays a schematic representation of Equation 2.8.

Note that the term h_L includes the losses due to resistance to flow (friction loss) as well as other losses such as those due to cross-sectional changes. If the energy

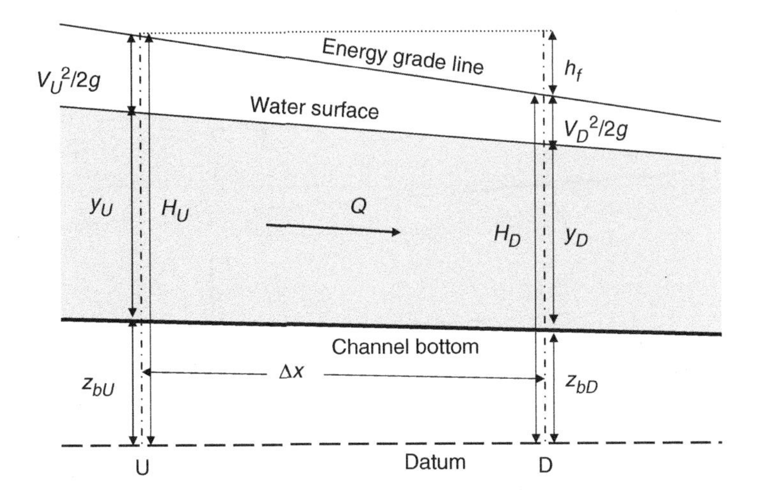

FIGURE 2.3 Energy principle for steady flow

loss is due to the friction alone, we can replace h_L in Equation 2.8 by h_f= friction loss. Then the average friction slope, S_f, between the two sections is expressed as $S_f = h_f/\Delta x$.

We often refer to z_b as *elevation head*, y as *pressure head*, and $V^2/2g$ as *velocity head*. The sum of these three terms is called *total energy head*, and is denoted by H. In other words

$$H = z_b + y + \frac{V^2}{2g} \tag{2.9}$$

A line connecting the energy head at various sections along a channel is called the *energy grade line*. The sum of the elevation head and the pressure head is often called the *hydraulic head* or *piezometric head*, and it is denoted by h. That is

$$h = z_b + y \tag{2.10}$$

A line connecting the hydraulic head at various sections along the channel is called the *hydraulic grade line* or *piezometric line*. For most open-channel flow situations, the water surface elevation above the datum is the same as the hydraulic head, and the water surface itself represents the hydraulic grade line. We sometimes refer to the water surface elevation as the *stage*.

The *specific energy*, denoted by E, is defined as the energy head relative to the channel bottom. Therefore, at any channel section

$$E = y + \frac{V^2}{2g} \tag{2.11}$$

Figure 2.4 displays a schematic representation of various heads in open-channel flow.

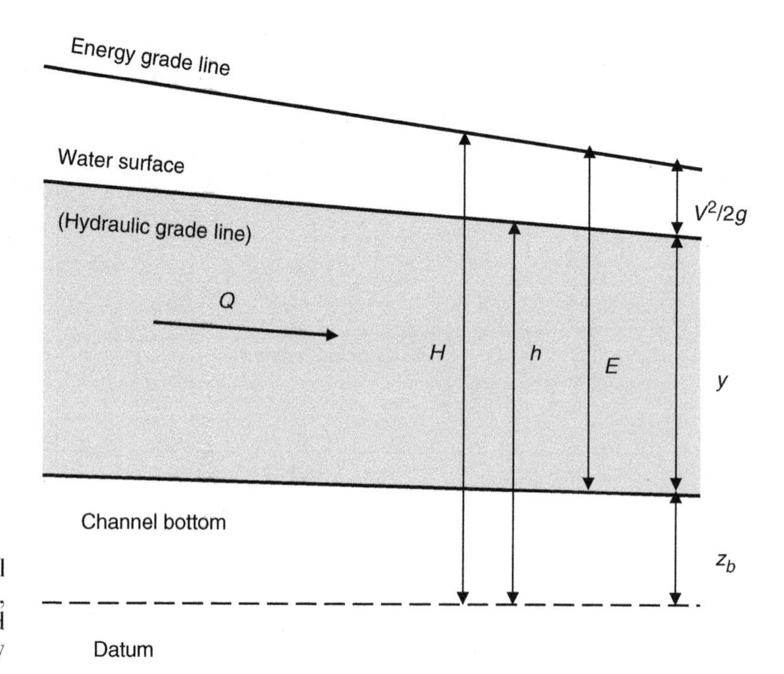

FIGURE 2.4 Total energy head, hydraulic head and specific energy

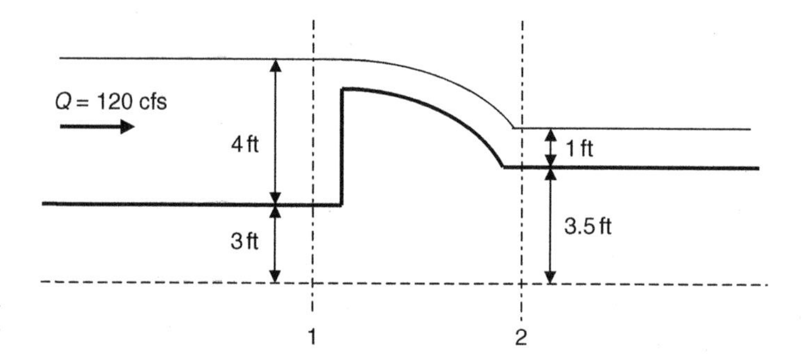

FIGURE 2.5 Definition sketch for Example 2.3

EXAMPLE 2.3 The channel shown in Figure 2.5 is rectangular in cross-section, and it is 10 ft wide. Determine the total energy head, hydraulic head, and specific energy at Sections 1 and 2. Also determine the head loss between Sections 1 and 2 due to the spillway.

From the problem statement, we have $b = 10$ ft, $Q = 120$ cfs, $z_{b1} = 3$ ft, $y_1 = 4$ ft, $z_{b2} = 3.5$ ft, and $y_2 = 1.0$ ft. Then $A_1 = (10)(4) = 40\text{ ft}^2$ and $V_1 = (120)/(40) = 3.0$ fps. Likewise, $A_2 = (10)(1) = 10\text{ ft}^2$ and $V_2 = (120)/(10) = 12$ fps. Now by using Equations 2.9, 2.10, and 2.11, respectively,

$$H_1 = 3.0 + 4.0 + \frac{(3.0)^2}{2(32.2)} = 7.14\text{ ft}$$

$$h_1 = 3.0 + 4.0 = 7.0\text{ ft}$$

$$E_1 = 4.0 + \frac{(3.0)^2}{2(32.2)} = 4.14\text{ ft}$$

Likewise

$$H_2 = 3.5 + 1.0 + \frac{(12.0)^2}{2(32.2)} = 6.74\,\text{ft}$$

$$h_2 = 3.5 + 1.0 = 4.5\,\text{ft}$$

$$E_2 = 1.0 + \frac{(12.0)^2}{2(32.2)} = 3.24\,\text{ft}$$

Finally, by using Equation 2.8, $h_L = H_1 - H_2 = 7.14 - 6.74 = 0.40$ ft.

2.2.2 SPECIFIC ENERGY DIAGRAM FOR CONSTANT DISCHARGE

The specific energy was defined as the energy head relative to the channel bottom, and it was expressed in terms of the flow depth and the velocity head in Equation 2.11. Noting that $V = Q/A$, we can rewrite Equation 2.11 in terms of the discharge Q and the flow area A as

$$E = y + \frac{Q^2}{2gA^2} \tag{2.12}$$

For steady flow, Q is constant. Also, the flow area A can be expressed in terms of the flow depth y and the channel cross-sectional dimensions (see Tables 1.1 and 2.1).

Therefore, for a fixed discharge and a given channel section, a plot of y versus E can be prepared as shown qualitatively in Figure 2.6. Such a plot is called a *specific energy diagram.*

The specific energy diagram reveals that the flow needs the minimum specific energy, E_{min}, to pass a channel section at critical depth. We can show this mathematically, by taking the derivative of Equation 2.12 with respect to y, noting $T = dA/dy$ (see Figure 2.7), and setting the derivative equal to zero as

$$\frac{dE}{dy} = 1 - \frac{Q^2}{2g}\frac{2(dA/dy)}{A^3} = 1 - \frac{Q^2}{g}\frac{T}{A^3} = 1 - \frac{V^2}{g}\frac{T}{A} = \frac{V^2}{gD} = 1 - F_r^2 = 0$$

Thus, when the specific energy is minimum, the Froude number is equal to unity, and the flow depth is equal to the critical depth (as first mentioned in Section 2.1). As we can see from Figure 2.6, the specific energy diagram has two limbs. The upper limb represents subcritical flow, since the flow depths on this limb are greater than the critical depth. The lower limb represents supercritical flow. Clearly at a given channel section, for a given discharge, two flow depths are possible for the same specific energy. These depths, denoted by y_1 and y_2 in Figure 2.6, are called the *alternate depths* (Chow 1959). Other

TABLE 2.1 Relationships between y, A, and Y_C for various channel sections

Section type	Area A	Product AY_C
Rectangular 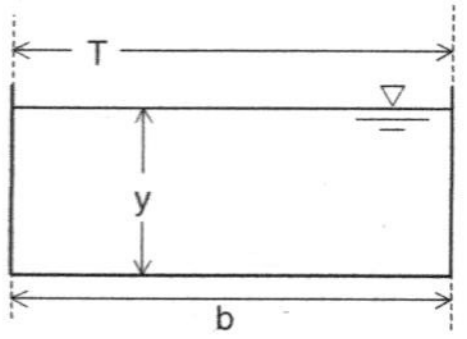	by	$b\frac{y^2}{2}$
Trapezoidal 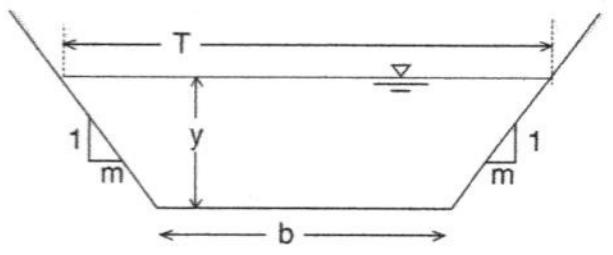	$(b+my)y$	$\frac{y^2}{6}(2my+3b)$
Triangular 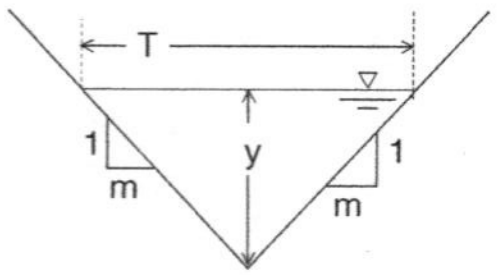	my^2	$\frac{my^3}{3}$
Circular 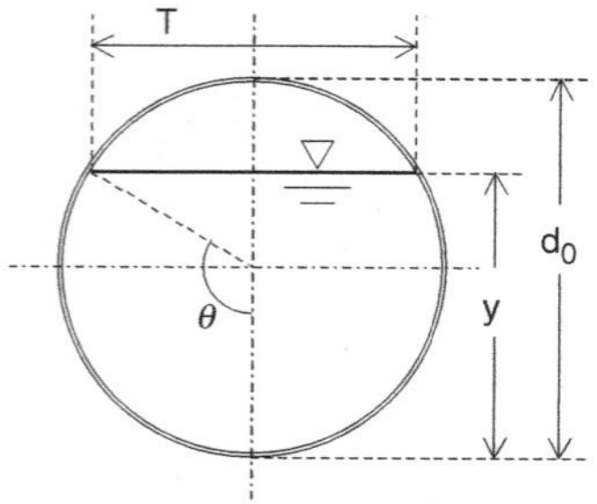	$\frac{1}{8}(2\theta - \sin 2\theta)d_0^2$ $\theta = \pi - \text{arc } \cos[(y - d_0/2)/(d_0/2)]$	$\frac{d_0^3}{24}(3\sin\theta - \sin^3\theta - 3\theta\cos\theta)$ $\theta = \pi - \text{arc } \cos[(y - d_0/2)/(d_0/2)]$

Y_C = vertical distance from the free surface to centroid of flow section (see Chapter 1).

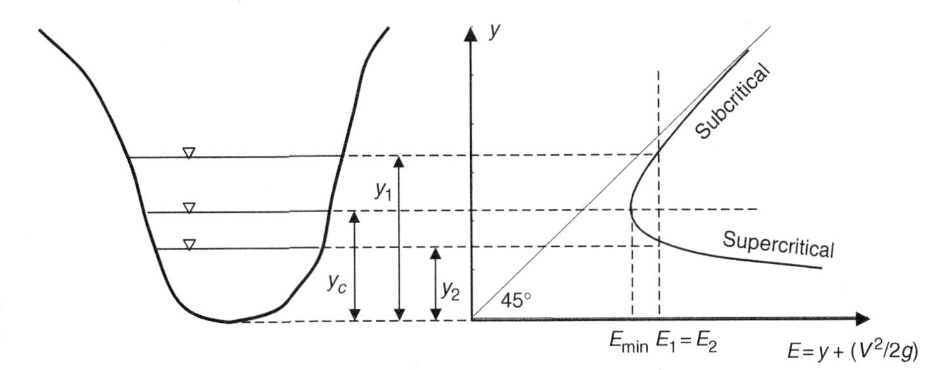

FIGURE 2.6 Specific energy diagram

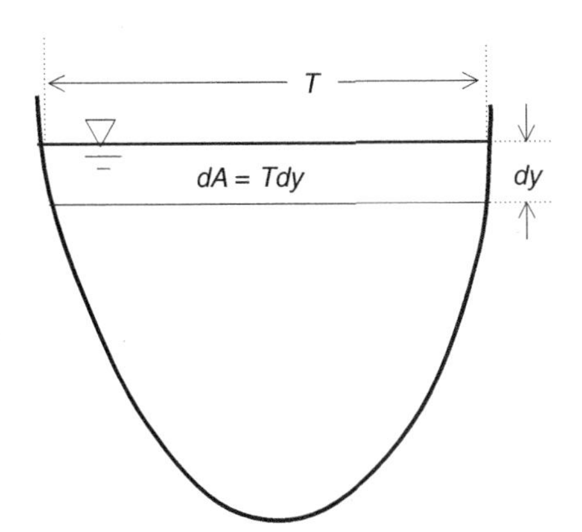

FIGURE 2.7 Expression for top width T

factors, which will be discussed later, will govern which of these two depths will actually occur.

Let us now take the derivative of both sides of Equation 2.12 with respect to x, where x is distance in flow direction. By manipulating the resulting expression we obtain

$$\frac{dy}{dx} = \frac{dE/dx}{(1 - F_r^2)} \tag{2.13}$$

This relationship shows that, for subcritical flow ($F_r < 1.0$), the flow depth increases in the flow direction with increasing specific energy. However, for supercritical flow ($F_r > 1.0$), the flow depth decreases in the flow direction with increasing specific energy. This is a good example of different hydraulic behavior under subcritical and supercritical conditions.

EXAMPLE 2.4 The trapezoidal channel considered in Example 2.1 has a bottom width of $b = 6$ ft and side slopes of $m = 2$ (1V:2H), and it carries a discharge of $Q = 290$ cfs. Calculate and plot the specific energy diagram for this

TABLE 2.2 Specific energy and Froude number calculations for trapezoidal channel

y (ft)	A (ft^2)	E (ft)	T (ft)	F_r
1.00	8.00	21.40	10.00	7.14
1.25	10.63	12.82	11.00	4.89
1.50	13.50	8.67	12.00	3.57
1.75	16.63	6.47	13.00	2.72
2.00	20.00	5.26	14.00	2.14
2.25	23.63	4.59	15.00	1.72
2.50	27.50	4.23	16.00	1.42
2.75	31.63	4.06	17.00	1.18
3.00	36.00	4.01	18.00	1.00
3.25	40.63	4.04	19.00	0.86
3.50	45.50	4.13	20.00	0.74
3.75	50.63	4.26	21.00	0.65
4.00	56.00	4.42	22.00	0.57
4.25	61.63	4.59	23.00	0.51
4.50	67.50	4.79	24.00	0.45
4.75	73.63	4.99	25.00	0.40
5.00	80.00	5.20	26.00	0.36
5.25	86.63	5.42	27.00	0.33
5.50	93.50	5.65	28.00	0.30
6.00	108.00	6.11	30.00	0.25
7.00	140.00	7.07	34.00	0.18
8.00	176.00	8.04	38.00	0.13
9.00	216.00	9.03	42.00	0.10
10.00	260.00	10.02	46.00	0.08

channel. Also, calculate and plot the specific energy diagrams for the same channel for $Q = 135$ cfs and $Q = 435$ cfs.

The specific energy calculations are summarized in Table 2.2. The values of y in column 1 are first picked, then the values of A in column 2 are calculated using the expression for A in Table 1.1 (or 2.1) for trapezoidal channels. For instance, for $y = 1.5$ ft,

$$A = (b + my)y = [6.0 + (2)(1.5)](1.5) = 13.5\ \text{ft}^2$$

Then the values of E in column 3 are calculated using Equation 2.12. For $y = 1.5$ ft,

$$E = y + \frac{Q^2}{2gA^2} = 1.5 + \frac{(290)^2}{2(32.2)(13.5)^2} = 8.67\ \text{ft}$$

The specific energy diagram, a plot of E versus y for $Q = 290$ cfs, is shown in Figure 2.8. Table 2.2 also includes the Froude number calculations. The top width

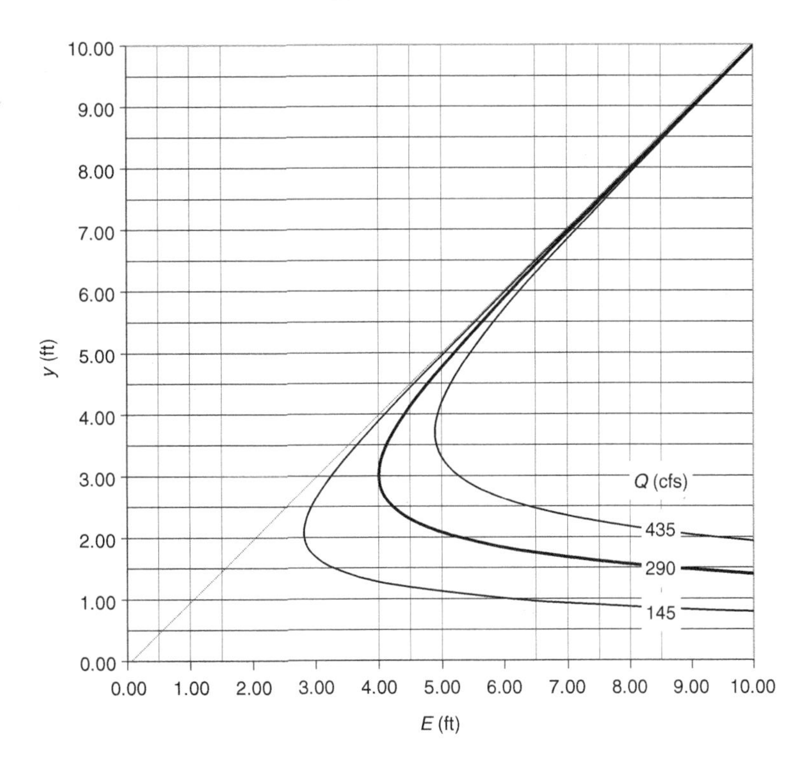

FIGURE 2.8
Example specific energy diagrams for a trapezoidal channel

T in column 4 is calculated using the expression in Table 1.1. For $y = 1.5$ ft, for instance,

$$T = b + 2my = 6.0 + (2)(2)(1.5) = 12.0 \text{ ft}$$

Then the Froude number, F_r, in column 5 is calculated using Equation 2.1 as

$$F_r = \frac{Q}{\sqrt{g(A^3/T)}} = \frac{290}{\sqrt{32.2(13.5)^3/(12.0)}} = 3.57$$

An inspection of Table 2.2 reveals that the minimum value of the specific energy is 4.01 ft, and the corresponding depth and the Froude number are, respectively, 3.0 ft and 1.0. Therefore, as expected, the minimum specific energy occurs at critical flow, and the critical depth is 3.0 ft. Also, for flow depths less than the critical depth, the Froude number is greater than 1.0, the flow is supercritical, and these depths represent the lower limb of the specific energy diagram. The depths on the upper limb are greater than the critical depth with Froude numbers smaller than 1. Therefore, the upper limb represents the subcritical flow range.

The specific energy diagrams for $Q = 145$ cfs and $Q = 435$ cfs are calculated in a similar manner. The results are displayed in Figure 2.8.

We can make some observations from Figure 2.8. A larger specific energy is needed to pass a larger discharge through a channel section at the same flow depth, regardless of whether the flow is subcritical or supercritical. This seems quite logical, since for a fixed depth (that is fixed flow area) a larger velocity would be required to pass a larger discharge. This increase in velocity would account for an increase in the specific energy. However, Figure 2.8 also reveals that to pass a larger discharge through a channel section at the same specific energy, the flow depth needs to be smaller in subcritical flow but larger in supercritical flow.

EXAMPLE 2.5 The storm sewer considered in Example 2.2 has a diameter of $d_0 = 36$ inches and carries a discharge of $Q = 30$ cfs. Calculate and plot the specific energy diagram for this channel. Also calculate the specific energy diagrams for $Q = 10$ cfs and Q = 20 cfs.

Table 2.3 summarizes the specific energy calculations. The values of y in column 1 are first picked, then the values of θ in column 2 are calculated using the expression for θ in Table 2.1 for circular channels. For instance, for $y = 1.2$ ft,

$$\theta = 3.1416 - \arccos\left(\frac{1.2 - 3.0/2}{3.0/2}\right) = 1.369 \text{ rad.}$$

Then, we calculate the A values in column 3 using the expression given for A in Table 2.1. For $y = 1.2$ ft and $2\theta = (2)(1.369) = 2.738$ rad.

$$A = \frac{1}{8}[2.738 - \sin(2.738)](3.0)^2 = 2.64 \text{ ft}^2$$

Finally, the E values in column 4 are calculated by using Equation 2.12. For $y = 1.2$ ft,

$$E = 1.2 + \frac{(30)^2}{2(32.2)(2.64)^2} = 3.20 \text{ ft}$$

TABLE 2.3 Specific energy calculations for a circular channel

y (ft)	θ (rad)	A (ft^2)	E (ft)
0.8	1.085	1.51	6.88
1.0	1.231	2.06	4.27
1.2	1.369	2.64	3.20
1.4	1.504	3.23	2.73
1.6	1.638	3.83	2.55
1.8	1.772	4.43	2.51
2.0	1.911	5.01	2.56
2.2	2.056	5.56	2.65
2.4	2.214	6.06	2.78
2.6	2.394	6.51	2.93
2.8	2.619	6.87	3.10

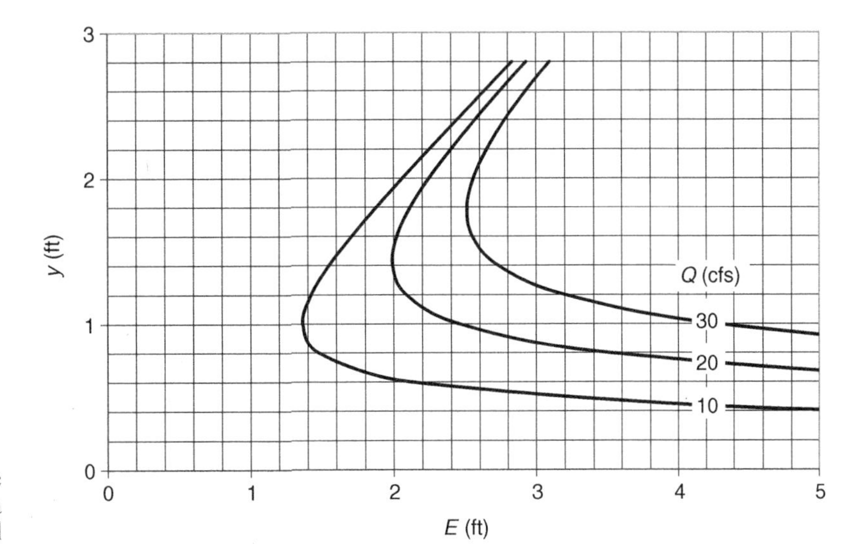

FIGURE 2.9 Specific energy diagrams for a circular channel

The calculated values of the specific energy are plotted in Figure 2.9. An inspection of the figure reveals that the minimum specific energy for $Q = 30$ cfs occurs at $y_c = 1.77$ ft. This concurs with Example 2.2.

The specific energy diagrams for $Q = 10$ cfs and $Q = 20$ cfs are calculated in the same manner. The results are displayed in Figure 2.9. The relative positioning of the specific energy diagrams with respect to the discharge in this figure is similar to that of Figure 2.8.

EXAMPLE 2.6 The trapezoidal channel considered in Examples 2.1 and 2.4 has a bottom width of $b = 6$ ft and side slopes of $m = 2$ (1V: 2H), and it carries a discharge of $Q = 290$ cfs. Suppose the channel is nearly horizontal, except that there is a smooth, short step rise in the channel bottom as shown in Figure 2.10. The height of the step is $\Delta z = 1.0$ ft. The energy loss due to this step is negligible. The depth at section A just before the step is 5.33 ft. Determine the flow depth over the step, that is at section B.

Noting that $A = (b + my)y$ and $T = b + 2my$, for $y_A = 5.33$ ft, we obtain $A_A = [6.0 + (2)(5.33)](5.33) = 88.8 \text{ ft}^2$, and $T = 6.0 + 2(2.0)(5.33) = 27.3$ ft Then, by using Equation 2.12

$$E_A = 5.33 + \frac{(290)^2}{2(32.2)(88.8)^2} = 5.5 \text{ ft}$$

Also, by using Equation 2.1,

$$F_{rA} = \frac{290}{\sqrt{32.2(88.8)^3/27.3}} = 0.32$$

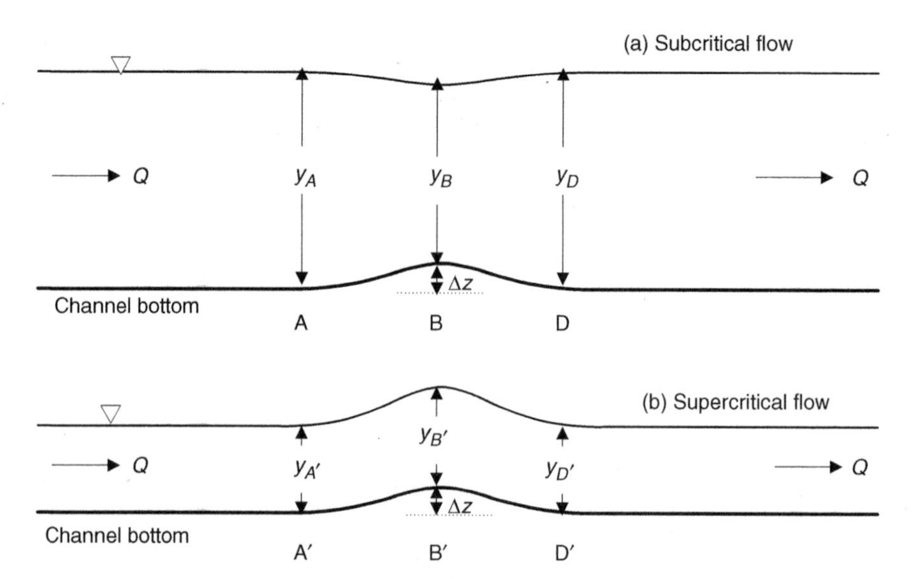

FIGURE 2.10 Flow over a smooth step

The flow is then subcritical at section A. The head loss between A and B is negligible, meaning that $H_A = H_B$. Then $E_B = E_A - \Delta z = 5.5 - 1.0 = 4.5$ ft. Noting that $A = (b + my)y$, we will now use Equation 2.12 to determine y_B as

$$4.5 = y_B + \frac{(290)^2}{2(32.2)\{[6.0 + (2.0)(y_B)](y_B)\}^2}$$

This equation has two positive roots for y_B, which are 4.10 ft and 2.30 ft. The former of the two is a subcritical depth and the latter is supercritical. We will need to examine the specific energy diagram that was calculated in Example 2.4 to determine which of the two depths will occur at section B. The diagram is replotted in Figure 2.11. Note that because the specific energy diagram depends on the discharge and the cross-sectional geometry (and not the elevation of the channel bottom), the diagram shown in Figure 2.11 represents both of sections A and B, and any other section between the two.

In Figure 2.11, point A represents the section A of the channel. Also shown in the figure are the values of E_A and E_B. Mathematically, for $E_B = 4.5$ ft, there are two possible flow conditions represented by points B and B′ in the figure. As the flow moves from section A to B in the channel, the point representing the flow on the diagram should move from A to either B or B′. Because the diagram represents all the flow sections between section A and B, the points representing the flow between the two sections should stay on the diagram. Any point that is not on the diagram would correspond to a discharge different from 290 cfs (see Figure 2.8). Thus the available paths from points A to B and A to B′ are around the curve. The path from A to B represents a gradual decrease in

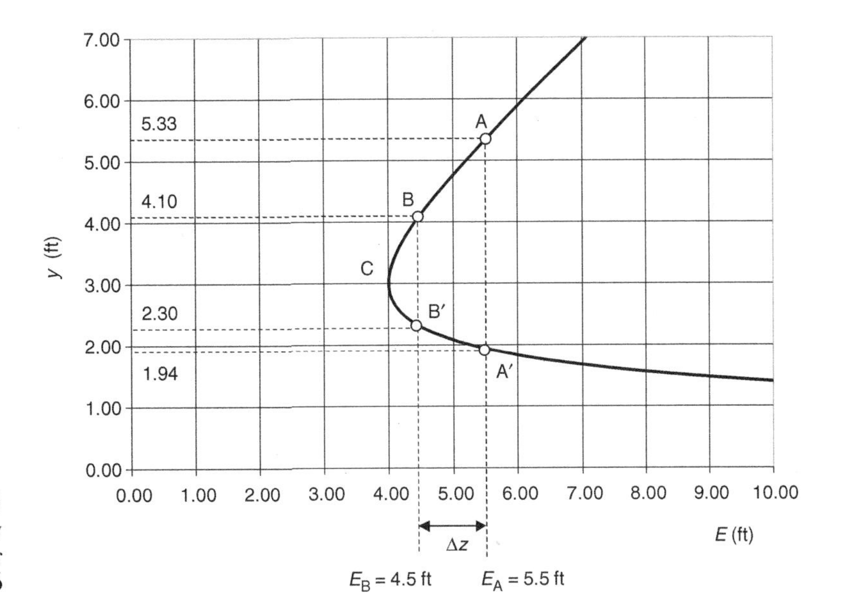

FIGURE 2.11 Specific energy diagram for Example 2.6

the specific energy, and is consistent with the given situation. The available path from A to B′, however, includes a segment from C to B′ representing an increase in the specific energy. However, such an increase in specific energy anywhere between sections A and B is not possible in the given channel situation. Therefore point B′ is not accessible from point A, and the solution to the problem is represented by point B on the diagram. As a result, we determine that $y_B = 4.1$ ft, and the flow is subcritical at B. The water surface elevation at B is then $h_B = 1.0 + 4.1 = 5.1$ ft. Note that this is lower than 5.33 ft, indicating that the water surface actually drops over the step (See Figure 2.10a). This can be explained by the fact that the subcritical limb of the specific energy diagram is steeper than the 45° line shown in Figure 2.6. In other words, $\Delta y > \Delta E$ when $\Delta E = \Delta z$.

We should also note that, in the absence of energy loss, the flow depth after the step (section D in Figure 2.10) will be the same as the depth at section A.

EXAMPLE 2.7 Suppose the flow depth at section A of the channel considered in Example 2.6 is 1.94 ft. Determine the flow depth in section B.

Section A is represented by point A′ in Figure 2.11. Mathematically, the flow depths represented by points B′ and B appear to be two possible solutions. However, a discussion similar to that of Example 2.6 can be given to show that point B is not accessible from A′, therefore point B′ will represent the actual solution. Accordingly, the flow depth at section B will be 2.30 ft. Note that, in this case, the flow is supercritical both at sections A and B. Also, the flow depth increases over the step and the water surface rises (see Figure 2.10b).

We can generalize the findings of Examples 2.6 and 2.7 by stating that the flow at section B will be subcritical if the flow at A is subcritical. If the flow at A is supercritical, then it will also be supercritical at B.

2.2.3 DISCHARGE DIAGRAM FOR CONSTANT SPECIFIC ENERGY

The specific energy diagram was defined as a graphical representation of Equation 2.12 for constant discharge, Q. Equation 2.12 can be rearranged also as

$$Q = \sqrt{2gA^2(E-y)} \tag{2.14}$$

A graphical representation of this equation in the form of y versus Q for constant specific energy E is called a *discharge diagram*. Figure 2.12 displays a discharge diagram qualitatively.

Figure 2.12 reveals that, for constant specific energy, two flow depths are possible for the same discharge. In the figure, these depths are denoted by y_1 (subcritical) and y_2 (supercritical). Also, for a given specific energy, a channel section will pass the maximum discharge at critical depth. The reader may recall that this was one of the conditions describing critical flow in section 2.1. Another observation is that, for constant specific energy, a smaller depth is required to pass a larger discharge under subcritical flow conditions. However, for supercritical flow, at constant specific energy, a larger discharge will cause a larger flow depth.

EXAMPLE 2.8 The trapezoidal channel considered in Example 2.1 has a bottom width of $b = 6$ ft and side slopes of $m = 2$ (1V: 2H). Calculate and plot the discharge diagrams for this channel for $E = 3$ ft, 4 ft, and 5 ft.

To calculate the discharge diagram, we will pick values for y and calculate the corresponding values of Q by using Equation 2.14. The calculations for $E = 3$ ft

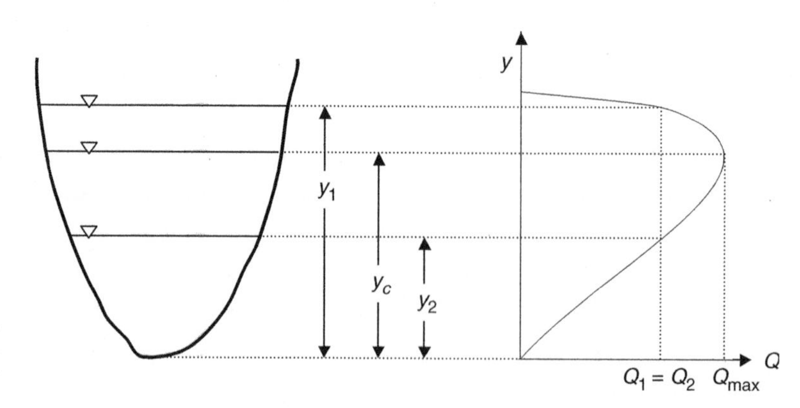

FIGURE 2.12 Discharge diagram for constant specific energy

TABLE 2.4 Discharge diagram calculations for Example 2.8

y (ft)	A (ft^2)	Q (cfs)
0.00	0.00	0.00
0.25	1.63	21.63
0.50	3.50	44.41
0.75	5.63	67.71
1.00	8.00	90.79
1.25	10.63	112.80
1.50	13.50	132.69
1.75	16.63	149.16
2.00	20.00	160.50
2.25	23.63	164.19
2.50	27.50	156.05
2.75	31.63	126.89
3.00	36.00	0.00

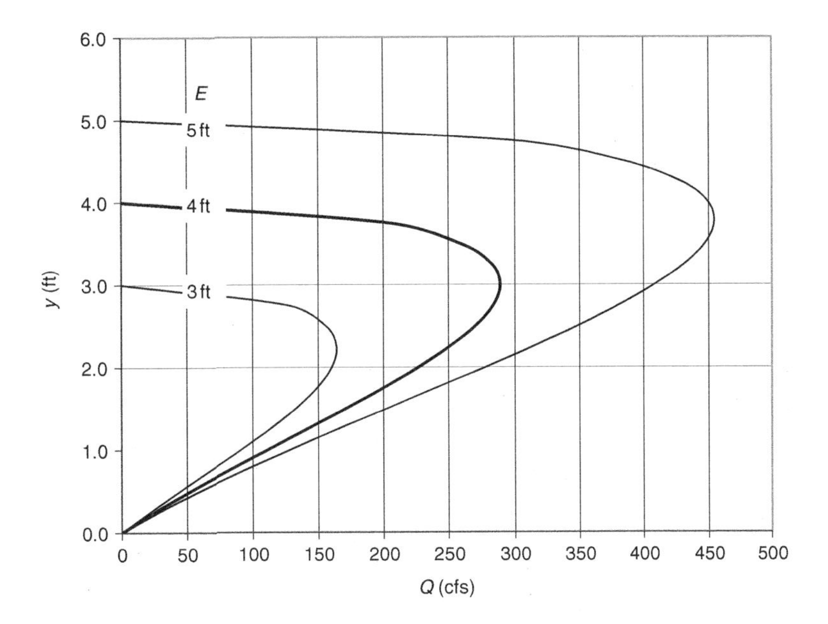

FIGURE 2.13 Discharge diagrams for a trapezoidal channel

are summarized in Table 2.4. Similar calculations are performed for $E = 4$ ft and 5 ft. The results are displayed in Figure 2.13.

2.2.4 SPECIFIC ENERGY IN RECTANGULAR CHANNELS

For rectangular channels, it is convenient to work with the unit width of the channel. Defining $q = Q/b =$ discharge per unit width, Equation 2.12 can be written for rectangular channels as

$$E = y + \frac{q^2}{2gy^2} \tag{2.15}$$

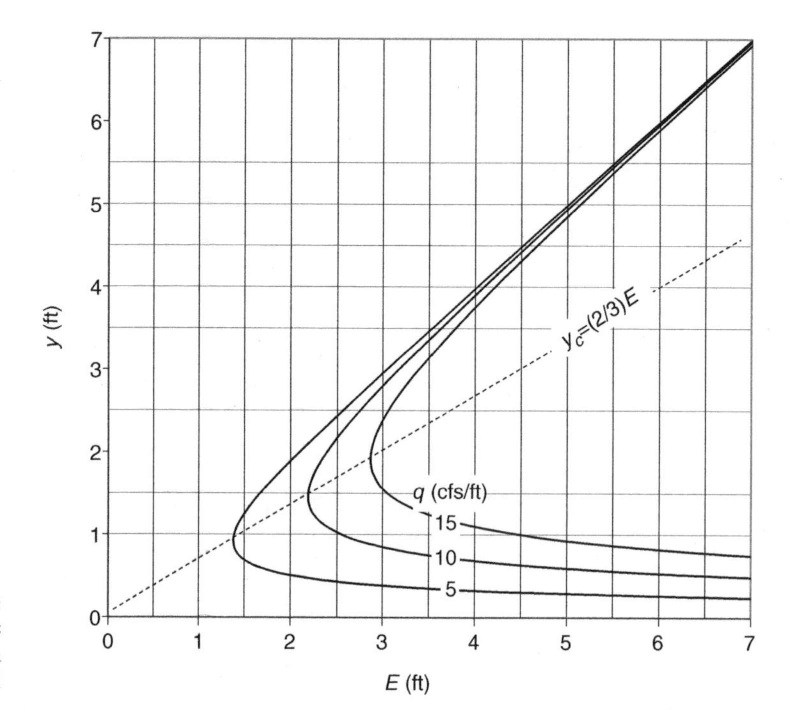

FIGURE 2.14
Example specific energy diagrams for a rectangular channel

This can be rearranged as

$$q = \sqrt{2gy^2(E - y)} \tag{2.16}$$

Substituting Equation 2.3 into Equation 2.15, for critical flow in rectangular channels, we obtain a simple relationship between the critical depth, y_c, and critical flow specific energy, E_c, as

$$y_c = \frac{2}{3}E_c \tag{2.17}$$

We can calculate the specific energy and the discharge diagrams for rectangular channels by using the same procedures we employed for trapezoidal channels in the preceding sections. Figure 2.14 shows example specific energy diagrams for various q values. Likewise, Figure 2.15 displays example discharge diagrams obtained for various values of E.

EXAMPLE 2.9 A nearly horizontal rectangular channel is 12 ft wide, and it carries 60 cfs. The width is smoothly contracted to 6 ft as shown in Figure 2.16. Determine the flow depth at section B if the depth at section A is (a) 2.50 ft, and (b) 0.43 ft.

(a) There is no head loss due to the contraction and the channel is nearly horizontal between sections A and B. Therefore, we have $H_A = H_B$, and $E_A = E_B$. The discharge per unit width at section A is $q_A = 60/12 = 5$ cfs/ft, and that at

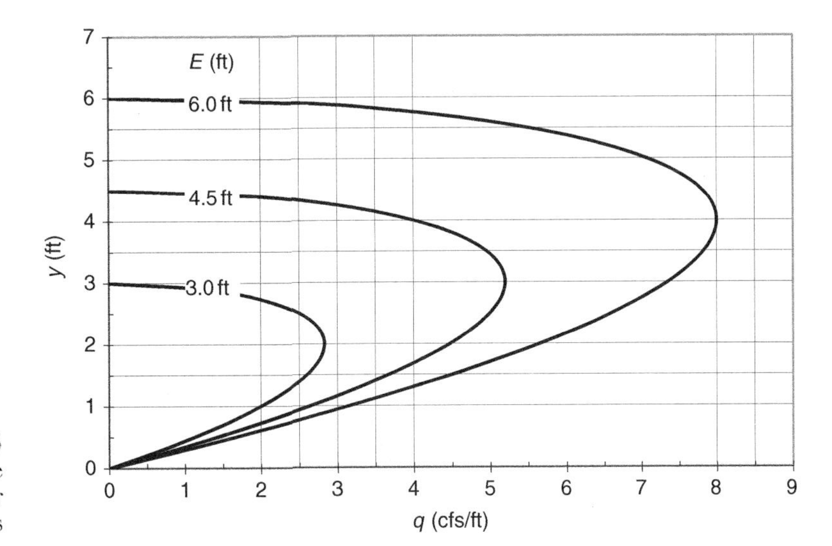

FIGURE 2.15 Example discharge diagrams for rectangular channels

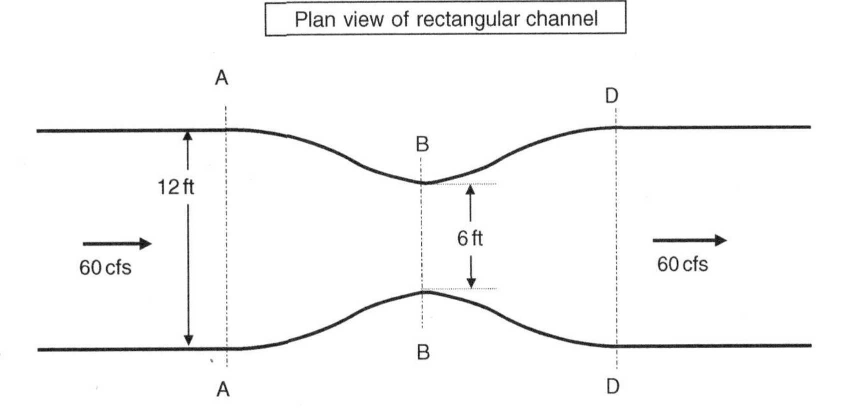

FIGURE 2.16 Contracted channel section of Example 2.9

section B is, $q_B = 60/6 = 10$ cfs/ft. For $y_A = 2.5$ ft, by using Equation 2.2 we calculate the Froude number as $F_{rA} = 5.0/[32.2(2.50)^3]^{1/2} = 0.22$ indicating that, at section A, the flow is subcritical. We can now calculate the specific energy at A using Equation 2.15 as

$$E_A = 2.50 + \frac{(5)^2}{2(32.2)(2.50)^2} = 2.56\ \text{ft}$$

Now, because $E_A = E_B$ we can write Equation 2.15 for section B as

$$2.56 = y_B + \frac{(10)^2}{2(32.2)y_B^2}$$

This equation has two positive roots for y_B, which are 2.26 ft and 1.0 ft. The former of the two is a subcritical depth and the latter is supercritical. We will need to examine the specific energy diagrams for $q = 5$ cfs/ft and

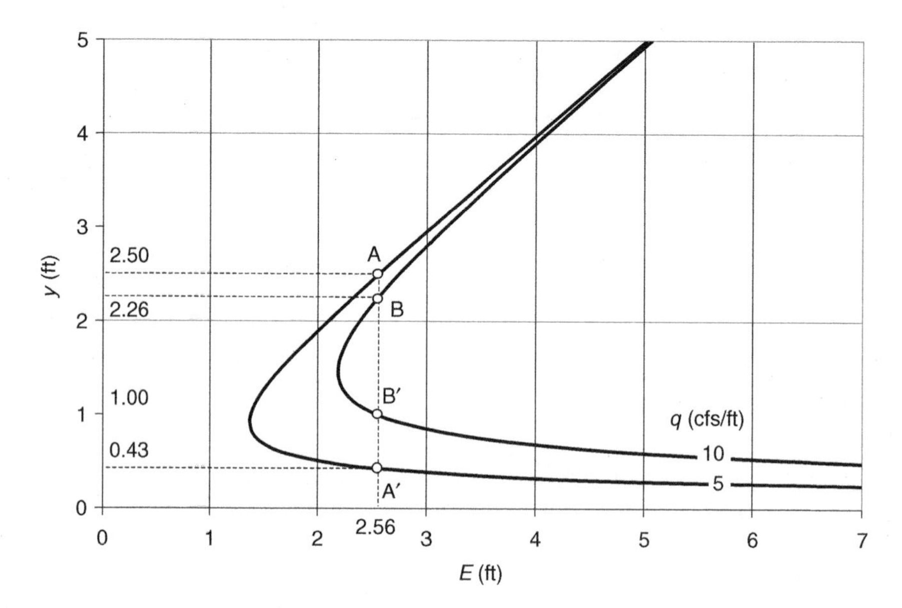

FIGURE 2.17 Specific energy diagrams for Example 2.9

$q = 10$ cfs/ft shown in Figure 2.17 to determine which of the two depths will occur at section B. In the figure, point A represents the section A of the channel. Also marked in the figure is $E_A = E_B = 2.56$ ft. Mathematically, for $E_B = 2.56$ ft, there are two possible flow conditions, which are represented by point B and B′ in the figure.

As the flow moves from section A to B in the channel, the point representing the flow in the figure should move from A to either B or B′. Because the specific energy is constant between sections A and B in the channel, the only path from A to B or B′ in Figure 2.17 is along the vertical line connecting these points. Moving from point A to B along this line, we will cross other specific energy diagrams (not shown in the figure) for q values increasing from 5 cfs/ft to 10 cfs/ft gradually between section A and B. This is consistent with the problem statement. Therefore point B is accessible from point A. However, part of the path from A to B′ (the part from B to B′) would imply that q is greater than 10 cfs/ft at some flow sections in the channel between section A and B (see Figure 2.14). Because such an increase in q is not possible, the point B′ in the figure is not accessible from point A. Therefore, point B represents the only possible solution, and $y_B = 2.26$ ft. Note that both y_A and y_B are subcritical, and the flow depth decreases in the contracted section. This can be generalized to other similar situations involving subcritical flow.

(b) The flow is supercritical if $y_A = 0.43$ ft, and E_A is calculated as being 2.56 ft. Section A of the channel in this case is represented by point A′ in Figure 2.17. Between points B and B′ in the figure, only B′ is accessible from A′. Thus $y_B = 1.0$ ft. Note that both y_A and y_B are supercritical in this case, and the flow depth increases in the contracted section. This can be generalized to other similar situations involving supercritical flow.

2.2.5 CHOKING OF FLOW

It should be clear from the preceding sections that a channel section cannot pass a discharge of any magnitude with any specific energy. For example at least 2 ft of specific energy is needed to pass 20 cfs through the circular channel section considered in Figure 2.9. Likewise, Figure 2.13 indicates that the trapezoidal section investigated can pass a maximum of 290 cfs at a specific energy of 4.0 ft.

If the flow does not have the minimum specific energy needed to pass a channel section, it will adjust itself to increase the specific energy, decrease the discharge, or both. This is often referred to as *choking*. We will defer the situations involving changes in the discharge to Chapter 4. The examples included here will discuss the situations in which the flow is backed up to increase the specific energy and maintain the same discharge.

Note that we are also deliberately deferring the case of flow choking in supercritical flow to Chapter 4. The reader has not been given adequate information to cover this topic yet.

EXAMPLE 2.10 The trapezoidal channel considered in Example 2.6 has a bottom width of $b = 6$ ft and side slopes of $m = 2$ (1V:2H), and it carries a discharge of $Q = 290$ cfs. Suppose the channel is nearly horizontal except that there is a smooth, short upward step in the channel bottom as shown in Figure 2.10. Suppose the height of the step is $\Delta z = 2.25$ ft (as opposed to 1.0 ft in Example 2.6). The energy loss due to this step is still negligible. As in Example 2.6, the depth at section A just before the step is 5.33 ft. Determine the flow depth over the step, that is at section B.

We may be tempted to solve this problem just like we solved Example 2.6. If we proceeded in the same manner, we would obtain $E_A = 5.5$ ft, $E_B = 5.5 - 2.25 = 3.25$ ft. Then we would set up the expression

$$3.25 = y_B + \frac{(290)^2}{2(32.2)\{[6.0 + (2.0)(y_B)](y_B)\}^2}$$

However, trying to solve this equation for y_B would become frustrating because no solutions are available. Indeed, we can see this from Figure 2.18, in which point A represents the channel section A. Also marked on the figure is $E_A = 5.50$ ft, and $E_B = 3.25$ ft is also marked; however, the vertical line representing $E_B = 3.25$ ft does not intersect the specific energy diagram, implying that 290 cfs cannot pass through section B if the available specific energy is only 3.25 ft. So choking will occur, and the flow will need to adjust.

To solve problems of this kind mathematically, without calculating and plotting a specific energy diagram, we should first check if choking will occur. We have already determined that without any flow adjustment, $E_B = 3.25$ ft. We should now calculate the minimum specific energy required at section B to

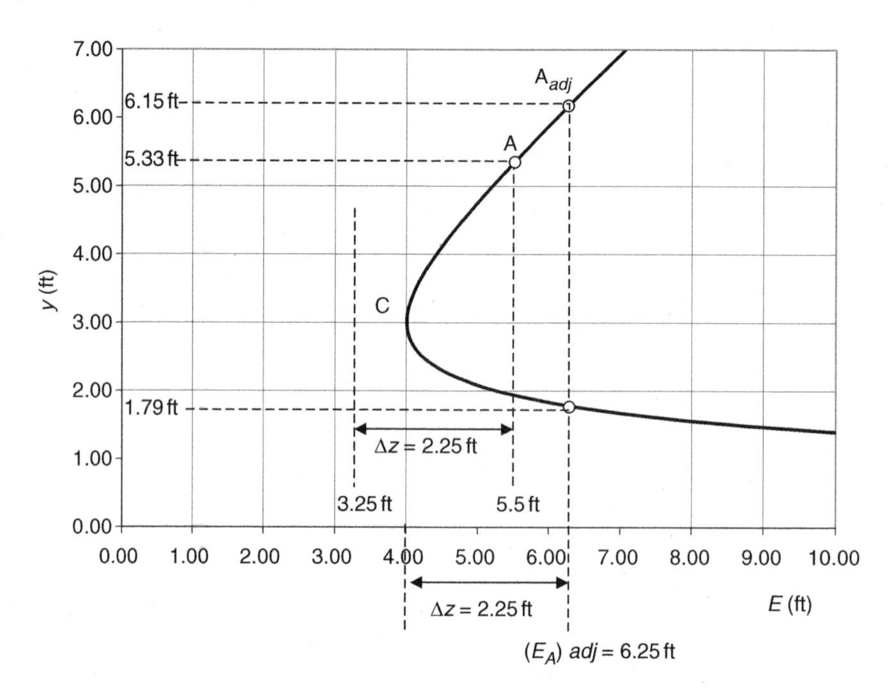

FIGURE 2.18
Example of choking due to step rise

pass 290 cfs. We know that the minimum specific energy occurs at critical depth. In Example 2.1, we calculated that for this channel $y_{cB} = 3.0$ ft. Then by using Equation 2.12

$$(E_B)_{\min} = y_{cB} + \frac{Q^2}{2gA_{cB}^2} = 3.0 + \frac{(290)^2}{2(32.2)\{[(6.0) + (2)(3.0)](3.0)]\}^2} = 4.0 \text{ ft}$$

Noting that 3.25 ft < 4.0 ft, we can conclude that, without any adjustment, the flow does not have the minimum required specific energy at section B – that is, $E_B < (E_B)_{\min}$. If E_B was greater than $(E_B)_{\min}$, we could solve this problem just like we solved Example 2.6. However, in this case choking occurs, and we should determine how the flow will adjust to it.

The minimum specific energy required at section B is 4.0 ft. Then, the specific energy needed at section A is 4.0 + 2.25 = 6.25 ft. The flow will back up and adjust itself to attain this specific energy of $(E_A)_{adj} = 6.25$ ft, where the subscript *adj* stands for adjusted. The corresponding adjusted flow depth at section A can be calculated by using Equation 2.12 as

$$6.25 = y_{Aadj} + \frac{(290)^2}{2(32.2)\{[6.0 + (2.0)(y_{Aadj})](y_{Aadj})\}^2}$$

By trial-and-error, we obtain $y_{Aadj} = 6.15$ ft. The second positive root for the equation is 1.79 ft, a supercritical depth.

Therefore, in this problem, the flow depth will back up to 6.15 ft at section A to maintain the discharge of 290 cfs. The flow at section B will be critical with a depth of 3.0 ft. The depth at section D can be either 6.15 ft or 1.79 ft, depending on the conditions further downstream (as we will see in Chapter 4). Figure 2.18 depicts a graphical solution to the problem.

EXAMPLE 2.11 The nearly horizontal rectangular channel considered in Example 2.9 and shown in Figure 2.16 is 12 ft wide; it carries 60 cfs, and the flow depth at section A is 2.5 ft. Determine whether choking occurs if the channel width is contracted to 4.0 ft (as opposed to 6.0 ft as in Example 2.9) at section B. Also determine the flow depth at section B.

In Example 2.9 we calculated that for $y_A = 2.5$ ft the flow is subcritical, and $E_A = 2.56$ ft. Without any flow adjustment, the available specific energy at B would also be 2.56 ft. The critical depth corresponding to $q_B = 60/4 = 15$ cfs/ft is $y_{cB} = [(15.0)^2/32.2)]^{1/3} = 1.91$ ft. Because the channel is rectangular, the minimum specific energy required at section B is $(E_B)_{\min} = (3/2)y_{cB} = 2.87$ ft. The minimum required specific energy is greater than that which would be available without any flow adjustment. Therefore choking will occur.

If the available specific energy without flow adjustment were greater than the minimum required specific energy, we would solve this problem the way we solved Example 2.9. However, in this case the available specific energy is not adequate, and therefore the flow will back up to attain a specific energy of 2.87 ft at section A. Then, noting that $q_A = 60/12 = 5$ cfs/ft, and using Equation 2.15,

$$2.87 = y_{Aadj} + \frac{(5)^2}{2(32.2)(y_{Aadj}]^2}$$

We can solve this equation by trial and error to obtain a subcritical depth of $y_{Aadj} = 2.82$ ft. A supercritical depth of 0.40 ft also satisfies the equation.

Then under the adjusted flow conditions, the depth at section A will be 2.82 ft. The flow at B will be critical at depth 1.91 ft. The depth at D can be either 0.40 ft or 2.82 ft depending on the downstream condition, as we will see in Chapter 4.

A graphical representation of this example is depicted in Figure 2.19.

2.3 APPLICATIONS OF MOMENTUM PRINCIPLE FOR STEADY FLOW

2.3.1 MOMENTUM EQUATION

We derived the conservation of momentum equation in Chapter 1 (see Equation 1.53). Assuming $\beta =$ momentum coefficient $= 1.0$, for simplicity, we can rewrite the equation between an upstream channel section U and a downstream section D as

$$\left(\frac{Q^2}{gA_U} + Y_{CU}A_U\right) - \frac{F_f}{\gamma} - \frac{F_e}{\gamma} + \Delta x S_0 \frac{A_D + A_U}{2} = \left(\frac{Q^2}{gA_D} + Y_{CD}A_D\right) \qquad (2.18)$$

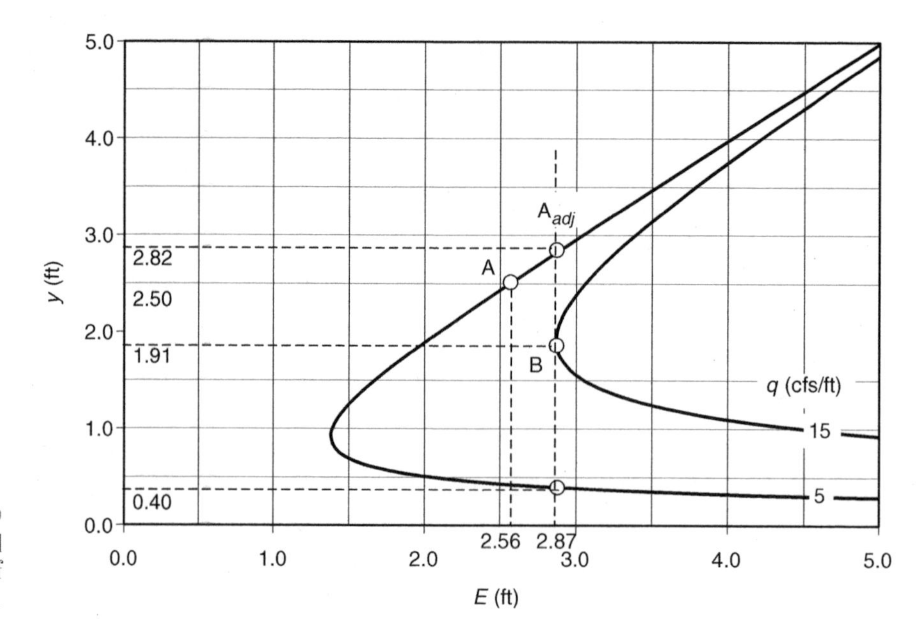

FIGURE 2.19 Graphical representation of Example 2.11

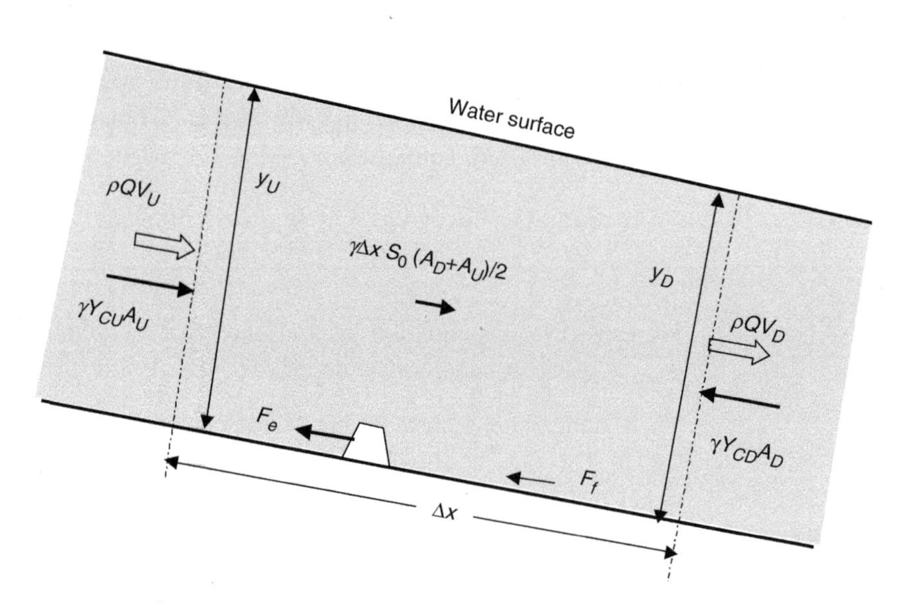

FIGURE 2.20 Momentum equation

where Q = discharge, Y_C = distance from free surface to centroid of the flow section, A = flow area, g = gravitational acceleration, F_f = friction force resisting to flow, F_e = sum of all external forces (other than hydrostatic pressure, friction, and gravity forces) assumed to act in the direction opposite to flow direction, γ = specific weight of water, Δx = distance between the two sections, and S_0 = longitudinal bottom slope of the channel. In this equation, $\gamma(\Delta x)(S_0)(A_D + A_U)/2$ represents the component of weight of water between the two sections in the flow direction. Likewise, $(\gamma Y_C A)$ represents the hydrostatic pressure force, and $\gamma Q^2/gA = \rho QV$ represents the rate of momentum transfer. Figure 2.20 displays a schematic representation of the momentum equation.

EXAMPLE 2.12 The channel shown in Figure 2.5 is rectangular in cross-section, and it is 10 ft wide. Suppose the friction forces and the weight component in the flow direction are negligible. Determine the magnitude and the direction of the force exerted by flow on the spillway.

We can simplify Equation 2.18 for this case as

$$\left(\frac{Q^2}{gA_U}+Y_{CU}A_U\right)-\frac{F_e}{\gamma}=\left(\frac{Q^2}{gA_D}+Y_{CD}A_D\right)$$

or

$$\left(\frac{(120)^2}{(32.2)(4.0)(10.0)}+(2.0)(4.0)(10.0)\right)-\frac{F_e}{62.4}=\left(\frac{(120)^2}{(32.2)(1)(10)}+(0.5)(1.0)(10.0)\right)$$

Solving this equation for F_e, we obtain $F_e = 2587$ lb. Note that the positive value indicates that the assumed direction is correct. Therefore the force exerted by the spillway on the flow is to the left (opposing the flow direction), and then the force exerted by the flow on the spillway is 2587 lb and it points to the right.

2.3.2 SPECIFIC MOMENTUM DIAGRAM FOR CONSTANT DISCHARGE

Specific momentum is defined as

$$M=\left(\frac{Q^2}{gA}+Y_CA\right) \tag{2.19}$$

and a plot of flow depth versus the specific momentum for a constant discharge is called the *specific momentum diagram*. As displayed qualitatively in Figure 2.21, a specific momentum diagram indicates that the same discharge can pass through a channel section at two different flow depths corresponding to the same specific momentum. These depths, marked as y_1 (supercritical) and y_2 (subcritical) in the figure, are called the *conjugate depths* (Henderson 1966). The minimum

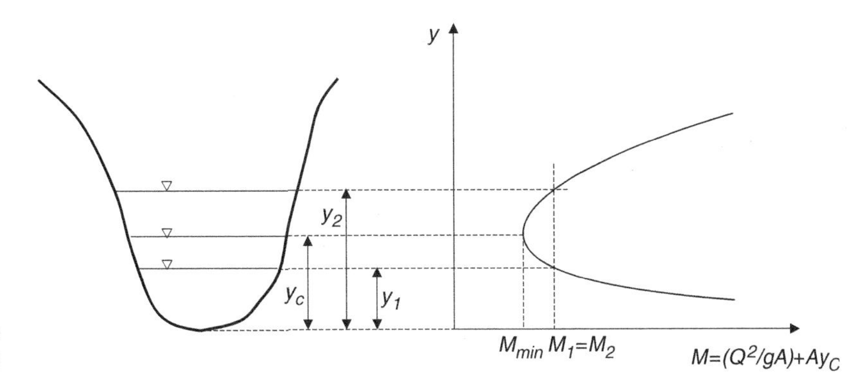

FIGURE 2.21 Specific momentum diagram

momentum required to pass a given discharge through the section occurs at critical depth (see conditions for critical depth in section 2.1). The upper limb of the diagram is for subcritical flow, and the lower limb represents supercritical flow.

EXAMPLE 2.13 The trapezoidal channel considered in Examples 2.1, 2.4 and 2.6 has a bottom width of $b = 6$ ft and side slopes of $m = 2$ (1V:2H), and it carries a discharge of $Q = 290$ cfs. Calculate and plot the specific momentum diagram for this channel. Also, calculate and plot the specific energy diagrams for the same channel for $Q = 135$ cfs and $Q = 435$ cfs.

The calculations for $Q = 290$ cfs are summarized in Table 2.5. The values of y in column 1 are first picked, then the expressions given in Table 2.1 for A and AY_C are used, respectively, to calculate the entries in columns 2 and 3. Finally, we use Equation 2.19 to determine the entries in column 4. The calculations for $Q = 145$ cfs and 435 cfs are performed in the same manner. The results are plotted in Figure 2.22.

TABLE 2.5 Specific momentum diagram calculations for a trapezoidal channel

y (ft)	A (ft^2)	AY_C (ft^3)	M (ft^3)
0.50	3.50	0.83	747.06
0.60	4.32	1.22	605.81
0.80	6.08	2.26	431.83
0.90	7.02	2.92	374.97
1.00	8.00	3.67	330.14
1.25	10.63	5.99	251.81
1.50	13.50	9.00	202.47
1.75	16.63	12.76	169.86
2.00	20.00	17.33	147.92
2.25	23.63	22.78	133.33
2.50	27.50	29.17	124.14
2.75	31.63	36.55	119.14
3.00	36.00	45.00	117.55
3.25	40.63	54.57	118.86
3.50	45.50	65.33	122.74
3.75	50.63	77.34	128.93
4.00	56.00	90.67	137.31
4.25	61.63	105.36	147.75
4.50	67.50	121.50	160.19
4.75	73.63	139.14	174.61
5.00	80.00	158.33	190.98
5.25	86.63	179.16	209.31
5.50	93.50	201.67	229.60
6.00	108.00	252.00	276.18
7.00	140.00	375.67	394.32
8.00	176.00	533.33	548.17

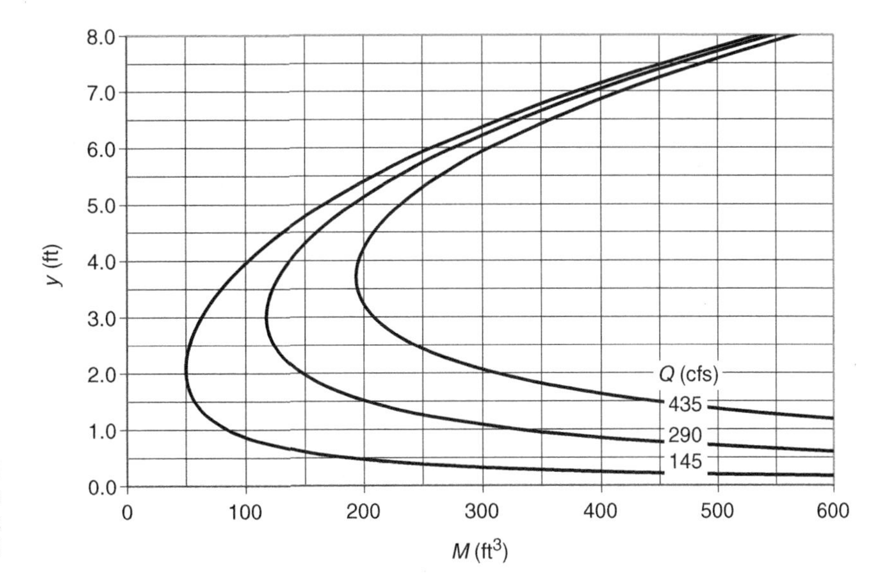

FIGURE 2.22
Specific momentum diagrams for a trapezoidal channel

An inspection of Figure 2.22 for $Q = 290$ cfs reveals that the critical depth for the channel is 3.0 ft. This is consistent with the results of Examples 2.1, 2.4, and 2.6. Also, for this channel, a specific momentum of at least 117.55 ft^3 is required (corresponding to the critical depth) to pass 290 cfs. The reader should note the relative positions of the specific momentum diagrams with respect to the increasing discharge in Figure 2.22.

EXAMPLE 2.14 The circular storm sewer considered in Examples 2.2 and 2.5 has a diameter of $d_0 = 36$ inches and carries a discharge of $Q = 30$ cfs. Calculate and plot the specific momentum diagram for this channel. Also, calculate and plot the specific momentum diagrams for $Q = 10$ cfs and $Q = 20$ cfs.

The calculations for $Q = 30$ cfs are summarized in Table 2.6. First, the y values in column 1 are picked. Then, the θ values in column 2 are obtained by using the expression given for θ in Table 2.1. Next, we use the expressions given for A and AY_C for circular channels in Table 2.1 to calculate the entries, respectively, in columns 3 and 4. Finally, by using Equation 2.19, we calculate the entries in column 5 for M.

The calculations are performed in the same manner for $Q = 10$ cfs and 20 cfs. The results are plotted in Figure 2.23. An inspection of the diagrams in Figure 2.23 reveals that for $Q = 30$ cfs, the critical depth is equal to 1.77 ft. This is consistent with the results of Examples 2.2 and 2.5. The corresponding specific momentum is 9.75 ft^3. In other words, a specific momentum of at least 9.75 ft^3 is needed to pass 30 cfs in this circular channel. Also note that the relative positions of the diagrams with respect to the increasing discharge are similar to those of a trapezoidal channel shown in Figure 2.22. This trend is the same for all types of channels.

TABLE 2.6 Specific momentum diagram calculations for a circular channel

y (ft)	θ (rad.)	A (ft^2)	$A\bar{Y}_C$ (ft^3)	M (ft^3)
0.20	0.52	0.20	0.02	138.12
0.40	0.75	0.56	0.09	49.98
0.60	0.93	1.01	0.25	28.02
0.80	1.09	1.51	0.50	18.97
1.00	1.23	2.06	0.85	14.41
1.20	1.37	2.64	1.32	11.91
1.40	1.50	3.23	1.91	10.55
1.60	1.64	3.83	2.62	9.91
1.77	1.75	4.34	3.31	9.75
1.80	1.77	4.43	3.44	9.76
2.00	1.91	5.01	4.39	9.97
2.20	2.06	5.56	5.45	10.48
2.40	2.21	6.06	6.61	11.22
2.60	2.39	6.51	7.87	12.16
2.80	2.62	6.87	9.21	13.28
2.90	2.77	7.00	9.90	13.89
2.95	2.88	7.04	10.25	14.22

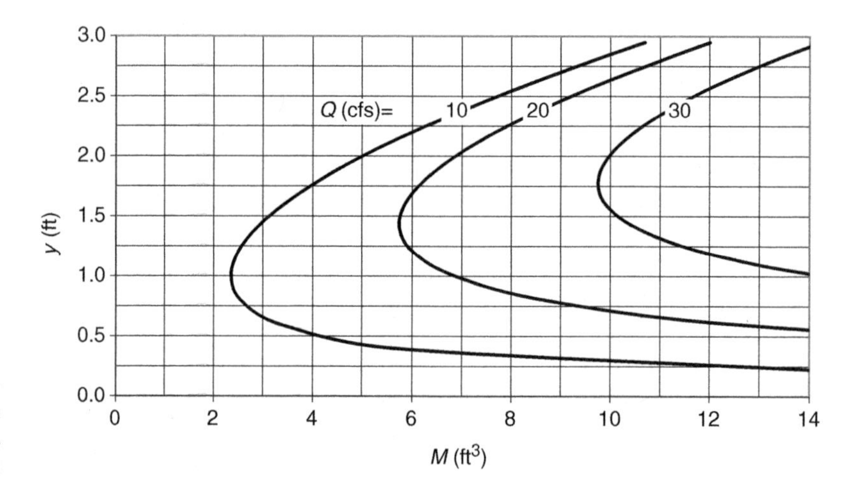

FIGURE 2.23 Specific momentum diagrams for a circular channel

EXAMPLE 2.15 The trapezoidal channel considered in Example 2.6 has a bottom width of $b = 6$ ft and side slopes of $m = 2$ (1V:2H), and it carries a discharge of $Q = 290$ cfs. The channel is nearly horizontal except that there is a smooth, short step in the channel bottom as shown in Figure 2.10. The height of the step is $\Delta z = 1.0$ ft. The depth at section A just before the step is 5.33 ft, and the depth over the step at section B is 4.10 ft. Determine the force exerted by the flow on the step between sections A and B.

We could solve this problem mathematically using the same procedure as in Example 2.12. However, because we have already calculated the specific

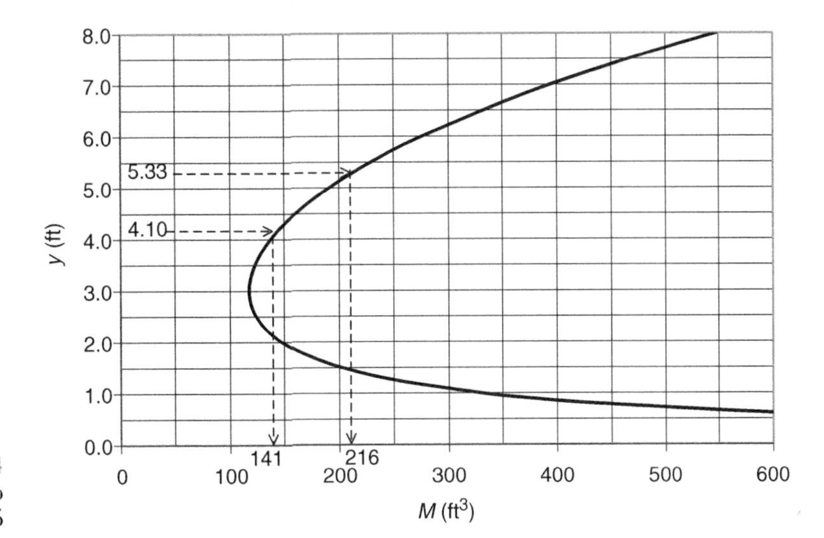

FIGURE 2.24 Solution to Example 2.15

momentum diagram for this channel in Example 2.13, we can also use a semi-graphical approach. The specific energy diagram for 290 cfs is replotted in Figure 2.24.

The specific momentum, M_A, corresponding to $y_A = 5.33$ ft at section A is determined directly from Figure 2.24 as being 216 ft^3. Likewise, at section B, $M_B = 141$ ft^3 for $y_B = 4.10$ ft. Now, by using Equation 2.18 with $F_f \approx 0$ and $S_0 \approx 0$, we obtain

$$216 - \frac{F_e}{62.4} = 141$$

Then $F_e = 62.4\ (216 - 141) = 4680$ lb. This is the force exerted by the step on the flow between sections A and B, and it is in the direction opposing the flow. The force exerted by the flow on the step has the same magnitude, but the direction is the same as the flow direction.

2.3.3 DISCHARGE DIAGRAM FOR CONSTANT SPECIFIC MOMENTUM

The specific momentum diagrams discussed in the preceding section are graphical representations of Equation 2.19 for a constant discharge, Q. We can calculate and plot similar diagrams displaying the variation of flow depth y with discharge Q for constant values of the specific momentum M. Figure 2.25 displays the discharge diagrams for a trapezoidal channel having a bottom width of 6.0 ft and side slopes of $m = 2$ (1V:2H).

The discharge diagrams demonstrate that the maximum discharge a channel section can pass depends on the specific momentum the flow has. For example, if the specific momentum is 50 ft^3, a trapezoidal section having $b = 6$ ft and $m = 2$

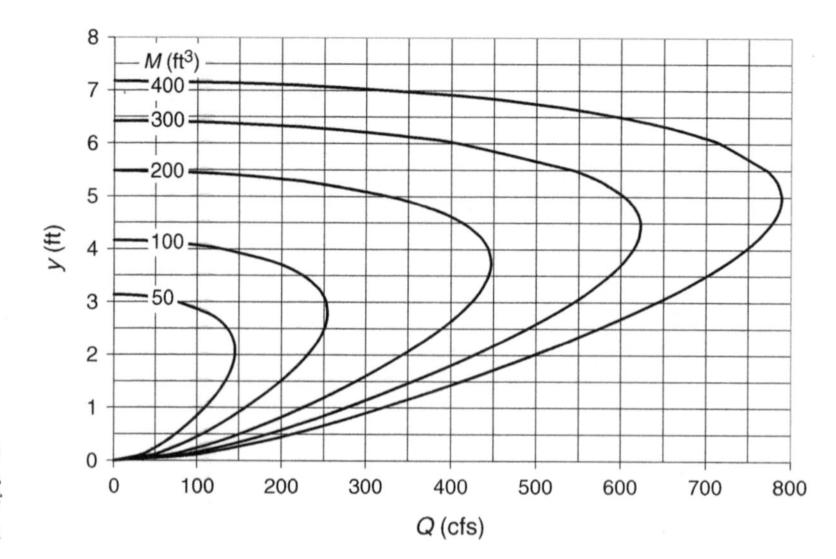

FIGURE 2.25
Discharge diagrams for constant values of specific momentum

can pass a maximum of 145 cfs as seen from Figure 2.25. The depth corresponding to 145 cfs is 2 ft, and the reader can confirm (by calculating the Froude number) that at this depth the flow is in critical state. Indeed, we can state that for a given specific momentum, a channel section passes the maximum discharge at the critical depth.

2.3.4 HYDRAULIC JUMP

Flow in a channel can change from subcritical state to supercritical and *vice versa* due to changes in the channel characteristics or boundary conditions, or the presence of hydraulic structures. The changes from subcritical to supercritical state usually occur rather smoothly via the critical depth. However, the change from supercritical to subcritical state occurs abruptly through a hydraulic jump, as shown in Figure 2.26. A hydraulic jump is highly turbulent, with complex internal flow patterns, and it is accompanied by considerable energy loss. In Figure 2.26, the flow is supercritical at depth y_{J1} just before the jump, and it is subcritical at depth y_{J2} just after the jump.

In most open-channel flow problems involving hydraulic jumps, one of the two depths y_{J1} or y_{J2} would be known, and we would need to calculate the second one. Because the energy loss due the hydraulic jump is usually significant and unknown, we cannot use the energy equation to determine the unknown depth. However, usually the friction force between sections J1 and J2 is negligible. Also, if the channel is nearly horizontal, the component of the weight in the flow direction is negligible. Then, in the absence of any other external forces (other than pressure forces), the momentum equation, Equation 2.18, can be written for the situation of Figure 2.26 as

$$\left(\frac{Q^2}{gA_{J1}} + Y_{CJ1}A_{J1}\right) = \left(\frac{Q^2}{gA_{J2}} + Y_{CJ2}A_{J2}\right) \tag{2.20}$$

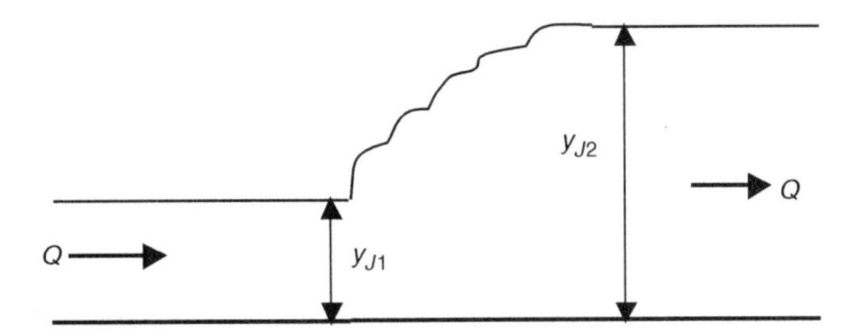

FIGURE 2.26
Hydraulic jump

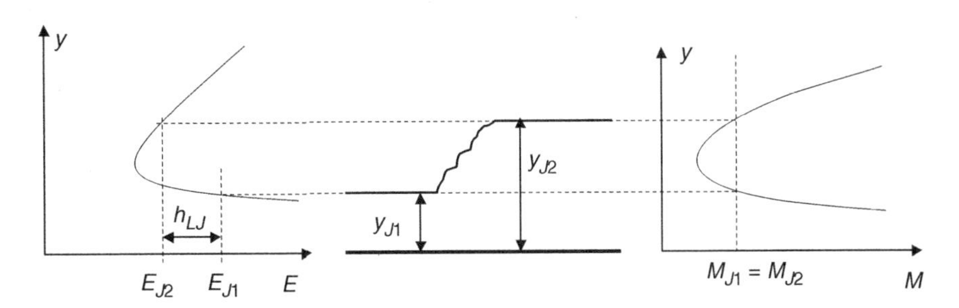

FIGURE 2.27
Hydraulic jump and specific energy and momentum diagrams (adopted from Mays 2001 with permission John Wiley & Son, Inc.)

if the pressure distribution is hydrostatic at Sections J1 and J2. Recalling the definition of the specific momentum (Equation 2.19), we can simplify Equation 2.20 as

$$M_{J1} = M_{J2} \tag{2.21}$$

Equation 2.21 is valid for any cross-sectional shape. Once this equation is solved for the unknown depth, the energy equation can be used to calculate the head loss due to the hydraulic jump. Figure 2.27 demonstrates the relationship between the flow depths before and after the jump, the specific momentum, the specific energy, and the energy loss due to the jump. In the figure, h_{LJ} stands for the head loss due to the jump. Similar figures were previously presented by Henderson (1966) and Mays (2001).

For rectangular channels, an explicit solution is available for Equation 2.21. Further discussion of hydraulic jumps in rectangular channels is given in the subsequent section.

For most other types, the solution requires either a trial and error procedure or construction of the momentum diagrams. Figures 2.28, 2.29, and 2.30 provide pre-determined solutions to the hydraulic jump equation for trapezoidal, circular and triangular channels, respectively.

Although the use of these charts is quite straightforward, an observation in Figure 2.29 is worth noting. Unlike the other types of channel sections considered, a circular section has a closing top. Therefore, it is possible that the flow will not have a free surface at the downstream side of a hydraulic jump. In other words, we may have a supercritical open-channel flow before a hydraulic

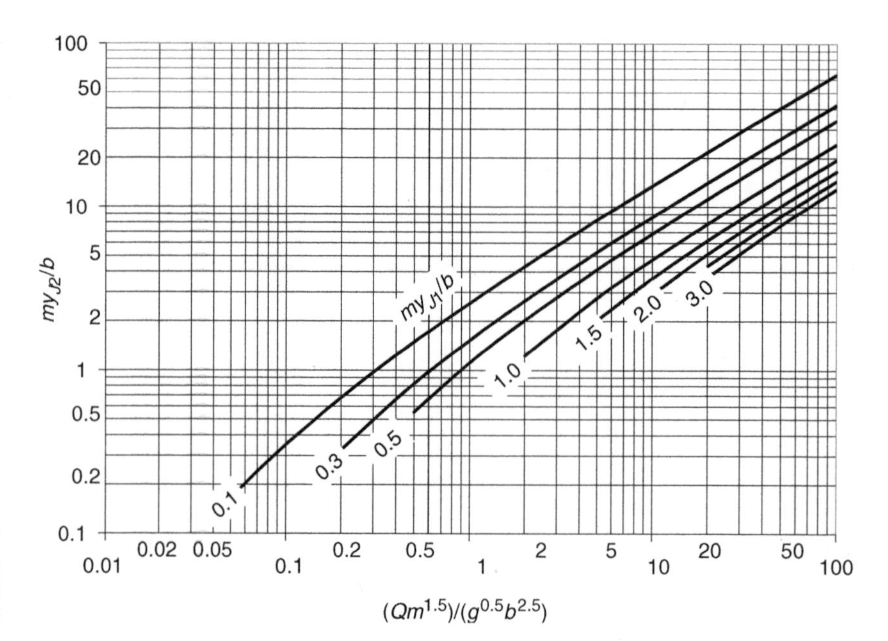

FIGURE 2.28 Hydraulic jump chart for trapezoidal channels

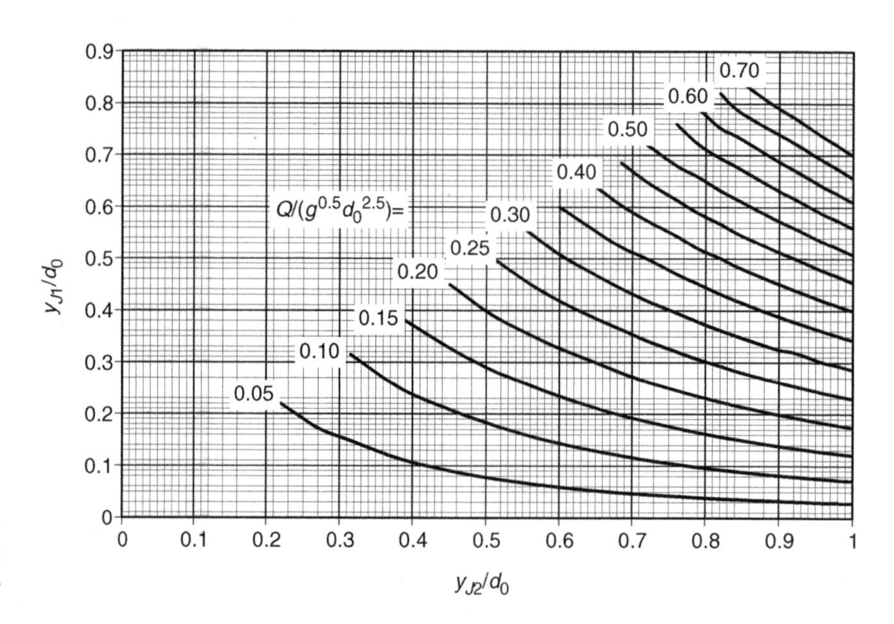

FIGURE 2.29 Hydraulic jump chart for circular channels

jump and full pipe flow after the jump. Figure 2.29 reflects this possibility. For instance, for $Q/(g^{0.5}d_0^{2.5}) = 0.25$ and $y_{J1}/d_0 = 0.3$, we can obtain $y_{J2}/d_0 = 0.8$ from the figure. Thus the flow will have a free surface after the hydraulic jump, and the depth will be equal to 0.8 times the diameter. However, for the same discharge, if $y_{J1}/d_0 = 0.2$, an inspection of Figure 2.30 will reveal that y_{J2}/d_0 is off the chart. We can then conclude that, in this case, the circular channel will flow full downstream of the jump.

EXAMPLE 2.16 The trapezoidal channel considered in Example 2.15 has a bottom width of $b = 6$ ft and side slopes of $m = 2$ (1V:2H), and it carries a

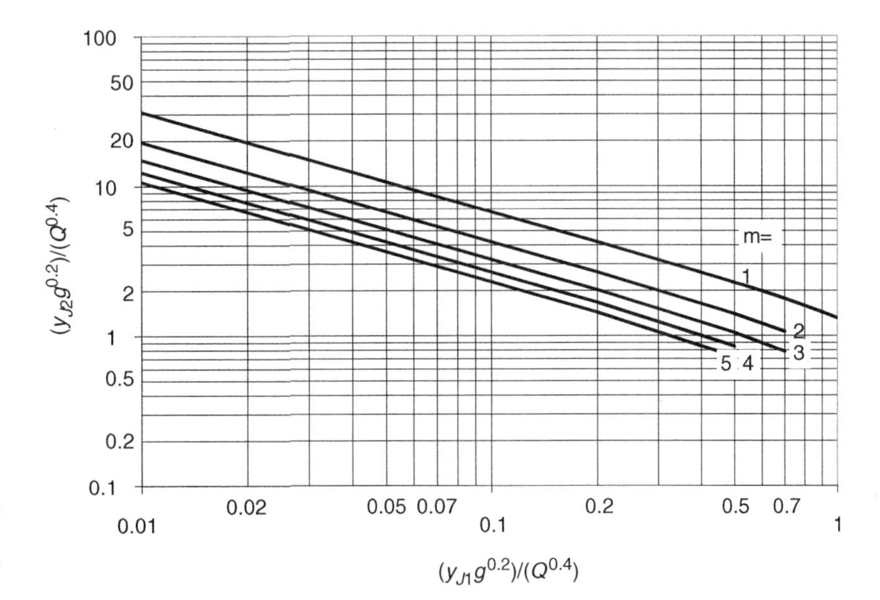

FIGURE 2.30 Hydraulic jump chart for triangular channels

discharge of $Q = 290$ cfs. A hydraulic jump occurs in this channel. The flow depth just before the jump is $y_{J1} = 0.9$ ft. Determine the depth after the jump.

We can solve this problem in several ways. The first and the most precise way is the mathematical approach. We will first calculate the specific momentum, M_{J1}. Substituting the expressions given in Table 2.1 for A and AY_C for trapezoidal channels into Equation 2.19,

$$M_{J1} = \frac{Q^2}{g(b + my_{J1})y_{J1}} + \frac{y_{J1}^2}{6}(2my_{J1} + 3b)$$

$$M_{J1} = \frac{(290)^2}{32.2[6.0 + (2.0)(0.9)](0.9)} + \frac{(0.9)^2}{6}[2(2.0)(0.9) + 3(6.0)] = 375 \text{ ft}^3$$

Now, because $M_{J1} = M_{J2}$, we can write

$$375 = \frac{(290)^2}{32.2[6.0 + (2.0)(y_{J2})](y_{J2})} + \frac{(y_{J2})^2}{6}[2(2.0)(y_{J2}) + 3(6.0)]$$

Solving this equation by trial and error, we obtain, $y_{J2} = 6.85$ ft.

Alternatively, we can solve this problem by first constructing the specific momentum diagram for the channel for $Q = 290$ cfs. This specific momentum diagram has already been calculated and plotted in Figure 2.24. For $y = y_{J1} = 0.9$ ft, we obtain $M = M_{J1} = 375 \text{ ft}^3$ from the figure. The corresponding subcritical depth, y_{J2}, is read directly from the figure as being about 6.85 ft.

We can also use Figure 2.28 to find a quick solution. Let us first evaluate

$$\frac{Qm^{1.5}}{g^{0.5}b^{2.5}} = \frac{(290)(2.0)^{1.5}}{(32.2)^{0.5}(6.0)^{2.5}} = 1.64$$

and

$$\frac{my_{J1}}{b} = \frac{(2.0)(0.9)}{6.0} = 0.3$$

Then, from Figure 2.28, we obtain $(my_{J2}/b) = 2.3$. Thus, $y_{J2} = (2.3)(6.0)/(2.0) = 6.9$ ft. This result is close to but slightly different from that of the mathematical approach due to reading errors. When precision is important, the chart in Figure 2.28 and the mathematical approach may be used together. The result obtained from the chart would be the first (and a very good) trial value in the trial-and-error solution.

2.3.5 SPECIFIC MOMENTUM IN RECTANGULAR CHANNELS

For rectangular channels, we can simplify the momentum equation (Equation 2.18) by writing it for a unit width of the channel. Noting that $q = Q/b$ = discharge per unit width, $A = by$, and $Y_C = y/2$ for a rectangular section, we can divide both sides of Equation 2.18 by b and simplify to obtain

$$\left(\frac{q^2}{gy_U} + \frac{y_U^2}{2}\right) - \frac{F_f}{b\gamma} - \frac{F_e}{b\gamma} + \Delta x S_0 \frac{y_D + y_U}{2} = \left(\frac{q^2}{gy_D} + \frac{y_D^2}{2}\right) \tag{2.22}$$

Likewise, the specific momentum for a rectangular section, M_r, is defined as

$$M_r = \left(\frac{q^2}{gy} + \frac{y^2}{2}\right) \tag{2.23}$$

We should note that the specific momentum, M_r, for rectangular channels is defined per unit width and has a dimension of (length)^2. Writing Equation 2.23 for an upstream section U and a downstream Section D, and substituting into Equation 2.22, we obtain

$$M_{rU} - \frac{F_f}{b\gamma} - \frac{F_e}{b\gamma} + \Delta x S_0 \frac{y_D + y_U}{2} = M_{rD} \tag{2.24}$$

For rectangular channels, specific momentum diagrams display the variation of y with M_r for constant q. These diagrams are graphical representations of Equation 2.23. Figure 2.31 shows the specific momentum diagrams calculated for various values of q. The procedure used to calculate these diagrams is similar to that of Example 2.13. However, here, for a constant q, we pick different y values and calculate the corresponding values of M_r from Equation 2.23. Of course it is also possible to prepare the discharge diagrams for constant M_r, but such diagrams are of limited use.

It is important to note that Equations 2.22 and 2.24 can be used only if the width, b, of the rectangular channel is the same (and therefore q is same) at Sections U and D. If the channel width varies, these equations cannot be used even if the channel is rectangular.

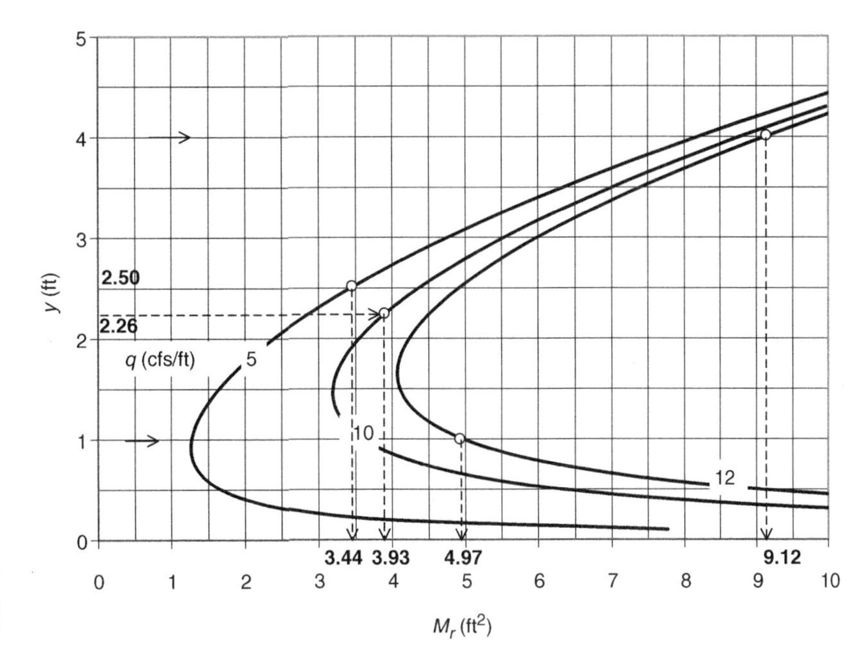

FIGURE 2.31 Specific momentum diagrams for $q = 5$, 10, and 12 cfs/ft

EXAMPLE 2.17 Redo Example 2.12 using the simplified momentum equation for rectangular channels.

Because the width of the rectangular channel is constant at 10 ft between Sections 1 and 2, we can use Equation 2.22 in this problem. The friction force and the component of weight in the flow direction are negligible. Therefore, with $q = Q/b = 120/10 = 12$ cfs/ft, we can write Equation 2.22 as

$$\left(\frac{q^2}{gy_1} + \frac{y_1^2}{2}\right) - \frac{F_e}{b\gamma} = \left(\frac{q^2}{gy_2} + \frac{y_2^2}{2}\right)$$

or

$$\left(\frac{(12)^2}{(32.2)(4.0)} + \frac{(4.0)^2}{2}\right) - \frac{F_e}{(10.0)(62.4)} = \left(\frac{(12)^2}{(32.2)(1.0)} + \frac{(1.0)^2}{2}\right)$$

Solving for F_e, we obtain $F_e = 2587$ lb. This is the force exerted by the spillway on the flow, and it is in the direction opposing the flow. The force exerted by the flow on the spillway is equal to this force in magnitude, but it is in the same direction as the flow direction.

We could also use the specific momentum diagram of the channel (if it has already been calculated and constructed as in Figure 2.31) to solve this problem. From the diagram for $q = 12$ cfs/ft, we obtain $M_{r1} = 9.12$ ft^2 for $y_1 = 4$ ft and $M_{r2} = 4.97$ ft^2 for $y_2 = 1$ ft. Neglecting the friction force and the component of the weight in the flow direction, Equation 2.24 can be written for this case as

$$M_{r1} - \frac{F_e}{b\gamma} = M_{r2}$$

or

$$9.12 - \frac{F_e}{(10)(62.4)} = 4.97$$

Solving for F_e, we obtain $F_e = 2590$ lb. The result is slightly different due to the reading errors.

EXAMPLE 2.18 Consider the rectangular channel investigated in Example 2.9. The channel is nearly horizontal and it carries 60 cfs. The width of the channel smoothly contracts from 12 ft at Section A to 6 ft at Section B (see Figure 2.16). The flow depth at A is 2.50 ft, and in Example 2.9 the flow depth at B was calculated as being 2.26 ft. Determine the force exerted on the flow by the segment of the channel walls between Sections A and B. Assume the friction force is negligible.

Because the width of the channel varies, we cannot use Equation 2.22 in this problem. Instead we will use Equation 2.18. Dropping the terms involving the friction force and the component of weight of water in the flow direction, Equation 2.18 can be written for Sections A and B as

$$\left(\frac{Q^2}{gA_A} + Y_{CA}A_A\right) - \frac{F_e}{\gamma} = \left(\frac{Q^2}{gA_B} + Y_{CB}A_B\right)$$

or

$$\left(\frac{(60)^2}{(32.2)(12.0)(2.50)} + \frac{2.50}{2}(12.0)(2.50)\right) - \frac{F_e}{62.4}$$
$$= \left(\frac{(60)^2}{(32.2)(6.0)(2.26)} + \frac{2.26}{2}(6.0)(2.26)\right)$$

Solving this equation for F_e, we obtain $F_e = 1102$ lb. The force is in the direction opposing the flow.

We could also use the specific momentum diagrams to solve this problem. Note that, in terms of the specific momentum, the momentum equation for this case is

$$M_A - \frac{F_e}{\gamma} = M_B$$

Recalling that $M = bM_r$, we can write

$$b_A M_{rA} - \frac{F_e}{\gamma} = b_B M_{rB}$$

Here $b_A = 12$ ft and $b_B = 6$ ft. Accordingly, $q_A = 60/12 = 5$ cfs/ft and $q_B = 60/6 = 10$ cfs/ft. From Figure 2.31, for $y_A = 2.5$ ft and $q_A = 5$ cfs/ft, we obtain $M_{rA} = 3.44$ ft^2.

Likewise, for $y_B = 2.26$ ft and $q_B = 10$ cfs/ft, we obtain $M_{rB} = 3.93$ ft^2. Substituting these in the equation above,

$$(12.0)(3.44) - \frac{F_e}{62.4} = (6.0)(3.93)$$

This will yield $F_e = 1104$ lb. Due to the reading errors, this result is slightly different.

2.3.6 HYDRAULIC JUMP IN RECTANGULAR CHANNELS

As discussed in the preceding section, for rectangular channels the momentum equation can be written for unit width of the channel (Equation 2.22). This equation applies to a hydraulic jump occurring in any rectangular channel. However, if the channel is horizontal ($S_0 = 0$), the friction force is negligible ($F_f = 0$), and there is no other external force acting on the flow other than the pressure forces ($F_e = 0$), Equation 2.22 reduces to

$$\left(\frac{q^2}{gy_{J1}} + \frac{y_{J1}^2}{2}\right) = \left(\frac{q^2}{gy_{J2}} + \frac{y_{J2}^2}{2}\right) \tag{2.25}$$

where $J1$ represents the flow section just upstream of the jump, and $J2$ represents the section just downstream. We can manipulate Equation 2.25 mathematically to obtain

$$y_{J2} = \frac{y_{J1}}{2}\left(\sqrt{1 + 8F_{rJ1}^2} - 1\right) \tag{2.26}$$

and

$$y_{J1} = \frac{y_{J2}}{2}\left(\sqrt{1 + 8F_{rJ2}^2} - 1\right) \tag{2.27}$$

Equation 2.26 is useful to calculate the flow depth just downstream of the jump if the flow conditions are known upstream. If the conditions are known downstream of the jump and the flow depth upstream is sought, then we can use Equation 2.27. We should recall that, in Equations 2.26 and 2.27, F_r stands for the Froude number, and for rectangular channels it can be calculated by using Equation 2.2.

Once we determine the flow depths upstream and downstream of the hydraulic jump, we can use the energy equation to calculate the head loss due to the jump as

$$h_{LJ} = \left(y_{J1} + \frac{q^2}{2gy_{J1}^2}\right) - \left(y_{J2} + \frac{q^2}{2gy_{J2}^2}\right) \tag{2.28}$$

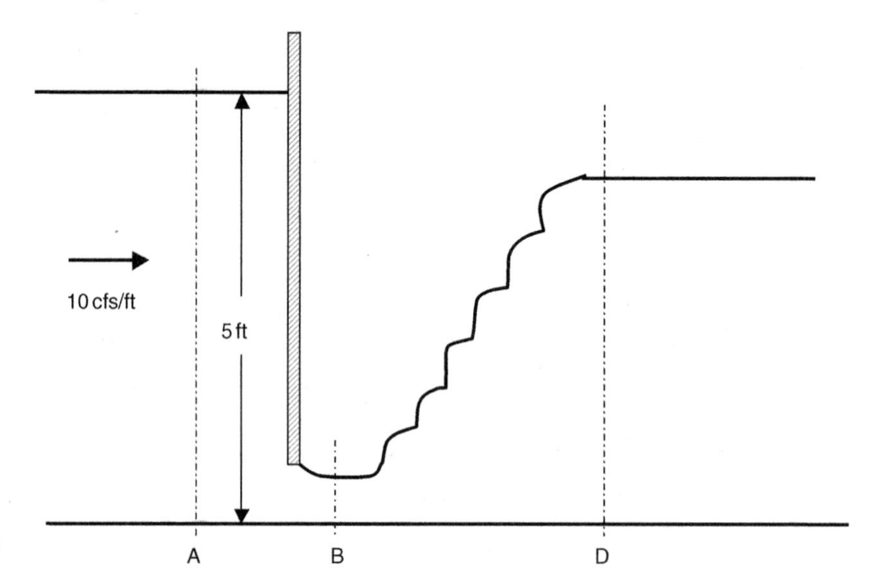

FIGURE 2.32
Example 2.19

This equation can be manipulated to obtain

$$h_{LJ} = \frac{(y_{J2} - y_{J1})^3}{4y_{J1}y_{J2}} \tag{2.29}$$

EXAMPLE 2.19 The rectangular channel shown in Figure 2.32 is nearly horizontal, and it carries $q = 10$ cfs/ft. The flow depth upstream of the sluice gate is 5 ft. A hydraulic jump occurs on the downstream side of the sluice gate. Determine the flow depth at Sections B and D, and the head loss due to the hydraulic jump.

We first need to calculate the flow depth at Section B. The sluice gate applies a force on the flow in the direction opposite to the flow. This force is unknown. Therefore, we can not use the momentum equation to calculate y_B given q and y_A, since the equation would include two unknowns: y_B and F_e. On the other hand, we can neglect the energy loss due to the sluice gate and write the energy equation between Sections A and B as

$$\left(y_A + \frac{q^2}{2gy_A^2}\right) = \left[5.0 + \frac{(10)^2}{2(32.2)(5.0)^2}\right] = 5.06 = \left(y_B + \frac{q^2}{2gy_B^2}\right)$$

This equation will yield two positive values for y_B; 5.0 ft and 0.59 ft. The former is a subcritical depth and the latter is a supercritical depth. Since the flow in Section B is supercritical (otherwise a jump could not occur), $y_B = 0.59$ ft. Now, we can use Equation 2.26 to calculate the depth at Section D. First, let us calculate the Froude number at Section B using Equation 2.2 as

$$F_{rB} = \frac{10}{\sqrt{(32.2)(0.59)^3}} = 3.89$$

Then by using Equation 2.26 with Section B in place of $J1$ and D in place of $J2$,

$$y_D = \frac{0.59}{2}\left(\sqrt{1 + (8)(3.89)^2} - 1\right) = 2.96\,\text{ft}$$

Finally, by using Equation 2.29

$$h_{LJ} = \frac{(2.96 - 0.59)^3}{(4)(0.59)(2.96)} = 1.91\,\text{ft}$$

2.3.7 CHOKING AND MOMENTUM PRINCIPLE

We discussed the problem of choking in terms of energy in Section 2.2.5. Basically, if the flow does not have the minimum required specific energy to pass a certain discharge through a contracted section, it has to back up to acquire the required energy. The critical flow condition at the contracted section determines the minimum required specific energy. This is demonstrated in Example 2.10, where the flow adjusts itself to a new depth at Section A to acquire the required energy to pass the discharge through Section B. In this example, we neglect the energy loss between the two sections.

We can also discuss the problem of choking in terms of momentum. We have already seen in Section 2.3.2 that a certain specific momentum is required to pass a specified discharge through a specified channel section. The minimum required specific momentum corresponds to the critical flow condition. Let us consider a channel contraction in a rectangular section, as shown in Figure 2.16. Let, for the given discharge, the critical depth at section B be y_{cB}. Then the minimum required specific momentum at Section B is

$$(M_B)_{\min} = \left(\frac{Q^2}{gA_B} + Y_{CB}A_B\right)_c \tag{2.30}$$

where the subscript c on the right-hand side denotes critical flow. For a rectangular channel, Equation 2.30 becomes

$$(M_B)_{\min} = \frac{Q^2}{gb_By_{cB}} + \frac{b_B}{2}y_{cB}^2 \tag{2.31}$$

Given the discharge and the channel width at Section B, we can easily evaluate Equation 2.31. However, we need to find the corresponding minimum required specific momentum at Section A to determine whether choking occurs. For this purpose, we can use Equation 2.18 with $F_f = 0$ and $S_0 = 0$ to obtain

$$(M_A)_{\min} - \frac{F_e}{\gamma} = (M_B)_{\min} \tag{2.32}$$

where F_e = external force exerted by the channel walls on the flow between Sections A and B. This external force is not always negligible, and it needs to be somehow evaluated to apply the momentum principle to the problem of choking. We will revisit the momentum approach in Chapter 7, when we discuss the problem of choking in more detail at bridge structures.

PROBLEMS

P.2.1 Determine if the flow is subcritical or supercritical in the channels tabulated below.

Section type	b (m)	Q (m^3/s)	d_0 (m)	m	y (m)
Rectangular	0.8	1.6		0	1.1
Trapezoidal	1.0	4.0		1.5	1.2
Triangular	0	2.0		2	0.9
Circular		1.2	1.0		0.8

P.2.2 Determine if the flow is subcritical or supercritical in the channels tabulated below.

Section type	b (ft)	Q (cfs)	d_0 (ft)	m	y (ft)
Rectangular	5.0	100		0	2.0
Trapezoidal	6.0	200		2	2.8
Triangular	0	45		1.5	2.0
Circular		100	5		3.2

P.2.3 Calculate and plot the specific energy diagrams for each of the channels in Problem P.2.1 for the given discharge, and verify your answers to Problem P.2.1.

P.2.4 Calculate and plot the specific energy diagrams for each of the channels in Problem P.2.2 for the given discharge, and verify your answers to Problem P.2.2.

P.2.5 Is the pressure distribution on the vertical face of the spillway in Figure 2.5 hydrostatic? Explain your answer.

P.2.6 Derive Equation 2.13 given Equation 2.12.

P.2.7 If the same scale is used for the Y and E axes in Figure 2.6, show that the straight line asymptotical to the upper limb makes a 45° angle with the E axis. Also show that the E axis will be asymptotical to the lower limb.

P.2.8 In Example 2.6, what is the minimum specific energy needed at Section B to pass 290 cfs through this section? What is the corresponding specific energy at A? Does the flow have adequate specific energy at A to sustain 290 cfs?

P.2.9 A trapezoidal channel has a bottom width of b = 30 ft and side slopes m = 2, and it carries Q = 5100 cfs.

(a) Calculate and plot the specific energy diagram for this channel. Use a depth range of 2 to 24 ft.
(b) Three piers, each 2 ft wide, support a bridge spanning the channel at a bridge section. Assume that at this location the channel section is trapezoidal with $m=2$ and $b=30-3(2)=24$ ft. Calculate and plot the specific energy diagram at the bridge section.
(c) Determine the flow depth at the bridge section if the depth upstream is 16 ft.

P.2.10 Suppose the nearly horizontal, 12-ft wide rectangular channel shown in Figure 2.16 carries 60 cfs at a depth 3.0 ft. The width is contracted to 6 ft at Section B. In addition, there is a smooth step rise of Δz at the contracted section. Determine the flow depth at A and B if

(a) $\Delta z=0.5$ ft.
(b) $\Delta z=1.0$ ft.

P.2.11 Verify the specific momentum diagrams given for $Q=145$ cfs and $Q=435$ cfs in Example 2.13 by calculating values of M for $y=1$, 2, 3, and 4 ft.

P.2.12 Verify the specific momentum diagrams given for $Q=10$ cfs and 20 cfs in Example 2.14 by calculating the values of M for $y=0.5$, 1.0, 1.5, and 2.0 ft.

P.2.13 A hydraulic jump occurs in a 36-inch storm sewer carrying 20 cfs. The flow depth just upstream of the jump is 1.0 ft. Determine the flow depth downstream of the jump.

P.2.14 A storm sewer with a diameter of 1.0 m carries a discharge of 0.75 m^3/s. A hydraulic jump occurs in this sewer, and the flow depth upstream of the jump is 0.30 m. Find the depth downstream of the jump.

P.2.15 A hydraulic jump occurs in a trapezoidal channel having $b=6$ ft, $m=2$, and $Q=200$ cfs. The flow depth just before the jump is 1 ft.

(a) Determine the flow depth after the jump.
(b) Determine the head loss due to the jump.

P.2.16 A trapezoidal channel having $b=10$ m and $m=1.5$ carries a discharge of $Q=1320$ m^3/s. A hydraulic jump occurs in this channel, and flow depth just after the jump is 13.3 m.

(a) Determine the flow depth before the jump.
(b) Determine the head loss due to the jump.

P.2.17 Derive Equation 2.26 given Equation 2.25.

P.2.18 Derive Equation 2.29 given Equation 2.28.

P.2.19 Equation 2.26 is obtained for horizontal rectangular channels. Everything else remaining the same, how would y_{J2} be affected if the channel sloped down in the flow direction?

P.2.20 Equation 2.26 is obtained assuming that the friction forces are negligible. Everything else remaining the same, how would y_{J2} be affected if the friction forces were significant?

REFERENCES

Chow, V. T. (1959). *Open-Channel Hydraulics*. McGraw-Hill Book Co., New York, NY.
Henderson, F. M. (1966). *Open Channel Flow*. Prentice Hall, Upper Saddle River, NJ.
Mays, L. W. (2001). *Water Resources Engineering*. John Wiley & Sons, Inc., New York, NY.

3 Normal flow

Flow in an open channel is called *uniform flow* or *normal flow* if the depth, flow area, and velocity remain constant at every cross-section along the channel. Strictly speaking, normal flow is possible only in prismatic channels, and it rarely occurs naturally. However, the flow tends to become normal in very long channels in the absence of flow controls such as hydraulic structures. The normal flow equations to be presented in this chapter appear to be satisfied even in irregular channels in the absence of hydraulic structures. Moreover, the concept of normal flow is central to the analysis and design procedures for open channels.

3.1 Flow Resistance

Resistance to flow can be explained in terms of the external or internal friction forces. External friction forces are encountered on the channel boundary, and are included in the momentum equation as we discussed in Chapter 1. The internal friction forces, however, occur due to velocity gradients within a flow cross-section. The energy equation derived in Chapter 1 includes the energy losses due to the internal friction. It is more convenient, and traditional, to explain the flow resistance in terms of the boundary friction.

In Section 1.6.2, we defined the *friction slope*, S_f, as the boundary friction force per unit weight of water present in the channel. For a channel segment of length ΔX, flow area A, and wetted perimeter P,

$$S_f = \frac{F_f}{A \Delta X \gamma} \tag{3.1}$$

where γ = specific weight of water and F_f = friction force on the channel bed. The friction force acts over the bed area of $P\Delta X$. Defining τ_0 = average friction force per unit area on the channel bed or the average shear stress, Equation 3.1 is written as

$$S_f = \frac{\tau_0 P \Delta X}{A \Delta X \gamma} \tag{3.2}$$

Noting that R = hydraulic radius = A/P, the expression for S_f becomes

$$S_f = \frac{\tau_0}{\gamma R} \tag{3.3}$$

Equation 3.3 is not convenient for determining the friction slope in practice. However, various more practical, empirical, and semi-empirical friction slope equations are available, as we will see later in this chapter.

3.1.1 BOUNDARY LAYER AND FLOW RESISTANCE

A brief review of the boundary layer concept may be useful in understanding how boundary roughness affects the flow resistance. When a fluid flows over a flat solid plate, the fluid particles in contact with the plate remain at rest while the particles above the plate have a finite velocity parallel to it. Therefore, the solid surface creates a transverse velocity gradient within the flow, as shown in Figure 3.1. The boundary shear stress, τ_w, is proportional to the velocity gradient at the plate surface, and can be evaluated as

$$\tau_w = \mu \frac{dv}{dy_w} \tag{3.4}$$

at $y_w = 0$, where μ = viscosity of the fluid, v = point velocity parallel to the plate, and y_w = distance from the plate.

If the flow over the plate is laminar, the effect of the plate on the flow velocity is limited to a layer called the *laminar boundary layer*. The thickness of this layer depends on the viscosity of water and the velocity outside the boundary layer, and it grows with distance along the surface.

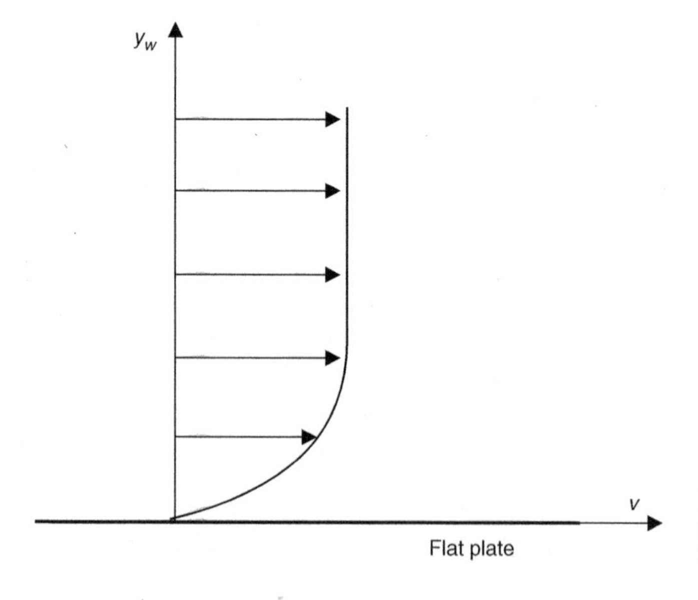

FIGURE 3.1 Velocity distribution above a flat plate

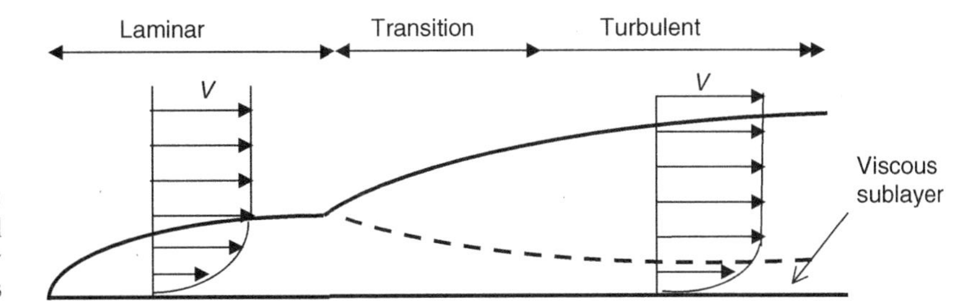

FIGURE 3.2
Laminar and turbulent boundary layers

If the flow over the flat plate is turbulent, the boundary layer may first be laminar near the leading edge of the plate, but soon a transition will occur and the boundary layer will become turbulent, as shown schematically in Figure 3.2. The thickness of the boundary layer grows much more rapidly once it becomes turbulent. Although the velocity increases with distance from the surface throughout the *turbulent boundary layer*, the velocity gradients are sharpest within a thin layer called the *viscous sublayer* near the wall. Sharper gradients lead to higher wall stresses in turbulent flow than in laminar flow.

When water enters a channel, say from a reservoir, a boundary layer will form near the channel bed in a similar way. As in the case of flow over a flat plate, the boundary layer may first be laminar before a transition to the turbulent state. After it becomes turbulent, the boundary layer will grow rapidly in thickness to encompass the entire flow depth. There will still be a viscous sublayer adjacent to the channel bed (Chow, 1959).

The hydraulic behavior of open-channel flow is affected by the thickness of the viscous sublayer and the surface roughness of the channel bed. The surface roughness is commonly characterized by *roughness height*, k_s, a length measure of roughness. Suggested values of k_s are 0.001 ft for very smooth cemented-plastered surfaces, 0.01 ft for straight earth channels, and 0.02 ft for rubble masonary (ASCE Task Force, 1963). These k_s values do not represent the actual heights of the roughness elements on a surface; rather, they indicate the equivalent sand-grain diameter. For example, the surface roughness of rubble masonry ($k_s = 0.02$ ft) is equivalent to that of a surface uniformly coated with sand grains of 0.02 ft in diameter.

As we will see in the subsequent sections, the flow resistance is calculated by using different expressions depending on whether the flow is laminar or turbulent. We further classify the turbulent flow into hydraulically smooth, transitional, and fully rough flows.

When the roughness elements of the channel bed are buried within the viscous sublayer, the flow is said to be *hydraulically smooth*. With increasing Reynolds number, the viscous sublayer shrinks and the flow enters a transitional state as the roughness elements break through this sublayer. At larger Reynolds numbers, with further shrinkage of the viscous sublayer, the roughness elements dominate the flow behavior.

As classified by Henderson (1966), the flow is hydraulically smooth if

$$\frac{V_* k_s}{\nu} < 4 \tag{3.5}$$

transitional if

$$4 < \frac{V_* k_s}{\nu} < 100 \tag{3.6}$$

and fully rough if

$$100 < \frac{V_* k_s}{\nu} \tag{3.7}$$

where $\nu = \mu/\rho$ is kinematic viscosity of water and $V_* =$ shear velocity, defined as

$$V_* = \sqrt{\frac{\tau_0}{\rho}} = \sqrt{gRS_f} \tag{3.8}$$

3.1.2 THE DARCY–WEISBACH EQUATION

The Darcy–Weisbach equation was originally developed for pipe flow (Chow, 1959). It is adopted for open-channel flow by replacing the pipe diameter d_0 with $4R$, where R is hydraulic radius. (Note that for a pipe with flow $A = \pi d_0^2/4$, $P = \pi d_0$, and thus $R = A/P = d_0/4$.) The Darcy–Weisbach equation for open-channel flow is

$$S_f = \frac{f}{R}\frac{V^2}{8g} \tag{3.9}$$

where f is a dimensionless factor called the *friction factor*. The friction factor is evaluated differently depending on whether the flow is laminar, turbulent and hydraulically smooth, transitional, or fully rough turbulent. A chart, called the Moody diagram, can be found in many fluid mechanics books to determine the friction factor for pipe flow. Although a Moody diagram for open-channel flow has not been reported, there are semi-empirical equations to calculate the friction factor (Henderson, 1966).

For laminar flow

$$f = \frac{64}{R_e} \tag{3.10}$$

For hydraulically smooth flow with $R_e < 100\,000$

$$f = \frac{0.316}{R_e^{0.25}} \tag{3.11}$$

while for hydraulically smooth flow with $R_e > 100\,000$

$$\frac{1}{\sqrt{f}} = -2\log\left(\frac{2.5}{R_e\sqrt{f}}\right) \tag{3.12}$$

For transitional flow

$$\frac{1}{\sqrt{f}} = -2\log\left(\frac{k_s}{12R} + \frac{2.5}{R_e\sqrt{f}}\right) \tag{3.13}$$

and for fully rough turbulent flow

$$\frac{1}{\sqrt{f}} = -2\log\left(\frac{k_s}{12R}\right) \tag{3.14}$$

Although the Darcy–Weisbach formula has some theoretical basis, it is rarely used in practice for open-channel flow. Perhaps the main reason is that the use of the equations given for f requires a trial-and-error procedure (we need to know R and/or R_e to find f, but R and R_e depend on f). However, these equations clearly demonstrate that viscosity is the dominant factor in flow resistance at low Reynold numbers, while the surface roughness affects the flow resistance in fully rough flow. Also, even for fully rough flow, unlike pipes flowing full, the friction factor is not constant for a given open channel; it depends on the hydraulic radius as well as the channel roughness.

3.1.3 THE CHEZY EQUATION

Chezy, a French engineer, introduced the expression

$$V = C\sqrt{RS_f} \tag{3.15}$$

for turbulent open-channel flow as early as 1769, where C = Chezy coefficient (Henderson, 1966). This coefficient has the dimensions of $(\text{length})^{1/2}/(\text{time})$. Although the Chezy equation appears to be simple, it has limited use in practice since the Chezy coefficient depends on the flow conditions as well as the channel roughness, and it is difficult to evaluate. To demonstrate the dependence of C on flow conditions, let us rewrite the Chezy equation as

$$S_f = \frac{V^2}{RC^2} \tag{3.16}$$

Comparing Equations 3.9 and 3.16, we can see that there is a direct relationship between the Darcy–Weisbach friction factor, f, and the Chezy coefficient, C, as

$$\frac{C}{\sqrt{8g}} = \frac{1}{\sqrt{f}} \tag{3.17}$$

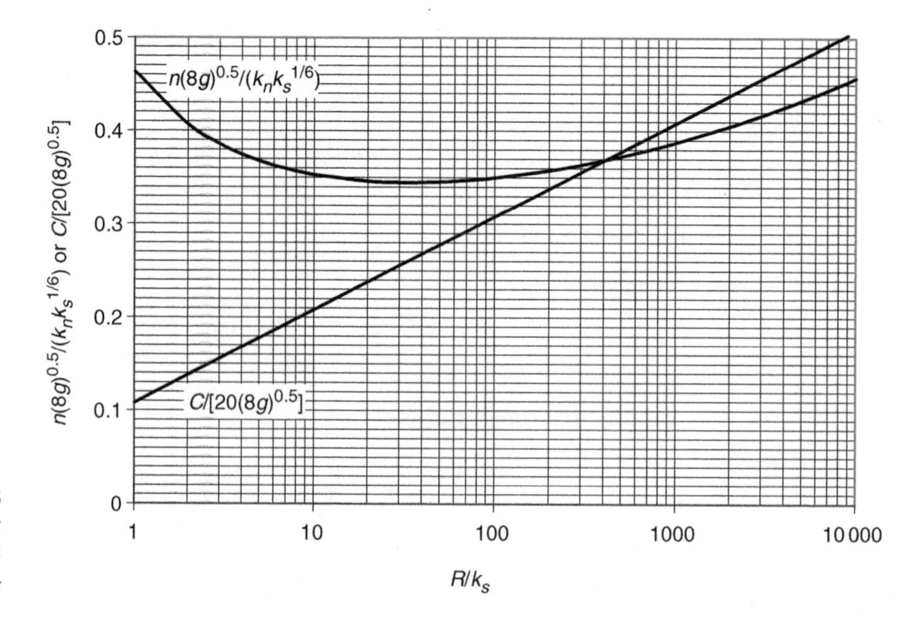

FIGURE 3.3 Variation of Chezy C and Manning n with R/k_s

Therefore, Equations 3.11 to 3.14 given for f in the preceding section can also be utilized for evaluating C. For example, for fully rough flow,

$$\frac{C}{\sqrt{8g}} = -2\log\left(\frac{k_s}{12R}\right) \tag{3.18}$$

Clearly, the Chezy coefficient depends on both the roughness height, k_s, and the hydraulic radius, R. Variation of C with the R/k_s ratio is demonstrated in Figure 3.3.

3.1.4 THE MANNING FORMULA

The Manning formula, also known as Strickler's equation, was first introduced in 1891 by Flamant (Henderson, 1966). It has found widespread use in engineering practice. The Manning formula, meant for fully rough turbulent flow, is written as

$$V = \frac{k_n}{n} R^{2/3} S_f^{1/2} \tag{3.19a}$$

or

$$Q = \frac{k_n}{n} A R^{2/3} S_f^{1/2} \tag{3.19b}$$

where $k_n = 1.0\ \text{m}^{1/3}/\text{s} = 1.49\ \text{ft}^{1/3}/\text{s}$, and n = Manning roughness factor. In practice, for a given channel, the Manning roughness factor is assumed not to vary with the flow conditions.

We can demonstrate the validity of this assumption. Let us first rewrite the Manning formula as

$$S_f = \frac{V^2 n^2}{k_n^2 R^{4/3}} \tag{3.20}$$

For fully rough turbulent flow, from Equations 3.9, 3.14, and 3.20, we can obtain the relationship

$$\frac{n\sqrt{8g}}{k_n k_s^{1/6}} = \frac{(R/k_s)^{1/6}}{2\log(12(R/k_s))} \tag{3.21}$$

Note that in this expression, g, k_n, and k_s are constant. Therefore, if the left-hand side of the expression remains constant, we can conclude that n is also constant. Figure 3.3 displays a graphical representation of Equation 3.21. An inspection of Figure 3.3 reveals that although n varies with R/k_s, the variations are less than $\pm 5\%$ over the average value within the range $4 < (R/k_s) < 600$. Therefore, we can assume that the Manning roughness factor for a given channel is constant within this range. Similar observations were previously reported by Yen (1992), Hager (2001), and Sturm (2001). Most practical open-channel flow situations fall within this range. For example, for a trapezoidal earth channel ($k_s = 0.01$ ft) with a bottom width of 5 ft and side slopes of $m = 3$ (3H : 1V), the corresponding flow depth range is about 0.04 ft to 16 ft. We should note that this justification for using a constant Manning roughness factor is based on the assumption that the flow is fully rough. By using Equation 3.7, the reader can easily show that the flow is indeed fully rough for most practical open-channel flow situations.

The Manning roughness factor is well documented and published in the literature. Chow (1959) presented an extensive table of minimum, normal, and maximum n values for a variety of channel materials. Chow's table was also reported by French (1985), Sturm (2001), and the US Army Corps of Engineers (2002). The Federal Highway Administration (Chen and Cotton, 1988) is the main source for the Manning roughness factors listed here in Table 3.1. However, these values are in general agreement with those of Chow (1959) and Henderson (1966). The lower values in this table are recommended for depths greater than 2.0 ft or 60 cm.

Selecting a Manning's n for a natural stream is not easy unless some field data are available to determine the roughness factor by calibration. Chow (1959), Barnes (1967), and Sturm (2001) presented photographs of various streams with calibrated n values. Table 3.2 summarizes the characteristics of selected streams calibrated and reported by Barnes (1967), where d_{50} = mean diameter of the streambed material. Cowan (1956) presented a procedure to account for the surface irregularities, variations in channel shape and size, obstructions, vegetation, and meandering in selecting a roughness factor. This procedure was reviewed and expanded later by Arcement and Schneider (1989).

TABLE 3.1 Manning roughness factor

Channel material	Manning roughness factor n
Concrete	0.013–0.015
Grouted riprap	0.028–0.040
Soil cement	0.020–0.025
Asphalt	0.016–0.018
Bare soil	0.020–0.023
Rock cut	0.025–0.045
Fiberglass roving	0.019–0.028
Woven paper net	0.015–0.016
Jute net	0.019–0.028
Synthetic mat	0.021–0.030

TABLE 3.2 Manning roughness factor for various streams and rivers

Location	Bed material and condition	Depth (ft)	d_{50} (mm)	n
Salt Creek at Roca, Nebraska	Sand and clay	6.3		0.030
Rio Chama near Chamita, New Mexico	Sand and gravel	3.5		0.032
Salt River below Stewart Mountain Dam, Arizona	Smooth cobbles, 4 to 10 inch diameter	1.8		0.032
West Fork Bitterroot River near Conner, Montana	Gravel and boulders	4.7	172	0.036
Middle Fork Vermilion River near Danville, Illinois	Gravel and small cobbles	3.9		0.037
Wenatchee River at Plain, Washington	Boulders	11.1	162	0.037
Etowah River near Dawsonville, Georgia	Sand and gravel with several fallen trees in the reach	9.0		0.039
Tobesofkee Creek near Macon Georgia	Sand, gravel and few rock outcrops	8.7		0.041
Middle Fork Flathed River near Essex, Montana	Boulders	8.4	142	0.041
Beaver Creek near Newcastle, Wyoming	Sand and silt	9.0		0.043
Murder Creek near Monticello, Georgia	Sand and gravel	4.2		0.045
South Fork Clearwater River near Grangeville, Idaho	Rock and boulders	7.9	250	0.051
Missouri Creek near Cashmere, Washington	Angular shaped boulders as large as 1 ft in diameter	1.5		0.057
Haw River near Banja, North Carolina	Coarse sand and a few outcrops	4.9		0.059
Rock Creek near Darby, Montana	Boulders	3.1	220	0.075

3.2 NORMAL FLOW EQUATION

Normal flow refers to steady open-channel flow in which the flow depth, area, and velocity remain constant at every cross-section along the channel. The momentum and energy equations for steady flow were derived in Chapter 1 (see Equations 1.49 and 1.50). If the velocity and the depth do not vary along the flow direction, then Equations 1.49 and 1.50 are, respectively, reduced to

$$S_f = S_0 \tag{3.22}$$

and

$$S_e = S_0 \tag{3.23}$$

where S_f = friction slope and S_e = energy slope. Also, as discussed in Section 1.6.6, S_f and S_e are interchangeable for practical purposes, and the term 'friction slope' refers to either. Indeed, in Section 2.2.1 we indicated that S_f, represents the slope of the energy grade line. Then, for normal flow, the energy grade line is parallel to the channel bottom. This also implies that the water surface is parallel to the channel bottom, since the flow depth and velocity are both constant.

Substituting Equation 3.22 into Equation 3.1, and simplifying, we obtain

$$F_f = \gamma(\Delta X)AS_0 \tag{3.24}$$

The left-hand side of this equation is the friction force acting on a channel segment that has a length ΔX, flow area A, and a bottom slope S_0. The right-hand side is the component of the weight of water (gravitational force) in the flow direction. Therefore, normal flow occurs when the gravitational force component in the flow direction is balanced by the flow resistance.

A qualitative inspection of Equation 3.24 will also reveal that, with everything else remaining the same, the flow area A (and therefore the depth y) will increase with increasing F_f. Therefore, the normal flow depth will be greater in rougher channels. Likewise, with everything else remaining the same, the flow area A (and depth y) will decrease with increasing S_0. In other words, with everything else remaining the same, the normal flow depth is smaller in steeper channels.

The Manning formula is the most commonly used flow-resistance equation for open-channel flow calculations. Substituting Equation 3.22 into Equation 3.19, the Manning formula for normal flow becomes

$$V = \frac{k_n}{n} R^{2/3} S_0^{1/2} \tag{3.25}$$

or

$$Q = \frac{k_n}{n} A R^{2/3} S_0^{1/2} \tag{3.26}$$

Two types of problems are encountered in analyzing channels under normal flow conditions. The first involves the calculation of normal flow velocity and discharge given the normal flow depth and the channel characteristics. This is a simple problem to solve. We first calculate A and R using the expressions in

Table 1.1, and then determine V and Q from Equations 3.25 and 3.26, respectively. The second type of problem involves the determination of normal flow depth given the discharge and channel characteristics. This is more difficult to solve, because it may involve a trial-and-error procedure.

EXAMPLE 3.1 A concrete, trapezoidal channel has a bottom slope of $S_0 = 0.0009$ and a Manning roughness factor of $n = 0.013$. The bottom width of the channel is $b = 2.5$ m, and the side slopes are $m = 2$ – that is, 2H:1V. Determine the velocity and discharge when the flow is normal at a depth of 1.8 m.

For trapezoidal channels, from Table 1.1

$$A = (b + my)y$$

$$P = b + 2y\sqrt{1 + m^2}$$

and

$$R = \frac{(b + my)y}{b + 2y\sqrt{1 + m^2}}$$

Therefore, for the given channel,

$$A = [2.5 + (2.0)(1.8)](1.8) = 10.98\,\text{m}^2$$

$$P = 2.5 + 2(1.8)\sqrt{1 + (2.0)^2} = 10.55\,\text{m}$$

$$R = \frac{10.98}{10.55} = 1.04\,\text{m}$$

Substituting these into Equations 3.25 and 3.26, and noting that $k_n = 1.0$ for the unit system used, we obtain

$$V = \frac{1.0}{0.013}(1.04)^{2/3}(0.0009)^{1/2} = 2.37\,\text{m/s}$$

$$Q = \frac{1.0}{0.013}(10.98)(1.04)^{2/3}(0.0009)^{1/2} = 26.00\,\text{m}^3/\text{s}$$

3.3 NORMAL DEPTH CALCULATIONS IN UNIFORM CHANNELS

The *normal depth* is the flow depth that satisfies Equations 3.25 and 3.26, and is denoted by y_n. We often need to calculate the normal depth given the discharge and the channel properties. For uniform channels, that is for prismatic channels made of uniform channel material, we can assume that the Manning roughness factor is constant. Also, for such channels, the cross-sectional relationships are available as presented in Table 1.1.

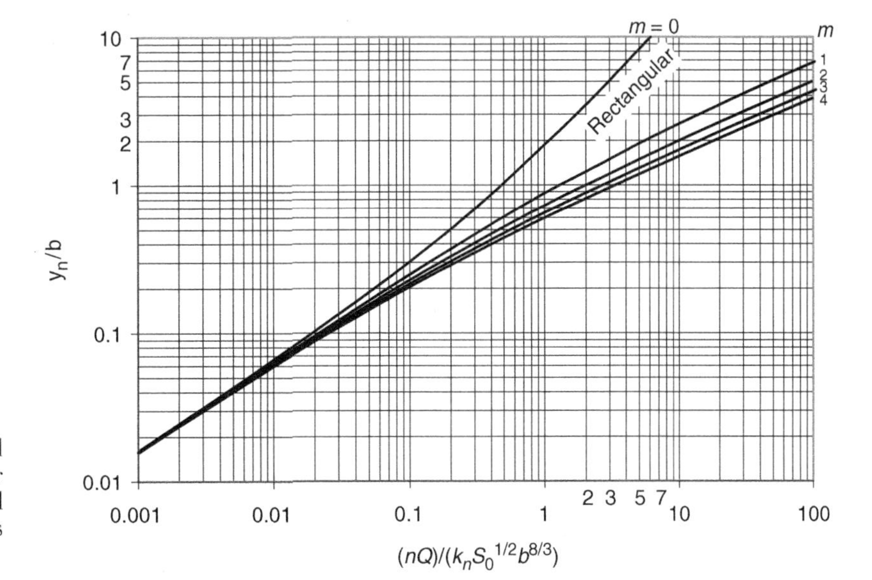

FIGURE 3.4 Normal depth chart for rectangular and trapezoidal channels

An explicit expression can be derived for the normal depth in triangular channels. For example, for a triangular channel having the same side slope on both sides we can substitute the expressions for A and R (in terms of y) from Table 1.1 into Equation 3.26 and rearrange it to obtain

$$y_n = \left(\frac{nQ}{k_n S_0^{1/2}}\right)^{3/8} \frac{\left(2\sqrt{1+m^2}\right)^{1/4}}{m^{5/8}} \tag{3.27}$$

For most other cross-sectional shapes an explicit expression for y_n is not available, and a trial-and-error procedure is needed to calculate the normal depth mathematically. Given the discharge and the channel properties, we first write Equation 3.26 as

$$AR^{2/3} = \frac{nQ}{k_n S_0^{1/2}} \tag{3.28}$$

where all the terms on the right-hand side of Equation 3.28 are given. The left-hand side is then expressed in terms of the unknown, y_n, using the expressions in Table 1.1. For example, noting that $AR^{2/3} = A^{5/3}/P^{2/3}$, for a trapezoidal channel of known bottom width b and side slopes m, Equation 3.28 becomes

$$\frac{[(b+my_n)y_n]^{5/3}}{[b+2y_n\sqrt{1+m^2}]^{2/3}} = \frac{nQ}{k_n S_0^{1/2}} \tag{3.29}$$

The only unknown in Equation 3.29 is y_n. However, the equation is implicit in y_n, and it needs to be solved by trial and error. Alternatively, we can use Figure 3.4, which presents predetermined solutions for Equation 3.29 in dimensionless form for normal depth in rectangular and trapezoidal channels. Because the graphical approach involves reading errors, the solution obtained by solving Equation 3.29

mathematically would be more precise than the one obtained graphically from Figure 3.4.

Likewise, for a circular channel of diameter d_0, from Table 1.1

$$A = \frac{1}{8}(2\theta - \sin 2\theta)d_0^2$$

$$P = \theta d_0$$

and

$$R = \frac{A}{P} = \frac{1}{4}\left(1 - \frac{\sin 2\theta}{2\theta}\right)d_0$$

Noting that $AR^{2/3} = A^{5/3}/P^{2/3}$, we can write Equation 3.28 as

$$\frac{\left[(d_0^2/8)(2\theta - \sin 2\theta)\right]^{5/3}}{(\theta d_0)^{2/3}} = \frac{nQ}{k_n S_0^{1/2}} \tag{3.30}$$

with

$$\theta = \pi - \arccos\left[(y_n - d_0/2)/(d_0/2)\right] \tag{3.31}$$

Equations 3.30 and 3.31 are implicit in y_n, and a trial-and-error procedure is needed to solve these equations mathematically. Alternatively, we can use the chart presented in Figure 3.5 to determine the normal depth graphically. When precision is required, the mathematical solution of Equations 3.30 and 3.31 is preferred.

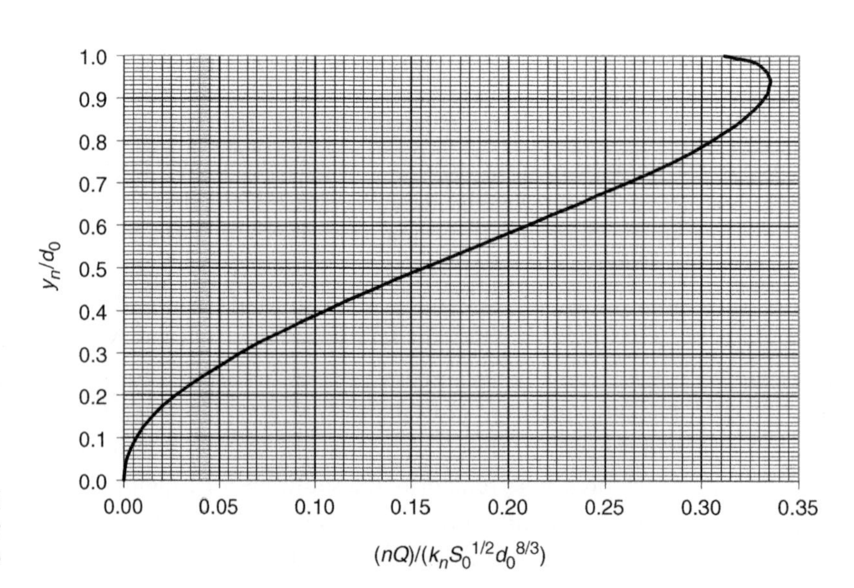

FIGURE 3.5 Normal depth chart for circular channels

EXAMPLE 3.2 A trapezoidal channel has a bottom slope of $S_0 = 0.0001$ and a Manning roughness factor of $n = 0.016$. The bottom width of the channel is $b = 5.0$ ft, and the side slopes are $m = 3$ – that is, 3H : 1V. Determine the normal depth in this channel for $Q = 136$ cfs.

To solve this problem, we first substitute all the known values into Equation 3.29:

$$\frac{[(5.0 + 3y_n)y_n]^{5/3}}{[5.0 + 2y_n\sqrt{1 + 3^2}]^{2/3}} = \frac{(0.016)(136)}{(1.49)(0.0001)^{1/2}}$$

or

$$\frac{[(5.0 + 3y_n)y_n]^{5/3}}{[5.0 + 6.32y_n]^{2/3}} = 146$$

Thus, we need to determine the value of y_n that will make the left-hand side of this expression equal to 146. To achieve this we try different values for y_n until the left-hand side becomes 146 with the results summarized as

Trial value for y_n (ft)	Left-hand side of expression
3.00	61.07
4.00	116.64
5.00	195.50
4.41	145.98

Therefore, the normal depth for this channel is 4.41 ft.

Alternatively, we can use Figure 3.4 to solve this problem. Let us first evaluate

$$\frac{nQ}{k_n S_0^{1/2} b^{8/3}} = \frac{(0.016)(136)}{(1.49)(0.0001)^{1/2}(5.0)^{8/3}} = 2.0.$$

Using this value and $m = 3$, we obtain $y_n/b = 0.88$ from Figure 3.4. Therefore, $y_n = (0.88)(5.0) = 4.40$ ft.

EXAMPLE 3.3 A circular storm sewer has a diameter of $d_0 = 1.0$ m, slope of $S_0 = 0.004$, and Manning roughness factor of $n = 0.013$. Determine the normal depth when the discharge is $Q = 1.33\ \mathrm{m^3/s}$.

To solve this problem, we will first substitute all the known values into Equation 3.30:

$$\frac{\left[((1.0)/8)^2(2\theta - \sin 2\theta)\right]^{5/3}}{[\theta(1.0)]^{2/3}} = \frac{(0.013)(1.33)}{(1.0)(0.004)^{1/2}}$$

or

$$\frac{(2\theta - \sin 2\theta)^{5/3}}{\theta^{2/3}} = 8.75$$

Now, substituting Equation 3.31 into this expression for θ with $d_0 = 1.0$, we obtain

$$\frac{\left(2\left[\pi - \arccos\frac{y_n - (1.0/2)}{(1.0/2)}\right] - \sin 2\left[\pi - \arccos\frac{y_n - (1.0/2)}{(1.0/2)}\right]\right)^{5/3}}{\left(\pi - \arccos\frac{y_n - (1.0/2)}{(1.0/2)}\right)^{2/3}} = 8.75$$

By trial and error, we find $y_n = 0.73$ m satisfies this expression.

Alternatively, we can use Figure 3.5 to solve this problem. Let us first evaluate

$$\frac{nQ}{k_n S_0^{1/2} d_0^{8/3}} = \frac{(0.013)(1.33)}{(1.00)(0.004)^{1/2}(1.0)^{8/3}} = 0.273$$

Then, from Figure 3.5, we obtain $y_n/d_0 = 0.73$. Finally, $y_n = (0.73)(1.0) = 0.73$ m.

EXAMPLE 3.4 A triangular channel has side slopes of $m = 2.5$, a Manning roughness factor of $n = 0.013$, and a bottom slope of $S_0 = 0.002$. The channel carries a discharge of $Q = 95$ cfs. Determine the normal depth.

An explicit equation (Equation 3.27) is available to calculate the normal depth in triangular channels. Substituting all the givens into Equation 3.27, we obtain

$$y_n = \left(\frac{nQ}{k_n S_0^{1/2}}\right)^{3/8} \frac{\left(2\sqrt{1+m^2}\right)^{1/4}}{m^{5/8}}$$

$$= \left(\frac{(0.013)(95)}{(1.49)(0.002)^{1/2}}\right)^{3/8} \frac{\left(2\sqrt{1+(2.5)^2}\right)^{1/4}}{(2.5)^{5/8}} = 2.57\text{ ft}$$

3.4 Normal Depth Calculations in Grass-Lined Channels

The assumption of a constant Manning factor does not apply when the channel bed is covered by vegetation. Part of the flow occurs through the vegetation on the channel bed at slower velocities. The overall roughness factor for the channel section varies depending on the magnitude of flow through the vegetation relative to the total flow in the whole section.

Based on experimental data, the Manning roughness factor for grass-lined channels can be expressed as (Chen and Cotton, 1988):

$$n = \frac{(RK_v)^{1/6}}{C_n + 19.97\log[(RK_v)^{1.4} S_0^{0.4}]} \tag{3.32}$$

where R = hydraulic radius, K_v = unit conversion factor = 3.28 m^{-1} = 1.0 ft^{-1}, S_0 = bottom slope, and C_n = dimensionless retardance factor. The retardance factor is given in Table 3.3 for five different retardance classes into which common grass types are grouped. Note that the same type of grass can belong to

TABLE 3.3 Retardance classes for vegetative covers (after Chen and Cotton, 1988)

Retardance class	Cover	Condition	C_n
A	Weeping lovegrass	Excellent stand, tall (average 30 in, 76 cm)	15.8
	Yellow bluestem ischaemum	Excellent stand, tall (average 36 in, 91 cm)	
B	Kudzu	Very dense growth, uncut	23.0
	Bermuda grass	Good stand, tall (average 12 in, 30 cm)	
	Native grass mixture (little bluestem, bluestem, blue gamma, and other long and short midwest grasses)	Good stand, unmowed	
	Weeping lovegrass	Good stand, tall (average 24 in, 61 cm)	
	Lespedeza sericea	Good stand, not woody, tall (average 19 in, 48 cm)	
	Alfalfa	Good stand, uncut (average 11 in, 28 cm)	
	Weeping lovegrass	Good stand, unmowed (average 13 in, 33 cm)	
	Kudzu	Dense growth, uncut	
	Blue gamma	Good stand, uncut (average 13 in, 33 cm)	
C	Crabgrass	Fair stand, uncut (10–48 in, 25–120 cm)	30.2
	Bermuda grass	Good stand, mowed (average 6 in, 15 cm)	
	Common lespedeza	Good stand, uncut (average 11 in, 28 cm)	
	Grass–legume mixture – summer (orchard grass, redtop, Italian ryegrass, and common lespedeza)	Good stand, uncut (6–8 in, 15–20 cm)	
	Centipedegrass	Very dense cover (average 6 in, 15 cm)	
	Kentucky bluegrass	Good stand, headed (6–12 in, 15–30 cm)	
D	Bermuda grass	Good stand, cut to 2.5 in (6 cm) height	34.6
	Common lespedeza	Excellent stand, uncut (average 4.5 in, 11 cm)	
	Buffalo grass	Good stand, uncut (3–6 in, 8–15 cm)	
	Grass–legume mixture – fall, spring (orchard grass, redtop, Italian ryegrass, and common lespedeza)	Good stand, uncut (4–5 in, 10–13 cm)	
	Lespedeza sericea	After cutting to 2 in (5 cm) height, very good stand before cutting	
E	Bermuda grass	Good stand, cut to 1.5 in (4 cm) height	37.7
	Bermuda grass	Burned stubble	

various retardance classes, depending on the maturity of the grass. In Table 3.3, class A represents the highest and class E the lowest degree of retardance.

To determine the normal depth in a grass-lined channel, we need a trial-and-error procedure:

1. Guess the normal depth, y_n
2. Calculate A and R using the expressions given in Table 1.1
3. Calculate n by using Equation 3.32
4. Calculate Q by using Equation 3.26. If the calculated Q is equal to the given Q, the guessed value of y_n is correct. Otherwise try another value for y_n.

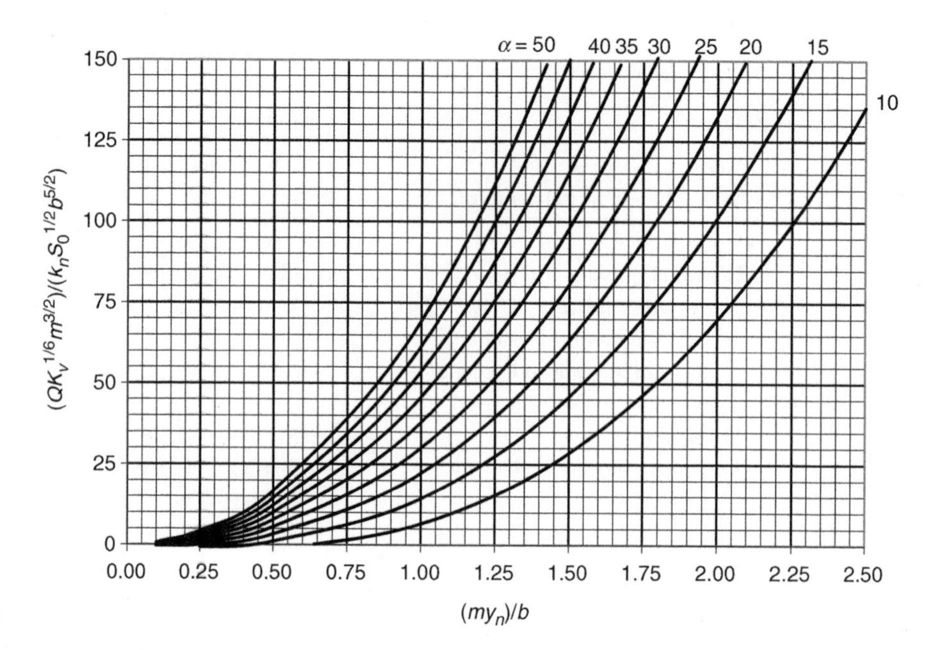

FIGURE 3.6 Approximate normal depth chart for grass-lined channels

The use of Figure 3.6 to pick the first trial value of y_n facilitates the trial-and-error procedure significantly. This figure, constructed using the procedure developed by Akan and Hager (2001), presents pre-determined solutions to Equations 3.26 and 3.32 for trapezoidal channels in terms of dimensionless parameters. The normal depth obtained from the figure should be exactly the same as that which we would calculate mathematically (except for reading errors) if $m = 3.0$. The figure will overestimate the normal depth slightly for $m > 3$ and underestimate it slightly for $m < 3$. In Figure 3.6, the dimensionless parameter α is defined as

$$\alpha = C_n + 19.97 \log\left(\frac{K_v^{1.4} b^{1.4} S_0^{0.4}}{m^{1.4}}\right) \tag{3.33}$$

EXAMPLE 3.5 A trapezoidal channel is lined with uncut buffalo grass that has a good stand. The channel has a bottom width of $b = 2.0$ m, side slopes of $m = 2.5$, and a longitudinal slope of $S_0 = 0.001$. The channel carries a discharge of $Q = 0.85$ m^3/s. Determine the normal depth.

First, the uncut buffalo grass belongs to retardance class D in Table 3.3, and the retardance coefficient is $C_n = 34.6$. Next, noting that $k_n = 1.0$ m$^{1/3}$/s and $K_v = 3.28$ m^{-1} in the metric unit system, we evaluate the dimensionless parameter α, using Equation 3.33, as

$$\alpha = C_n + 19.97 \log\left(\frac{K_v^{1.4} b^{1.4} S_0^{0.4}}{m^{1.4}}\right)$$

$$= 34.6 + 19.97 \log\left(\frac{(3.28)^{1.4}(2.0)^{1.4}(0.001)^{0.4}}{(2.5)^{1.4}}\right) = 22.3$$

Next, we evaluate the dimensionless parameter

$$\frac{QK_v^{1/6}m^{3/2}}{k_n S_0^{1/2} b^{5/2}} = \frac{(0.85)(3.28)^{1/6}(2.5)^{3/2}}{(1.0)(0.001)^{1/2}(2.0)^{5/2}} = 22.9$$

Then from Figure 3.6 we obtain $my_n/b = 0.93$ and $y_n = (0.93)(2.0)/2.5 = 0.74$ m.

We can now check whether this depth satisfies the Manning formula. Using the expressions given for a trapezoidal channel in Table 1.1,

$$A = (b + my)y = [2.0 + (2.5)(0.74)](0.74) = 2.85 \text{ m}$$

$$P = b + 2y\sqrt{1 + m^2} = 2.0 + 2(0.74)\sqrt{1 + (2.5)^2} = 5.99 \text{ m}.$$

Then $R = A/P = (2.85)/(5.99) = 0.48$ m. Substituting the known values into Equation 3.33,

$$n = \frac{(RK_v)^{1/6}}{C_n + 19.97 \log(RK_v)^{1.4} S_0^{0.4}}$$

$$= \frac{[(0.48)(3.28)]^{1/6}}{34.6 + 19.97 \log\{[(0.48)(3.28)]^{1.4}(0.001)^{0.4}\}} = 0.067$$

Therefore, by using Equation 3.26,

$$Q = \frac{k_n}{n} AR^{2/3} S_0^{1/2} = \frac{1.0}{0.067}(2.85)(0.48)^{2/3}(0.001)^{0.5} = 0.82 \text{ m}^3/\text{s}$$

The calculated Q is different from and slightly lower than the given $Q = 0.85$ m^3/s. We should now try another normal depth value slightly higher than 0.74 m. If we pick $y_n = 0.75$ m, we can follow the same procedure to show that $y_n = 0.75$ m will result in $R = 0.48$ m and $n = 0.066$, and will satisfy the Manning formula. As expected, the approximate normal depth obtained by using Figure 3.6 is very close to the actual normal depth.

3.5 Normal Depth Calculations in Riprap Channels

There have been several studies reported in the literature to find a relationship between the Manning roughness factor and the stone size of riprap lining. Some of these studies suggest a simple relationship in the form of

$$n = C_m(K_v d_{50})^{1/6} \tag{3.34}$$

where C_m = constant coefficient, K_v = unit conversion factor = 3.28 m^{-1} = 1.0 ft^{-1}, and d_{50} = mean stone diameter. This equation was first suggested by Strickler, with $C_m = 0.034$ for gravel-bed streams as reported by Henderson (1966). Other suggested values for C_m are 0.039 (Hager, 2001) and 0.038 (Maynord, 1991).

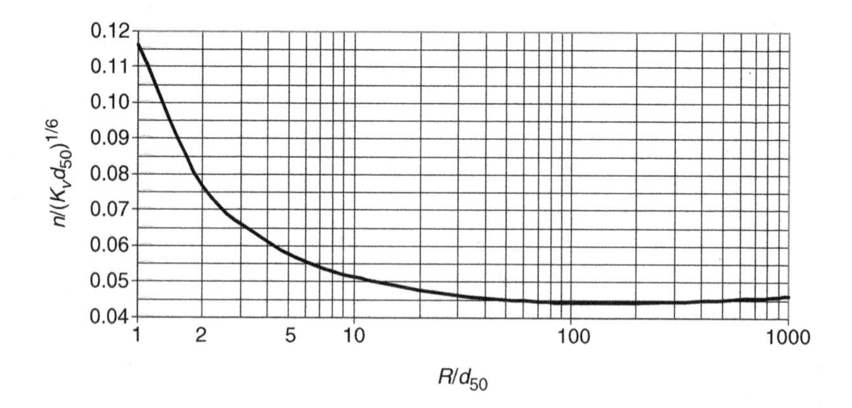

FIGURE 3.7 Manning roughness factor for riprap channels

Based on the findings of Blodgett and McConaughy (1985), Chen and Cotton (1988) adopted the relationship

$$n = \frac{(K_v R)^{1/6}}{8.60 + 19.98 \log(R/d_{50})} \tag{3.35}$$

Figure 3.7 displays a graphical representation of this relationship. This figure reveals that a relationship of the form of Equation 3.34 is valid for $500 > R/d_{50} > 50$ with $C_m = 0.045$. Comparing this value of C_m to those reported previously, we may conclude that Equation 3.35 overestimates the Manning roughness factor. However, this equation accounts for the expected variability of the Manning roughness factor with R/d_{50} for smaller values of R/d_{50}. Therefore, Equation 3.35 is recommended here, particularly for channels with small R/d_{50} ratios, despite its complexity.

Equation 3.34 assumes that the Manning roughness factor is constant for a given riprap size. If we adopt this equation, after determining the Manning roughness factor from Equation 3.34, we can use the procedures discussed in Section 3.3 to calculate the normal depth. If Equation 3.35 is adopted, however, a trial-and-error procedure will be needed:

1. Guess the normal depth, y_n
2. Calculate A and R using the expressions given in Table 1.1
3. Calculate n by using Equation 3.35
4. Calculate Q by using Equation 3.26. If the calculated Q is equal to the given Q, the guessed value of y_n is correct. Otherwise try another value for y_n.

Figure 3.8 can be used to determine the first trial value of y_n. The value obtained from this figure should be very close to the actual result.

EXAMPLE 3.6 A trapezoidal channel that has a bottom width of $b = 3.0$ ft, side slopes of $m = 3$ (1V:3H), and a longitudinal slope of $S_0 = 0.01$ carries a discharge of $Q = 40$ cfs. The channel is lined with riprap of $d_{50} = 0.5$ ft. Determine the normal depth in this channel.

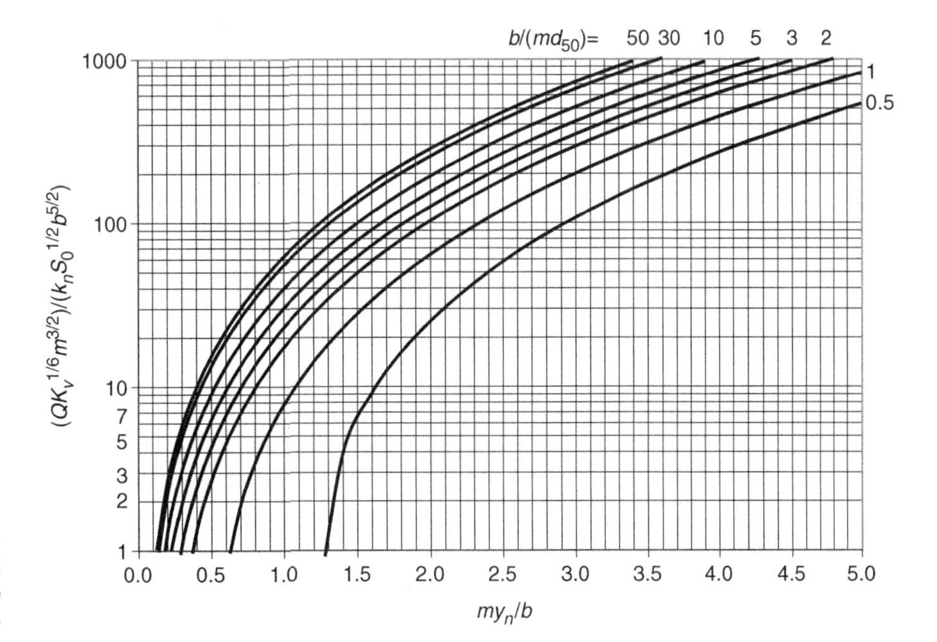

FIGURE 3.8 Approximate normal depth in riprap channels

Noting that $k_n = 1.49\,\text{ft}^{1/3}/\text{s}$ and $K_v = 1.0\,\text{ft}^{-1}$ for the unit system used, let us calculate the dimensionless parameters of Figure 3.8 as

$$\frac{b}{md_{50}} = \frac{3.0}{3(0.50)} = 2.0$$

and

$$\frac{QK_v^{1/6}m^{3/2}}{k_n S_0^{1/2} b^{5/2}} = \frac{(40.0)(1.0)(3)^{3/2}}{1.49(0.01)^{1/2}(3.0)^{5/2}} = 89.5$$

Using these values, we obtain $(my_n/b) = 1.90$ from Figure 3.8. Therefore, $y_n = (1.90)(3.0)/3 = 1.90\,\text{ft}$.

Let us now check if this value satisfies the Manning formula. First by using the expressions given for trapezoidal channels in Table 1.1:

$$A = (b + my)y = [3.0 + (3)(1.90)](1.90) = 16.53\,\text{ft}^2$$
$$P = b + 2y\sqrt{1 + m^2} = 3.0 + 2(1.90)\sqrt{1 + 3^2} = 15.02\,\text{ft}$$

and

$$R = A/P = 16.53/15.02 = 1.10\,\text{ft}$$

Now, by using Equation 3.35,

$$n = \frac{(K_v R)^{1/6}}{8.60 + 19.98\log(R/d_{50})} = \frac{[(1.0)(1.10)]^{1/6}}{8.60 + 19.98\log(1.10/0.50)} = 0.0658$$

Then, by using Equation 3.26,

$$Q = \frac{k_n}{n} A R^{2/3} S_0^{1/2} = \frac{1.49}{0.0658}(16.53)(1.10)^{2/3}(0.01)^{0.5} = 39.9\,\text{cfs}$$

The calculated value of Q, 39.9 cfs, is very close to the given value, 40.0 cfs. Therefore, we can conclude that $y_n = 1.90$ ft.

3.6 Normal Flow in Composite Channels

The channel roughness may be different on different parts of the wetted perimeter. For example, it is possible to use different types of lining materials on the sides and the bottom of a drainage channel. Likewise, a laboratory flume may have a metal bottom and glass sidewalls. Such channels are called *composite channels*. Different parts of the perimeter of a composite channel are then represented by different Manning roughness factors. This may cause different average velocities in various parts of a composite channel section. However, in what we categorize as composite channels these velocity differences are small, and the whole section can be represented by one cross-sectional average velocity. The channels having significantly different velocities in different parts of a section are called *compound channels*, and are treated differently as we will see in Section 3.7.

We usually define an *equivalent roughness factor* or *composite roughness factor* for composite channels for flow calculations. As summarized by Chow (1959), there are various formulas to evaluate the equivalent roughness factor. Two of these formulas are included herein. Either one of these formulas is acceptable.

Suppose the channel perimeter is made of N distinct segments having different values of Manning roughness factor. Suppose P_i = length of the i-th segment and n_i = Manning roughness factor for the i-th segment. If we assume that the velocities corresponding to the different segments are equal, we obtain

$$n_e = \left[\frac{\sum_{i=1}^{N} (P_i n_i^{1.5})}{\sum_{i=1}^{N} P_i} \right]^{2/3} \tag{3.36}$$

where n_e = equivalent roughness factor. Equation 3.36 is attributed to Horton (Chow, 1959). If we use the condition that the total force resisting to flow is equal to the sum of forces resisting to flow over the different segments of the perimeter, we obtain

$$n_e = \left[\frac{\sum_{i=1}^{N} (P_i n_i^{2})}{\sum_{i=1}^{N} P_i} \right]^{1/2} \tag{3.37}$$

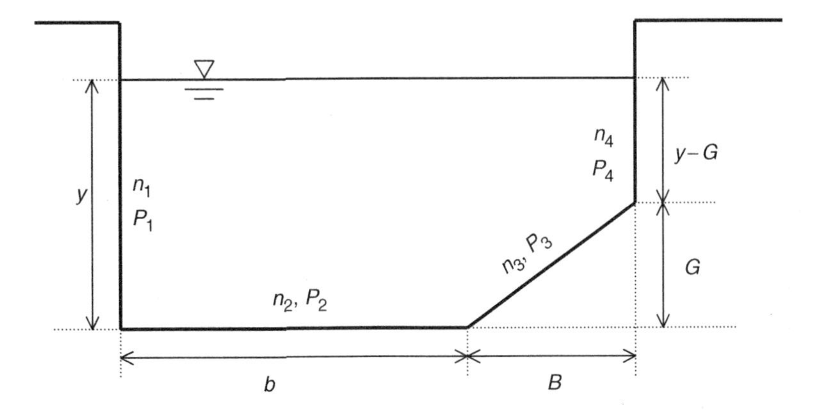

FIGURE 3.9
Example composite channel

Equation 3.37 is attributed to Einstein and Banks (Chow 1959). If we want to calculate the discharge for a given normal flow depth, we can first determine the equivalent roughness using Equation 3.36 or 3.37. Then we apply the Manning formula to calculate the discharge. However, if the discharge is given and the depth is unknown, a trial-and-error procedure will be needed.

EXAMPLE 3.7 For the composite channel shown in Figure 3.9, $b=5$ ft, $B=3$ ft, $G=2$ ft, $n_1=0.016$, $n_2=0.020$, $n_3=0.022$, and $n_4=0.016$. The channel has a slope of $S_0=0.0004$. Determine the discharge if the normal flow depth is 4.5 ft.

From the cross-sectional geometry, we have $P_1=y=4.5$ ft, $P_2=b=5$ ft, $P_3=(B^2+G^2)^{1/2}=(3^2+2^2)^{1/2}=3.61$ ft, and $P_4=y-G=4.5-2=2.5$ ft. We can now evaluate the equivalent roughness. Let us use Equation 3.37 to determine the composite roughness factor as

$$n_e=\left[\frac{\sum_{i=1}^{N}(P_i n_i^2)}{\sum_{i=1}^{N}P_i}\right]^{1/2}$$

$$=\left[\frac{4.5(0.016)^2+5(0.020)^2+3.61(0.022)^2+2.5(0.016)^2}{4.5+5.0+3.61+2.5}\right]^{1/2}=0.019$$

Now we calculate the flow area as $A=by+0.5(y+y-G)B=(4.5)(5.0)+0.5(4.5+2.5)3=33.0\text{ ft}^2$. Likewise, the total perimeter becomes $P=4.5+5.0+3.61+2.50=15.61$ ft. The hydraulic radius of the whole section is $R=33.0/15.61=2.11$ ft. Substituting these into Equation 3.26,

$$Q=\frac{k_n}{n}AR^{2/3}S_0^{1/2}=\frac{1.49}{0.019}(33.0)(2.11)^{2/3}(0.0004)^{1/2}=85.2\text{ cfs}$$

EXAMPLE 3.8 Determine the normal flow depth for the channel considered in Example 3.7 if the discharge is $Q=150$ cfs.

In this problem we cannot determine the equivalent roughness factor directly, since we cannot evaluate P_1 and P_4 without knowing the flow depth. Thus we

TABLE 3.4 Summary calculations for Example 3.8

y_n (ft)	P_1 (ft)	P_2 (ft)	P_3 (ft)	P_4 (ft)	P (ft)	n_e	A (ft^2)	R (ft)	Q (cfs)
5.50	5.50	5.00	3.61	3.50	17.61	0.019	41.00	2.33	115.81
6.00	6.00	5.00	3.61	4.00	18.61	0.018	45.00	2.42	131.26
6.50	6.50	5.00	3.61	4.50	19.61	0.018	49.00	2.50	147.00
6.60	6.60	5.00	3.61	4.60	19.81	0.018	49.80	2.51	150.18

need to use a trial-and-error procedure. We pick a trial value for the normal flow depth, then, using this depth, we calculate the equivalent roughness factor and the discharge as in Example 3.7. If the calculated discharge turns out to be 150 cfs, the tried value of normal depth is correct. Otherwise, we try different depths until we obtain 150 cfs. Table 3.4 summarizes the calculations for the various normal depths tried. The normal depth for this problem is found to be $y_n = 6.60$ ft.

3.7 NORMAL FLOW IN COMPOUND CHANNELS

The foregoing sections were devoted to prismatic channels having constant cross-sectional shapes and dimensions. Most natural channels have irregular shapes that vary with distance along the channel. Moreover, many natural channels have a main part and one or two overbank areas, as shown in Figure 3.10. The average flow velocity in the main channel and the overbank areas can be significantly different because of different flow depths and roughness factors. We need to take these differences into account in our calculations.

Let us first rewrite the Manning formula (Equation 3.19b) as

$$Q = KS_f^{1/2} \tag{3.38}$$

where K = conveyance, defined as

$$K = \frac{k_n}{n}AR^{2/3} \tag{3.39}$$

For a compound channel section made of several subsections of different characteristics (like the main channel subsection and two overbank flow subsections of Figure 3.10), we can define the conveyance of each individual subsection as

$$K_i = \frac{k_n}{n_i}A_iR_i^{2/3} \tag{3.40}$$

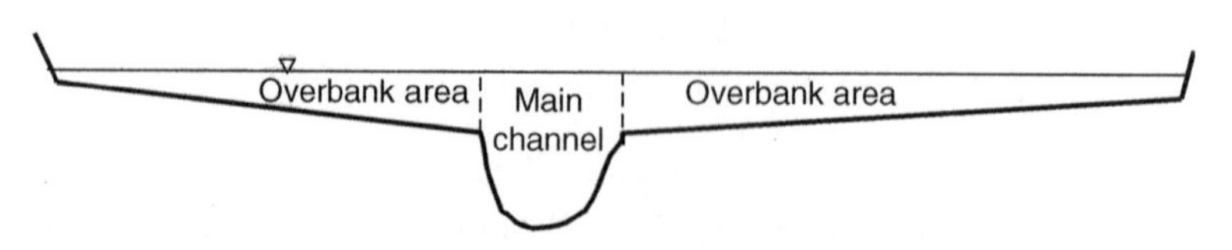

FIGURE 3.10 Compound channel

where i = index referring to the i-th subsection. The total discharge Q in the compound section is equal to the sum of the subsection discharges, Q_i. Assuming that S_f is the same for all subsections (that is, assuming the energy head does not vary across the compound section), we can write that

$$Q = \left(\sum K_i\right) S_f^{1/2} \tag{3.41}$$

Equation 3.41 is applicable to gradually-varied open-channel flow as well as normal flow. However, we can simplify Equation 3.41 for normal flow, noting that $S_f = S_0$, as

$$Q = \left(\sum K_i\right) S_0^{1/2} \tag{3.42}$$

The normal flow depth for a given discharge in a compound channel is the depth that satisfies Equation 3.42.

Given the normal flow depth, we can determine the corresponding discharge explicitly from Equation 3.42. However, in most applications we would need to determine the normal flow depth given the discharge. This requires a trial-and-error procedure. We evaluate the right-hand side of Equation 3.42 by trying different values for the normal flow depth until the equation is balanced.

Strict normal flow is unlikely to occur in a natural compound channel. However, the concept of normal flow is used in a variety of practical applications, such as flood plain encroachment calculations. *Encroachment* (or narrowing of the flood plain) occurs due to construction of levees to confine flood flows, earth-fills in the flood plain, and construction of building sites. Obviously, narrowing the channel will affect the flow depth. Typically, encroachments are allowed as long as the flow depth during a 100-year flood will not increase more than a specified height above the depth that would have occurred in the original channel (Roberson *et al.*, 1997).

EXAMPLE 3.9 The cross-section of a stream can be approximated by the compound channel shown in Figure 3.11. The bottom slope is $S_0 = 0.0009$. The Manning roughness factor is $n = 0.025$ for the main channel and $n = 0.035$ for the overbank areas. Determine the normal depth for a discharge of 57 000 cfs.

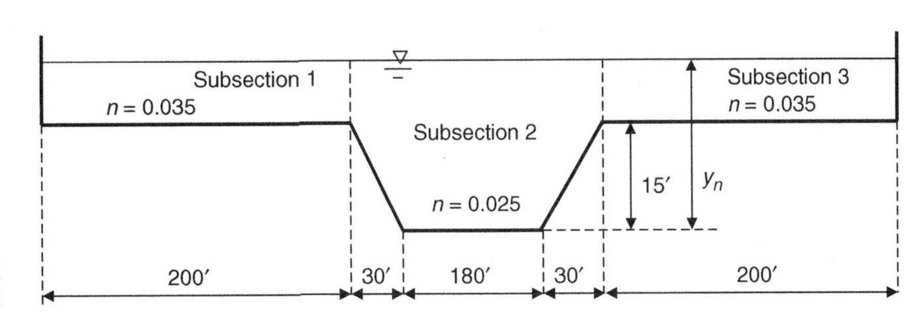

FIGURE 3.11 Example compound channel

We need to use a trial-and-error procedure to solve this problem. First we will try $y_n = 18$ ft. Then, for Subsection 1:

$$A_1 = (18 - 15)200 = 600\ \text{ft}^2$$

$$P_1 = (18 - 15) + 200 = 203\ \text{ft}$$

$$R_1 = \frac{600}{203} = 2.96\ \text{ft}$$

$$K_1 = \frac{1.49}{0.035} 600(2.96)^{2/3} = 52\,660\ \text{cfs}$$

Because Subsections 1 and 3 are identical, we also have $K_3 = 52\,660$ cfs. For the main channel, that is for Subsection 2,

$$A_2 = (180 + 240)\frac{15}{2} + (240)(18 - 15) = 3870\ \text{ft}^2$$

$$P_2 = 180 + 2\sqrt{30^2 + 15^2} = 247\ \text{ft}$$

$$R_2 = \frac{3870}{247} = 15.67\ \text{ft}$$

$$K_2 = \frac{1.49}{0.025} 3870(15.67)^{2/3} = 1\,444\,450\ \text{cfs.}$$

Then the total conveyance becomes

$$\sum K_i = 52\,660 + 444\,350 + 52\,660 = 1\,549\,670\ \text{cfs}$$

and the discharge is calculated as

$$Q = (1\,549\,670)(0.0009)^{1/2} = 46\,490\ \text{cfs}$$

Because this discharge is different from the given 57 000 cfs, we will try another depth. Using the same procedure, we obtain $Q = 75\,000$ cfs for $y_n = 22$ ft and $Q = 57\,000$ for $y_n = 19.58$ ft. We therefore conclude that $y_n = 19.58$ ft.

EXAMPLE 3.10 Determine the momentum coefficient β and the energy coefficient α for the compound channel considered in Example 3.9 under normal flow conditions.

In the previous example, we determined that $y_n = 19.58$ ft. Therefore, for Subsections 1 and 3,

$$A_1 = A_3 = (19.58 - 15.0)(200) = 916\ \text{ft}^2$$

$$P_1 = P_3 = (19.58 - 15.0) + 200 = 204.58\ \text{ft}$$

$$R_1 = R_3 = (916)/(204.58) = 4.48\ \text{ft}$$

$$K_1 = K_3 = \frac{1.49}{0.035}(916)(4.48)^{2/3} = 105\,980\ \text{cfs}$$

Then for $S_0 = 0.0009$,

$$Q_1 = Q_3 = (105\,980)(0.0009)^{1/2} = 3180\,\text{cfs}$$

$$V_1 = V_3 = (3180)/(916) = 3.47\,\text{fps}$$

For Subsection 2, we have

$$A_2 = (180 + 240)\frac{15}{2} + (240)(19.58 - 15.00) = 4250\,\text{ft}^2$$

$$P_2 = 180 + 2\sqrt{(30)^2 + (15)^2} = 247\,\text{ft}$$

$$R_2 = (4250)/(247) = 17.20\,\text{ft}$$

$$K_2 = \frac{1.49}{0.025}(4250)(17.2)^{2/3} = 1\,687\,900\,\text{cfs}$$

$$Q_2 = (1\,687\,900)(0.0009)^{1/2} = 50\,640\,\text{cfs}$$

$$V_2 = (50\,640)/(4250) = 11.91\,\text{fps}$$

For the whole section, we have

$$Q = 2(3180) + 50\,640 = 57\,000\,\text{cfs}$$

$$A = 2(916) + 4250 = 6082\,\text{ft}^2$$

$$V = (57\,000)/(6082) = 9.37\,\text{fps}$$

Now, using Equation 1.15,

$$\beta = \frac{V_1^2 A_1 + V_2^2 A_2 + V_3^2 A_3}{V^2 A} = \frac{(3.47)^2(916) + (11.91)^2(4250) + (3.47)^2(916)}{(9.37)^2(6082)} = 1.17$$

Likewise, by using Equation 1.21,

$$\alpha = \frac{V_1^3 A_1 + V_2^3 A_2 + V_3^3 A_3}{V^3 A} = \frac{(3.47)^3(916) + (11.91)^3(4250) + (3.47)^3(916)}{(9.37)^3(6082)} = 1.45$$

EXAMPLE 3.11 The width of the channel considered in Example 3.9 will be reduced; however, this reduction must not cause an increase of more than 1 ft in the flow depth for the discharge of 57 000 cfs. The encroachment will be over a long distance, and we can assume that normal flow will occur throughout the encroached portion of the channel. Determine the minimum allowable channel width, B, shown in Figure 3.12.

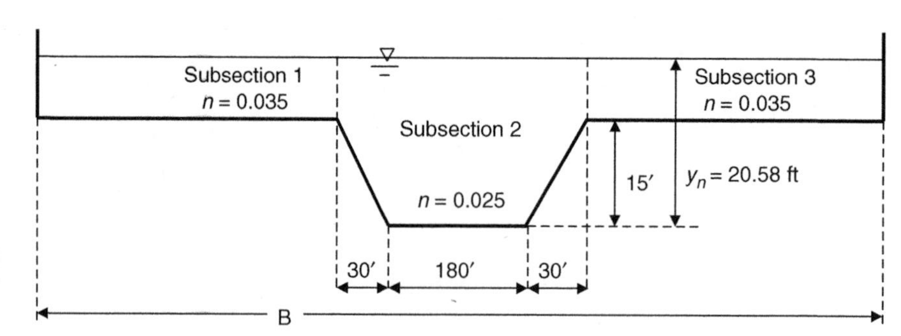

FIGURE 3.12 Definition sketch for Example 3.11

The allowable depth is $19.58 + 1.0 = 20.58$ ft. Then, with reference to Figure 3.12 for Subsections 1 and 3,

$$A_1 = A_3 = (20.58 - 15)\frac{(B - 240)}{2} = 2.79B - 669.6$$

$$P_1 = P_3 = (20.58 - 15) + \frac{B - 240}{2} = 0.5B - 114.42$$

$$R_1 = R_3 = \frac{2.79B - 669.6}{0.5B - 114.42}$$

$$K_1 = K_3 = \frac{1.49}{0.035}\frac{(2.79B - 669.6)^{5/3}}{(0.5B - 114.42)^{2/3}}$$

For the main channel, that is for Subsection 2,

$$A_2 = (180 + 240)\left(\frac{15.0}{2}\right) + 240(20.58 - 15.0) = 4489\ \text{ft}^2$$

$$P_2 = 180 + 2\sqrt{30^2 + 15^2} = 247\ \text{ft}$$

$$R_2 = \frac{4489}{247} = 18.17\ \text{ft}$$

$$K_2 = \frac{1.49}{0.025}4489(18.17)^{2/3} = 1\,849\,300\ \text{cfs}$$

Thus for the whole section,

$$\left(\sum K_i\right)S_0^{1/2} = \left[(2)\frac{1.49}{0.035}\frac{(2.79B - 669.6)^{5/3}}{(0.5B - 114.42)^{2/3}} + 1\,849\,300\right](0.0009)^{1/2} = 57\,000$$

or, simplifying,

$$2.554\frac{(2.79B - 669.6)^{5/3}}{(0.5B - 114.42)^{2/3}} + 55\,479 = 57\,000$$

By trial and error, $B = 314.5\ \text{ft} = 315\ \text{ft}$.

PROBLEMS

P.3.1 Suppose $b = 2$ ft, $m_1 = 2.0$, $m_2 = 2.0$, $m = 1.5$ for the channels shown in Figure P.1.1 of Chapter 1. Determine the discharge in each channel for a normal flow depth of 2.5 ft if $n = 0.016$ and $S_0 = 0.0009$. Also determine whether the normal flow is subcritical or supercritical.

P.3.2 Determine the normal flow depth for each channel considered in Problem P.3.1 for $Q = 90$ cfs.

P.3.3 Suppose $b = 1$ m, $m_1 = 1.0$, $m_2 = 2.0$, $m = 1.5$ for the channels shown in Figure P.1.1 of Chapter 1. If $n = 0.016$ and $S_0 = 0.0009$, determine the discharge in each channel for a normal flow depth of 1.0 m. Also, determine whether the normal flow is subcritical or supercritical.

P.3.4 Determine the normal flow depth for each channel considered in Problem P.3.3 for $Q = 4.5\ m^3/s$.

P.3.5 The normal flow depth is $y_n = 2.6$ ft in a trapezoidal, earthen channel that has a bottom width $b = 10$ ft and side slopes of $m = 2$. The bottom slope of the channel is $S_0 = 0.0004$. Assuming that $k_s = 0.01$ ft for this channel, and the water temperature is 60°F, determine whether the flow is fully rough. Note that for normal flow, $S_f = S_0$.

P.3.6 Consider a length segment of $\Delta x = 10$ ft of the channel discussed in Example 3.1.

(a) Determine the weight of water stored in the length segment.
(b) Determine the component of the weight in the flow direction.
(c) Determine the friction force resisting to the flow. Compare this with your answer for part (b).

P.3.7 As part of a drainage improvement project, a drainage ditch will be straightened and cleaned. As a result, the length of the ditch will decrease from 1800 ft to 1400 ft and the Manning roughness factor will be reduced to 0.018 from 0.022. Determine the percentage increase in the discharge the ditch can accommodate at the same normal depth.

P.3.8 A circular storm sewer to be laid on a slope of 0.0009 will be sized to carry 38 cfs. It is required that the normal depth to diameter ratio be less than 0.8. Select the minimum acceptable diameter among the available sizes of 12 in., 15 in., 18 in., 21 in., 24 in., 27 in., 30 in., 36 in., 42 in., 48 in, 54 in., and 60 in. Use a Manning roughness factor of 0.016.

P.3.9 A 24-in storm water sewer laid on a slope of 0.0009 has a Manning roughness factor of 0.013. Determine the discharge the sewer carries at normal flow depths of 0.8 ft, 1.6 ft, 1.88 ft, and 1.98 ft. Discuss your results.

P.3.10 A circular storm sewer has a diameter of $d_0 = 0.5$ m, Manning roughness factor of $n = 0.013$, and a bottom slope of $S_0 = 0.0004$. What is the maximum discharge this sewer can accommodate at normal flow if the normal depth-to-diameter ratio is not to exceed 0.70?

P.3.11 What is the maximum discharge the sewer in Problem P.3.10 can accommodate under normal flow conditions if there is no limit on the normal depth-to-diameter ratio?

P.3.12 A trapezoidal channel to be excavated into the ground will be sized to carry $Q = 200$ cfs at normal flow. The bottom slope is $S_0 = 0.0004$ and the

Manning roughness factor is $n = 0.020$. The land surface also has a slope of 0.0004. If cost of excavation is the main concern, which of the alternatives listed below is the best? Neglect free board.

b (ft)	m
10	1.5
10	2.0
10	2.5

P.3.13 A trapezoidal drainage canal is being considered to carry $10\,m^3/s$. The slope of the canal will be $S_0 = 0.0002$. To avoid erosion, the canal will be lined with asphalt for which $n = 0.017$. If cost of lining is the dominant factor, determine which of the following alternatives is the best. Neglect free board.

b (m)	m
2.5	1.0
2.0	1.5
1.5	2.0

P.3.14 A trapezoidal channel lined with Bermuda grass has a bottom width of $b = 7$ ft, side slopes of $m = 2.5$ and a longitudinal bottom slope of $S_0 = 0.0009$. Determine the normal flow depth in the channel for $Q = 70$ cfs if:

(a) the grass has a good stand and is about 30 cm tall
(b) the grass is cut to about 4 cm.

P.3.15 The normal flow depth in a trapezoidal channel lined with uncut crabgrass is 1.0 m. The channel has a bottom width of $b = 2.0$ m, side slopes of $m = 2.5$, and a longitudinal bottom slope of $S_0 = 0.002$. Determine the discharge.

P.3.16 The normal flow depth in a trapezoidal channel lined with uncut crabgrass is 2.5 ft. The channel has a bottom width of $b = 6$ ft, side slopes of $m = 2.5$, and a longitudinal bottom slope of $S_0 = 0.001$. Determine the discharge.

P.3.17 A trapezoidal channel lined with Bermuda grass has a bottom width of $b = 2.2$ m, side slopes of $m = 2.5$, and a longitudinal bottom slope of $S_0 = 0.001$. Determine the normal flow depth for $Q = 2.5\,m^3/s$:

(a) if the grass has a good stand and is about 30 cm tall
(b) if the grass is cut to about 4 cm.

P.3.18 A trapezoidal channel has a bottom width of $b = 4$ ft, side slopes of $m = 3$, and a bottom slope of $S_0 = 0.0009$. The channel is lined with riprap of $d_{50} = 0.5$ ft. The normal flow depth is $y_n = 3$ ft.

(a) Is Equation 3.34 or Equation 3.35 more suitable for this channel?
(b) Determine the discharge at a normal depth of 3 ft.

P.3.19 A trapezoidal channel has a bottom width of $b=1.0$ m, side slopes of $m=3$, and a bottom slope of $S_0=0.01$. The channel is lined with riprap of $d_{50}=15$ cm. The normal flow depth is $y_n=0.60$ m.

(a) Is Equation 3.34 or Equation 3.35 more suitable for this channel?
(b) Determine the discharge at a normal depth of 0.60 m.

P.3.20 A trapezoidal channel has a bottom width of $b=2.5$ ft., side slopes of $m=3$, bottom slope of $S_0=0.001$, and is lined with riprap of $d_{50}=0.4$ ft. Determine the normal depth and Froude number for $Q=43$ cfs.

P.3.21 A trapezoidal channel lined with $d_{50}=10$ cm has a bottom width of $b=1.2$ m, side slopes of $m=3$, and a longitudinal bottom slope of $S_0=0.008$. Determine the normal depth and the Froude number for $Q=1.4\ \text{m}^3/\text{s}$.

P.3.22 Figure 3.9 represents the cross-sectional shape of various composite channels, the dimensions of which are tabulated below. Determine the discharge in each channel if $y_n=3$ ft and $S_0=0.001$.

b (ft)	*B* (ft)	*G* (ft)	n_1	n_2	n_3	n_4
5.0	2.0	3.0	0.016	0.016	0.012	0.012
4.0	1.0	2.0	0.012	0.012	0.018	0.018
3.0	3.0	2.0	0.015	0.016	0.020	0.015

P.3.23 Figure 3.9 represents the cross-sectional shape of various composite channels, the dimensions of which are tabulated below. Determine the discharge in each channel if $y_n=1.2$ m and $S_0=0.008$.

b (m)	*B* (m)	*G* (m)	n_1	n_2	n_3	n_4
2.0	1.2	1.0	0.016	0.016	0.012	0.012
1.0	3.0	1.2	0.012	0.012	0.018	0.018
0.5	3.5	1.0	0.015	0.016	0.020	0.015

P.3.24 Suppose the composite channel shown in Figure 3.9 has $b=3$ ft, $B=1$ ft, $G=3$ft, $n_1=0.012$, $n_2=0.016$, $n_3=0.016$, $n_4=0.012$, and $S_0=0.001$. Determine the normal depth for $Q=80$ cfs.

P.3.25 Suppose the composite channel shown in Figure 3.9 has $b=1.2$ m, $B=0.3$ m, $G=1.0$ m, $n_1=0.012$, $n_2=0.016$, $n_3=0.016$, $n_4=0.012$ and $S_0=0.001$. Determine the normal depth for $Q=2.9\ \text{m}^3/\text{s}$.

P.3.26 The dimensions of the compound channel shown in Figure P.3.1 are $b_1=50$ ft, $b_2=150$ ft, $n_1=0.020$, $n_2=0.035$, and $Z=10$ ft. The slope is $S_0=0.001$. If the normal depth is $y_n=12$ ft:

(a) Determine the discharge
(b) Determine the energy and momentum coefficients α and β.

P.3.27 Suppose the 100-year flood for the compound channel described in Problem P.3.26 is 9200 cfs. Determine the normal flow depth for this discharge.

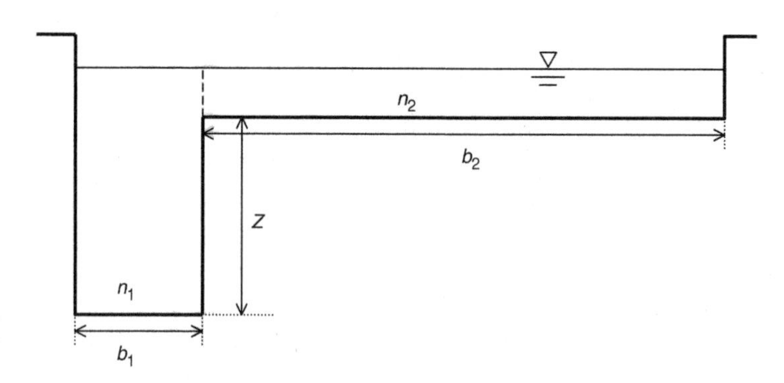

FIGURE P.3.1 Compound channel for Problems P.3.26–P.3.29

P.3.28 Suppose the compound channel shown in Figure P.3.1 has the dimensions of $b_1 = 15$ m, $b_2 = 50$ m, $n_1 = 0.020$, $n_2 = 0.035$, and $Z = 3$ m. The bottom slope is $S_0 = 0.001$. Determine the discharge, Q, the energy coefficient, α, and the momentum coefficient, β, for the normal depth $y_n = 4$ m.

P.3.29 Suppose the 100-year flood for the compound channel described in Problem P.3.28 is 300 m^3/s. Determine the normal flow depth for the 100-year flood.

REFERENCES

Akan, A. O. and Hager, W. W. (2001). Design aid for grass-lined channels. *Journal of Hydraulic Engineering, ASCE*, **127(3)**, 236–237.

Arcement, G. J. and Schneider, V. R. (1989). *Guide for Selecting Manning's Roughness Coefficient for Natural Channels and Floodplains*. US Geological Service, Water Supply Paper 2339, Washington, DC.

ASCE Task Force (1963). Friction Factors in Open Channels, Progress Report of the Task Force in Open Channels of the Committee on Hydromechanics. *Journal of the Hydraulics Division, ASCE*, **89(HY2)**, 97–143.

Barnes, H. H. (1967). *Roughness Characteristics of Natural Channels*. US Geological Survey, Water Supply Paper 1849, Washington, DC.

Blodgett, J. C. and McConaughy, C. E. (1985). *Evaluation of Design Practices for Riprap Protection of Channels near Highway Structures*. US Geological Survey, prepared in cooperation with the Federal Highway Administration Preliminary Draft, Sacramento, CA.

Chen, Y. H. and Cotton, G. K. (1988). *Design of Roadside Channels with Flexible Linings*. Hydraulic Engineering Circular No. 15, Publication No. FHWA-IP-87-7, US Department of Transportation, Federal Highway Administration, McLean, VA.

Chow, V. T. (1959). *Open-Channel Hydraulics*. McGraw-Hill Book Co., New York, NY.

Cowan, W. L. (1956). Estimating hydraulic roughness coefficients. *Agricultural Engineering*, **37(7)**, 473–475.

French, R. H. (1985). *Open-Channel Hydraulics*. McGraw-Hill Book Co., New York, NY.

Hager, W. W. (2001). *Wastewater Hydraulics: Theory and Practice*. Springer-Verlag, New York, NY.

Henderson, F. M. (1966). *Open Channel Flow*. Prentice Hall, Upper Saddle River, NJ.

Maynord, S. T. (1991). Flow resistance of riprap. *Journal of Hydraulic Engineering, ASCE*, **117(6)**, 687–695.

Roberson, J. A., Cassidy, J. J. and Chaudhry, M. H. (1997). *Hydraulic Engineering*. John Wiley and Sons, Inc., New York, NY.

Sturm, T. W. (2001). *Open Channel Hydraulics*. McGraw-Hill Book Co., New York, NY.

US Army Corps of Engineers (2002). HEC-RAS River Analysis System. *Hydraulic Reference Manual*, Hydrologic Engineering Center, Davis, CA.

Yen, B. C. (1992). Hydraulic resistance in open channels. In: B. C. Yen (ed.), *Channel Flow Resistance: Centennial of Manning's Formula*. Water Resources Publications, Littleton, CO.

4 Gradually-varied flow

A flow control is any feature that imposes a relationship between the flow depth and discharge in a channel. A critical flow section, for instance, is a flow control, since at this section $F_r = 1.0$. Likewise, various hydraulic structures such as weirs and gates will control the flow. Normal flow may be viewed as a flow control also because a normal flow equation like Equation 3.26 describes a depth-discharge relationship. In the absence of other flow controls, the flow in an open channel tends to become normal. However, where present, the other controls will pull the flow away from the normal flow conditions. The flow depth varies between two flow controls. Such a non-uniform flow is called *gradually-varied flow* if the changes in the flow depth are gradual. This chapter is devoted to steady, gradually-varied flow.

To obtain an expression for gradually-varied flow, let us recall Equation 2.9, defining the total energy head, H, as

$$H = z_b + y + \frac{V^2}{2g} \tag{4.1}$$

where z_b = elevation of the channel bottom, y = flow depth, V = average cross-sectional velocity, and g = gravitational acceleration. Now, recalling the definition of specific energy given in Equation 2.11 as

$$E = y + \frac{V^2}{2g} \tag{4.2}$$

Equation 4.1 can be expressed as

$$H = z_b + E \tag{4.3}$$

Let us differentiate both sides of Equation 4.3 with respect to x to obtain

$$\frac{dH}{dx} = \frac{dz_b}{dx} + \frac{dE}{dx} \tag{4.4}$$

where x is the displacement in the flow direction. By definition, $S_f = -dH/dx$, and $S_0 = -dz_b/dx$. By substituting these into Equation 4.4 and rearranging, we obtain one form of the gradually-varied flow equation as

$$\frac{dE}{dx} = S_0 - S_f \tag{4.5}$$

We can obtain another form of the gradually-varied flow equation by expanding the left-hand side of Equation 4.5 to

$$\frac{dE}{dx} = \frac{dy}{dx} + \frac{d(V^2/2g)}{dx} = \frac{dy}{dx} + \frac{V}{g}\frac{dV}{dx} = \frac{dy}{dx} + \frac{V}{g}\frac{dV}{dy}\frac{dy}{dx}$$

$$= \frac{dy}{dx} + \frac{V}{g}\frac{dy}{dx}\frac{d(Q/A)}{dy} = \frac{dy}{dx} + \frac{V}{g}\frac{dy}{dx}\frac{Qd(1/A)}{dy}$$

where Q = constant discharge and A = area. Further mathematical manipulation by using the definitions T = top width = dA/dy, D = hydraulic depth = A/T, and F_r = Froude number = $V/(gD)^{0.5}$ will lead to

$$\frac{dE}{dx} = \frac{dy}{dx} + \frac{V}{g}\frac{dy}{dx}\frac{Qd(1/A)}{dy} = \frac{dy}{dx} - \frac{V}{g}\frac{dy}{dx}\frac{Q(dA/dy)}{A^2}$$

$$= \frac{dy}{dx} - \frac{V}{g}\frac{dy}{dx}\frac{Q(T)}{A^2} = \frac{dy}{dx}\left(1 - \frac{V^2}{gD}\right) = \frac{dy}{dx}(1 - F_r^2)$$

Therefore,

$$\frac{dE}{dx} = \frac{dy}{dx}(1 - F_r^2) \tag{4.6}$$

By substituting Equation 4.6 into 4.5 and rearranging, we obtain

$$\frac{dy}{dx} = \frac{S_0 - S_f}{(1 - F_r^2)} \tag{4.7}$$

4.1 Classification of Channels for Gradually-Varied Flow

Open channels are classified as being mild, steep, critical, horizontal, and adverse in gradually-varied flow studies. If for a given discharge the normal depth of a channel is greater than the critical depth, the channel is said to be *mild*. If the normal depth is less than the critical depth, the channel is called *steep*. For a *critical* channel, the normal depth and the critical depth are equal. If the bottom slope of a channel is zero, the channel is called *horizontal*. A channel is said to have an *adverse* slope if the channel bottom rises in the flow direction. In summary:

Mild channels	$y_n > y_c$
Steep channels	$y_n < y_c$
Critical channels	$y_n = y_c$
Horizontal channels	$S_0 = 0$
Adverse channels	$S_0 < 0$

where y_n = normal depth and y_c = critical depth.

4.2 Classification of Gradually-Varied Flow Profiles

A *gradually-varied flow profile* or *gradually-varied water surface profile* is a line indicating the position of the water surface. It is a plot of the flow depth as a function of distance along the flow direction. A sound understanding of possible profiles under different flow situations is essential before we can obtain numerical solutions to gradually-varied flow problems. A qualitative investigation of Equation 4.7 will serve this purpose.

Consider a mild channel as shown in Figure 4.1. By definition, $y_n > y_c$. The channel bottom, the critical depth line, and the normal depth line divide the channel into three zones in the vertical dimension, namely M1, M2, and M3 (M stands for mild). The solid lines in the figure represent the shapes of the possible flow profiles in these three zones. Obviously, the normal depth line itself would represent the water surface if the flow in the channel were normal. In zone M1, the water surface is above the normal depth line. Therefore, in this zone $y > y_n$ and consequently $S_f < S_0$. Also, $y > y_c$ and thus $F_r < 1.0$ in zone M1. Therefore, both the numerator and the denominator of Equation 4.7 are positive quantities, and $(dy/dx) > 0$. In other words, the flow depth must increase in the flow direction in zone M1. We can examine the zones M2 and M3 in a similar manner, and conclude that $(dy/dx) < 0$ in zone M2 and $(dy/dx) > 0$ in zone M3.

The behavior of the water surface profile near the zone boundaries can also be examined. From Equation 4.7, as $y \rightarrow \infty$ we can see that $F_r \rightarrow 0$ and $S_f \rightarrow 0$. Thus $(dy/dx) \rightarrow S_0$, meaning the water surface will approach a horizontal line asymptotically as $y \rightarrow \infty$. Likewise, as $y \rightarrow y_n$, by definition $S_f \rightarrow S_0$ and thus $(dy/dx) \rightarrow 0$. Therefore, the surface profile approaches the normal depth line asymptotically. Near the critical depth line, $y \rightarrow y_c$ and $F_r \rightarrow 1.0$. Thus $(dy/dx) \rightarrow \infty$, and the water surface will approach the critical depth line at an angle close to a right-angle. Near the bottom of the channel, as $y \rightarrow 0$, both $S_f \rightarrow \infty$, and $F_r \rightarrow \infty$. Therefore, the water surface will approach the channel

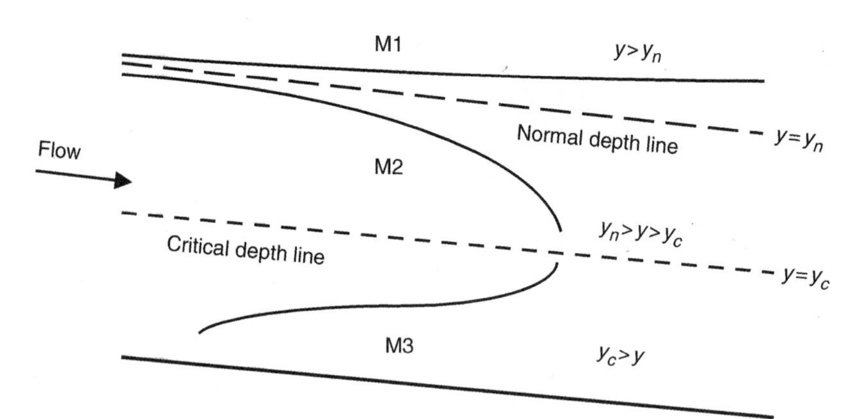

FIGURE 4.1 Flow profiles in mild channels

bottom at a finite positive angle. The magnitude of this angle depends on the friction formula used and the specific channel section.

Based on this qualitative examination of Equation 4.7 near the zone boundaries, we conclude that in zone M1 the water surface profile is asymptotical to the normal depth line as $y \rightarrow y_n$ and is asymptotical to a horizontal line as $y \rightarrow \infty$. The M2 profile is asymptotical to the normal depth line, and it makes an angle close to a right-angle with the critical depth line. The M3 profile makes a positive angle with the channel bottom and an angle close to a right-angle with the critical depth line. The water surface profiles sketched in Figure 4.1 reflect these considerations.

We should note that a flow profile does not have to extend from one zone boundary to another. For example, an M2 profile does not have to begin at the normal depth line and end at the critical depth line. It is possible that an M2 profile begins at a point below the normal depth line and ends at a point above the critical depth line.

For a steep channel, $y_n > y_c$ by definition. The channel bottom, the normal depth line, and the critical depth line divide the channel into three zones in the vertical dimension, namely S1, S2, and S3 (S stands for steep) as shown in Figure 4.2. As before, the solid lines in the figure represent the shapes of the possible flow profiles in these three zones. If the flow were normal in this channel, the normal depth line itself would represent the water surface. In zone S1 the water surface is above the critical depth line, therefore in this zone $y > y_c$ and thus $F_r < 1.0$. Also, $y > y_c > y_n$, and consequently $S_f < S_0$. Therefore, both the numerator and the denominator of Equation 4.7 are positive quantities, and in zone S1 $(dy/dx) > 0$. In other words, the flow depth must increase in the flow direction. We can examine the zones S2 and S3 in a similar manner, and conclude that $(dy/dx) < 0$ in zone S2 and $(dy/dx) > 0$ in zone S3.

The behavior of the surface profile near the zone boundaries examined for mild channels is valid for steep channels as well, since Equation 4.7 is applicable to both steep and mild channels. Accordingly, the S1 profile makes an angle close to the right-angle with the critical depth line, and it approaches to a horizontal line

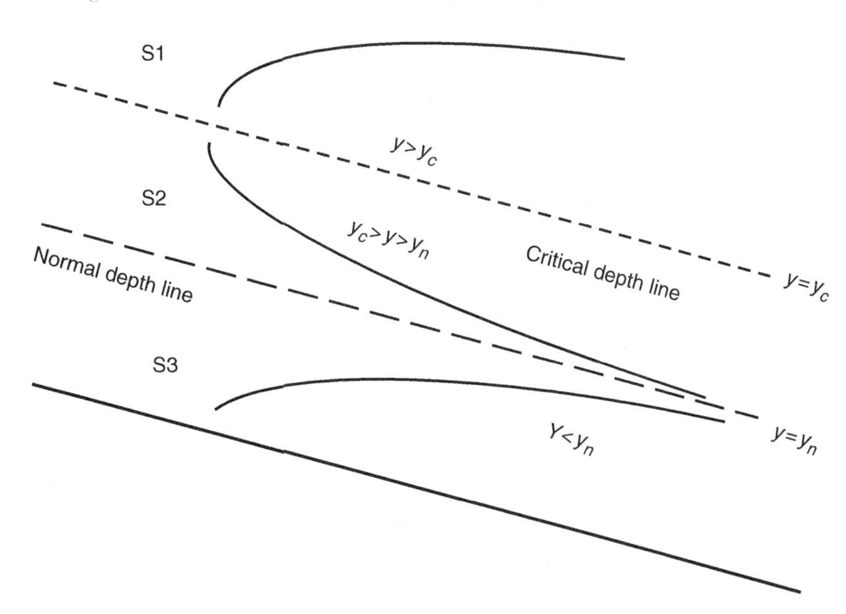

FIGURE 4.2 Flow profiles in steep channels

asymptotically as $y \to \infty$. The S2 profile makes an angle close to the right-angle with the critical depth line, and it approaches the normal depth line asymptotically. The S3 profile will make a positive angle with the channel bottom, and it will approach the normal depth line asymptotically.

The possible profile types that can occur in horizontal, adverse, and critical channels are shown in Figure 4.3. These profiles are sketched by examining the sign of (dy/dx) with the help of Equation 4.7, and considering the behavior of the profile near the zone boundaries. Note that for horizontal and adverse channels normal flow is not possible, thus y_n is not defined, and zones H1 and A1 do not exist. Likewise, for critical channels $y_n = y_c$, and therefore zone C2 does not exist. It is also worth noting that the flow is subcritical in zones M1, M2, S1, H2, A2, and C1, and it is supercritical in zones M3, S2, S3, H3, A3, and C3.

4.3 Significance of Froude Number in Gradually-Varied Flow Calculations

The gradually-varied flow equation (Equation 4.6 or 4.7) is a differential equation, and we need a boundary condition to solve it. Mathematically, the flow depth at any given flow section can be used as a boundary condition. However, for correct representation of open-channel flow the boundary condition will be prescribed at either the upstream or the downstream end of the channel, depending on whether the flow in the channel is supercritical or subcritical. The following observation is presented to explain the reason for this.

A pebble thrown into a large still body of water will create a disturbance, which will propagate outward in the form of concentric circles as shown in Figure 4.4a.

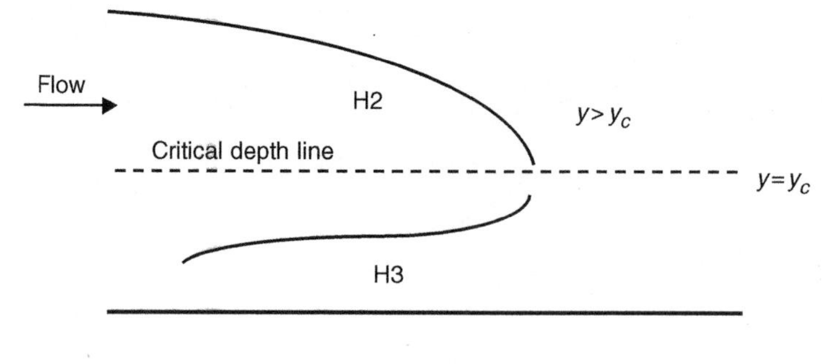

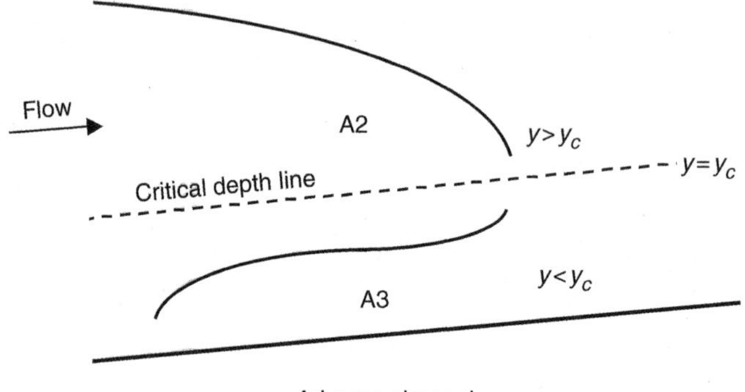

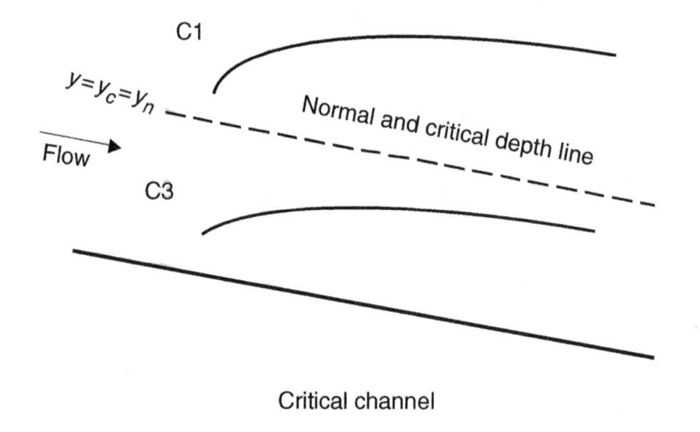

FIGURE 4.3 Flow profiles in horizontal, adverse, and critical channels

The speed with which the disturbance propagates is called *celerity* (or celerity of gravity waves in shallow water), and it is evaluated as

$$c = \sqrt{gD} \tag{4.8}$$

where c = celerity, g = gravitational acceleration, and D = hydraulic depth. If the pebble is thrown into a body of water moving with a velocity V the wave propagation will no longer be in the form of concentric circles. Recalling the definition of Froude number (Equation 1.24), we can write that

$$F_r = \frac{V}{\sqrt{gD}} = \frac{V}{c}$$

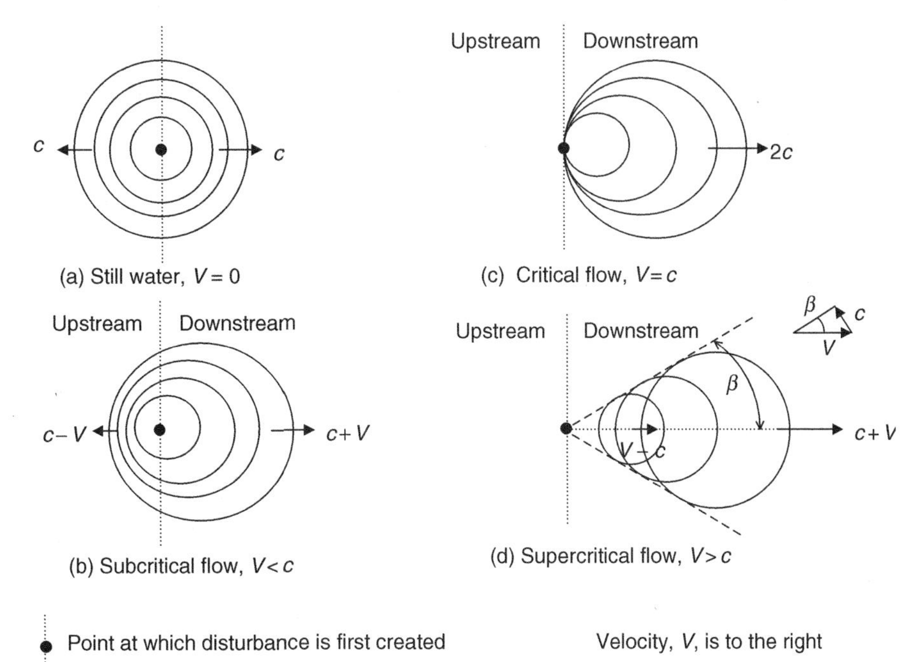

FIGURE 4.4 Effect of Froude number on propagation of a disturbance in channel flow (after Chow, 1959, with permission of Estate of Ven Te Chow)

For subcritical flow, $F_r < 1$ and $V < c$. On the other hand, for supercritical flow, $F_r > 1$ and $V > c$. Obviously, for critical flow $V = c$. Therefore, if the flow is subcritical, the disturbances will propagate upstream at a speed $(c - V)$ and downstream at a speed $(c + V)$, as shown in Figure 4.4b. If the flow is critical, then the upstream edge of the wave will be stationary while the downstream propagation will be at a speed $2c$, as shown in Figure 4.4c. If the flow is supercritical, then the propagation will be in the downstream direction only as shown in Figure 4.4d, with the back and front edges moving with speeds $(V - c)$ and $(V + c)$, respectively. It is important for us to remember that a disturbance in subcritical flow will propagate upstream as well as downstream to affect the flow in both further upstream and downstream sections. However, in supercritical flow the propagation will be only in the downstream direction and the flow at upstream sections will not be affected. Also, as shown in Figure 4.4d, in the case of supercritical flow the lines tangent to the wave fronts lie at an angle $\beta = \arcsin(c/V) = \arcsin(1/F_r)$ with the flow direction.

Because the disturbances can propagate upstream in subcritical flow, the conditions at the downstream end of a channel affect flow in the channel. In other words, subcritical flow is subject to downstream control. Therefore, a downstream boundary condition is needed to solve the gradually-varied flow equations for subcritical flow profiles. On the other hand, because disturbances in supercritical flow cannot propagate upstream, supercritical flow in a channel is not affected by the conditions at the downstream end as long as the flow remains supercritical. Therefore, supercritical flow is subject to upstream control, and we need an upstream boundary condition to solve the gradually-varied flow equations.

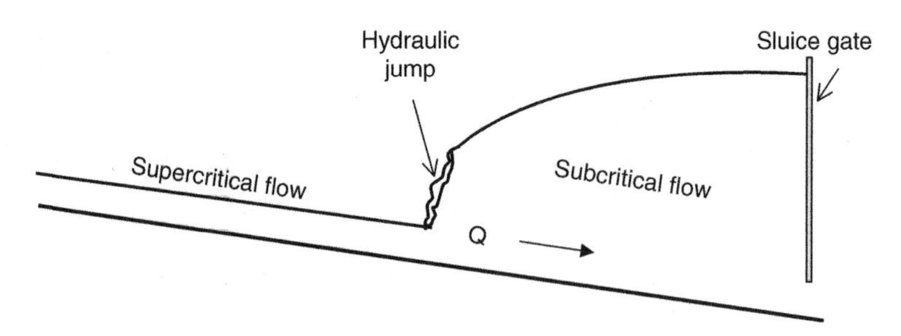

FIGURE 4.5 Hydraulic jump caused by a sluice gate

By stating that subcritical flow is subject to downstream control we do not mean that the flow is not affected by the conditions at the upstream end. The flow enters the channel at the upstream end, and the discharge represents the conditions upstream of the channel. We should also emphasize that the supercritical flow depths are not affected by what is downstream as long as the flow remains supercritical. However, a hydraulic structure placed at the downstream end of a channel, like the sluice gate in Figure 4.5, may cause the flow to change from the supercritical state to the subcritical state through a hydraulic jump. In that event, upstream of the jump the flow will still be supercritical and the flow depths will remain unaffected by the downstream hydraulic structure. However, on the downstream side of the jump the flow will be subcritical and the depths will be affected by the hydraulic structure.

4.4 QUALITATIVE DETERMINATION OF EXPECTED GRADUALLY-VARIED FLOW PROFILES

As discussed in the preceding section, to solve the gradually-varied flow equations we need a boundary condition. To specify this boundary condition, however, we first need to determine, qualitatively, the types of profiles that will occur. This task is not too difficult if we remember some general rules:

1. Subcritical flow is subject to downstream control
2. Supercritical flow is subject to upstream control
3. In the absence of (or far away from) flow controls, flow tends to become normal in prismatic channels
4. The only possible shapes for the water surface profile occurring in the different zones labeled as M1, S2, etc., are those shown in Figures 4.1, 4.2, and 4.3
5. Normal flow in a mild channel is subcritical and that in a steep channel is supercritical
6. Flow is subcritical upstream of a sluice gate and supercritical downstream
7. When subcritical flow is present in a channel terminating at a free fall, the depth at the free fall will be equal to the critical depth (this is an assumption, and a good one, since the critical depth actually occurs a short distance, about $4y_c$, upstream of the free fall)
8. The change from supercritical to subcritical flow is possible only through a hydraulic jump in a prismatic channel.

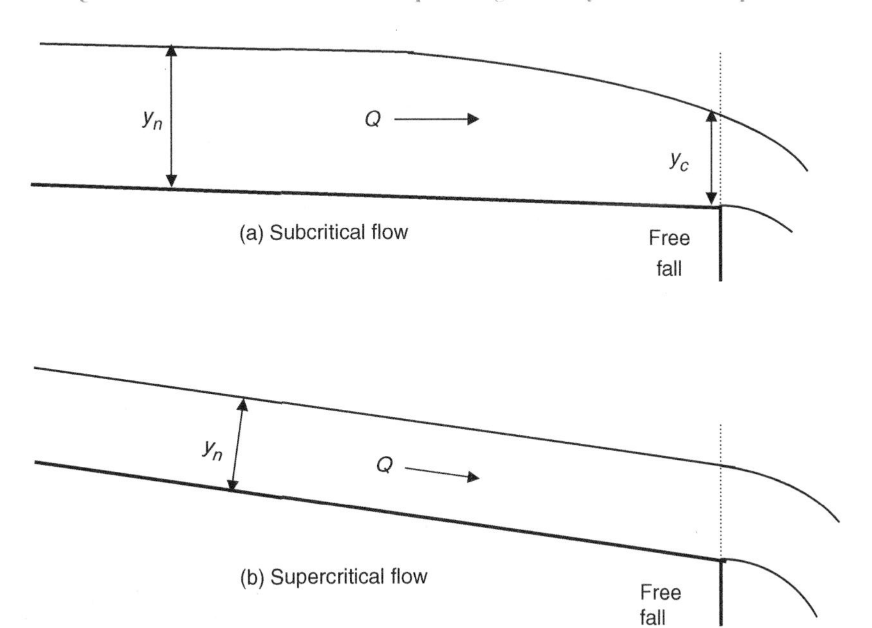

FIGURE 4.6 Subcritical and supercritical flows approaching a free fall

For example, let us consider a very long mild channel that terminates at a free fall as in Figure 4.6a. Far away from the free fall the flow depth will be nearly equal to normal depth (although we do not know the exact location where the flow is normal). Therefore, the flow is subcritical and the depth at the free fall should be critical. Then, the water surface will be positioned between the normal depth line and the critical depth line, and an M2 profile will occur with a boundary condition $y = y_c$ at the downstream end. However, if the channel is steep then the normal flow is supercritical and will not be affected by the free fall (Figure 4.6b).

The problem may be a bit more involved when we deal with composite profiles. Let us consider the situation in Figure 4.7. Suppose the two channels are very long, and identical except for the slope. Let both channels be mild, but channel 1 is milder. Then $y_{n1} > y_{n2}$. Also, y_c is the same in both channels, since the critical depth does not depend on the slope. We can now sketch the normal depth and critical depth lines in both channels. The flow depth will change from nearly y_{n1} to y_{n2}. It may not always be obvious how this change will occur, and in this event we can just sketch some profiles that seem logical at first and then investigate whether they can actually occur. For example, let us consider the sketched profile in Figure 4.7a. This profile cannot occur, since the M1 curve for channel 2 does not have the correct shape (see Figure 4.1). The profile in Figure 4.7b cannot occur either, for the same reason. The only possible profile is the one shown in Figure 4.7c. A subcritical M2 profile will occur in channel 1 with a downstream boundary condition $y = y_{n2}$. Note that y_{n2} is known (or can be calculated based on the known discharge and channel 2 characteristics). The flow will be normal everywhere in channel 2, and therefore gradually-varied flow calculations will be needed for channel 1 only.

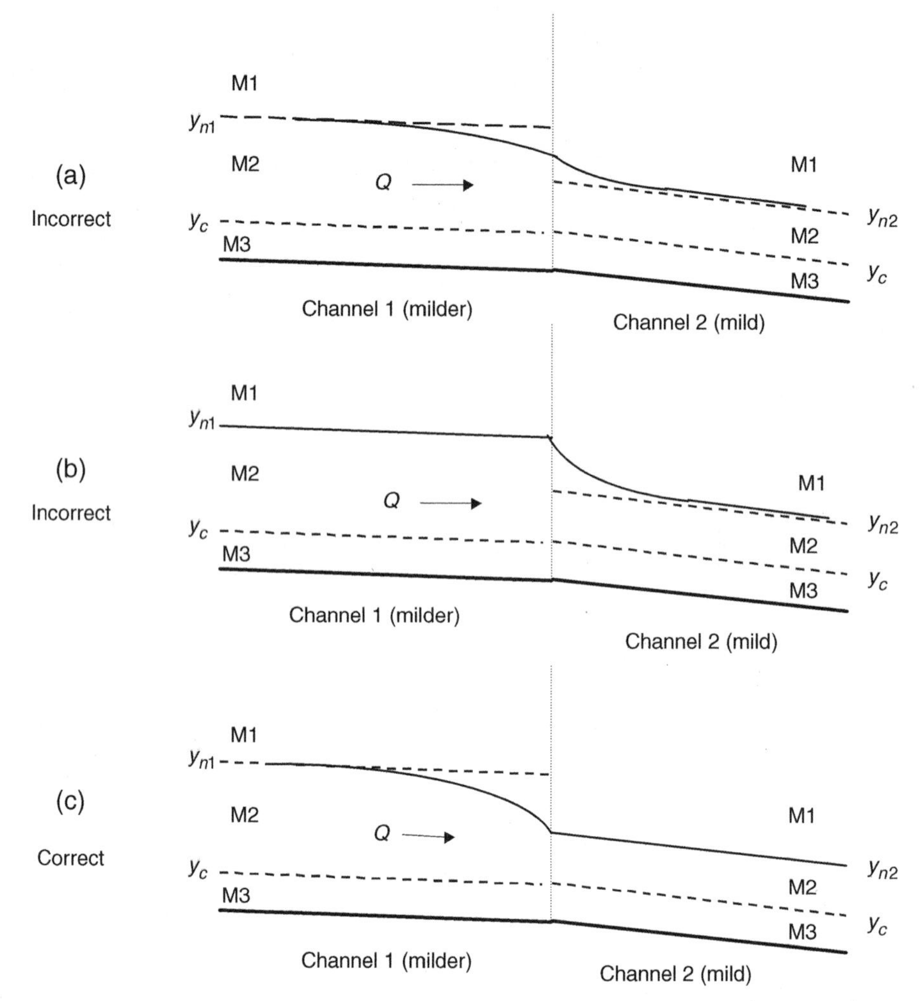

FIGURE 4.7
Composite profile in milder and mild channels

Several other composite profile situations are displayed in Figure 4.8. In all these situations both channels 1 and 2 are very long, and they are identical except for the slope. In Figure 4.8a, where both channels are mild but channel 2 is milder, a subcritical M1 profile will occur in channel 1 with a downstream boundary condition $y = y_{n2}$. The flow in channel 2 will be normal. In Figure 4.8b, both channels are steep but channel 1 is steeper. The normal flow depth is smaller than the critical depth in both channels, and therefore the flow will remain supercritical. Supercritical flow is not subject to downstream control, so in channel 1 the normal flow is maintained. There will be a S3 profile in channel 2 with an upstream boundary condition of $y = y_{n1}$. Note that, because S3 is supercritical, an upstream boundary condition is appropriate. A downstream boundary condition is not available anyhow, since we do not know exactly at what distance the S3 profile will reach the normal depth line. In Figure 4.8c, both channels are steep but channel 2 is steeper. Again, supercritical flow will be maintained in both channels. The flow will be normal in channel 1, and an S2 profile will occur in channel 2 with an upstream boundary condition $y = y_{n1}$. In Figure 4.8d, channel 1 is mild with a subcritical normal depth and channel 2 is

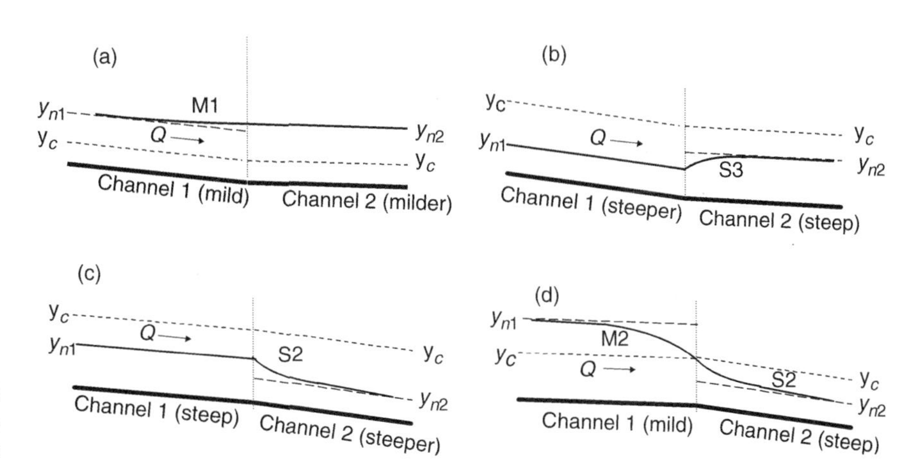

FIGURE 4.8 Various composite flow profiles

steep with a supercritical normal depth. Thus the flow will change from subcritical to supercritical through the critical depth where the two channels join. A subcritical M2 profile will occur in channel 1 with a downstream boundary condition $y = y_c$. In channel 2, a supercritical S2 profile will occur with an upstream boundary condition $y = y_c$. Note that y_c is either known or can be calculated given the discharge and channel cross-sectional properties.

Figure 4.9 displays a situation in which both channels are very long, channel 1 is steep, and channel 2 is mild. The normal depth in channel 1 is supercritical and that in channel 2 is subcritical. Therefore the flow has to change from the supercritical state to subcritical state, and this is possible only through a hydraulic jump. The profiles shown in Figures 4.9a and 4.9b are both possible, qualitatively. However, for a given situation with a specified discharge and channel characteristics, only one of these profiles will occur (See examples 4.1 and 4.2). In Figure 4.9a, the hydraulic jump occurs in channel 1. The profile will be of type S1 in channel 1 after the jump. The flow will remain normal in channel 2. For the S1 profile, the downstream boundary condition is $y = y_{n2}$. In Figure 4.9b, the hydraulic jump occurs in channel 2. The normal flow is maintained in channel 1. In channel 2, an M3 profile will occur before the jump, and the flow will be normal after the jump. The M3 profile is supercritical with an upstream boundary condition $y = y_{n1}$.

The reader may have noticed that, in all the examples discussed above, all the boundary depths are either known or can be determined without any gradually-varied flow calculations. However, this may not always be the case. For example, in Figure 4.10, channel 1 is very long, but channel 2 has a finite length. Both channels are mild, but channel 1 is milder. A qualitative analysis will indicate that an M2 profile will occur in channel 2 with a downstream boundary condition $y = y_c$. The critical depth, y_c, is either known or can be readily determined. This information allows us to begin the gradually-varied flow calculation in channel 2. The profile in channel 1 will also be of type M2. The downstream boundary condition for this profile is $y = y_D$. However, the value of y_D, which is also the depth at the upstream end of channel 2, is not known

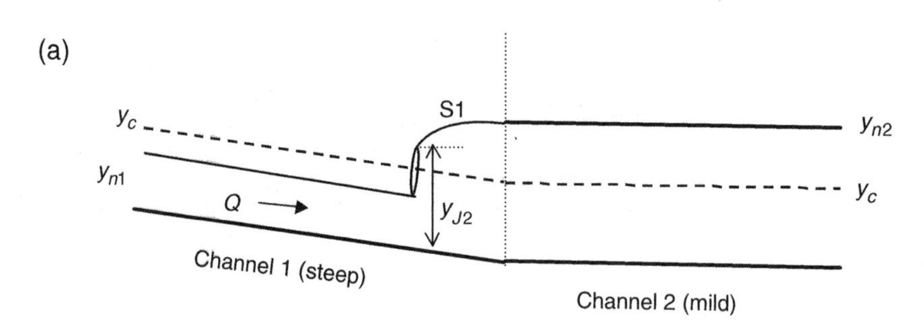

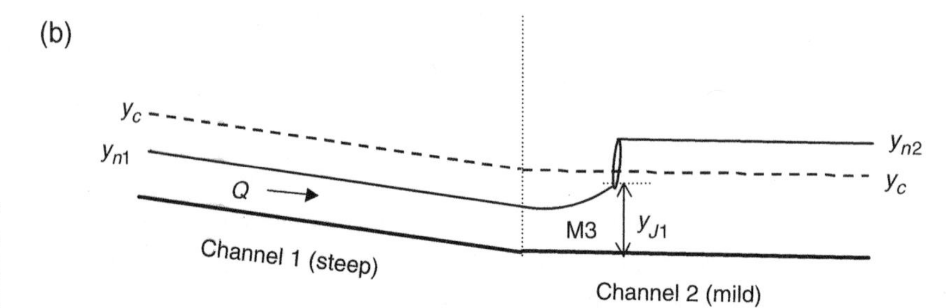

FIGURE 4.9 Flow profiles in a steep channel followed by a mild channel

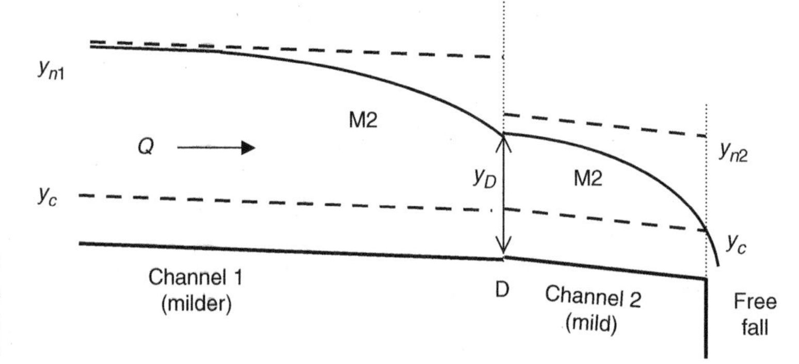

FIGURE 4.10 Example composite profiles involving a short channel

a priori; it will become available only when the gradually-varied flow calculations are completed for channel 2.

EXAMPLE 4.1 A very long rectangular channel (channel 1) has a width of $b = 10$ft, Manning roughness factor of $n = 0.020$, and a bottom slope of $S_0 = 0.02$. It carries a discharge of $Q = 300$ cfs. This channel joins another channel (channel 2) downstream, as in Figure 4.9, that has identical properties except for a slope of $S_0 = 0.005$. Determine the type of water surface profile occurring in these two channels.

We will first determine if these two channels are mild or steep by calculating and comparing the critical and normal depths. The critical depth is the same in both channels. Because the channels are rectangular, we can use Equation 2.3:

$$y_c = \sqrt[3]{\frac{Q^2}{gb^2}} = \sqrt[3]{\frac{(300)^2}{(32.2)(10)^2}} = 3.03 \text{ ft}$$

To calculate the normal depth in channel 1, noting that $P=A/R$, we will write the Manning formula, Equation 3.26 as

$$Q=\frac{k_n}{n}AR^{2/3}S_0^{1/2}=\frac{k_n}{n}\frac{A^{5/3}}{P^{2/3}}S_0^{1/2}=\frac{k_n}{n}\frac{(by_{n1})^{5/3}}{(b+2y_{n1})^{2/3}}S_0^{1/2}$$

or

$$300=\frac{1.49}{0.02}\frac{(10y_{n1})^{5/3}}{(10+2y_{n1})^{2/3}}(0.02)^{1/2}$$

By trial and error, $y_{n1}=2.16$ ft. We can find the normal depth in channel 2, in a similar way, to be $y_{n2}=3.51$ ft. Comparing the normal and critical depths, we can conclude that channel 1 is steep and channel 2 is mild. Therefore, this situation is similar to those illustrated in Figure 4.9 and two different profiles need to be considered.

Let us first check if the profile of Figure 4.9a is possible for this case, where a hydraulic jump occurs in channel 1 from a depth $y_{j1}=y_{n1}$ to a depth y_{j2} that is smaller than y_{n2}. Here, $y_{J1}=2.16$ ft and

$$F_{rj1}=\frac{Q}{A\sqrt{gD}}=\frac{Q}{(by_{J1})\sqrt{g(by_{J1})/b}}=\frac{300}{(10)(2.16)\sqrt{32.2(10)(2.16)/10}}=1.67$$

Now, by using Equation 2.26,

$$y_{J2}=\frac{1}{2}y_{J1}\left(\sqrt{1+8F_{rj1}^2}-1\right)=\frac{1}{2}(2.16)\left(\sqrt{1+8(1.67)^2}-1\right)=4.13\text{ ft}$$

The calculated y_{j2} is greater than y_{n2}. This is not possible, since along an S1 profile the flow depths should increase, not decrease. Accordingly, Figure 4.9a does not represent the solution in this problem, and the jump will not occur in channel 1. At this point, without further calculations, we can reach the conclusion that the jump will be in channel 2 as in Figure 4.9b, since a jump has to occur somewhere for the flow to change from the supercritical to the subcritical state. We can easily confirm this conclusion if needed. In channel 2 the hydraulic jump would occur from a depth y_{J1}, greater than y_{n1}, to a depth $y_{J2}=y_{n2}$. For $y_{J2}=3.51$ ft:

$$F_{rJ2}=\frac{Q}{(by_{J2})\sqrt{g(by_{J2})/b}}=\frac{300}{(10)(3.51)\sqrt{32.2(10)(3.51)/10}}=0.80$$

Then, by using Equation 2.27,

$$y_{J1}=\frac{1}{2}y_{J2}\left(\sqrt{1+8F_{rJ2}^2}-1\right)=\frac{1}{2}(3.51)\left(\sqrt{1+8(0.80)^2}-1\right)=2.59\text{ ft}$$

This depth is greater than y_{n1} and smaller than y_c. Therefore, an M3 profile in channel 2 is possible followed by a hydraulic jump, as shown in Figure 4.9b.

EXAMPLE 4.2 Suppose in Example 4.1 the slope of channel 2 were $S_0 = 0.001$. Determine whether the hydraulic jump would occur in channel 1 or channel 2.

In Example 4.1, we determined that channel 1 is steep since $y_c = 3.03$ ft and $y_{n1} = 2.16$ ft. Here, using the same procedure as in Example 4.1, we can calculate $y_{n2} = 6.41$ ft for $S_0 = 0.001$. Then, channel 2 is mild ($y_c = 3.03$ ft as in channel 1) and a hydraulic jump has to occur since the flow will change from a supercritical to a subcritical state.

Let us check whether the jump will occur in channel 1 followed by an S1 curve. In Example 4.1, we determined that if the jump occurred in channel 1, the depth right after the jump would be $y_{J2} = 4.13$ ft. This depth is smaller than $y_{n2} = 6.41$ ft, and hence an S1 profile is possible. Therefore, in this case the jump will be in channel 1.

The difference between Examples 4.1 and 4.2 is that the slope of channel 2 in Example 4.2 is milder. This results in a larger y_{n2}. The increase in the hydrostatic pressure force due to the larger depth will push the hydraulic jump back to channel 1.

4.5 GRADUALLY-VARIED FLOW COMPUTATIONS

We can solve either Equation 4.5 or Equation 4.7 in order to determine the gradually-varied flow depths at different sections along a channel. However, we find Equation 4.5 more convenient for this purpose. As we pointed out before, this is a differential equation; a boundary condition is required for solution. It is very important to remember that subcritical flow is subject to downstream control. Therefore, if flow in the channel is subcritical, then a downstream boundary condition must be used to solve Equation 4.5 given Q. Conversely, supercritical flow is subject to upstream control, and an upstream boundary condition is needed to solve Equation 4.5 for supercritical flow. By boundary condition, we generally mean a known flow depth associated with a known discharge.

Analytical solutions to Equation 4.5 are not available for most open-channel flow situations typically encountered. In practice, we apply a finite difference approach to calculate the gradually-varied flow profiles. In this approach, the channel is divided into short reaches and computations are carried out from one end of the reach to the other.

Consider the channel reach shown in Figure 4.11 having a length of ΔX. Sections U and D denote the flow sections at the upstream and downstream ends of the reach, respectively. Using the subscripts U and D to denote the upstream and downstream sections, we can write Equation 4.5 for this reach in finite difference form as

$$\frac{E_D - E_U}{\Delta X} = S_0 - S_{fm} \tag{4.9}$$

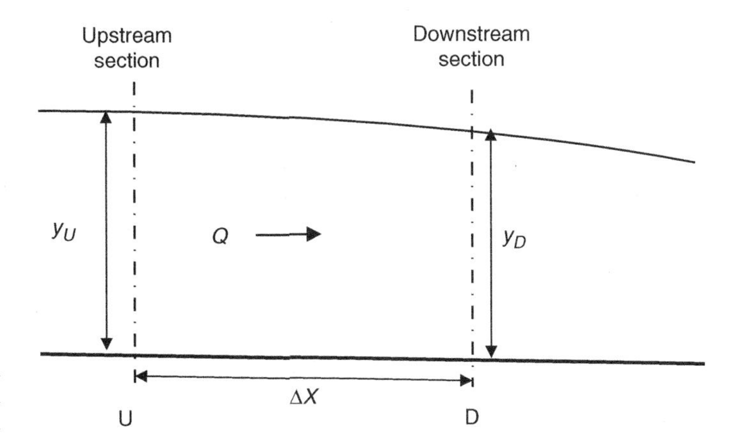

FIGURE 4.11
Definition sketch for gradually-varied flow formulation

where S_{fm} = average friction slope in the reach, approximated as

$$S_{fm} = \frac{1}{2}(S_{fU} + S_{fD}) \tag{4.10}$$

By rearranging the Manning formula, the friction slopes at sections U and D are obtained as

$$S_{fU} = \frac{n^2}{k_n^2}\frac{V_U^2}{R_U^{4/3}} \tag{4.11}$$

and

$$S_{fD} = \frac{n^2}{k_n^2}\frac{V_D^2}{R_D^{4/3}} \tag{4.12}$$

The two most common methods used to perform the gradually-varied flow calculations are the *direct step method* and the *standard step method.*

4.5.1 DIRECT STEP METHOD

In the direct step method, we write Equation 4.9 as

$$\Delta X = \frac{E_D - E_U}{S_0 - S_{fm}} = \frac{(y_D + V_D^2/2g) - (y_U + V_U^2/2g)}{S_0 - S_{fm}} \tag{4.13}$$

In a typical subcritical flow problem, the condition at the downstream section D is known. In other words, y_D, V_D, and S_{fD} are given. We pick an appropriate value (depending on the type of flow profile we have predicted) for y_U, and calculate the corresponding V_U, S_{fU}, and S_{fm}. Then we calculate ΔX from Equation 4.13. Conversely, where supercritical flow is involved, conditions at section U are known. In this case, we pick a value for y_D to calculate the reach length. This method is called the direct step method, since the reach length is obtained directly from Equation 4.13 without any trial and error. These calculations are repeated for the subsequent reaches to determine the water surface profile.

For subcritical flow calculations, we start from the downstream end of a channel and proceed in the upstream direction. In other words, the first reach considered is at the downstream end of the channel, and the downstream section of this reach coincides with the downstream extremity of the channel. At the downstream extremity, y_D is known from the boundary condition. Using the known discharge and the cross-sectional properties, we first calculate V_D and S_{fD}. Next we pick a value for y_U and calculate the corresponding V_U and S_{fU}. Then, from Equation 4.13, we determine the channel reach ΔX. This process is repeated for further upstream reaches until the entire length of the channel is covered. Note that y_U of any reach becomes y_D for the reach considered next. Also, we must be careful in picking the values for y_U. These values depend on the type of the profile that will occur in the channel. For example, if an M2 profile is being calculated, y_U must satisfy the inequalities $y_U > y_D$ and $y_n > y_U > y_c$. Likewise, for an S1 profile, $y_U < y_D$ and $y_U > y_c > y_n$.

For supercritical profiles, we start at the upstream end and proceed in the downstream direction. For the first reach, y_U is known from the upstream boundary condition. We choose a value for y_D and calculate the reach length, ΔX, using Equation 4.13. This process is repeated for further downstream reaches until the length of the channel is covered. The y_D of any reach becomes y_U of the subsequent reach. The values of y_D must be chosen carefully in the process. For instance, for M3 profiles, $y_D > y_U$ and $y_D < y_c < y_n$. Likewise, for S2 profiles, $y_D < y_U$ and $y_n < y_D < y_c$.

In certain situations, the flow depths at both ends of a surface profile will be known and we can perform the calculations to determine the total length of the profile. In such a case we can start from either the upstream end or the downstream end, regardless of whether the flow is subcritical or supercritical. However, a downstream boundary condition is always known for subcritical flow, and an upstream boundary condition is always known for supercritical flow. Therefore, it is reasonable to adopt the general rule that subcritical flow calculations start at the downstream end, and supercritical flow calculations start at the upstream end.

EXAMPLE 4.3 A very long trapezoidal canal has $b = 18$ ft, $m = 2.0$, $S_0 = 0.001$, and $n = 0.020$, and it carries $Q = 800$ cfs. The canal terminates at a free fall. Calculate the water surface profile.

To solve this problem, we need to predict the type of profile. First, we should calculate y_c and y_n and determine whether the channel is mild, critical or steep. Using the procedures discussed in Chapters 2 and 3, we obtain $y_n = 5.16$ ft and $y_c = 3.45$ ft. Because, $y_n > y_c$, the channel is mild.

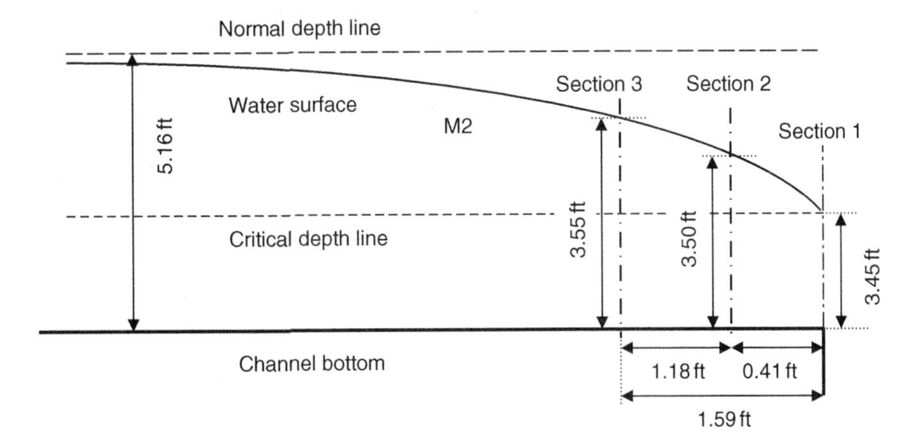

FIGURE 4.12 Direct step method example for subcritical flow

We know that far away from the free fall, the flow will tend to be normal. In a mild channel, the normal flow is subcritical. It is known that when subcritical flow approaches a free fall, critical depth occurs near the brink. For practical purposes, we assume that y_c occurs at the brink. Thus in this example an M2 profile will occur, and the flow depth will change between 5.16 ft and 3.45 ft as shown in Figure 4.12.

The calculations are best performed in tabular form as shown in Table 4.1, where the entries in the first column denote the section numbers. Because the flow is subcritical, the calculations will start at the downstream end of the channel and will proceed upstream. Referring to Figure 4.12, let us consider the most downstream reach – the reach between Sections 1 and 2. For this reach, we already know that $y_D = y_c = 3.45$ ft. This value is entered in column 2 of Table 4.1 for Section 1. Then, by using the expressions given for trapezoidal channels in Table 1.1,

$$A_D = (b + my_D)y_D = [18.0 + (2.0)(3.45)](3.45) = 85.91\ \text{ft}^2$$

$$P_D = b + 2y_D\sqrt{1+m^2} = 18.0 + 2(3.45)\sqrt{1+(2.0)^2} = 33.43\ \text{ft}$$

$$R_D = A_D/R_D = (85.91)/(33.43) = 2.57\ \text{ft}$$

Also,

$$V_D = Q/A_D = (800)/(85.91) = 9.313\ \text{fps}$$

$$S_{fD} = \frac{n^2}{k_n^2}\frac{V_D^2}{R_D^{4/3}} = \frac{(0.020)^2(9.313)^2}{(1.49)^2(2.57)^{4/3}} = 0.00444$$

$$E_D = y_D + \frac{V_D^2}{2g} = (3.45) + \frac{(9.313)^2}{2(32.2)} = 4.79666\ \text{ft}$$

We are now ready to choose a value for y_U. We know that for an M2 curve, $y_U > y_D$ and $y_n > y_U > y_c$. Thus we choose $y_U = 3.50$ ft. This value is entered in column 2 as the flow depth for Section 2. Knowing the flow depth, we can calculate all the flow characteristics at the upstream

TABLE 4.1 Summary of direct step method calculations for subcritical flow

	Variables for section i						Variables for reach between sections i and $i-1$					
i	y (ft)	A (ft^2)	P (ft)	R (ft)	V (fps)	E (ft)	$\Delta E = E_D - E_U$ (ft)	S_f	S_{fm}	$S_0 - S_{fm}$	ΔX (ft)	$\Sigma \Delta X$ (ft)
1	3.45	85.905	33.429	2.570	9.313	4.79666		0.00444				0
2	3.50	87.500	33.652	2.600	9.143	4.79801	−0.00135	0.00421	0.00433	−0.00333	0.41	0.41
3	3.55	89.105	33.876	2.630	8.978	4.80167	−0.00366	0.00400	0.00411	−0.00311	1.18	1.59
4	3.60	90.720	34.100	2.660	8.818	4.80750	−0.00583	0.00380	0.00390	−0.00290	2.01	3.60
5	3.65	92.345	34.323	2.690	8.663	4.81538	−0.00788	0.00361	0.00371	−0.00271	2.91	6.51
6	3.70	93.980	34.547	2.720	8.512	4.82518	−0.00980	0.00344	0.00353	−0.00253	3.88	10.39
7	3.75	95.625	34.771	2.750	8.366	4.83680	−0.01162	0.00327	0.00336	−0.00236	4.93	15.32
8	3.80	97.280	34.994	2.780	8.224	4.85014	−0.01334	0.00312	0.00320	−0.00220	6.08	21.40
9	3.85	98.945	35.218	2.810	8.085	4.86509	−0.01495	0.00297	0.00304	−0.00204	7.32	28.71
10	3.90	100.620	35.441	2.839	7.951	4.88158	−0.01649	0.00283	0.00290	−0.00190	8.67	37.38
11	3.95	102.305	35.665	2.869	7.820	4.89951	−0.01793	0.00270	0.00277	−0.00177	10.14	47.52
12	4.00	104.000	35.889	2.898	7.692	4.91881	−0.01930	0.00258	0.00264	−0.00164	11.76	59.28
13	4.05	105.705	36.112	2.927	7.568	4.93941	−0.02060	0.00246	0.00252	−0.00152	13.53	72.81
14	4.10	107.420	36.336	2.956	7.447	4.96124	−0.02183	0.00236	0.00241	−0.00141	15.48	88.29
15	4.15	109.145	36.559	2.985	7.330	4.98423	−0.02299	0.00225	0.00230	−0.00130	17.64	105.93
16	4.20	110.880	36.783	3.014	7.215	5.00833	−0.02410	0.00215	0.00220	−0.00120	20.03	125.96
17	4.25	112.625	37.007	3.043	7.103	5.03347	−0.02515	0.00206	0.00211	−0.00111	22.70	148.66

18	4.30	114.380	37.230	3.072	6.994	5.05962	−0.02614	0.00197	0.00202	−0.00102	25.70	174.36
19	4.35	116.145	37.454	3.101	6.888	5.08670	−0.02709	0.00189	0.00193	−0.00093	29.07	203.43
20	4.40	117.920	37.677	3.130	6.784	5.11469	−0.02799	0.00181	0.00185	−0.00085	32.89	236.32
21	4.45	119.705	37.901	3.158	6.683	5.14354	−0.02884	0.00174	0.00177	−0.00077	37.27	273.59
22	4.50	121.500	38.125	3.187	6.584	5.17320	−0.02966	0.00167	0.00170	−0.00070	42.31	315.90
23	4.55	123.305	38.348	3.215	6.488	5.20363	−0.03044	0.00160	0.00163	−0.00063	48.17	364.07
24	4.60	125.120	38.572	3.244	6.394	5.23481	−0.03117	0.00153	0.00157	−0.00057	55.08	419.15
25	4.65	126.945	38.795	3.272	6.302	5.26668	−0.03188	0.00147	0.00150	−0.00050	63.32	482.47
26	4.70	128.780	39.019	3.300	6.212	5.29924	−0.03255	0.00141	0.00144	−0.00044	73.32	555.79
27	4.75	130.625	39.243	3.329	6.124	5.33243	−0.03319	0.00136	0.00139	−0.00039	85.69	641.48
28	4.80	132.480	39.466	3.357	6.039	5.36623	−0.03380	0.00131	0.00133	−0.00033	101.37	742.85
29	4.85	134.345	39.690	3.385	5.955	5.40062	−0.03439	0.00126	0.00128	−0.00028	121.88	864.74
30	4.90	136.220	39.913	3.413	5.873	5.43557	−0.03495	0.00121	0.00123	−0.00023	149.84	1014.58
31	4.95	138.105	40.137	3.441	5.793	5.47105	−0.03548	0.00116	0.00119	−0.00019	190.13	1204.71
32	5.00	140.000	40.361	3.469	5.714	5.50704	−0.03599	0.00112	0.00114	−0.00014	253.20	1457.91
33	5.05	141.905	40.584	3.497	5.638	5.54351	−0.03648	0.00108	0.00110	−0.00010	365.83	1823.74
34	5.10	143.820	40.808	3.524	5.563	5.58046	−0.03694	0.00104	0.00106	−0.00006	623.94	2447.68
35	5.15	145.745	41.032	3.552	5.489	5.61785	−0.03739	0.00100	0.00102	−0.00002	1820.87	4268.55

section, Section 2, just as we did for the downstream section. We find that $A_U = 87.50\,ft^2$, $P_U = 33.65$ ft, $R_U = 2.60$ ft, $V_U = 9.143$ fps, $(S_f)_U = 0.00421$, and $E_U = 4.79801$ ft as shown in the table. Next, by using Equation 4.10, $S_{fm} = (0.00444 + 0.00421)/2 = 0.00433$, as shown in column 10 of the table for the reach between Sections 1 and 2. For the same reach, $(S_0 - S_{fm}) = 0.001 - 0.00433 = -0.00333$ and $\Delta E = E_D - E_U = 4.79801 - 4.79666 = -0.00135$, as shown in columns 11 and 8 respectively. Finally, $\Delta X = \Delta E/(S_0 - S_{fm}) = (-0.00135)/(-0.00333) = 0.41$ ft. Therefore the distance between Section 1 (where the depth is 3.45 ft) and Section 2 (where the depth is 3.50 ft) is 0.41 ft.

We can now move on to the reach between Sections 2 and 3. For this reach, Section 2 is the downstream section and 3 is the upstream section. Referring to Figure 4.12, $y = 3.50$ ft now becomes the downstream depth, y_D, and the corresponding values of E_D and S_{fD} have already been calculated. Now we choose $y_U = 3.55$ ft, for Section 3, and perform the calculations in a similar way to obtain $\Delta X = 1.18$ ft for the reach between Sections 2 and 3. Obviously, $y = 3.55$ ft is located at a distance $0.41 + 1.18 = 1.59$ ft from the brink, as shown in column 13. The positions of other selected depths can be calculated in the same way.

EXAMPLE 4.4 Flow enters a long, rectangular flume at its upstream end from under a sluice gate. The flume has $b = 3$ ft, $n = 0.013$, and $S_0 = 0.02$. The flow depth at the entrance is 1.30 ft and the discharge is 30 cfs. Determine the water surface profile.

Using the procedures discussed in foregoing sections, we obtain $y_n = 0.91$ and $y_c = 1.46$ ft; the channel is steep because $y_c > y_n$. The upstream depth, 1.30 ft, is between y_c and y_n. Also, the flow will tend to become normal away from the upstream control. Therefore, an S2 profile will occur.

We perform the calculations in tabular form as shown in Table 4.2. Because the flow is supercritical, the calculations will start at the upstream end of the channel and proceed in the downstream direction. Referring to Figure 4.13, let us consider the most upstream reach between Sections 1 and 2. The value 1.30 ft is entered in column 2 for Section 1 as the known upstream depth, y_U, for this reach. We now choose $y_D = 1.28$ ft, and enter this value in column 2 for Section 2. Note that the selected y_D is smaller than y_U and greater than y_n in accordance with the shape of an S2 profile. The calculations are similar to those of Example 4.8, and ΔX is obtained from Equation 4.13 as being 0.71 ft. As we move to the second reach, the reach between Sections 2 and 3, the flow depth of 1.28 ft at Section 2 becomes the upstream depth, y_U. We pick $y_D = 1.26$, and calculate the distance between Sections 2 and 3 as being $\Delta X = 0.85$ ft. The same procedure is repeated for further downstream reaches. The values in column 13 of Table 4.2 represent the distance from the upstream end of the channel.

Table 4.2 Summary of direct step method calculations for supercritical flow

	Variables for section i						Variables for reach between sections i and $i-1$					
i	y (ft)	A (ft^2)	P (ft)	R (ft)	V (fps)	E (ft)	$\Delta E = E_D - E_U$ (ft)	S_f	S_{fm}	$S_0 - S_{fm}$	ΔX (ft)	$\Sigma \Delta X$ (ft)
1	1.30	3.900	5.600	0.696	7.692	2.21881		0.00730				
2	1.28	3.840	5.560	0.691	7.813	2.22775	0.00894	0.00761	0.00745	0.01255	0.71	0.71
3	1.26	3.780	5.520	0.685	7.937	2.23808	0.01033	0.00794	0.00778	0.01222	0.85	1.56
4	1.24	3.720	5.480	0.679	8.065	2.24988	0.01181	0.00830	0.00812	0.01188	0.99	2.55
5	1.22	3.660	5.440	0.673	8.197	2.26326	0.01338	0.00868	0.00849	0.01151	1.16	3.71
6	1.20	3.600	5.400	0.667	8.333	2.27833	0.01507	0.00908	0.00888	0.01112	1.35	5.07
7	1.18	3.540	5.360	0.660	8.475	2.29519	0.01686	0.00951	0.00929	0.01071	1.57	6.64
8	1.16	3.480	5.320	0.654	8.621	2.31398	0.01879	0.00996	0.00973	0.01027	1.83	8.47
9	1.14	3.420	5.280	0.648	8.772	2.33483	0.02085	0.01045	0.01021	0.00979	2.13	10.60
10	1.12	3.360	5.240	0.641	8.929	2.35788	0.02305	0.01097	0.01071	0.00929	2.48	13.08
11	1.10	3.300	5.200	0.635	9.091	2.38330	0.02542	0.01154	0.01126	0.00874	2.91	15.99
12	1.08	3.240	5.160	0.628	9.259	2.41127	0.02797	0.01214	0.01184	0.00816	3.43	19.42
13	1.06	3.180	5.120	0.621	9.434	2.44198	0.03071	0.01278	0.01246	0.00754	4.07	23.49
14	1.04	3.120	5.080	0.614	9.615	2.47565	0.03366	0.01348	0.01313	0.00687	4.90	28.39
15	1.02	3.060	5.040	0.607	9.804	2.51250	0.03685	0.01423	0.01386	0.00614	6.00	34.39
16	1.00	3.000	5.000	0.600	10.000	2.55280	0.04030	0.01504	0.01464	0.00536	7.51	41.91
17	0.98	2.940	4.960	0.593	10.204	2.59682	0.04403	0.01592	0.01548	0.00452	9.74	51.65
18	0.96	2.880	4.920	0.585	10.417	2.64489	0.04807	0.01687	0.01639	0.00361	13.33	64.97
19	0.94	2.820	4.880	0.578	10.638	2.69735	0.05246	0.01790	0.01738	0.00262	20.05	85.02
20	0.92	2.760	4.840	0.570	10.870	2.75459	30.05724	0.01902	0.01846	0.00154	37.14	122.16

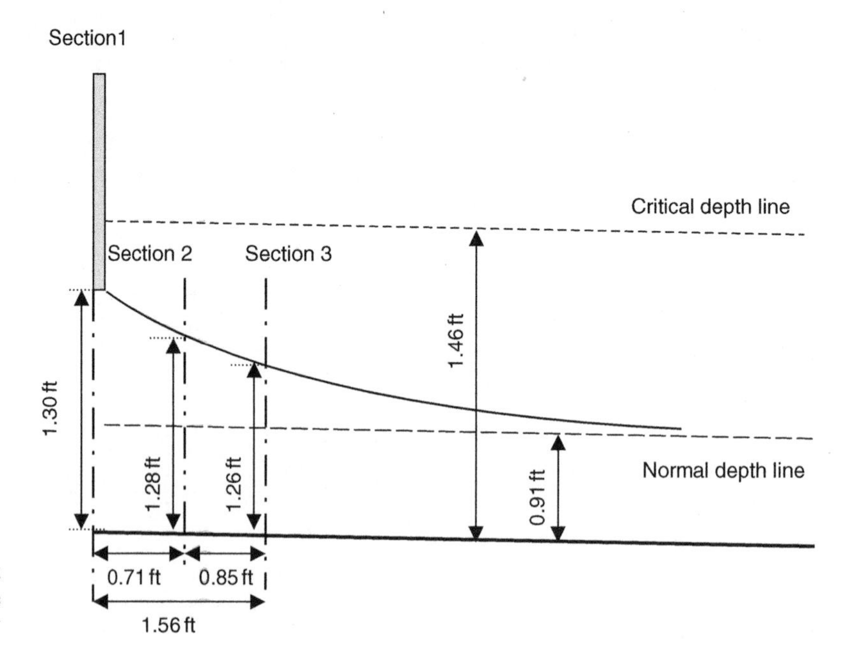

FIGURE 4.13 Direct step method example for supercritical flow

4.5.2 STANDARD STEP METHOD

In the standard step method, the flow depths are calculated at specified locations. As in the direct step method, we know the flow depth and velocity at one end of a channel reach. We then choose the reach length, ΔX, and calculate the depth at the other end of the reach.

For subcritical flow, the conditions at the downstream section will be known. For this case, to facilitate the calculations, we will rearrange Equation 4.13 as

$$y_U + \frac{V_U^2}{2g} - \frac{1}{2}(\Delta X)S_{fu} = y_D + \frac{V_D^2}{2g} + \frac{1}{2}(\Delta X)S_{fD} - (\Delta X)S_0 \tag{4.14}$$

For a constant discharge, we can express V_U and $(S_f)_U$ in terms of y_U. Therefore, the only unknown in Equation 4.14 is y_U. However, the expression is implicit in y_U, and we can solve it by use of an iterative technique. We try different values for y_U until Equation 4.14 is satisfied. Because of the iterative nature of the procedure the standard step method is not suitable for calculation by hand, and we normally employ a computer program. However, in the absence of such a program we can improve the guessed values of y_U in each iteration using

$$(y_U)_{k+1} = (y_U)_k - \Delta y_k \tag{4.15}$$

with

$$\Delta y_k = \frac{(LHS)_k - (RHS)}{\left(1 - F_{rU}^2 + 3(\Delta X)S_{fU}/2R_U\right)_k} \tag{4.16}$$

where $(y_U)_k$ is the guessed value of y_U for the k-th trial, $(\text{LHS})_k$ is the left-hand side of Equation 4.14 evaluated using $(y_U)_k$, and F_{rU} is the Froude number corresponding to y_U (Henderson, 1966).

For supercritical flow, the conditions at the upstream section will be known. Then, we will need to solve Equation 4.14 iteratively to determine y_D. The guessed values of y_D can be improved for each iteration using

$$(y_D)_{k+1} = (y_D)_k - \Delta y_k \tag{4.17}$$

where

$$\Delta y_k = \frac{(RHS)_k - LHS}{\left(1 - F_{rD}^2 - 3(\Delta X)S_{fD}/2R_D\right)_k} \tag{4.18}$$

The terms *RHS* and *LHS* again refer to the right-hand side and left-hand side of Equation 4.14.

A spreadsheet program can be used to perform the standard step calculations. The 'goal seek' or similar functions built into the standard spreadsheet programs perform the iterations rapidly without much effort from the user.

The standard step method can be used for non-prismatic channels as well, with some modifications. Section 4.8 of this chapter discusses the use of the standard step method for non-prismatic channels.

EXAMPLE 4.5 A very long trapezoidal canal has $b = 18$ ft, $m = 2.0$, $S_0 = 0.001$, and $n = 0.020$, and it carries $Q = 800$ cfs. The canal terminates at a free fall. Calculate the water surface profile in this channel using the standard step method and a constant space increment of $\Delta X = 10$ ft.

Using the procedures discussed in Chapters 2 and 3, we determine that $y_n = 5.16$ ft and $y_c = 3.45$ ft. Thus the channel is mild, and we expect an M2 profile. We will start the calculations at the downstream end of the channel, with 3.45 ft as the depth at the brink.

Table 4.3 summarizes the iterative procedure to determine the flow depth 10 ft upstream of the brink. For the first iteration, $k = 1$, suppose we guess that the flow depth at the upstream end of the 10-ft reach ($\Delta X = 10$ ft) is 3.65 ft. Then we calculate the depth correction by using Equation 4.16 as being -0.058 ft, as listed in column 12 of the table. The upstream depth for iteration $k = 2$ becomes $3.65 - (-0.058) = 3.708$ ft. We repeat the same procedure until the value in column 10 or 12 is zero (or tolerably small). Note that in this case we were able to find the upstream depth in three iterations.

The water surface calculations for the other reaches of this channel are summarized in Table 4.4 without reporting the iterations. Again, a constant

TABLE 4.3 Iteration in standard step method

k	Section	y (ft)	A (ft^2)	R (ft)	V (fps)	S_f	RHS (ft)	LHS (ft)	$LHS-RHS$ (ft)	Fr	Δy_k (ft)
	D	3.450	85.905	2.570	9.313	0.00444	4.809				
1	U	3.650	92.345	2.690	8.663	0.00361		4.797	−0.012	0.907	−0.058
2	U	3.708	94.259	2.725	8.487	0.00341		4.810	0.001	0.883	0.005
3	U	3.704	94.103	2.723	8.501	0.00343		4.809	0.000	0.885	0.000

TABLE 4.4 Summary of standard step method calculations

i	$\Sigma\Delta X$ (ft)	y (ft)	A (ft^2)	P (ft)	R (ft)	V (cfs)	S_f	RHS (ft)	LHS (ft)
1	0	3.45	85.91	33.429	2.570	9.313	0.00444	4.80885	
2	10	3.70	94.07	34.560	2.722	8.504	0.00343	4.83294	4.80866
3	20	3.79	97.11	34.972	2.777	8.238	0.00313	4.85439	4.83306
4	30	3.86	99.38	35.276	2.817	8.050	0.00293	4.87390	4.85456
5	40	3.92	101.24	35.524	2.850	7.902	0.00278	4.89196	4.87411
6	50	3.97	102.84	35.736	2.878	7.779	0.00266	4.90883	4.89220
7	60	4.01	104.26	35.923	2.902	7.673	0.00256	4.92473	4.90911
8	70	4.05	105.54	36.091	2.924	7.580	0.00248	4.93978	4.92503
9	80	4.08	106.71	36.244	2.944	7.497	0.00240	4.95409	4.94010
10	90	4.11	107.79	36.384	2.963	7.422	0.00233	4.96776	4.95443
11	100	4.14	108.79	36.514	2.980	7.353	0.00227	4.98083	4.96811
12	110	4.17	109.72	36.633	2.995	7.291	0.00222	4.99318	4.98099
13	120	4.19	110.58	36.745	3.009	7.234	0.00217	5.00499	4.99329
14	130	4.22	111.43	36.854	3.024	7.179	0.00212	5.01678	5.00554
15	140	4.24	112.19	36.951	3.036	7.131	0.00208	5.02759	5.01675
16	150	4.26	112.92	37.044	3.048	7.085	0.00205	5.03799	5.02753
17	160	4.28	113.60	37.131	3.059	7.042	0.00201	5.04800	5.03788
18	170	4.30	114.25	37.214	3.070	7.002	0.00198	5.05762	5.04782
19	180	4.31	114.87	37.293	3.080	6.964	0.00195	5.06686	5.05736
20	190	4.33	115.46	37.367	3.090	6.929	0.00192	5.07573	5.06651
21	200	4.35	116.02	37.438	3.099	6.895	0.00190	5.09423	5.07527

space increment of $\Delta X = 10$ ft is used. The flow depth at the most downstream section ($i = 1$) is 3.45 ft, and 10 ft upstream at Section 2 ($i = 2$) the depth is 3.70 ft. Note that for this reach, the RHS of Equation 4.14 is evaluated at Section 1 ($i = 1$) and the LHS is evaluated at Section 2 ($i = 2$). For the next reach, Section 2 becomes the downstream section and Section 3 the upstream section. Then for this reach the RHS is evaluated at Section 2 and the LHS at Section 3. The calculations for further upstream reaches are performed in a similar way.

4.6 APPLICATIONS OF GRADUALLY-VARIED FLOW

4.6.1 LOCATING HYDRAULIC JUMPS

The hydraulic jump equations were discussed in Chapter 2. To determine the jump location in a channel, we need to use the jump equation along with the gradually-varied flow calculations. The jump length is usually negligible compared to the length of a channel. Therefore, we often perform these calculations assuming that the jump occurs vertically. The flow depths, y_{J1} and y_{J2}, just upstream and downstream of the jump should satisfy the jump equation. If there is gradually-varied flow upstream of the jump, y_{J1} should also satisfy the gradually varied equations upstream. Likewise, if there is gradually varied flow downstream, then y_{J2} should also satisfy the downstream gradually-varied flow equations.

EXAMPLE 4.6 Determine the distance between the hydraulic jump and the downstream end of channel 1 in Example 4.2.

We have already determined, in Example 4.2, that a hydraulic jump will occur in channel 1 followed by an S1 profile as in Figure 4.9a. We also determined that, based on the hydraulic jump equation, $y_{J2} = 4.13$ ft, and therefore the S1 curve extends from an upstream depth of 4.13 ft from the jump location to a depth of 6.41 ft at the downstream end of channel 1.

To determine the jump location, we perform the gradually-varied flow calculations starting with a downstream depth of 6.41 ft. We will continue until we reach the upstream depth of 4.13 ft. It is easier and more convenient to use the direct step method. The calculations are summarized in Table 4.5. The channel characteristics used in the calculations are $b = 10$ ft, $m = 0$ (rectangular), $n = 0.020$, $S_0 = 0.02$, and $Q = 300$ cfs. A review of the results will show that the distance from the downstream end of channel 1 to the jump is 98.73 ft.

EXAMPLE 4.7 The flow enters a rectangular channel from under a sluice gate, as shown in Figure 4.14, at a depth of 1.75 ft. The channel has a width of $b = 4$ ft, a Manning roughness factor of $n = 0.013$, and a bottom slope of $S_0 = 0.001$. The discharge is $Q = 133$ cfs. The channel is 200 ft long, and it terminates at free fall. Calculate the free surface profile.

Using the procedures discussed in the previous sections, we first calculate $y_n = 6.87$ ft and $y_c = 3.25$ ft. Therefore the channel is mild. Because the flow enters the channel at a supercritical depth, an M3 curve should occur just downstream of the sluice gate. Two possible water surface profiles for this problem are shown in Figure 4.14. The profile of Figure 4.14a is possible only if flow remains supercritical throughout the length of the channel. In other words, when we perform the gradually varied flow calculations for the M3 curve,

TABLE 4.5 Calculations for Example 4.6

	Variables for section i						Variables for reach between sections i and $i-1$					
i	y (ft)	A (ft²)	P (ft)	R (ft)	V (fps)	E (ft)	$\Delta E = E_D - E_U$ (ft)	S_f	S_{fm}	$S_0 - S_{fm}$	ΔX (ft)	$\Sigma\Delta X$ (ft)
1	6.41	64.100	22.820	2.809	4.680	6.75013		0.00100				0
2	6.21	62.100	22.420	2.770	4.831	6.57239	0.17774	0.00108	0.00104	0.01896	9.37	9.37
3	6.01	60.100	22.020	2.729	4.992	6.39691	0.17548	0.00118	0.00113	0.01887	9.30	18.67
4	5.81	58.100	21.620	2.687	5.164	6.22400	0.17290	0.00129	0.00123	0.01877	9.21	27.88
5	5.61	56.100	21.220	2.644	5.348	6.05405	0.16995	0.00141	0.00135	0.01865	9.11	37.00
6	5.41	54.100	20.820	2.598	5.545	5.88749	0.16656	0.00155	0.00148	0.01852	8.99	45.99
7	5.21	52.100	20.420	2.551	5.758	5.72485	0.16264	0.00171	0.00163	0.01837	8.85	54.84
8	5.01	50.100	20.020	2.502	5.988	5.56678	0.15807	0.00190	0.00181	0.01819	8.69	63.53
9	4.81	48.100	19.620	2.452	6.237	5.41404	0.15274	0.00212	0.00201	0.01799	8.49	72.02
10	4.61	46.100	19.220	2.399	6.508	5.26759	0.14645	0.00238	0.00225	0.01775	8.25	80.27
11	4.41	44.100	18.820	2.343	6.803	5.12859	0.13900	0.00268	0.00253	0.01747	7.96	88.23
12	4.21	42.100	18.420	2.286	7.126	4.99848	0.13010	0.00304	0.00286	0.01714	7.59	95.82
13	4.13	41.300	18.260	2.262	7.264	4.94933	0.04916	0.00320	0.00312	0.01688	2.91	98.73

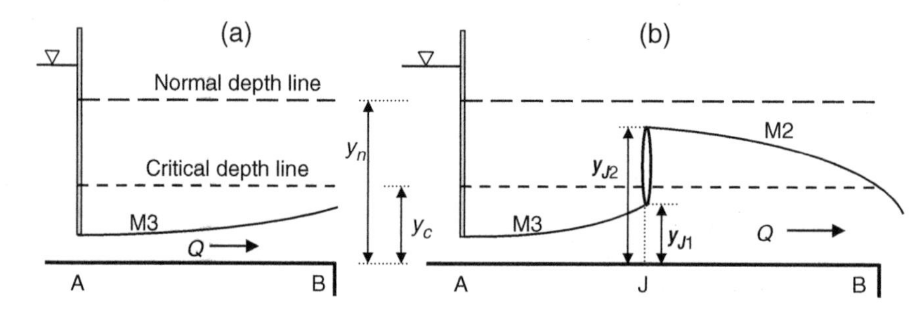

FIGURE 4.14 Possible profiles for Example 4.7

the flow depths should remain smaller than the critical depth throughout the entire length of the channel. Otherwise, the profile shown in Figure 4.14b will occur.

Let us first determine whether the channel is short enough for the profile shown in Figure 4.14a to occur. For this purpose we perform the gradually-varied flow calculations for the M3 curve using an upstream boundary depth of 1.75 ft and depth increments of 0.10 ft. The details of the calculations are omitted for brevity. The results show that the profile reaches the critical depth, $y_c = 3.25$ ft, at a distance 158 ft from the gate. Thus the flow would remain supercritical if the channel were shorter than 158 ft. In this problem the length of the channel is 200 ft, so the flow has to change from the supercritical to the subcritical state. Such a change is possible only through a hydraulic jump. The resulting flow profile will be similar to that of Figure 4.14b.

We can calculate the M3 and the M2 profiles easily by using the direct step or the standard step method. However, locating the hydraulic jump will require

Table 4.6 Example 4.7

M3 profile and sequent depths			M2 profile	
X_A (ft)	y_{M3} (ft)	y_{J2} (ft)	X_B (ft)	y_{M2} (ft)
0	1.75	5.45	0	3.25
10	1.80	5.33	10	3.58
20	1.86	5.22	20	3.70
30	1.92	5.10	30	3.79
40	1.98	4.99	40	3.87
50	2.04	4.87	50	3.94
60	2.10	4.76	60	4.00
70	2.17	4.65	70	4.05
80	2.24	4.53	80	4.10
90	2.31	4.42	90	4.15
100	2.39	4.30	100	4.19
110	2.47	4.18	110	4.23
120	2.56	4.05	120	4.26
130	2.66	3.92	130	4.30
140	2.78	3.77	140	4.33

X_A = distance from point A (sluice gate).
X_B = distance from point B (free overfall).

some additional work. We should first note that the flow depth just before the jump, y_{J1}, is on the M3 curve, and the depth right after the jump, y_{J2}, is on the M2 curve. Moreover, the two depths, y_{J1} and y_{J2}, should satisfy the hydraulic jump equation (Equation 2.26 for rectangular channels). Knowing the flow depth at A, let us first calculate the M3 profile using the standard step method with a constant space increment of, say, $\Delta X = 10$ ft. Then for every flow depth calculated on the M3 curve, we also calculate the sequent depth, y_{J2}, using the hydraulic jump equation. The results are summarized in the first three columns of Table 4.6, omitting the details of the standard step method. These results indicate that if, for example, a hydraulic jump occurred at $X_A = 20$ ft, the flow depth just upstream of the jump would be 1.86 ft and that just downstream of the jump would be 5.22 ft. Now, starting with the critical depth at B, we calculate the M2 profile by using the standard step method with a constant space increment of $\Delta X = 10$ ft. The results are summarized in the last two columns of Table 4.6.

Let us now plot the calculated M3 and M2 profiles as well as the sequent depths as shown in Figure 4.15. The point where the sequent depth line intersects the M2 profile will determine y_{J2}, since this depth should satisfy both the hydraulic jump equation and the M2 profile calculations. Because we are assuming that the jump occurs vertically, the point of intersection also determines the location of the hydraulic jump. As we can determine graphically from Figure 4.15, the hydraulic jump occurs at a distance 113 ft from the sluice gate. The flow jumps from a supercritical depth of 2.50 ft to a subcritical depth of 4.14 ft. The flow

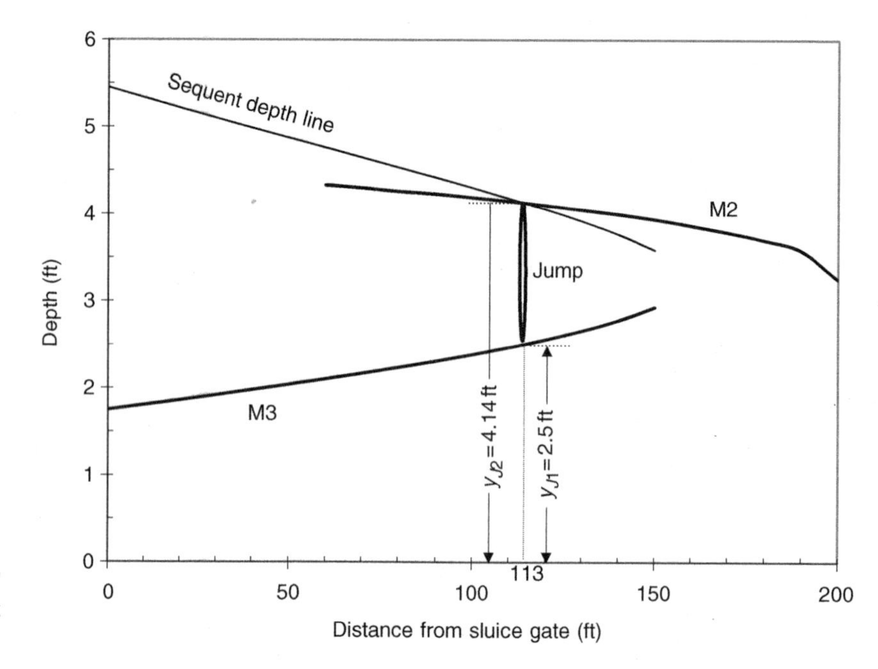

FIGURE 4.15 Jump location in Example 4.7

depth will increase from 1.75 ft to 2.5 ft along an M3 profile between the sluice gate and the hydraulic jump. Downstream of the jump an M2 profile will form along which the depth will decrease from 4.14 ft to 3.25 ft.

4.6.2 LAKE AND CHANNEL PROBLEMS

The discharge, Q, has been given in all the gradually-varied flow problems we have seen so far in this chapter. However, there are situations in which the discharge needs to be determined as part of the solution. For instance, we may need to determine the discharge and the water surface profile in a channel leading from a lake. The solution will depend on whether the channel is mild or steep. Of course, we need to know the discharge to determine with certainty whether the channel is mild or steep. Without the discharge given, if the channel slope is milder than, say, 0.001, we can assume that the channel is mild, solve the problem accordingly, but verify the assumption in the end. If the slope is steeper than, say, 0.02, we can first assume that the channel is steep, solve the problem accordingly, and verify the assumption in the end. For slopes between 0.001 and 0.02, we may have to try both alternatives.

4.6.2.1 Lake and mild channel

Let us consider an infinitely long channel leading from a lake, as shown in Figure 4.16a, and investigate whether a gradually-varied flow can occur in this channel. The normal depth line is above the critical depth line in a mild channel. However, it has to be below the lake water surface since the lake water surface represents the highest energy head available. Then, along any gradually-varied flow profile that seems possible, the water surface should drop gradually from

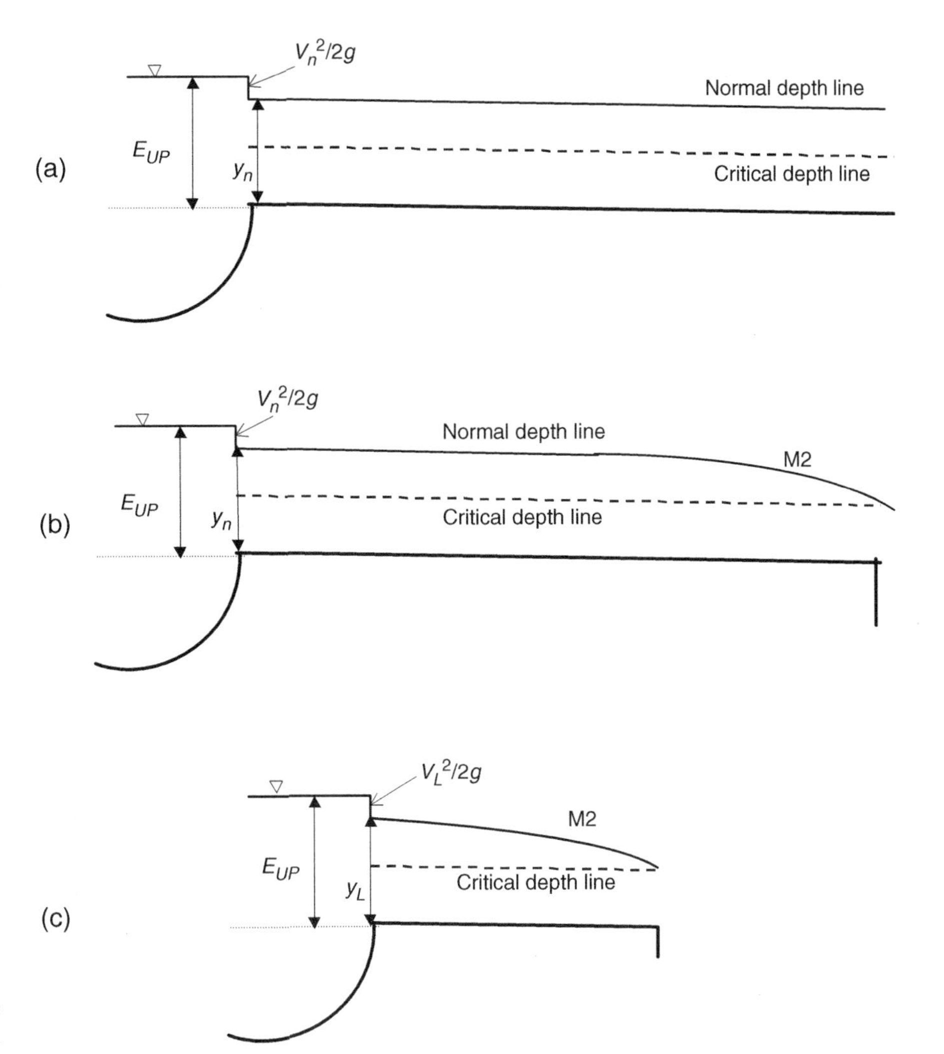

FIGURE 4.16 Lake and channel problem

near the lake water level towards the normal depth line. On the other hand, such a profile would be in zone M1 in which the flow depths can only increase. Therefore, a gradually-varied flow cannot occur in this channel. The flow will become normal immediately at the upstream end of the channel, and will remain normal. Therefore, at upstream end, both the normal flow equation and the energy equation should be satisfied. That is,

$$Q = \frac{k_n}{n} \frac{A_n^{5/3}}{P_n^{2/3}} S_0^{1/2} \tag{4.19}$$

and

$$E_{UP} = y_n + \frac{Q^2}{A_n^2 2g} \tag{4.20}$$

where E_{UP} = vertical distance between the lake water surface and the channel invert. Solving these two equations simultaneously, we can determine Q and y_n.

Suppose now that the channel terminates at a free fall, as shown in Figure 4.16b. The flow will be critical at the brink, forming an M2 curve. If the M2 curve reaches the normal depth line before reaching the lake, then the flow will be normal between this point and the lake. In this case, Equations 4.19 and 4.20 are still valid, and they can be solved to calculate Q and y_n. Such a channel is said to be *hydraulically long*.

If the channel is hydraulically short, the M2 curve reaches the lake at a depth less than the normal depth, as in Figure 4.16c. In that event we cannot use Equations 4.19 and 4.20, and the problem will require a trial-and-error procedure. Perhaps the easiest way is as follows:

- Assume the value of y_c and determine the corresponding Q
- Calculate the M2 profile, starting at the free fall and proceeding towards the lake; let y_L and V_L be, respectively, the flow depth and velocity so calculated at the upstream end of the channel
- Check whether the energy equation is satisfied:

$$E_{UP} = y_L + \frac{V_L^2}{2g} \tag{4.21}$$

If the equation is satisfied, the assumed critical depth is correct. Otherwise, assume another y_c and repeat the same procedure.

EXAMPLE 4.8 A trapezoidal channel leading from a lake is infinitely long. It has a bottom width of $b = 10$ ft, side slopes of $m = 2$, a bottom slope of $S_0 = 0.0004$, and a Manning roughness factor of 0.013. The lake water surface is 5.45 ft above the invert of the channel. Determine the discharge in the channel.

To solve this problem, let us assume that the channel is mild. Then, the flow will be normal throughout the channel, and Equations 4.19 and 4.20 will apply. Using the expressions for A and P from Table 1.1, we can write these two equations as

$$Q = \frac{k_n}{n}\frac{[(b+my_n)y_n]^{5/3}}{(b+2y_n\sqrt{1+m^2})^{2/3}}S_0^{1/2} = \frac{1.49}{0.013}\frac{[(10+2y_n)y_n]^{5/3}}{(10+2y_n\sqrt{1+2^2})^{2/3}}(0.0004)^{1/2}$$

and

$$E_{UP} = 5.45 = y_n + \frac{Q^2}{[(b+my_n)y_n]^2 2g} = \frac{Q^2}{[(10+2y_n)y_n]^2 2(32.2)}$$

Solving these two equations simultaneously, we obtain $Q = 500$ cfs and $y_n = 5.08$ ft. We can also calculate the corresponding critical depth as being $y_c = 3.38$ ft. Comparing the normal and the critical depths, we can verify that the channel is indeed mild as assumed.

EXAMPLE 4.9 Suppose the channel considered in Example 4.8 is 20 000 ft long and terminates at a free fall. Determine the discharge and the water surface profile.

Let us assume that the channel is hydraulically long. With this assumption, we can calculate the M2 profile using Q = 500 cfs and $y_c = 3.38$ ft (as obtained in Example 4.8). The direct step method with depth increments of $\Delta y = 0.05$ ft will show that the profile reaches 5.07 ft at about 11 500 ft from the brink. (Because the M2 curve approaches the normal depth asymptotically, we usually calculate the distance to a depth very close to the normal depth rather than exactly the normal depth. That is why 5.07 ft is used instead of 5.08 ft in these calculations.) The flow will be normal at the upstream end, and therefore Equations 4.19 and 4.20 are still valid. Then the discharge is 500 cfs, as obtained in Example 4.8.

EXAMPLE 4.10 Suppose the length of the channel considered in Example 4.9 is 800 ft. Determine the discharge in the channel and the water surface profile.

Referring to the results of Example 4.9, this channel is hydraulically short since the length, 800 ft, is much smaller than 11 500 ft. Therefore, Equations 4.19 and 4.20 are not applicable and instead we will use a trial-and-error procedure. After a few trials, we find that $y_c = 3.74$ ft and the corresponding discharge $Q = 596.5$ ft produces an M2 profile that satisfies Equation 4.21 at 800 ft upstream of free fall. Table 4.7 summarizes the gradually varied flow calculations. Note that for $Q = 596.5$ cfs, we can calculate $y_n = 5.55$ ft. Comparing the critical and normal depths, we can conclude that the channel is mild as it was initially assumed.

4.6.2.2 Lake and steep channel

The flow usually enters from a lake into a steep channel through the critical depth followed by an S2 curve, as shown in Figure 4.17. This can be demonstrated qualitatively by examining the gradually-varied flow equation (Equation 4.7)

$$\frac{dy}{dx} = \frac{S_0 - S_f}{(1 - F_r^2)}$$

rearranged as

$$\frac{dy}{dx}(1 - F_r^2) = S_0 - S_f \qquad (4.22)$$

In this equation, S_f represents the friction loss per unit flow length. Suppose the friction loss at the entrance of the channel is negligible – that is, $S_f = 0$. Also, $S_0 = -dz_b/dx$ = slope of the channel bed. As can be seen in Figure 4.17, the channel bed elevation is at a local maximum at the channel entrance where

TABLE 4.7 Calculations for Example 4.10

	Variables for section i						Variables for reach between sections i and $i-1$					
i	y (ft)	A (ft^2)	P (ft)	R (ft)	V (fps)	E (ft)	$\Delta E = E_D - E_U$ (ft)	S_f	S_{fm}	$S_0 - S_{fm}$	ΔX (ft)	$\Sigma\Delta X$ (ft)
1	3.74	65.375	26.726	2.446	9.124	5.03273		0.00192				0
2	3.79	66.628	26.949	2.472	8.953	5.03457	−0.00184	0.00183	0.00187	−0.00147	1.25	1.25
3	3.84	67.891	27.173	2.498	8.786	5.03869	−0.00412	0.00173	0.00178	−0.00138	2.99	4.24
4	3.89	69.164	27.397	2.525	8.624	5.04497	−0.00628	0.00165	0.00169	−0.00129	4.87	9.10
5	3.94	70.447	27.620	2.551	8.467	5.05329	−0.00831	0.00157	0.00161	−0.00121	6.89	15.99
6	3.99	71.740	27.844	2.577	8.315	5.06352	−0.01023	0.00149	0.00153	−0.00113	9.07	25.06
7	4.04	73.043	28.067	2.602	8.166	5.07556	−0.01204	0.00142	0.00145	−0.00105	11.42	36.49
8	4.09	74.356	28.291	2.628	8.022	5.08931	−0.01375	0.00135	0.00138	−0.00098	13.97	50.46
9	4.14	75.679	28.515	2.654	7.882	5.10468	−0.01537	0.00129	0.00132	−0.00092	16.72	67.18
10	4.19	77.012	28.738	2.680	7.746	5.12157	−0.01689	0.00123	0.00126	−0.00086	19.71	86.89
11	4.24	78.355	28.962	2.705	7.613	5.13991	−0.01834	0.00117	0.00120	−0.00080	22.97	109.86
12	4.29	79.708	29.185	2.731	7.484	5.15962	−0.01971	0.00112	0.00114	−0.00074	26.51	136.37
13	4.34	81.071	29.409	2.757	7.358	5.18062	−0.02101	0.00107	0.00109	−0.00069	30.38	166.75
14	4.39	82.444	29.633	2.782	7.235	5.20286	−0.02223	0.00102	0.00104	−0.00064	34.62	201.36
15	4.44	83.827	29.856	2.808	7.116	5.22626	−0.02340	0.00097	0.00100	−0.00060	39.28	240.64
16	4.49	85.220	30.080	2.833	7.000	5.25076	−0.02451	0.00093	0.00095	−0.00055	44.42	285.06
17	4.54	86.623	30.303	2.859	6.886	5.27632	−0.02556	0.00089	0.00091	−0.00051	50.11	335.17
18	4.59	88.036	30.527	2.884	6.776	5.30287	−0.02655	0.00085	0.00087	−0.00047	56.43	391.60
19	4.64	89.459	30.751	2.909	6.668	5.33037	−0.02750	0.00081	0.00083	−0.00043	63.49	455.09
20	4.69	90.892	30.974	2.934	6.563	5.35878	−0.02840	0.00078	0.00080	−0.00040	71.42	526.51
21	4.74	92.335	31.198	2.960	6.460	5.38804	−0.02926	0.00075	0.00076	−0.00036	80.38	606.90
22	4.79	93.788	31.422	2.985	6.360	5.41811	−0.03008	0.00072	0.00073	−0.00033	90.58	697.47
23	4.84	95.251	31.645	3.010	6.262	5.44897	−0.03085	0.00069	0.00070	−0.00030	102.26	799.73

$S_0 = -dz_b/dx = 0$. Therefore, for Equation 4.22 to be satisfied at this location, either dy/dx must be equal to zero or F_r must be equal to 1.0 since the flow is locally accelerating ($dy/dx \neq 0$). Then $F_r = 1.0$, and the flow is critical.

Therefore, at the upstream end of the channel, both the critical flow equation and the energy equation should be satisfied. That is,

$$F_r^2 = \frac{Q^2}{A_c^2 g D_c} = \frac{Q^2 T_c}{g A_c^3} = 1 \tag{4.23}$$

and

$$E_{UP} = y_c + \frac{Q^2}{A_c^2 2g} \tag{4.24}$$

where the subscript c stands for critical.

If the channel is infinitely long, the S2 profile will approach the normal depth line asymptotically as shown in Figure 4.17a. If the channel is short, and say

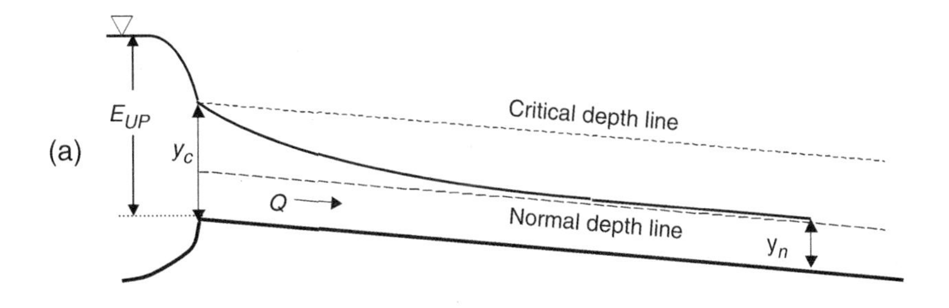

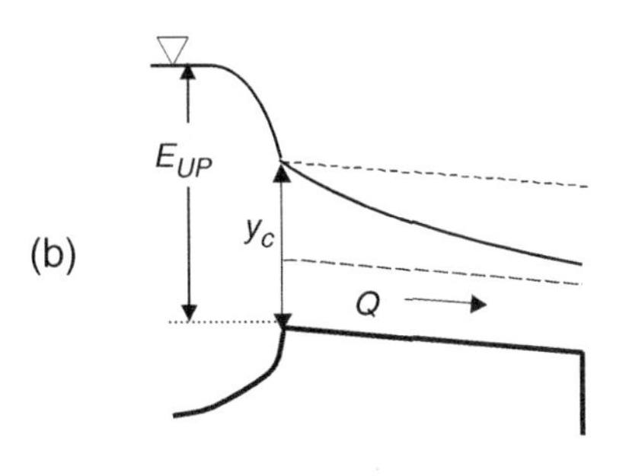

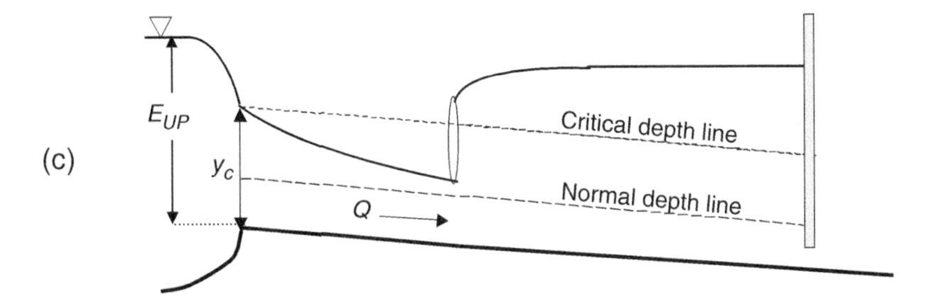

FIGURE 4.17 Lake and steep channel problem

terminates at a free fall as shown in Figure 4.17b, the S2 curve will remain unaffected, and Equations 4.23 and 4.24 will still be valid. If the flow is forced to change from the supercritical to subcritical state as shown in Figure 4.17c, the S2 profile near the lake will remain unaffected, and the discharge can still be determined by using Equations 4.23 and 4.24 unless the jump moves all the way up near to the lake location. The jump location can be determined using a procedure similar to that in Example 4.7.

EXAMPLE 4.11 The trapezoidal channel considered in Example 4.8 has a bottom width of $b = 10$ ft, side slopes of $m = 2$, bottom slope of $S_0 = 0.0004$, and Manning roughness factor of 0.013. The lake water surface is 5.45 ft above the invert of the channel. Determine the discharge and the normal depth in this channel if the bottom slope is $S_0 = 0.02$.

Let us first assume that the channel is steep. Then, using Equations 4.23 and 4.24,

$$\frac{Q^2 T_c}{gA_c^3} = \frac{Q^2(b + 2my_c)}{g[(b + my_c)y_c]^3} = \frac{Q^2[10 + 2(2)y_c]}{32.2[(10 + 2y_c)y_c]^3} = 1$$

and

$$E_{UP} = y_c + \frac{Q^2}{A_c^2 2g} = y_c + \frac{Q^2}{[(b+my_c)y_c]^2 2g} = y_c + \frac{Q^2}{[(10+2y_c)y_c]^2 2(32.2)} = 5.45$$

Solving these two equations simultaneously, we obtain $y_c = 4.05$ ft and $Q = 696$ cfs. The corresponding normal depth can be found as being $y_n = 2.19$ ft. Comparing the critical and the normal depths, we can verify our initial assumption that the channel is steep.

The reader should notice that the two situations investigated in Example 4.8 and 4.11 are identical except that the channel slope is $S_0 = 0.0004$ in Example 4.8 and $S_0 = 0.02$ in Example 4.11. The discharge is found to be $Q = 500$ cfs in Example 4.8 and $Q = 696$ cfs in Example 4.11. The larger discharge in Example 4.11 is what we would expect, because from Chapter 2 we recall that the discharge through a channel section for a given specific energy is maximum at critical depth. At the channel entrance the specific energy is 5.45 in both examples, while the flow depth is critical in Example 4.11, resulting in the maximum discharge.

4.6.3 TWO-LAKE PROBLEMS

When a channel connects two lakes as shown in Figure 4.18, the discharge in the channel may be affected by the variation in the water levels in the two lakes. Plots of discharge versus lake water levels are sometimes called *delivery curves*. In the discussions presented here, we will assume that the head loss at the channel entrance is negligible. Therefore, the water surface will drop by an amount equal to the velocity head as the flow enters the channel from the upper lake. In other words,

$$E_{UP} = y_{UP} + \frac{Q^2}{A_{UP}^2 2g} \tag{4.25}$$

where E_{UP} = depth of water in the upper lake above the channel invert, y_{UP} = flow depth at the upstream end of the channel, and A_{UP} = flow area at the upstream end of the channel.

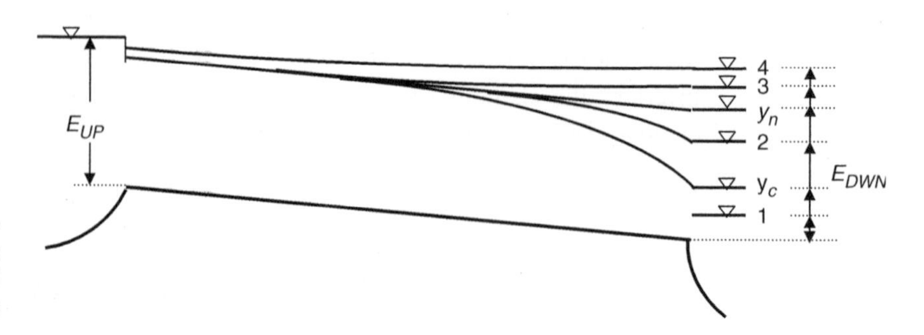

FIGURE 4.18 Flow profiles for a long, mild channel connecting two lakes

At the downstream end, where the channel joins the lake, the water surface in the channel is assumed to match the lake water surface as long as the lake water surface is above the critical depth. In other words,

$$y_{DWN} = E_{DWN} \tag{4.26}$$

where E_{DWN} = depth of water in the lake above the channel invert and y_{DWN} = flow depth at the downstream end of the channel. However, if the water level in the lower lake is below the critical depth and the channel flow is subcritical, then critical depth will occur at the downstream end of the channel (similar to the case of a free fall). That is, $y_{DWN} = y_c$.

4.6.3.1 Two lakes and a mild channel

Let us consider a very long, mild channel connecting two lakes. Suppose that E_{UP} is fixed and we are interested in how the discharge, Q, varies with E_{DWN}. Figure 4.18 shows various possible flow profiles. When, $E_{DWN} = y_n$, the flow in the channel will remain normal and the discharge is calculated by using Equations 4.19 and 4.20, as though the channel is infinitely long. Let us denote this discharge by Q_n. If $E_{DWN} = y_c$, then an M2 profile will form and the profile will meet the normal depth line before reaching the upper lake (since the channel is very long). In that event Equations 4.19 and 4.20 will still be valid, and the discharge will be equal to Q_n. If $E_{DWN} < y_c$, like the case labeled '1' in Figure 4.18, we will have $y_{DWN} = y_c$, and the discharge will be Q_n as before. If $y_n > E_{DWN} > y_c$, like the case labeled '2' in Figure 4.18, again an M2 profile will form. The profile will meet the normal depth line before reaching the upper lake (since the channel is very long), and the discharge will still be equal to Q_n. If $E_{DWN} > y_n$, then an M1 profile will form. If the M1 profile reaches the normal depth line before the upper lake, such as in case 3 in Figure 4.18, Equations 4.19 and 4.20 will still be valid and the discharge will be equal to Q_n. However, if the M1 profile is longer than the channel – that is, if the profile reaches the upper lake above the normal depth line, as in case 4 in Figure 4.18 – the discharge will satisfy Equation 4.25 and it will be less than Q_n. Figure 4.19 displays the relationship between the discharge, Q, and the lower lake water depth, E_{DWN}, calculated for a channel that is 12 000 ft long with $b = 10$ ft, $n = 0.013$, $S_0 = 0.0004$, and $m = 2$. The upper lake water depth is assumed to remain constant at $E_{UP} = 5.45$ ft. Note that $Q_n = 500$ cfs for this channel, and it is the maximum discharge that can occur for the given conditions. The numbered zones marked on the figure directly correspond to the cases labeled in Figure 4.18 with the same number.

The gradually-varied flow calculations are performed to develop the Q versus E_{DWN} relationship shown in Figure 4.19. All the profiles are subcritical and would normally be calculated starting from the downstream end. However, for the cases with M1 curves longer than the channel, this approach would require a trial-and-error solution to find the discharge for

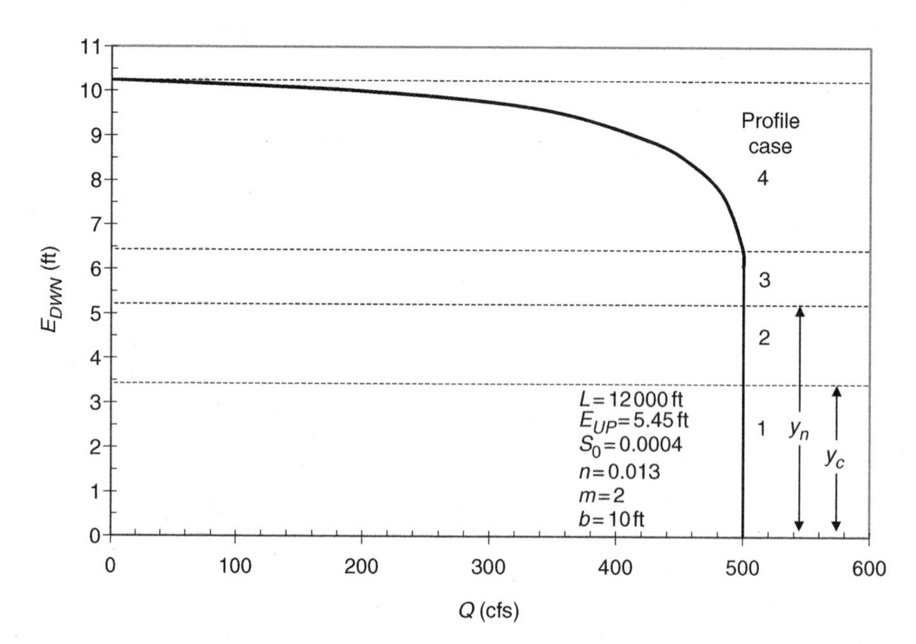

FIGURE 4.19
Delivery curve for a long, mild channel

a given E_{DWN}. After picking a value for E_{DWN}, we would have to try different values of Q until Equation 4.25 was satisfied at the lake location. However, we can avoid these lengthy trials if we start the calculations at the upstream end. We first pick a value for Q. Next we calculate y_{UP} using Equation 4.25 (making sure that we pick the subcritical value between the two positive roots). Then we perform the gradually-varied flow calculations, starting from the upstream end and proceeding in the downstream direction even though the flow is subcritical. The resulting flow depth at the downstream end of the channel will be E_{DWN}.

When a mild, short channel connects two lakes, the M2 and M1 profiles will be longer than the channel. In other words, the profiles will reach the upper lake before reaching the normal depth line. In this case, $Q_n = 500$ cfs will occur only if $E_{DWN} = 5.08$ ft $= y_n$. Otherwise, the discharge will be different from Q_n. The results displayed in Figure 4.20 are calculated for a channel identical to that of Figure 4.19, except in this case the channel is only 800 ft long. The maximum discharge corresponds to the case $y_c = E_{DWN} = 3.74$ ft. Although all the profiles are subcritical, the calculations are carried out starting from the upstream end to avoid trial-and-error solutions in developing the delivery curve of Figure 4.20.

Figure 4.21 depicts the delivery curves calculated for the same channel for constant $E_{DWN} = 4.0$ ft and variable E_{UP}. As before, the channel is 800 ft long with $b = 10$ ft, $n = 0.013$, $S_0 = 0.0004$, and $m = 2$. Normal flow occurs in this channel only if $y_n = E_{DWN} = 4.0$ ft and the corresponding discharge is 310 cfs. The maximum discharge, 680 cfs, occurs when $y_{DWN} = y_c = E_{DWN} = 4.0$ ft. The corresponding E_{UP} is obtained as being 5.81 ft through the M2

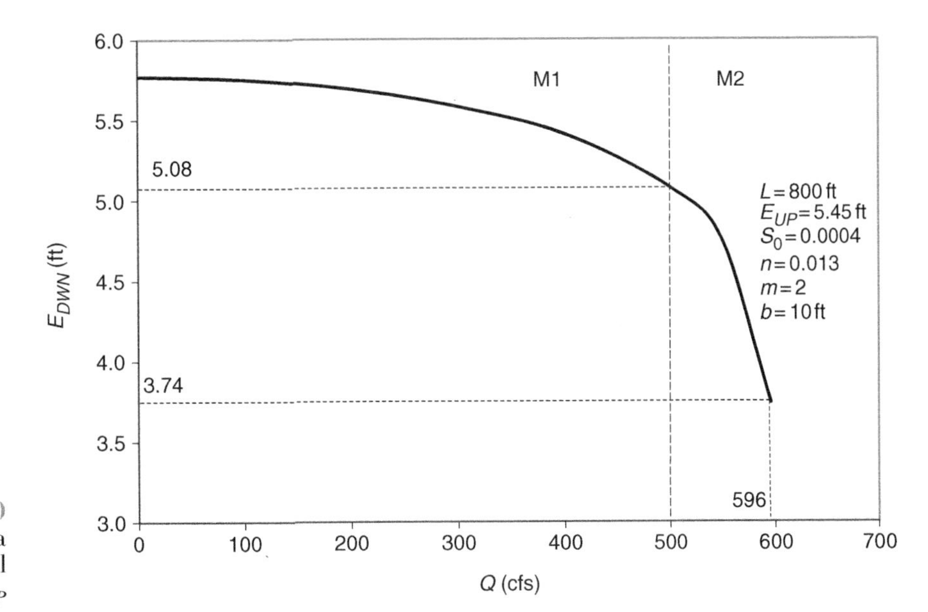

FIGURE 4.20 Delivery curve for a mild, short channel for constant E_{UP}

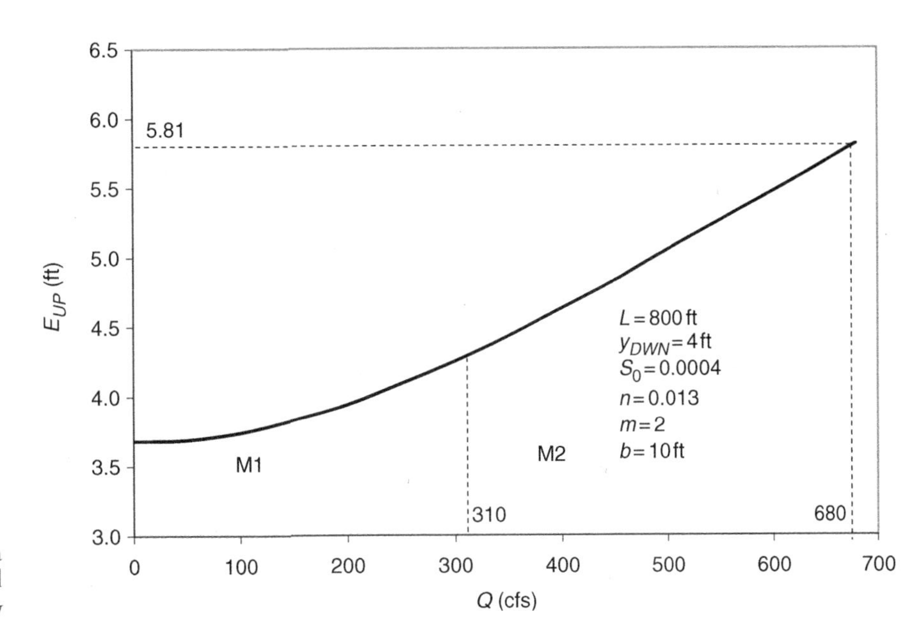

FIGURE 4.21 Delivery curve for a short, mild channel for constant E_{DWN}

profile calculations with a downstream boundary depth of 4.0 ft and discharge of 680 cfs.

If the water level in the upper lake rises above 5.81 ft, the discharge will increase beyond 680 cfs. However, Equation 4.26 will no longer be valid. The resulting water surface profile will rise above the profile calculated for 680 cfs, and it will terminate at a critical depth that is higher than $E_{DWN} = 4$ ft. Figure 4.22 displays the two profiles calculated for $Q = 680$ cfs ($E_{UP} = 5.81$ ft) and $Q = 848$ cfs ($E_{UP} = 6.45$ ft). The depth of 4.5 ft, at the downstream end of the higher profile, is the critical depth for $Q = 848$ cfs.

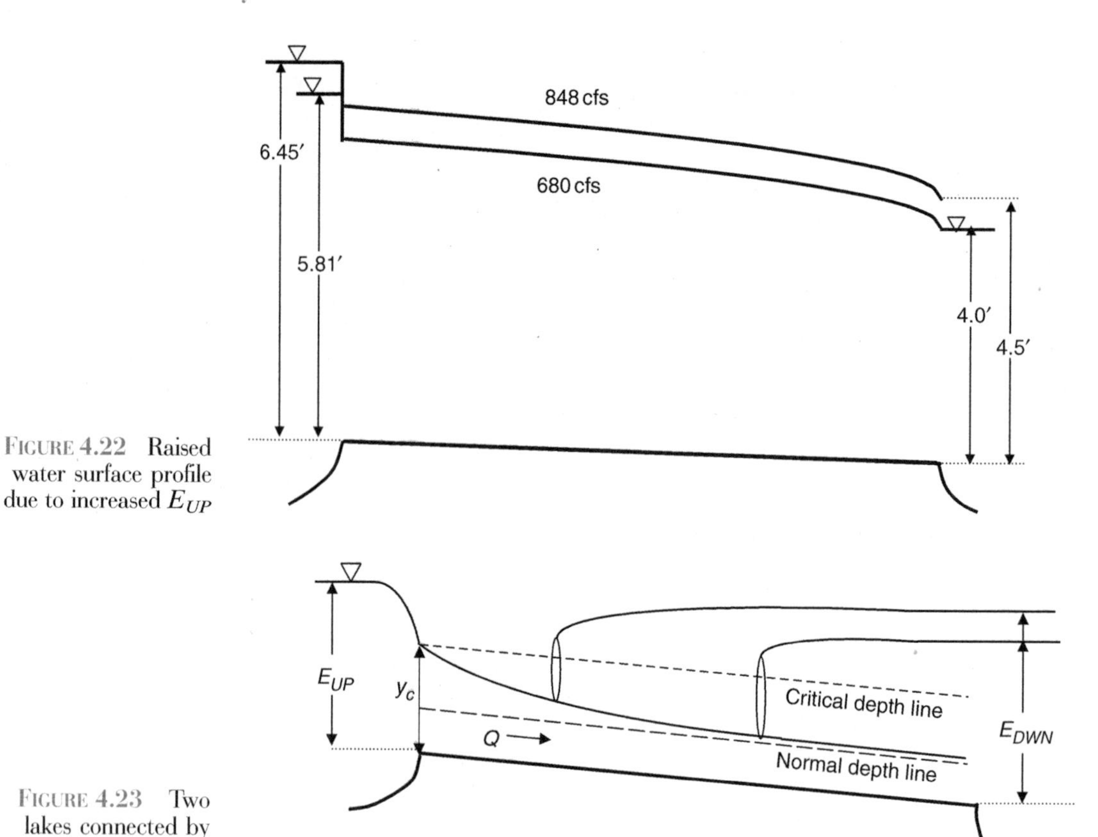

FIGURE 4.22 Raised water surface profile due to increased E_{UP}

FIGURE 4.23 Two lakes connected by a steep channel

4.6.3.2 Two lakes and a steep channel

When a steep channel connects two lakes, the water level in the upper lake will govern the discharge in the channel. A critical depth will occur at the upstream end of the channel satisfying the Equations 4.23 and 4.24, then an S2 profile will form approaching the normal depth line. Rising water levels in the lower lake may force the flow in the channel to change from a supercritical to subcritical state through a hydraulic jump, as shown in Figure 4.23. However, the discharge in the channel will still remain the same unless the hydraulic jump moves to the upstream end and submerges the channel entrance. In that event, the flow in the channel will be subcritical and the discharge will be determined by S1 curve calculations. Different values of Q will be tried until the results satisfy both Equations 4.25 and 4.26.

4.6.4 EFFECT OF CHOKING ON WATER SURFACE PROFILE

We discussed the problem of choking in Chapter 2. Choking occurs when the flow has to adjust to acquire the specific energy needed to pass a constricted section, a section where the width is reduced or the bottom is raised, or both. In Chapter 2, we investigated choking as a local phenomenon. We will now examine how choking affects the water surface profile upstream and downstream.

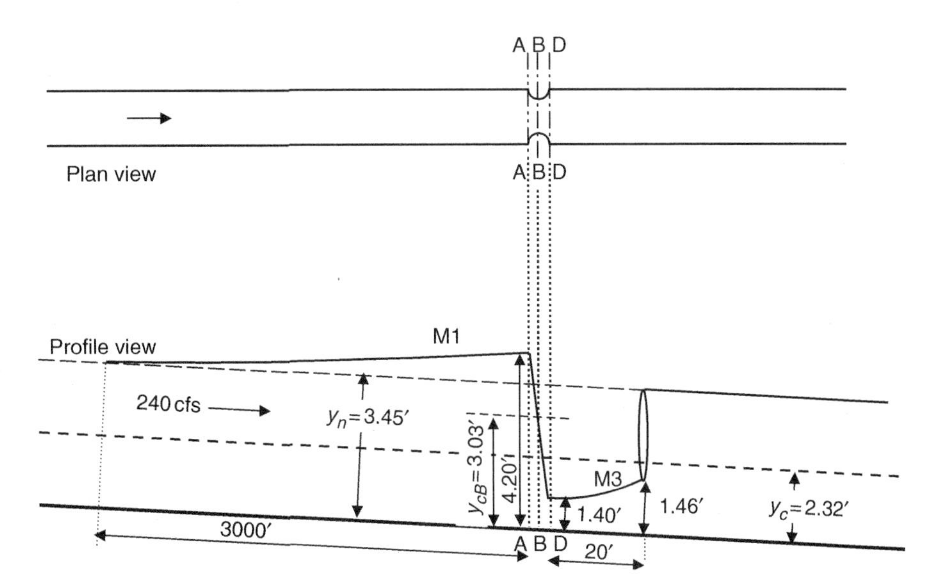

FIGURE 4.24 Choking in a long, mild channel

4.6.4.1 Choking in mild channels

As previously discussed, the flow tends to become normal in long channels. However, where choking occurs, the flow is pulled away from the normal conditions. Upstream of the choked section, the flow will build up to a depth greater than the normal depth. This will result in an M1 curve, as shown in Figure 4.24. If the length of the M1 curve is less than the length of the channel – that is, if the M1 curve meets the normal depth line – the channel is said to be long. In this case, the discharge in the channel will not be affected. However, if the M1 curve does not meet the normal depth line, then the channel is short and the flow rate will be affected by choking. Downstream of the constricted section, the flow will likely be supercritical. An M3 curve will form, followed by a hydraulic jump.

EXAMPLE 4.12 A very long rectangular channel has a bottom width of $b = 12$ ft, Manning roughness factor of $n = 0.013$, and longitudinal slope of $S_0 = 0.0009$. The channel carries $Q = 240$ cfs. The channel width is smoothly constricted to 8 ft at section B, as shown in Figure 4.24. Determine how this constriction affects the flow profile upstream and downstream. Assume the energy loss due to the channel constriction is negligible.

We can determine that for this channel $y_n = 3.45$ ft and $y_c = 2.32$ ft. Therefore, the channel is mild. Without the constriction the depth would be equal to 3.45 ft everywhere, including point A in Figure 4.24. The specific energy corresponding to this depth is

$$E_n = y_n + \frac{Q^2}{2gA_n^2} = 3.45 + \frac{(240)^2}{2(32.2)[(12.0)(3.45)]^2} = 3.97 \text{ ft}$$

We will now investigate whether this amount of specific energy is adequate to pass the 240 cfs through the constricted section. We know that minimum specific energy is required at the critical state. At section B, for $b = 8$ ft, the critical depth is $y_{cB} = 3.03$ ft. Then

$$(E_B)_{\min} = y_{cB} + \frac{Q^2}{2gA_{cB}^2} = 3.03 + \frac{(240)^2}{2(32.2)[(8.0)(3.03)]^2} = 4.55\text{ ft}$$

Assuming the energy loss between the sections A and B is negligible, the minimum specific energy required at A is also 4.55 ft. This is larger than the 3.97 ft that would be available at section A under the normal flow conditions. Therefore, as we recall from Chapter 2, choking will occur, and the flow will back up. The adjusted depth at section A will satisfy the expression

$$4.55 = y_{\text{Aadj}} + \frac{(240)^2}{2(32.2)[12.0(y_{\text{Aadj}})]^2}$$

By trial and error, we determine that the adjusted depth at section A is $y_{\text{Aadj}} = 4.20$ ft. This depth is higher than the normal depth. Therefore, an M1 profile will form upstream of the choked section. We can calculate the M1 profile using either the direct step method or the standard step method with a downstream boundary depth of 4.20 ft. Using a depth increment of $\Delta y = 0.05$ ft, the direct step method will indicate that the M1 curve meets the normal depth line at a distance approximately 3000 ft from the choked section. In other words, the effect of choking extends to 3000 ft upstream in this problem.

Both in Chapter 2 and here we have indicated that the flow needs additional energy to pass the constricted section. We have also indicated that the flow would acquire this additional energy by an increase in the flow depth just upstream of the choked section. However, we have not yet explained the source of this energy. In other words, where is this additional energy coming from? To answer this question, we should remember that the flow depths along an M1 curve are higher than the normal depth (as shown in Figure 4.24). Then, for the same discharge, the velocities are lower than the normal flow velocities, and the energy loss due to friction occurs at a slower rate (the friction slope, S_f, is milder than the bottom slope, S_0). As a result of this, compared to normal flow, less energy is dissipated and more energy is saved along an M1 curve. Therefore, when the flow is choked and backed up in the form of an M1 curve, the amount of energy saved upstream will be available to pass the constant discharge through the choked section.

We will now consider the situation downstream of the constricted section. Assuming that the energy loss is negligible between sections B and D, the flow depth, y_D, at D should satisfy

$$4.55 = y_D + \frac{(240)^2}{2(32.2)[12.0(y_D)]^2}$$

Two positive roots can be found; a subcritical depth of 4.20 ft (the same as y_{Aadj}) and a supercritical depth of 1.40 ft. The former is higher than the normal depth, and is in the M1 zone. However, an M1 profile is not possible downstream of the choked section. In this case, the flow depths here should decrease to approach the normal depth line ($y_n = 3.45$ ft), while along an M1 curve the depths can only increase. Therefore at section D the depth will be $y_D = 1.40$, followed by an M3 profile. Further down, the flow will change to the subcritical state through a hydraulic jump. The depth after the jump will be $y_{J2} = 3.45$ ft. To find the supercritical depth, y_{J1}, just before the jump we can use the hydraulic jump equation (Equation 2.27) for rectangular channels. Let us first determine the Froude number, F_{rJ2}, by using Equation 2.1:

$$F_{rJ2} = \frac{Q}{\sqrt{gA_{J2}^3/T_{J2}}} = \frac{240}{\sqrt{32.2[(12.0)(3.45)]^3/(12)}} = 0.55$$

Then, by using Equation 2.27,

$$y_{J1} = \frac{y_{J2}}{2}\left[\sqrt{1 + 8F_{rJ2}^2} - 1\right] = \frac{3.45}{2}\left[\sqrt{1 + 8(0.55)^2} - 1\right] = 1.47 \text{ ft}$$

We can now use the direct step method to calculate the M3 curve with an upstream boundary depth of 1.40 ft. Calculations performed with depth increments of 0.01 ft will show that the M3 curve reaches the depth of 1.47 ft at 20 ft from the constricted section.

It is important to note that in this example $y_D = 1.40$ ft turned out to be smaller than $y_{J1} = 1.47$ ft. Therefore, the flow depth changes from 1.40 ft to 1.47 ft through an M3 profile. This may not always be the case. Depending on the channel conditions, the calculations may yield a y_{J1} that is smaller than y_D. In that event, gradually-varied flow cannot occur downstream of point D. A drowned jump will occur at the constriction, and the flow will become normal immediately downstream.

EXAMPLE 4.13 Suppose the rectangular channel considered in Example 4.12 receives flow from a lake at a distance of 1080 ft from the constricted section. Let the water level in the lake be 3.97 ft. We are to determine how the channel constriction affects the discharge and the flow profile.

Let us first assume that the channel is hydraulically long. Then, we can show that for $b = 12$ ft, $n = 0.013$, and $S_0 = 0.0009$, a discharge of $Q = 240$ cfs and a normal depth of $y_n = 3.45$ ft satisfy Equations 4.19 and 4.20 with $E_{UP} = 3.97$ ft. Let us now proceed as in Example 4.12. We find that the critical depth $y_{cB} = 3.03$ ft will occur at section B, the flow depth will rise to 4.20 ft at section A, and an M1 curve will form upstream. If, as in Example 4.12, we use the direct step method to calculate the M1 profile with a downstream boundary depth of 4.20 ft and depth increments of $\Delta y = 0.05$ ft, the calculations will yield a flow depth of $y = 3.80$ ft and a specific energy of $E = 4.23$ ft at the lake location (1080 ft upstream of the choked section), as shown in Table 4.8. This means that, to maintain $Q = 240$ cfs, the water level in the lake should be 4.23 ft above

TABLE 4.8 Effect of choking on discharge

$Q=240$ cfs			$Q=220$ cfs		
y (ft)	E (ft)	X (ft)	y (ft)	E (ft)	X (ft)
4.20	4.55	0	3.96	4.29	0
4.15	4.51	110.75	3.91	4.25	109.37
4.10	4.47	226.19	3.86	4.21	223.60
4.05	4.43	347.10	3.81	4.17	343.52
4.00	4.39	474.43	3.76	4.13	470.22
3.95	4.35	609.42	3.71	4.09	605.09
3.90	4.31	753.70	3.66	4.05	750.02
3.85	4.27	909.48	3.61	4.01	907.61
3.80	4.23	1080.00	3.56	3.97	1080.00

the channel invert. Because the given lake surface is only 3.97 ft above the channel invert in this problem, the discharge has to be less than 240 cfs.

Finding the discharge in this problem requires a lengthy trial-and-error procedure. Let us try $Q=220$ cfs. We can determine that $y_c=2.19$ ft and $y_n=3.25$ ft for $Q=220$ cfs. The specific energy for the normal depth is $E_n=3.74$ ft. If the flow in the channel remained unaffected by the channel constriction, the specific energy just upstream of the constriction would then be 3.74 ft. Let us now calculate the minimum specific energy required to pass 220 cfs through the constricted section. The critical depth in the constricted section ($b=8$ ft) is 2.86 ft, and the corresponding specific energy is 4.29 ft. Because the energy loss due to constriction is neglected, the minimum specific energy required just upstream of the constriction is also 4.29 ft. This is higher than 3.74 ft, the specific energy corresponding to the normal depth. Therefore, choking will occur. Using the procedures in Example 4.12, we can determine that the depth will increase to 3.96 ft at section A to acquire the required specific energy of 4.29 ft. Now we can perform the M1 profile calculations using the direct step method with a downstream boundary depth of 3.96 ft and depth increments of 0.05 ft. The results summarized in Table 4.8 show that at 1080 ft upstream, the flow depth will be 3.56 ft and the specific energy will be 3.97 ft (equal to the given lake surface elevation). Therefore, we conclude that $Q=220$ cfs satisfies all the boundary conditions. If we obtained a specific energy different from 3.97 ft at the lake location, we would try another value for the flow rate and repeat the same procedure.

Once the discharge has been determined, we can follow the same procedure as in Example 4.12 to determine the profile downstream. In this case, the depth just downstream of the constriction will be 1.33 ft, followed by an M3 curve. The M3 curve will reach the depth of 1.38 ft at a distance of 14 ft from the constriction, and then a hydraulic jump will occur from the depth of 1.38 ft to 3.25 ft. Downstream of the jump, the flow will be normal.

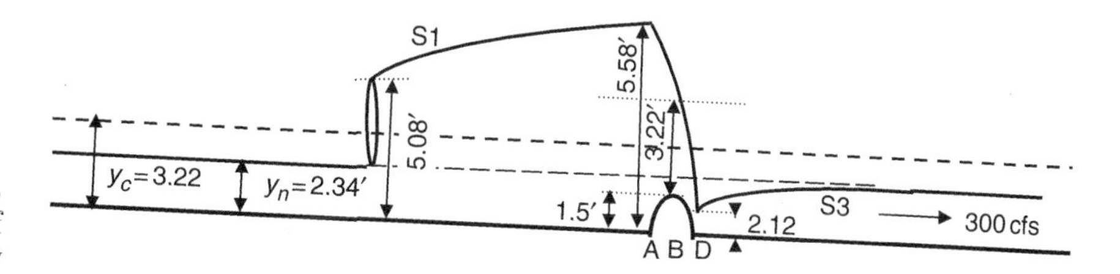

FIGURE 4.25 Choking of supercritical flow

4.6.4.2 Choking in steep channels

The discussion of the flow choking problem in Chapter 2 was limited to subcritical flow. We are now ready to discuss the case of choking of supercritical flow as well. We know that choking occurs when the flow does not have adequate specific energy to pass a horizontally constricted section or a section with a raised bottom. When choked, the flow adjusts itself to acquire the additional specific energy needed. As we recall from Figures 2.8, 2.9, and 2.14, for the same discharge, a higher specific energy corresponds to a larger depth in subcritical flow. However, if the flow is supercritical, for the same discharge, a higher specific energy occurs at a lower depth. Then, where supercritical flow approaches a choked section, in order to gain additional specific energy the flow depth needs to decrease if the flow is to remain supercritical. However, this is not possible since supercritical to subcritical state flow is not subject to downstream control, and the supercritical flow depths cannot be affected by what is downstream as long as the flow remains supercritical. Therefore, the only possibility is that the flow will change from supercritical through a hydraulic jump as shown in Figure 4.25. Then a subcritical S1 profile will form between the jump and the choked section, with flow depths increasing in the flow direction. Accordingly, the losses due to friction will be less, allowing the flow to maintain enough energy to compensate for the energy loss due to the hydraulic jump and to push the flow through the choked section.

Note that the profile is unaffected upstream of the jump. Therefore, as long as the length of the channel is greater than the length of the S1 curve, the discharge in the channel will not be affected by choking.

EXAMPLE 4.14 A long trapezoidal channel carrying a discharge of $Q=300$ cfs has $b=5$ ft, $m=2$, $n=0.013$, and $S_0=0.008$. There is a smooth step rise of 1.5 ft on the channel bottom, as shown in Figure 4.25. Determine the water surface profile in this channel.

We can determine that the normal depth is $y_n=2.34$ ft and the critical depth is $y_c=3.22$ ft. Therefore, the channel is steep and the normal flow is supercritical. Without choking, the flow would be normal upstream of the step with a specific energy of

$$E_n = y_n + \frac{Q^2}{2gA_n^2} = 2.34 + \frac{(300)^2}{2(32.2)\{[(5.0)+(2)(2.34)](2.34)\}^2} = 5.06\,\text{ft}$$

The minimum specific energy is required for the flow to pass section B if the flow at B is critical. The critical depth is $y_{cB} = 3.22$ ft. Therefore,

$$(E_B)_{\min} = y_{cB} + \frac{Q^2}{2gA_{cB}^2} = 3.22 + \frac{(300)^2}{2(32.2)\{[(5.0) + (2)(3.22)](3.22)\}^2} = 4.25 \text{ ft}$$

The channel bottom at section A is 1.5 ft below that at section B. Accordingly, the minimum specific energy required at A is $4.25 + 1.5 = 5.75$ ft. This is higher than the normal flow specific energy of 5.08 ft. Hence, the flow cannot remain normal at section A. It will adjust to acquire the required specific energy of 5.75 ft. To determine the adjusted flow depth at A, we solve the equation

$$(E_A)_{\min} = y_{\text{Aadj}} + \frac{Q^2}{2gA_{\text{Aadj}}^2} = 5.75 = y_{\text{Aadj}} + \frac{(300)^2}{2(32.2)\{[(5.0) + (2)y_{\text{Aadj}}]y_{\text{Aadj}}\}^2}$$

This equation yields two positive roots; 2.12 ft (supercritical) and 5.58 ft (subcritical). A supercritical depth at section A less than the normal depth (2.34 ft) is not possible. Such a depth would be in zone S3, in which the flow depths can only increase (not decrease from 2.34 ft to 2.12 ft). Moreover, if the flow remains supercritical, it should not be subject to downstream control. For these reasons we conclude that a depth of 2.12 ft is not possible, and the adjusted flow depth at section A will be 5.58 ft. The flow must change from the supercritical to the subcritical state through a hydraulic jump. As shown in Figure 4.25, the depth just before the jump, y_{J1}, is equal to the normal depth, 2.34 ft. We need to perform the hydraulic jump calculations to determine the depth after the jump, y_{J2}. Note that the channel is trapezoidal (not rectangular), and we cannot use Equation 2.26 or 2.27. However, by employing one of the procedures presented in section 2.3.4 and in Example 2.16, we determine that $y_{J2} = 4.25$ ft. Now we can perform the gradually-varied flow equations to determine the jump location. Using the direct step method with a downstream boundary depth of 5.58 ft and a depth increment of 0.05 ft, we find that the 4.25 ft occurs 141 ft upstream of section A.

We will now determine the flow depth at D. Assuming the energy loss is negligible over the step rise, the specific energy at D should be the same as that at A (5.75 ft) with the corresponding flow depths of 2.12 ft and 5.58 ft. The depth 5.58 ft would be in zone S1, in which the flow depths can only increase in the flow direction. However, in this case the profile should approach the normal depth at 2.34 ft. Hence, a depth of 5.58 ft is not possible at section D, so $y_D = 2.12$ ft followed by an S3 profile. The direct step method with an upstream boundary depth of 2.12 ft and depth increments of 0.02 ft shows that the S3 profile reaches the normal depth line at a distance 744 ft from point D.

4.7 Gradually-Varied Flow in Channel Systems

The gradually-varied flow calculations can also be used for channel systems. A schematic of a channel system is shown in Figure 4.26. If the flow in the system is subcritical, we first calculate the profile in channel D (most downstream channel) and proceed in the upstream direction. The calculations are performed for each channel individually. If the flow in the system is supercritical, we first calculate the profiles in channels E and F (most upstream channels) and proceed in the downstream direction. However, at channel junctions, where two or more channels meet, a junction equation should be satisfied to account for the interaction between the channels. The continuity equation should always be satisfied. For instance, for steady flow, $Q_D = Q_A + Q_B$ in Figure 4.26. Likewise, $Q_B = Q_G + Q_H$.

Also, either the energy or the momentum equation should be used at the junction. For subcritical flow, it is convenient to use the energy equation. Figure 4.27 displays a schematic of a junction where three channels meet. Let Section 3 represent the most upstream section of channel D, and Sections 1 and 2, respectively, represent the most downstream sections of channels A and B. Let the junction area between these three sections be negligible. Then the energy equation between Sections 1 and 3 can be written as

$$z_{b1} + y_1 + \frac{V_1^2}{2g} = z_{b3} + y_3 + \frac{V_3^2}{2g} + h_j \tag{4.27}$$

where h_j = head loss due to the junction. This head-loss is evaluated by using

$$h_j = k_j \left| \frac{V_1^2}{2g} - \frac{V_3^2}{2g} \right| \tag{4.28}$$

The junction loss coefficient, k_j, usually has a value of between 0 and 1. For simplicity, if we let $k_j = 0$, then Equation 4.27 becomes

$$z_{b1} + y_1 + \frac{V_1^2}{2g} = z_{b3} + y_3 + \frac{V_3^2}{2g} \tag{4.29}$$

Equation 4.29 implies that the energy grade line is continuous at the junction. If the channel bottom elevations at Sections 1 and 3 are the same, $z_{b1} = z_{b3}$,

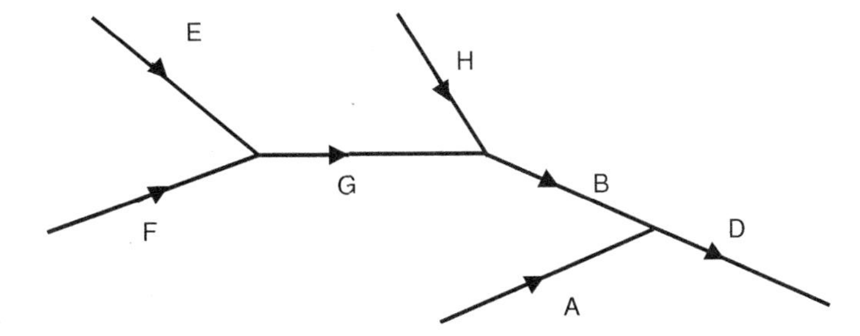

Figure 4.26 Schematic of a channel system

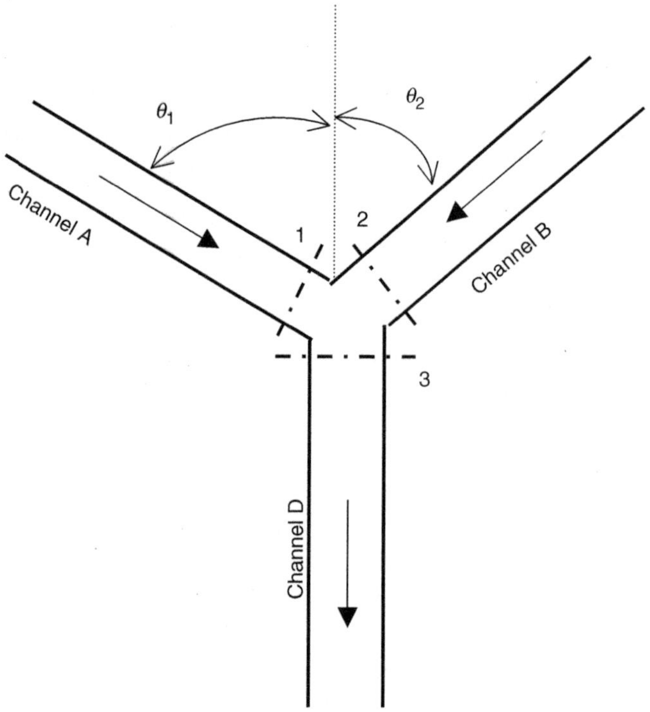

FIGURE 4.27
Schematic of a channel junction

Equation 4.29 is further simplified to yield $E_1 = E_3$, where E denotes the specific energy. If we set $k_j = 1$, then Equation 4.27 becomes

$$z_{b1} + y_1 = z_{b3} + y_3 \tag{4.30}$$

Equation 4.30 implies that the water surface is continuous. If the channel bottom elevations at Sections 1 and 3 are the same, $z_{b1} = z_{b3}$, Equation 4.30 will yield $y_1 = y_3$. In a typical subcritical flow situation, y_3 and V_3 would be known from the gradually-varied flow calculations of channel 3. Then the junction equation (one of Equations 4.27, 4.29, or 4.30) is used to determine the flow depth at Section 1. This becomes the downstream boundary condition for channel A. The flow depth at Section 2 of channel B is determined in a similar way.

Some channel junctions include a drop structure, as shown in Figure 4.28. In this case, Equation 4.27 is valid only if

$$z_{b1} + y_{1c} < z_{b3} + y_3 \tag{4.31}$$

Otherwise, $y_1 = y_{1c}$, where y_{1c} denotes the critical depth at Section 1.

If the flow in the channel system is supercritical, then in Figure 4.27 the conditions at Sections 1 and 2 would be known from the gradually-varied flow calculations for channels A and B. We would need to determine an upstream depth, y_3, for channel D. In this case, the momentum approach is more convenient. We will neglect the friction forces in the junction. We will also

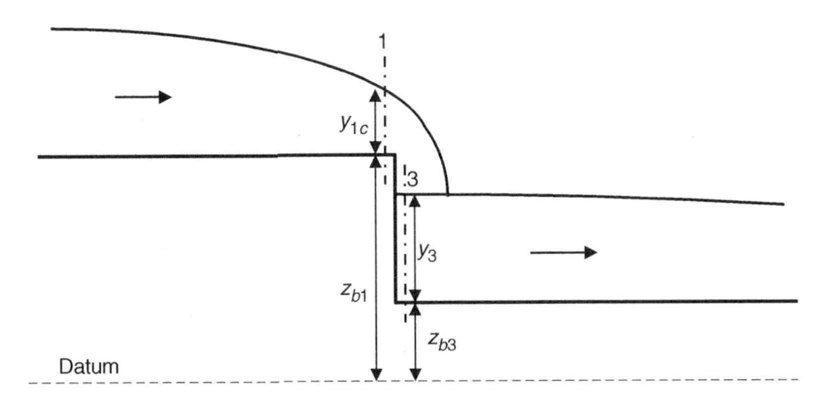

FIGURE 4.28 Drop at a junction

neglect the flow direction component of the weight of water occupying the junction. The momentum balance for the junction can then be written as

$$M_1 \cos\theta_1 + M_2 \cos\theta_2 = M_3 \tag{4.32}$$

where M denotes the specific momentum (see Section 2.3.2), θ_1 = angle between the flow directions of channels A and D, and θ_2 = angle between the flow directions of channels B and D. Substituting Equation 2.19 into Equation 4.32 for all three channels,

$$\left(\frac{Q_1^2}{gA_1} + Y_{C1}A_1\right)\cos\theta_1 + \left(\frac{Q_2^2}{gA_2} + Y_{C2}A_2\right)\cos\theta_2 = \left(\frac{Q_3^2}{gA_3} + Y_{C3}A_3\right) \tag{4.33}$$

where Y_C = the distance from the water surface to the centroid of the flow area. Everything on the left-hand side of Equation 4.33 would be known from the gradually-varied flow calculations for channels 1 and 2. The right-hand side can be expressed in terms of y_3. We can solve Equation 4.33 for y_3 and use this depth as the upstream boundary condition for the supercritical water surface profile calculations for channel D.

EXAMPLE 4.15 Consider the situation of flow around an island shown in Figure 4.29. Suppose the discharge in the channel downstream of section A is $30\ \text{m}^3/\text{s}$, and the gradually-varied flow calculations completed for this channel indicate that the specific energy at A is $E_A = 1.98$ m. Suppose we can approximate branch 1 as a trapezoidal channel that is 1000 m long with a bottom width of $b_1 = 2$ m, side slopes of $m_1 = 2$, Manning roughness factor of $n_1 = 0.016$, and a longitudinal bottom slope of $S_{01} = 0.00036$. Branch 2 is also trapezoidal in cross-section, with a bottom width of $b_2 = 3$ m and side slopes of $m_2 = 2$. This branch is 900 m long, and it has a Manning roughness factor of $n_2 = 0.013$ and a longitudinal bottom slope of $S_{02} = 0.0004$. Determine the discharge in each branch. Assume that the energy grade line is continuous at A and B.

The total discharge of $30\ \text{m}^3/\text{s}$ will split between the channels so that the specific energy at section B will be the same as that calculated from each channel. We can

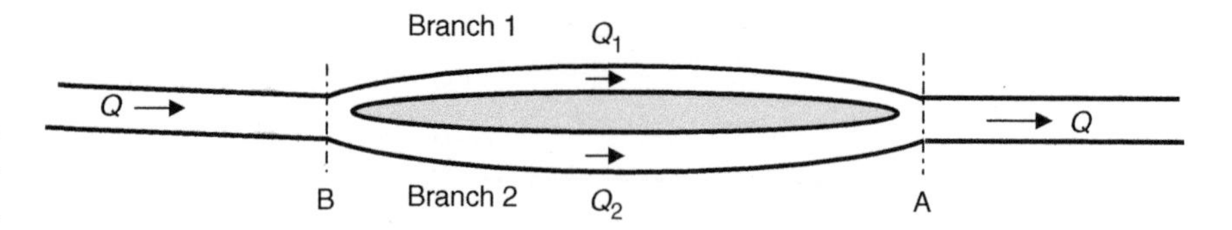

FIGURE 4.29
Example of flow around an island

TABLE 4.9 Summary of results for Example 4.15

	Branch 1			Branch 2		
Trial	Q_1 (m^3/s)	y_{1A} (m)	E_{1B} (m)	Q_2 (m^3/s)	y_{2A} (m)	E_{2B} (m)
1	10	1.94	1.836	20	1.85	1.976
2	11	1.93	1.877	19	1.86	1.944
3	12	1.92	1.916	18	1.88	1.916

determine this split only by trial and error. Since branch 2 is larger, smoother, and shorter, it will carry a larger discharge than branch 1.

Let us first try $Q_1 = 10\,\text{m}^3/\text{s}$ and $Q_2 = 20\,\text{m}^3/\text{s}$. Calculating the normal and critical depths in these channels as being $y_{n1} = 1.66$ m, $y_{c1} = 0.99$ m, $y_{n2} = 1.84$ m, and $y_{c2} = 1.25$ m, we determine that both channels are mild. Denoting the flow depth at the downstream end of branch 1 by y_{1A}, we can write

$$y_{1A} + \frac{(10.0)^2}{2(9.81)[2.0 + 2(y_{1A})]^2 y_{1A}^2} = 1.98$$

Solving this expression, we obtain $y_{1A} = 1.94$ m. This is larger than $y_{n1} = 1.66$ m. Therefore, an M1 curve will occur in branch 1. We can now perform the gradually-varied flow calculations using a downstream boundary depth of 1.94 m. The direct step method with a depth increment of 0.005 m will yield a specific energy of $E_{1B} = 1.839$ m at section B (1000 m upstream of A). Similarly, for $Q_2 = 20\,\text{m}^3/\text{s}$, we obtain $y_{2A} = 1.85$ m for branch 2. The M1 profile calculations in this channel will yield $E_{2B} = 1.976$ m at section B (900 m upstream of A). Because the calculated E_{2B} is larger than E_{1B}, we need to try another set of Q_1 and Q_2. Table 4.9 summarizes the results obtained for other trial values of Q_1 and Q_2, and reveals that the final solution is obtained with $Q_1 = 12\,\text{m}^3/\text{s}$ and $Q_2 = 18\,\text{m}^3/\text{s}$.

4.8 GRADUALLY-VARIED FLOW IN NATURAL CHANNELS

The foregoing sections have been devoted to gradually-varied flow calculations in prismatic channels having constant cross-sectional shapes and dimensions. Most natural channels have irregular shapes that vary with distance along the channel. Also, the channel slope is not well defined since the bottom of the channel may be far from being straight. Moreover, as discussed in Section 3.7, many natural channels have a main part and one or two overbank areas (Figure 3.10). The average flow velocity in the main channel and the overbank areas can be significantly different because of different flow depths and roughness factors.

We introduced the direct step method and the standard step method in the preceding sections to calculate the water surface profiles in prismatic channels. In the direct step method, we first pick a depth and then calculate the flow area, velocity, energy head, and the friction slope corresponding to this depth. Then, using this information, we determine the distance at which this depth occurs. In other words, in the direct step method at first we do not know the location where the selected depth will occur, yet we calculate the flow area, velocity, energy head, and friction slope at this unknown location. This is possible only if the cross-sectional characteristics, like the bottom width and side slopes, remain constant with distance along the channel. In natural channels the cross-sectional characteristics vary, and we cannot determine the flow area, say, for a given depth unless we know where this depth occurs and what the cross-sectional characteristics are at this location. Therefore, the direct step method is not applicable to non-prismatic, natural channels. In the standard step method, we first select the location (or the distance) along a channel and then determine the flow depth at this location using the known cross-sectional characteristics. The cross-sectional characteristics would be known at surveyed sections of natural channels, as well as the distance between these sections. Therefore, the standard step method is suitable for natural channels. However, some modifications are needed to account for the irregular channel sections.

For non-prismatic channels, the bottom slope is not well defined. Therefore, it is more convenient to write the gradually-varied flow equation in terms of the bottom elevations of the surveyed sections rather than the bottom slope. Also, in non-prismatic channels the flow velocity varies considerably within a channel section, particularly if the section is compound. We account for this variation by using the energy correction factors discussed in Chapter 1. Moreover, besides the friction losses, in non-prismatic channels additional energy losses occur due to changes in the cross-sectional shapes and sizes along a channel. We call these losses resulting from flow expansions and contractions the *eddy losses*.

Referring to Figure 4.11, the gradually-varied flow equation for natural, compound channels is written as

$$z_{bU} + y_U + \alpha_U \frac{V_U^2}{2g} - \frac{1}{2}(\Delta X) S_{fU} - h_e = z_{bD} + y_D + \alpha_D \frac{V_D^2}{2g} + \frac{1}{2}(\Delta X) S_{fD} \qquad (4.34)$$

where U and D, respectively, denote the upstream and downstream sections, and z_b = elevation of channel bottom above a horizontal datum, α = energy correction coefficient, and h_e = eddy loss. In this equation, V_U and V_D stand for the cross-sectional average velocities in the upstream and downstream sections, respectively.

For a compound channel section, as in Figure 3.10, the friction slope is evaluated by using Equation 3.41, rewritten here as

$$S_f = \left(\frac{Q}{\sum K_i}\right)^2 \tag{4.35}$$

where i = index referring to the i-th subsection of the compound channel section, and K = conveyance (calculated as in Equation 3.39):

$$K_i = \frac{k_n}{n_i} A_i R_i^{2/3} = \frac{k_n}{n_i} \frac{A_i^{5/3}}{P_i^{2/3}} \tag{4.36}$$

The discharge in the i-th subsection is

$$Q_i = K_i S_f^{1/2} \tag{4.37}$$

The energy coefficient is evaluated by using Equation 1.20 rewritten here as

$$\alpha = \frac{\sum V_i^3 A_i}{V^3 \sum A_i} = \frac{\left(\sum A_i\right)^2}{\left(\sum K_i\right)^3} \sum \frac{K_i^3}{A_i^2} \tag{4.38}$$

The eddy loss is evaluated by using

$$h_e = k_e \left| \alpha_U \frac{V_U^2}{2g} - \alpha_D \frac{V_D^2}{2g} \right| \tag{4.39}$$

where k_e = a coefficient. For gradually converging or diverging channels, we generally pick a value of between 0 and 0.50 for k_e.

For subcritical flow, for any channel reach between two surveyed sections, the downstream depth, y_D, would be known. With this we can calculate V_D, α_D, and S_{fD}. Next we guess the value of y_U and determine V_U, α_U, and S_{fU} as well as h_e. If the calculated values satisfy Equation 4.34, then the guessed value of y_U is acceptable. Otherwise we need to try another y_U until Equation 4.34 is satisfied. Equation 4.16 can still be used to improve the trial values of y_U in each iteration (Henderson, 1966).

We can also use Equation 4.34 for supercritical flow, although in this case we start from the upstream end and carry out the calculations in the downstream direction. In other words, for a channel reach between two surveyed sections, y_U would be known and y_D would be sought. Equation 4.18 can still be used to improve the trial value of y_D in each iteration.

We can model the stream junctions using the procedures discussed in Section 4.7. However, for natural streams we need to include the energy and momentum correction factors in the equations. Moreover, a stream junction can occupy a fairly large area. Accordingly, Equations 4.27 and 4.28 are modified as

$$z_{b1} + y_1 + \alpha_1 \frac{V_1^2}{2g} = z_{b3} + y_3 + \alpha_3 \frac{V_3^2}{2g} + h_j + L_{1-3} \frac{S_{f1} + S_{f3}}{2} \tag{4.40}$$

and

$$h_j = k_j \left| \alpha_1 \frac{V_1^2}{2g} - \alpha_3 \frac{V_3^2}{2g} \right| \tag{4.41}$$

where L_{1-3} = distance between Sections 1 and 3 through the junction. Likewise, Equation 4.33 is modified as

$$\left(\beta_1 \frac{Q_1^2}{gA_1} + Y_{C1}A_1 \right) \cos\theta_1 + \left(\beta_2 \frac{Q_2^2}{gA_2} + Y_{C2}A_2 \right) \cos\theta_2 - F_{fj} + W_j$$
$$= \left(\beta_3 \frac{Q_3^2}{gA_3} + Y_{C3}A_3 \right) \tag{4.42}$$

where β = momentum correction factor, F_{fj} = friction forces opposing the flow at the junction, and W_j = the component of the weight of water occupying the junction in the flow direction. Certain assumptions are needed to evaluate F_{fj} and W_j (US Army Corps of Engineers, 2002).

Generally speaking, gradually-varied flow calculations for natural, compound channels are too cumbersome to perform manually. Various graphical methods have been developed for this purpose, such as the Ezra, Grimm, and Escoffier methods (Chow, 1959; Henderson, 1966). However, these methods are no longer feasible, since we can perform the gradually-varied flow calculations much more quickly using personal computers. The HEC-RAS program (US Army Corps of Engineers, 2002) is the leading computer program available in the public domain.

EXAMPLE 4.16 The water surface profile is to be calculated for the channel shown schematically in Figure 4.30. The channel is straight, and the sections marked in Figure 4.30 are 600 m apart. The section numbers (0.7, 0.8, 1, 4 and 4.3) are simply identifiers without any other significance. The cross-section of the channel can be approximated, as shown in Figure 4.31. The Manning roughness factor is 0.025 for the main channel and 0.05 for the left overbank and the right overbank. The main channel bed elevation, z_b, at section 0.7 is 64 m. The channel has a fairly well-defined longitudinal bottom slope of 0.0005. Therefore the bed elevation at section 0.8 is $64 + (0.0005)(600) = 64.3$ m. Likewise, the bed elevations for Sections 1, 4, and 4.3 are 64.6 m, 64.9 m, and 65.2 m, respectively. The eddy loss coefficient, k_e, is 0.1 for contraction and 0.3 for expansion. Calculate the water surface profile for $Q = 250$ m^3/s if the water surface elevation at section 0.7 is 66.3 m.

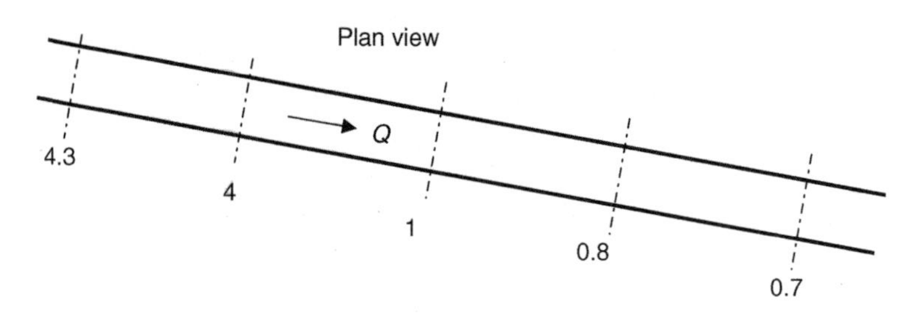

FIGURE 4.30 Schematic of channel reach for Example 4.16

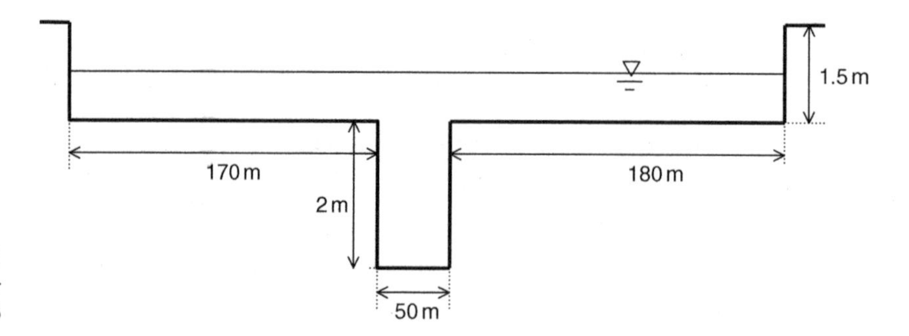

FIGURE 4.31 Channel section for Example 4.16

Given the bottom slope of 0.0005, the channel is more than likely mild, and the flow is subcritical (this can be verified by checking the Froude numbers later). Therefore, the computations will begin at section 0.7 and proceed in the upstream direction. Given the water surface elevation at section 0.7, we will first determine the water surface elevation at section 0.8, which will satisfy Equation 4.34. Then we will proceed to Sections 1, 4, and 4.3.

Let us first evaluate the cross-sectional properties at section 0.7. Reviewing Figure 4.31 with $z_b = 64.0$ m at section 0.7, we can see that a water surface elevation of 66.3 m corresponds to a depth of 2.3 m in the main channel and 0.3 m in the overbank sections. Therefore, for the main channel $A = (50)(2.3) = 115\ \text{m}^2$ and $P = 50 + 2 + 2 = 54$ m. For the left overbank, $A = (170)(0.3) = 51\ \text{m}^2$ and $P = 170 + 0.3 = 170.3$ m. Likewise, for the right overbank, $A = (180)(0.3) = 54\ \text{m}^2$ and $P = 180 + 0.3 = 180.3$ m.

Now we can evaluate the conveyances of the main channel and the overbank areas by using Equation 4.36. For the main channel with $k_n = 1.0$ for metric units,

$$K_i = \frac{k_n}{n_i}\frac{A_i^{5/3}}{P_i^{2/3}} = \frac{1.0}{0.025}\frac{(115.0)^{5/3}}{(54.0)^{2/3}} = 7614.3\ \text{m}^3/\text{s}$$

Likewise, we obtain $K = 456.6\ \text{m}^3/\text{s}$ and $483.5\ \text{m}^3/\text{s}$ for the left overbank and right overbank, respectively. Then, by using Equation 4.35,

$$S_f = \left(\frac{Q}{\sum K_i}\right)^2 = \left(\frac{250}{456.6 + 7614.3 + 483.5}\right)^2 = 0.000854$$

Now we can determine the energy correction factor by using Equation 4.38

$$\alpha = \frac{\left(\sum A_i\right)^2}{\left(\sum K_i\right)^3} \sum \frac{K_i^3}{A_i^2} = \frac{(456.6 + 7614.3 + 483.5)^2}{(51.0 + 115.0 + 54.0)^3} \left(\frac{456.6^3}{51.0^2} + \frac{7614.3^3}{115.0^2} + \frac{483.5^3}{54.0^2} \right)$$
$$= 2.59$$

Also, the average cross-sectional velocity is

$$V = \frac{Q}{\sum A_i} = \frac{250}{(51.0 + 115.0 + 54.0)} = 1.14 \text{ fps}$$

Although not needed for the solution of this example problem, we can calculate the discharge and the velocity in the three different segments of the compound channel. For the main channel, by using Equation 4.37, $Q = (7614.3)(0.000854)^{1/2} = 222.5 \text{ m}^3/\text{s}$. Then the velocity in the main channel is $(222.5)/(115.0) = 1.93$ m/s. Likewise, for the left overbank, we can determine that the discharge is $13.38 \text{ m}^3/\text{s}$ and the velocity is 0.26 m/s. For the right overbank, the discharge and velocity are found to be $14.14 \text{ m}^3/\text{s}$ and 0.26 m/s, respectively.

The right-hand side of Equation 4.34 becomes

$$z_{bD} + y_D + \alpha_D \frac{V_D^2}{2g} + \frac{1}{2}(\Delta X) S_{fD}$$
$$= 64.0 + 2.3 + 2.59 \frac{1.14^2}{2(9.81)} + \frac{1}{2}(600)(0.000854) = 66.73 \text{ m}$$

We now need to determine the water surface elevation at section 0.8 that will satisfy Equation 4.34. This is a trial-and-error process. Let us try a depth of $y_U = 2.35$ m at section 0.8. For this depth, we obtain $S_{fU} = 0.000754$, $\alpha_U = 2.72$, and $V_U = 1.04$ m/s. In addition, we can find the cross-sectional properties as being $A_U = 240 \text{ m}^3$, $P_U = 404.7$ m, $R_U = 0.59$ m, $T_U = 400$ m, and $F_{rU} = 0.43$. Noting that $z_b = 64.3$ m at section 0.8, and $k_e = 0.1$ for contracting flow, we can now evaluate the left-hand side of Equation 4.34 as

$$z_{bU} + y_U + \alpha_U \frac{V_U^2}{2g} - \frac{1}{2}(\Delta X) S_{fU} - h_e$$
$$= 64.3 + 2.35 + 2.72 \frac{1.04^2}{2(9.81)}$$
$$- \frac{1}{2}(600)(0.000754) - (0.1)\left| 2.72 \frac{1.04^2}{2(9.81)} - 2.59 \frac{1.14^2}{2(9.81)} \right|$$
$$= 66.57 \text{ m}$$

Because the calculated left-hand side is different from the right-hand side, we need to try another value for y_U. We will use Equation 4.16 to determine the

TABLE 4.10 Summary of iterations for Example 4.16

y_U (m)	A_U (m²)	P_U (m)	V_U (m/s)	α_U	S_{fU}	*LHS* (m)	R_U (m)	T_U (m)	D_U (m)	F_{rU}	Δy (m)
2.35	240	404.7	1.04	2.72	0.000754	66.57	0.59	400	0.60	0.43	−0.08
2.43	272	404.9	0.92	2.88	0.000618	66.67	0.67	400	0.68	0.36	−0.04
2.47	288	404.9	0.87	2.93	0.000560	66.71	0.71	400	0.72	0.33	−0.01
2.48	292	405.0	0.86	2.94	0.000547	66.72	0.72	400	0.73	0.32	−0.01
2.49	296	405.0	0.84	2.95	0.000534	66.73					

LHS = Left-hand side of Equation 4.34.

TABLE 4.11 Summary of results for Example 4.16

Section	Whole section			Elements: Left overbank	Elements: Main channel	Elements: Right overbank
0.7	WS elevation (m)	66.30	y (m)	0.30	2.30	0.30
	z_b (m)	64.0	A (m²)	51.0	115.0	54.0
	y (m)	2.30	P (m)	170.3	54.0	180.3
	V (m/s)	1.14	K (m³/s)	456.6	7614.4	483.5
	α	2.59	Q (m³/s)	13.36	222.50	14.14
	S_f	0.000854	V (m/s)	0.26	1.93	0.26
0.8	WS elevation (m)	66.89	y (m)	0.49	2.49	0.49
	z_b (m)	64.3	A (m²)	83.3	124.5	88.2
	y (m)	2.49	P (m)	170.49	54.0	180.49
	V (m/s)	0.85	K (m³/s)	1019.4	8668.3	1079.4
	α	2.95	Q (m³/s)	23.88	200.83	25.29
	S_f	0.000534	V (m/s)	0.29	1.64	0.29
1	WS elevation (m)	67.11	y (m)	0.51	2.51	0.51
	z_b (m)	64.6	A (m²)	86.24	125.37	91.32
	y (m)	2.51	P (m)	170.51	54.0	180.51
	V (m/s)	0.83	K (m³/s)	1095.0	8792.8	1159.6
	α	2.97	Q (m³/s)	24.78	198.98	26.24
	S_f	0.000512	V (m/s)	0.29	1.59	0.29
4	WS elevation (m)	67.41	y (m)	0.51	2.51	0.51
	z_b (m)	64.9	A (m²)	87.35	125.68	92.49
	y (m)	2.51	P (m)	170.51	54.0	180.51
	V (m/s)	0.82	K (m³/s)	1118.5	8829.1	1184.4
	α	2.97	Q (m³/s)	25.12	198.28	26.60
	S_f	0.000504	V (m/s)	0.29	1.58	0.29
4.3	WS elevation (m)	67.72	y (m)	0.52	2.52	0.52
	z_b (m)	65.2	A (m²)	87.72	125.79	92.88
	y (m)	2.52	P (m)	170.52	54.0	180.52
	V (m/s)	0.82	K (m³/s)	1126.5	8841.9	1192.8
	α	2.98	Q (m³/s)	25.23	198.05	26.72
	S_f	0.000502	V (m/s)	0.29	1.57	0.29

second guess. Substituting the known values into Equation 4.16,

$$\Delta y_k = \frac{(LHS)_k - (RHS)}{\left(1 - F_{rU}^2 + 3(\Delta X)S_{fU}/2R_U\right)_k}$$

$$= \frac{(66.57) - (66.72)}{(1 - (0.43)^2 + 3(600)(0.000754)/2(0.59))} = -0.08$$

Therefore, next we will use $y_u = 2.35 - (-0.08) = 2.43$ m. The results of the next few iteration cycles are summarized in Table 4.10. The depth at section 0.8 is found as being 2.49 m on the fifth iteration.

The reader will agree by now that the trial and error procedure is lengthy and is not suitable for manual calculations. We often use computer programs to calculate the water surface profile in compound channels. The HEC-RAS (US Army Corps of Engineers, 2002) computer program was used to complete this example. The results are presented in Table 4.11.

PROBLEMS

P.4.1 Classify the trapezoidal channels listed below as mild, steep, horizontal, critical, or adverse.

Channel	Q (cfs)	b (ft)	m	n	S_0
1	300	5.0	2.0	0.016	0.0004
2	350	4.5	2.0	0.022	0.0001
3	200	5.0	1.5	0.0	0
4	300	5.0	1.5	0.013	0.01
5	250	4.5	1.0	0.013	0.02
6	300	5.0	0	0.022	0.001

P.4.2 Classify the trapezoidal channels listed below as mild, steep, horizontal, critical, or adverse.

Channel	Q (m^3/s)	b (m)	m	n	S_0
1	30	5.0	1.0	0.013	0.0001
2	18	3.0	2.0	0.016	0.01
3	22	3.5	1.5	0.020	0
4	20	3.0	2.0	0.021	0.005
5	100	15.0	3.0	0.020	0.0002
6	50	8.0	2.0	0.016	0.004

P.4.3 Suppose that both channels shown in Figure P.4.1 are trapezoidal, infinitely long, and have a bottom width of $b = 5$ ft and side slopes of $m = 2$.

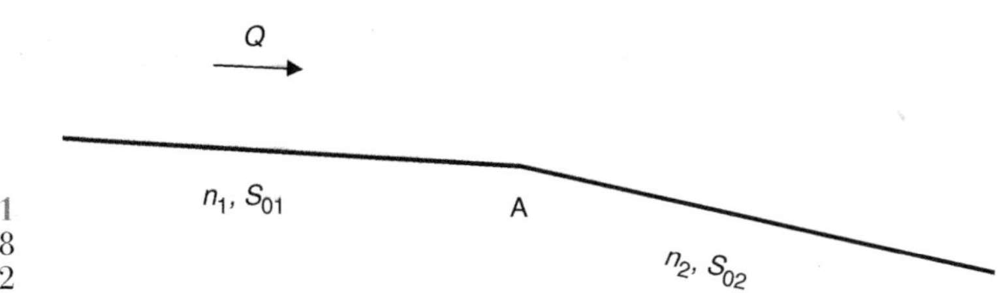

FIGURE P.4.1
Problems P.4.3–P.4.8
and P.4.11–P.4.12

The discharge is $Q=300$ cfs. Determine the type of the water surfaces profiles in these channels if:

(a) $n_1=n_2=0.013$, $S_{01}=0.0004$, and $S_{02}=0.0001$
(b) $n_1=n_2=0.013$, $S_{01}=0.0001$, and $S_{02}=0.0004$
(c) $n_1=n_2=0.013$, $S_{01}=0.0004$, and $S_{02}=0.01$
(d) $n_1=n_2=0.013$, $S_{01}=0.01$, and $S_{02}=0.02$
(e) $n_1=n_2=0.013$, $S_{01}=0.02$, and $S_{02}=0.01$
(f) $n_1=n_2=0.013$, $S_{01}=0.01$, and $S_{02}=0.0004$
(g) $S_{01}=S_{02}=0.0004$, $n_1=0.013$, and $n_2=0.022$
(h) $S_{01}=S_{02}=0.0004$, $n_1=0.022$, and $n_2=0.013$
(i) $S_{01}=S_{02}=0.01$, $n_1=0.013$, and $n_2=0.022$
(j) $S_{01}=S_{02}=0.01$, $n_1=0.022$, and $n_2=0.013$.

P.4.4 Suppose that both channels shown in Figure P.4.1 are trapezoidal, infinitely long, and have a bottom width of $b=2$ m and side slopes of $m=2$. The discharge is $Q=15\,m^3/s$. Determine the type of the water surfaces profiles in these channels if:

(a) $n_1=n_2=0.013$, $S_{01}=0.0004$, and $S_{02}=0.0001$
(b) $n_1=n_2=0.013$, $S_{01}=0.0001$, and $S_{02}=0.0004$
(c) $n_1=n_2=0.013$, $S_{01}=0.0004$, and $S_{02}=0.01$
(d) $n_1=n_2=0.013$, $S_{01}=0.01$, and $S_{02}=0.02$
(e) $n_1=n_2=0.013$, $S_{01}=0.02$, and $S_{02}=0.01$
(f) $n_1=n_2=0.013$, $S_{01}=0.01$, and $S_{02}=0.0004$
(g) $S_{01}=S_{02}=0.0004$, $n_1=0.013$ and $n_2=0.022$
(h) $S_{01}=S_{02}=0.0004$, $n_1=0.022$ and $n_2=0.013$
(i) $S_{01}=S_{02}=0.01$, $n_1=0.013$ and $n_2=0.022$
(j) $S_{01}=S_{02}=0.01$, $n_1=0.022$ and $n_2=0.013$.

P.4.5 Suppose that both channels shown in Figure P.4.1 are trapezoidal, infinitely long, and have a bottom width of $b=5$ ft and side slopes of $m=2$. The discharge is $Q=300$ cfs. Calculate the water surface profiles if $n_1=n_2=0.013$, $S_{01}=0.0004$, and $S_{02}=0.01$. Use the direct step method with depth increments of 0.10 ft.

P.4.6 Suppose that both channels shown in Figure 4.P.1 are trapezoidal, infinitely long, and have a bottom width of $b=2$ m and side slopes of $m=2$. The discharge is $Q=15\,m^3/s$. Calculate the water surface profiles if $n_1=n_2=0.013$,

$S_{01} = 0.0004$, and $S_{02} = 0.01$. Use the direct step method with depth increments of 0.03 m.

P.4.7 Suppose the infinitely long channels shown in Figure P.4.1 both have $n = 0.013$, $b = 5$ ft, $m = 2$, and $Q = 300$ cfs. Determine the flow depth 50 ft upstream of point A if $S_{01} = 0.0001$ and $S_{02} = 0.01$. Use the standard step method with $\Delta X = 25$ ft.

P.4.8 Suppose the infinitely long channels shown in Figure P.4.1 both have $n = 0.013$, $b = 2$ m, $m = 2$, and $Q = 15\ m^3/s$. Determine the flow depth 16 m upstream of point A if $S_{01} = 0.0001$ and $S_{02} = 0.01$. Use the standard step method with $\Delta X = 8$ m.

P.4.9 A very long trapezoidal channel carrying $Q = 400$ cfs has a bottom width of $b = 6$ ft, side slopes of $m = 1.5$, and a Manning roughness factor of $n = 0.013$. Determine the flow depth in the channel 60 ft upstream of the brink if:

(a) $S_0 = 0.0001$
(b) $S_0 = 0.02$.

P.4.10 A very long trapezoidal channel carrying $Q = 13\ m^3/s$ has a bottom width of $b = 2$ m, side slopes of $m = 1.5$, and a Manning roughness factor of $n = 0.013$. Determine the flow depth in the channel 20 m upstream of the brink if:

(a) $S_0 = 0.0001$
(b) $S_0 = 0.02$.

P.4.11 Suppose the two channels shown in Figure P.4.1 are very long and rectangular in cross-section, with a bottom width of $b = 10$ ft and a Manning roughness factor of $n = 0.013$. The longitudinal bottom slopes are $S_{01} = 0.02$ and $S_{02} = 0.0002$ and the discharge is $Q = 400$ cfs. Will a hydraulic jump occur? If your answer is 'no', explain it fully. If your answer is 'yes', find the distance between point A and the hydraulic jump.

P.4.12 Suppose the two channels shown in Figure P.4.1 are very long and rectangular in cross-section, with a bottom width of $b = 3$ m and a Manning roughness factor of $n = 0.013$. The longitudinal bottom slopes are $S_{01} = 0.02$ and $S_{02} = 0.0002$ and the discharge is $Q = 15\ m^3/s$. Will a hydraulic jump occur? If your answer is 'no', explain it fully. If your answer is 'yes', find the distance between point A and the hydraulic jump.

P.4.13 An infinitely long rectangular channel leading from an upstream lake has a bottom width of 20 ft, slope of S_0, and Manning roughness factor of 0.016. The lake water surface is 6.40 ft above the channel invert. Determine the discharge in the channel if:

(a) $S_0 = 0.01$
(b) $S_0 = 0.0002$.

P.4.14 An infinitely long rectangular channel leading from an upstream lake has a bottom width of 7 m, slope of S_0, and Manning roughness factor of 0.016.

The lake water surface is 2.0 m above the channel invert. Determine the discharge in the channel if:

(a) $S_0 = 0.01$
(b) $S_0 = 0.0002$.

P.4.15 A rectangular channel leading from an upstream lake has a bottom width of 20 ft, bottom slope of 0.0002, and a Manning roughness factor of 0.016. The lake water surface is 6.40 ft above the channel invert. The discharge depth relationship for a hydraulic structure built in the channel at distance L from the lake is

$$Q = 64(y - 4.6)^{1.5}$$

where Q is in cfs and y is in ft. Determine the discharge in the channel if:

(a) $L = 50\,000$ ft
(b) $L = 1000$ ft.

P.4.16 A rectangular channel leading from an upstream lake has a bottom width of 7 m, bottom slope of 0.0002, and a Manning roughness factor of 0.016. The lake water surface is 2.00 m above the channel invert. The discharge depth relationship for a hydraulic structure built in the channel at distance L from the lake is

$$Q = 20(y - 2.3)^{1.5}$$

where Q is in m^3/s and y is in m. Determine the discharge in the channel if:

(a) $L = 12\,000$ m
(b) $L = 460$ m.

P.4.17 Redo Problem P.4.15a using $S_0 = 0.01$ instead of 0.0002. Also, calculate the water surface profiles.

P.4.18 Redo Problem P.4.16a using $S_0 = 0.01$ instead of 0.0002. Also, calculate the water surface profiles.

P.4.19 The stormwater drainage system of an industrial park includes two ponds connected by a channel, as shown in Figure P.4.2. The channel is rectangular with a bottom width of $b = 15$ ft, Manning roughness factor of $n = 0.02$, and length of $L = 500$ ft. Determine the discharge in the channel and calculate the surface profile if:

(a) $H_{UP} = 16.5$ ft, $z_{bUP} = 12.0$ ft, $H_{DWN} = 15.0$ ft, and $z_{bDWN} = 10.0$ ft
(b) $H_{UP} = 18.8$ft, $z_{bUP} = 12.0$ ft, $H_{DWN} = 14.7$ ft, and $z_{bDWN} = 10.0$ ft.

Use the direct step method with depth increments of about 0.10 ft

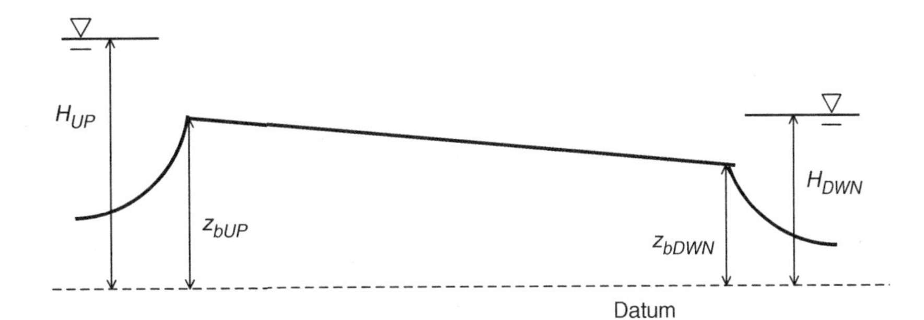

FIGURE P.4.2 Problems P.4.19–P.4.22

P.4.20 The stormwater drainage system of an industrial park includes two ponds connected by a channel, as shown in Figure P.4.2 The channel is rectangular with a bottom width of $b = 4.5$ m, Manning roughness factor of $n = 0.02$, and length of $L = 150$ m. Determine the discharge in the channel and calculate the surface profile if:

(a) $H_{up} = 11.9$ m, $z_{bu} = 10.6$ m, $H_{DWN} = 11.5$ m, and $z_{bDWN} = 10.0$ m
(b) $H_{up} = 12.77$ m, $z_{bu} = 10.6$ m, $H_{DWN} = 11.4$ m, and $z_{bDWN} = 10.0$ m.

Use the direct step method with depth increments of about 0.02 m.

P.4.21 What is the maximum discharge that would occur in the channel described in Problem P.4.19 if $H_{UP} = 18.8$ ft, $z_{bUP} = 12.0$ ft, and $z_{bDWN} = 10.0$ ft?

P.4.22 What is the maximum discharge that would occur in the channel described in Problem P.4.20 if $H_{UP} = 12.77$ m, $z_{bUP} = 10.6$ m, and $z_{bDWN} = 10.0$ m?

P.4.23 Channels A and B shown in Figure P.4.3 are very long. Both channels are rectangular. Channel A has a bottom width of 4 ft, longitudinal slope of 0.015, and Manning roughness factor of 0.016, and it carries 200 cfs. Channel B has a bottom width of 6 ft, longitudinal slope of 0.015, and Manning roughness factor of 0.016, and it carries 300 cfs. Channel C is 3000 ft long and it terminates at a free fall. It is rectangular, and has a bottom width of 14 ft, longitudinal slope of 0.015, and a Manning roughness factor of 0.016. Determine the flow depth at the upstream end of channel C.

P.4.24 Channels A and B shown in Figure P.4.3 are very long. Both channels are rectangular. Channel A has a bottom width of 1.2 m, longitudinal slope of 0.015, and Manning roughness factor of 0.016, and it carries 6.0 m^3/s. Channel B has a

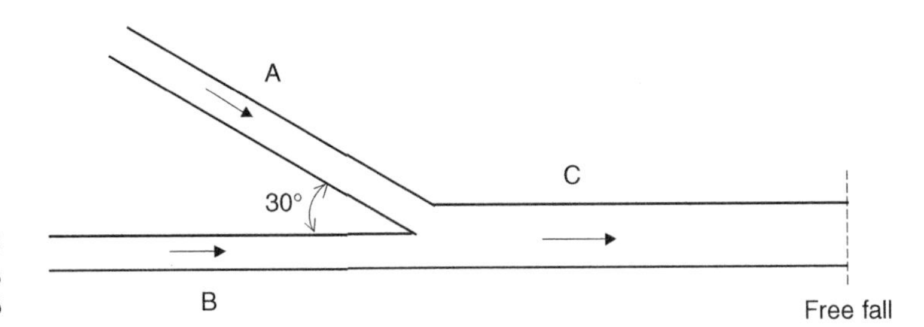

FIGURE P.4.3 Problems P.4.24–P.4.26

bottom width of 2.0 m, longitudinal slope of 0.015, and Manning roughness factor of 0.016, and it carries 9.0 m^3/s. Channel C is 1200 m long and it terminates at a free fall. It is rectangular, and has a bottom width of 4.2 m, longitudinal slope of 0.015, and a Manning roughness factor of 0.016. Determine the flow depth at the upstream end of channel C.

P.4.25 Redo Problem P.4.23 using a width of 8 ft for Channel C.

REFERENCES

Chow, V. T. (1959). *Open-Channel Hydraulics*. McGraw-Hill Book Co., New York, NY.
Henderson, F. M. (1966). *Open Channel Flow*. Prentice Hall, Upper Saddle River, NJ.
US Army Corps of Engineers (2002). HEC-RAS River Analysis System. *Hydraulic Reference Manual*, Hydrologic Engineering Center, Davis, CA.

5 Design of open channels

Open channels are designed to carry a design discharge in a safe and economical way. For flood control channels the design discharge represents the peak discharge expected to result from a flood event of a specified return period. Normally, the design discharge is obtained from the hydrologic study of upstream watersheds. For water distribution channels, however, such as those used in irrigation and water supply projects, the design discharge is determined on the basis of total delivery requirements. Open channels are usually designed for *uniform* or *normal* flow conditions.

Designing an open channel involves the selection of channel alignment, size and shape of the channel, longitudinal slope, and the type of lining material. Normally, we consider several hydraulically feasible alternatives, and compare them to determine the most cost-effective alternative. This chapter will emphasize the hydraulic considerations involved in channel design rather than economic analyses of different alternatives.

5.1 General Design Considerations

Selection of channel alignment is the first step in designing an open channel. Generally, the topography of the area, available width of right-of-way, and existing and planned adjacent structures and transportation facilities control the channel alignment. The topography also controls the invert elevations and bottom slope of the channel.

Most manmade surface channels are trapezoidal in cross-section, although triangular, parabolic and rectangular channels are also used. The primary concern in selecting a cross-sectional shape and size is the section's hydraulic capacity to accommodate the design discharge. There are, however, other factors to be considered. For instance, the depth of the channel may be limited due to a high water table in the underlying soil, or underlying bedrock. Also, large channel widths and mild side slopes will result in high costs of right-of-way and structures such as bridges. Small channel widths, on the other hand, may create construction difficulties. Likewise, steep side slopes can cause slope stability

TABLE 5.1 Steepest recommended side slopes for channels

Material	Side slope, m (run to rise ratio)
Rock	0–0.25
Earth with concrete lining	0.50
Stiff clay or earth	1.0
Soft clay	1.5
Loose sandy soil	2.0
Light sand, sandy loam	3.0

problems as well as high erosion rates in earthen channels. The steepest recommended side slopes for different types of channel materials are given in Table 5.1. This table is compiled from the information previously presented by Chow (1959), Chaudhry (1993), and Bankston and Baker (1995). In Table 5.1, m represents the run-to-rise ratio of the side slope. In other words, $m = 3$ means 3 horizontal over 1 vertical. If channel sides are to be mowed, slopes of $m = 3$ or milder ($m > 3$) are recommended.

For the most part, open channels are designed for subcritical flow. It is important to keep the Froude number sufficiently lower than the critical value of 1.0 under the design conditions. We must remember that the design discharge is only a single estimated value; the actual discharge occurring in a channel will vary possibly above and below the design discharge. Therefore, if the design Froude number is close to 1.0, there is a possibility that the actual flow might be fluctuating between subcritical and supercritical conditions. This fluctuation would be an unstable flow situation, and it should be avoided.

Channels are often lined to prevent the sides and the bottom of the channel from suffering erosion due to the shear stresses caused by the flow. The types of channel linings can be categorized into two broad groups: rigid and flexible. *Rigid lining* materials include cast-in-place concrete, cast-in-place asphaltic concrete, stone masonry, soil cement, and grouted riprap. Rigid linings can resist high shear stresses and provide a much higher conveyance capacity for the same cross-sectional size and channel slope than can a flexible lining. Where limited right-of-way is available, rigid linings may be the only alternative. They also reduce losses of water from the channel due to seepage. However, they are susceptible to failure from structural instability caused by freeze-thaw, swelling, and excessive soil pore pressures. When a rigid lining deteriorates, large broken slabs may be dislodged and displaced by the channel flow, resulting in significant erosion problems and slope and structure failures.

Flexible linings can be further classified into permanent and temporary linings. *Permanent flexible linings* include riprap, wire-enclosed riprap (although wire may corrode and break), vegetation lining, and gravel. *Temporary linings* are used for temporary protection against erosion until vegetation is established. Temporary linings include straw with net, curled wood mat, jute net, synthetic mat, and fiberglass roving. Flexible linings have several advantages compared to rigid linings. They are less susceptible to structural failure, because they can conform to the changes in the channel shape. They allow infiltration and exfiltration, and they provide habitat opportunities for local flora and fauna. The main disadvantage of flexible linings is that they can only sustain limited magnitudes of erosive forces. To accommodate the same design discharge safely, a channel section with a flexible lining would have to be considerably larger than a section lined with a rigid material. Therefore, flexible lining can lead to higher overall channel costs although the flexible lining materials are usually less expensive than the rigid lining materials in terms of construction costs.

Freeboard is the vertical distance between the top of the channel and the water surface that prevails under the design flow conditions. This distance should be sufficient to allow variations in the water surface due to wind-driven waves, tidal action, occurrence of flows exceeding the design discharge, and other causes. There are no universally accepted rules to determine a freeboard. In practice, freeboard selection is often a matter of judgment, or it is stipulated as part of the prevailing design standards. For preliminary estimates, the US Bureau of Reclamation (Chow, 1959) recommends that the unlined channel freeboard be computed as

$$F = \sqrt{Cy} \tag{5.1}$$

where F = freeboard, y = flow depth, and C = a coefficient. If F and y are in imperial units, C varies from 1.5 ft for a canal capacity of 20 cfs to 2.5 ft for a canal capacity of 3000 cfs or more. If metric units are used, with F and y in meters, C varies from 0.5 m for a flow capacity of 0.6 m^3/s to 0.76 m for a capacity of 85 m^3/s or more. Linear interpolation is acceptable to determine the intermediate values of C. For lined channels, the curves displayed in Figure 5.1 can be used to estimate the height of bank above water surface and the height of lining above water surface. This figure follows the US Bureau of Reclamation recommendations, and it is similar to figures presented previously by Chow (1959) and French (1985).

5.2 Design of Unlined Channels

The sides and bottoms of earthen channels are both erodible. The main criterion for earthen channel design is that the channel is not eroded under the design flow conditions. There are two approaches to erodible channel design, namely the maximum permissible velocity method and the tractive force method. Both are discussed in the following sections.

5.2.1 Maximum Permissible Velocity Method

This method is based on the assumption that a channel will not be eroded if the average cross-sectional velocity in the channel does not exceed the *maximum*

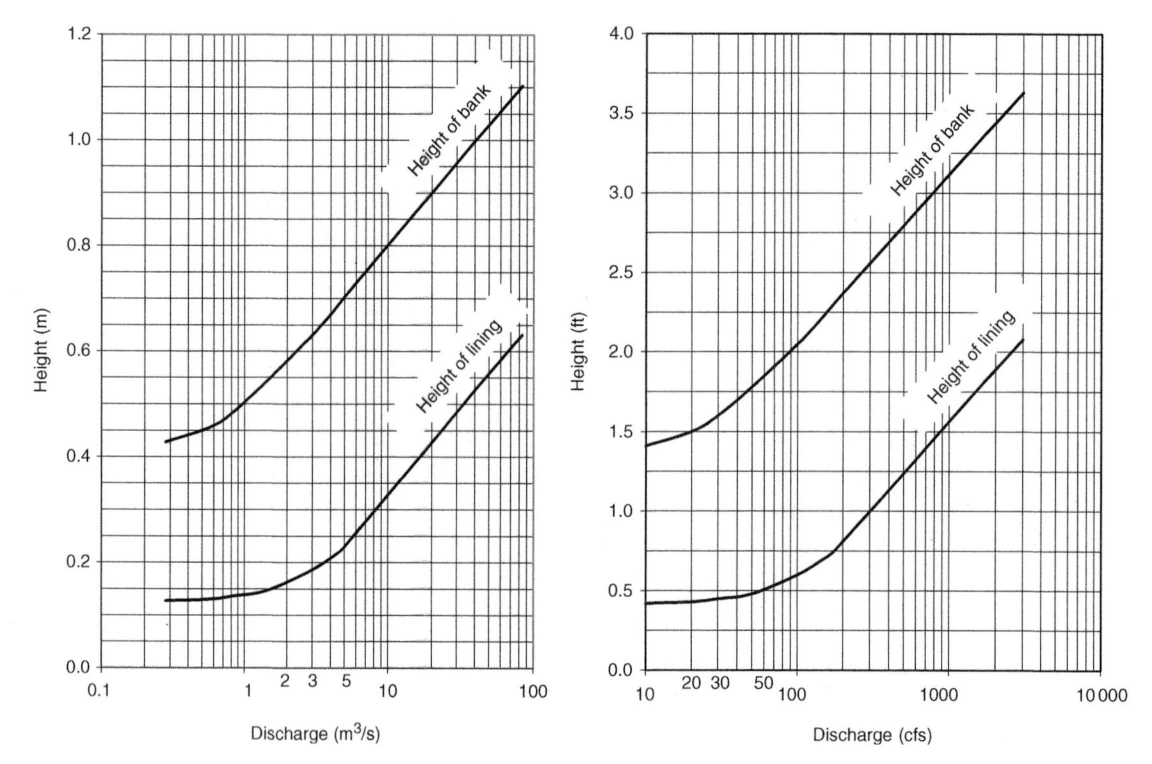

FIGURE 5.1 Suggested heights of lining and bank above water surface (source: Chow 1959 with permission Estate of Ven Te Chow)

permissible velocity. Therefore, a channel cross-section is designed so that, under the design flow conditions, the cross-sectional average velocity remains below the maximum permissible value. The magnitude of the maximum permissible velocity depends on the type of the material into which the channel is excavated, as well as the channel alignment. The maximum permissible velocities presented in Table 5.2 are adopted from the US Army Cops of Engineers (1991). These values are usually considered to be valid for straight channels having a flow depth of up to 3 ft or 1 m. Following Lane (1955), the values given in Table 5.2 can be reduced by 13% for moderately sinuous and 22% for very sinuous channels. Also, for flow depths exceeding 3 ft or 1 m, the maximum permissible velocities can be increased by about 0.50 fps or 0.15 m/s.

In a typical problem regarding sizing a channel section, the channel bottom slope, S_0, the design discharge, Q, and the type of the channel material would be known. The procedure to size the channel section consists of the following steps:

1. For the specified channel material, determine the Manning roughness factor n from Table 3.1, the side slope m from Table 5.1, and the maximum permissible velocity V_{MAX} from Table 5.2.
2. Compute the corresponding hydraulic radius, R, from the Manning formula, Equation 3.25, rearranged as

$$R = \left(\frac{n\ V_{MAX}}{k_n\ \sqrt{S_0}}\right)^{3/2} \tag{5.2}$$

TABLE 5.2 Suggested maximum permissible channel velocities (adapted from US Army Corps of Engineers, 1991)

Channel material	V_{MAX} (fps)	V_{MAX} (m/s)
Fine sand	2.0	0.6
Coarse sand	4.0	1.2
Fine gravel*	6.0	1.8
Sandy silt	2.0	0.6
Silt clay	3.5	1.0
Clay	6.0	1.8
Bermuda grass on sandy silt**	6.0	1.8
Bermuda grass on silt clay**	8.0	2.4
Kentucky bluegrass on sandy silt**	5.0	1.5
Kentucky bluegrass on silt clay**	7.0	2.1
Sedimentary rock	10.0	3.0
Soft sandstone	8.0	2.4
Soft shale	3.5	1.0
Igneous or hard metamorphic rock	20.0	6.0

*Applies to particles with d_{50} less than 0.75 in (20 mm).
**Velocities should be kept less than 5.0 fps (1.5 m/s) unless good cover and proper maintenance can be obtained. Slopes should be less than 5%.

where $k_n = 1.49\ \text{ft}^{1/3}/\text{s}$ for the conventional US unit system and $1.0\ \text{m}^{1/3}/\text{s}$ for the metric system.

3. Compute the required flow area as $A = Q/V_{MAX}$.
4. Compute the wetted perimeter as $P = A/R$.
5. Knowing the magnitudes of A and P and using the expressions for A and P given in Table 1.1, solve for the flow depth y and the bottom width b simultaneously.
6. Check the Froude number to ensure that it is not close to the critical value of 1.0.
7. Add a freeboard and modify the section for practical purposes.

Step (5) of this procedure requires the solution of two simultaneous equations. This can be facilitated for trapezoidal channels by using the following equations

$$W = \frac{Q}{R^2 V_{MAX}(2\sqrt{1+m^2} - m)} \tag{5.3}$$

$$y = \frac{RW}{2}\left(1 - \sqrt{1 - \frac{4}{W}}\right) \tag{5.4}$$

$$b = \frac{Q}{RV_{MAX}} - 2y\sqrt{1+m^2} \tag{5.5}$$

where W is an intermediate dimensionless parameter.

EXAMPLE 5.1 An unlined channel to be excavated in stiff clay will convey a discharge of $Q = 13.15\,\text{m}^3/\text{s}$ over a slope of $S_0 = 0.002$. Proportion the section dimensions using the maximum permissible velocity method.

Using Table 5.1 as a guide, we pick $m = 1.5$ (milder than the steepest recommended slope) for stiff clay, and from Table 3.1 we obtain $n = 0.020$. From Table 5.2, $V_{MAX} = 1.8\,\text{m/s}$. Using Equation 5.2 with $k_n = 1.00$,

$$R = \left[\frac{0.020(1.8)}{1.00\sqrt{0.002}}\right]^{3/2} = 0.72\,\text{m}$$

Also, $A = Q/V_{MAX} = 13.15/1.8 = 7.32\,\text{m}^2$. Hence $P = A/R = 7.32/0.72 = 10.17\,\text{m}$. Now, from expressions given in Table 1.1 and using $m = 1.5$,

$$A = (b + my)y = (b + 1.5y) = 7.32\,\text{m}^2$$

and

$$P = b + 2y\sqrt{1 + m^2} = b + 3.61y = 10.17\,\text{m}$$

We now have two equations with two unknowns – y and b. From the second equation, $b = 10.17 - 3.61y$. Substituting this into the first equation and simplifying,

$$2.11y^2 - 10.17y + 7.32 = 0$$

This equation has two roots: $y = 0.88\,\text{m}$ and $3.94\,\text{m}$. The first root results in a channel width of $b = 10.17 - 3.61(0.88) = 7.00\,\text{m}$; the second results in $b = 10.17 - 3.61(3.94) = -4.05\,\text{m}$. Obviously a negative channel width has no physical meaning, therefore $y = 0.88\,\text{m}$ will be used. Also note that there is no need to modify the value of V_{MAX} picked from Table 5.2, because $y = 0.88\,\text{m} < 1.0\,\text{m}$.

Alternatively, we could use Equations 5.3 to 5.5 to obtain y and b as

$$W = \frac{13.15}{(0.72)^2(1.8)(2\sqrt{1 + 1.5^2} - 1.5)} = 6.70$$

$$y = \frac{(0.72)(6.60)}{2}\left(1 - \sqrt{1 - \frac{4}{6.70}}\right) = 0.88\,\text{m}$$

$$b = \frac{13.15}{(0.72)(1.8)} - 2(0.88)\sqrt{1 + 1.5^2} = 7.00\,\text{m}$$

Next we will check whether the Froude number is close to the critical value of 1.0. From the expression given for the top width, T, in Table 1.1,

$$T = b + 2my = 7.0 + 2(1.5)0.88 = 9.64\,\text{m}$$

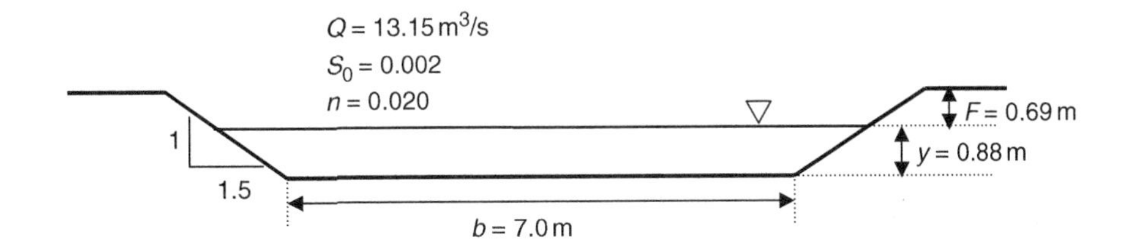

FIGURE 5.2 Channel section proportioned in Example 5.1

Then the hydraulic depth becomes $D = A/T = 7.32/9.64 = 0.76$ m, and

$$F_r = \frac{V}{\sqrt{gD}} = \frac{1.8}{\sqrt{9.81(0.76)}} = 0.66$$

This value indicates that under the design flow conditions the flow will not be near the critical state.

Finally we will determine a freeboard using Equation 5.1. It is known that C varies from 0.5 m for a channel capacity of 0.6 m^3/s to 0.76 m for a capacity of 85 m^3/s. Assuming this variation is linear, we determine C as being 0.54 m for $Q = 13.15$ m^3/s by interpolation. Then,

$$F = \sqrt{0.54(0.88)} = 0.69 \text{ m}$$

The total depth for the channel is $(0.88 + 0.69) = 1.57$ m. Then the width of the channel at freeboard is $b + 2m(y + F) = 7.00 + 2(1.5)(1.57) = 11.70$ m. The results of this design are summarized in Figure 5.2.

5.2.2 TRACTIVE FORCE METHOD

The forces acting on the soil particles comprising the channel bottom and sides are considered in this method. Flow in a channel exerts *tractive forces* (or *shear forces*) on the channel bed that are equal in magnitude but opposite in direction to the friction forces exerted by the channel bed on the flow. The tractive forces tend to move the particles on the channel bed in the flow direction. Erosion will occur if the tractive forces exceed the resistive forces preventing the movement of these particles. When we design an earthen channel, we proportion the channel section so that the particles will not move under the design flow conditions.

Assuming that the channel bottom is nearly level, the flow-induced tractive forces are the only forces tending to move the soil particles lying on the channel bottom. The flow exerts tractive forces on the sides of the channel as well. In addition, the particles on the sides of the channel tend to roll down the slope due to the effect of gravity. Therefore, the forces tending to move the particles on the sides of a channel are the resultant of the flow-induced tractive forces and the gravitational forces acting on the soil particles. For cohesive soils, however, the gravitational forces are much smaller than the cohesive forces keeping the soil particles together.

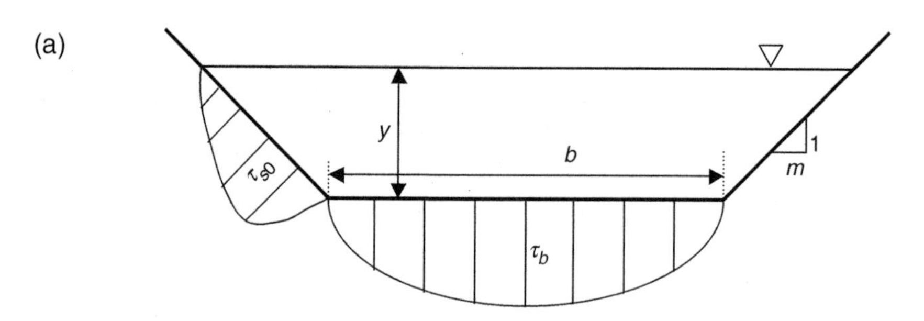

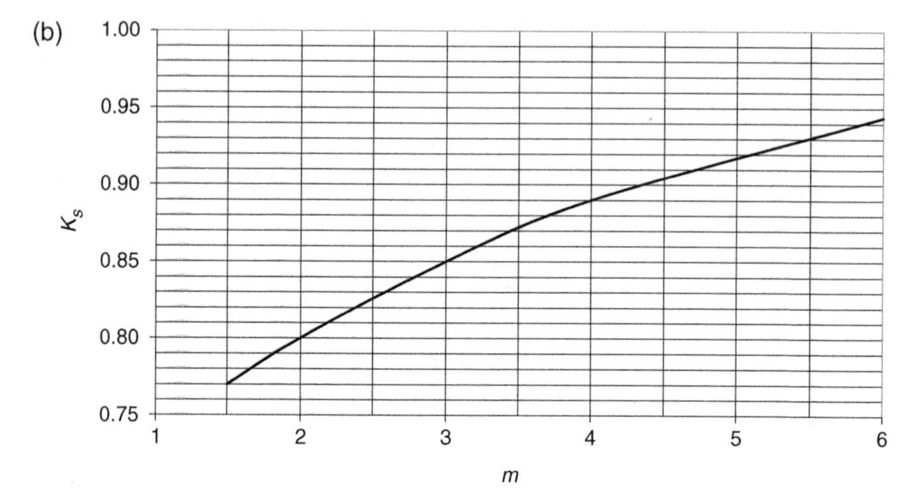

FIGURE 5.3 Shear stress distribution on channel bottom and sides

For design purposes, the forces acting on unit areas on the channel bottom and sides are considered rather than individual soil particles. For normal flow, the flow-induced average *unit tractive force*, or the average tractive force per unit area over the channel perimeter, is equal to $\gamma R S_0$ where γ = specific weight of water, R = hydraulic radius, and S_0 = bottom slope of the channel. However, the distribution of the unit tractive force over the channel perimeter is non-uniform, as shown in Figure 5.3.

Defining τ_b = maximum unit tractive force on the channel bottom, and τ_{s0} = maximum unit tractive force on the sides, following Lane (1955), we can express τ_b and τ_{s0} in terms of the flow depth as

$$\tau_b = K_b \gamma y S_0 \tag{5.6}$$

and

$$\tau_{s0} = K_s \gamma y S_0 \tag{5.7}$$

The dimensionless coefficients K_b and K_s depend on the side slope, m, and the bottom width to depth ratio, b/y. The largest values of K_b are near but below unity (Chaudhry, 1993). Therefore, we use $K_b = 1.0$ for simplicity. Figure 5.3 presents the suggested values of K_s for $1 < (b/y) < 6$ as a function of the side slope, m. Information presented by Lane (1955) and Anderson *et al.* (1970) was utilized in constructing Figure 5.3.

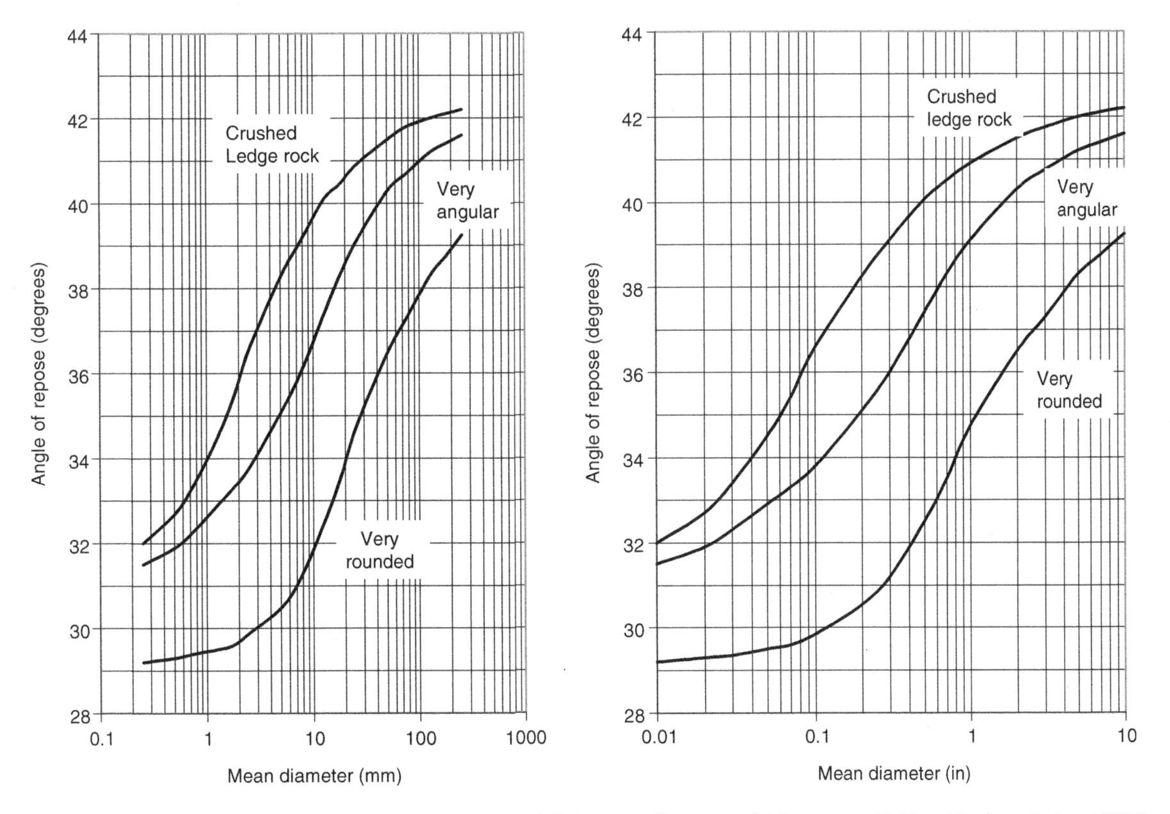

FIGURE 5.4 Angle of repose for non-cohesive material (source: Simon and Albertson, 1960, with permission ASCE)

Equation 5.7 expresses the maximum flow-induced tractive force on the sides of a trapezoidal channel. However, as mentioned previously, the soil particles on the channel sides also tend to roll down the slope due to the gravitational forces. The maximum unit force (force per unit area) tending to move the particles due to the flow-induced tractive force and the gravitational forces combined can be expressed $\tau_s = \tau_{s0}/K$, or

$$\tau_s = \frac{K_s \gamma y S_0}{K} \tag{5.8}$$

where τ_s = maximum unit force tending to move the particles, and K = tractive force ratio = a dimensionless parameter reflecting the tendency of soil particles to roll down the side slopes due to gravity. For cohesive soils, $K = 1.0$ – that is, the effect of the gravitational forces is negligible. For cohesionless (or non-cohesive) soils,

$$K = \sqrt{1 - \frac{1}{(1 + m^2)\sin^2 \alpha_R}} \tag{5.9}$$

where m = side slope of the channel, and α_R = angle of repose of the cohesionless channel material. The derivation of Equation 5.9 can be found elsewhere (Chow, 1959) and is omitted here for brevity. Figure 5.4 can be used to determine the angles of repose for non-cohesive soils. This figure was constructed

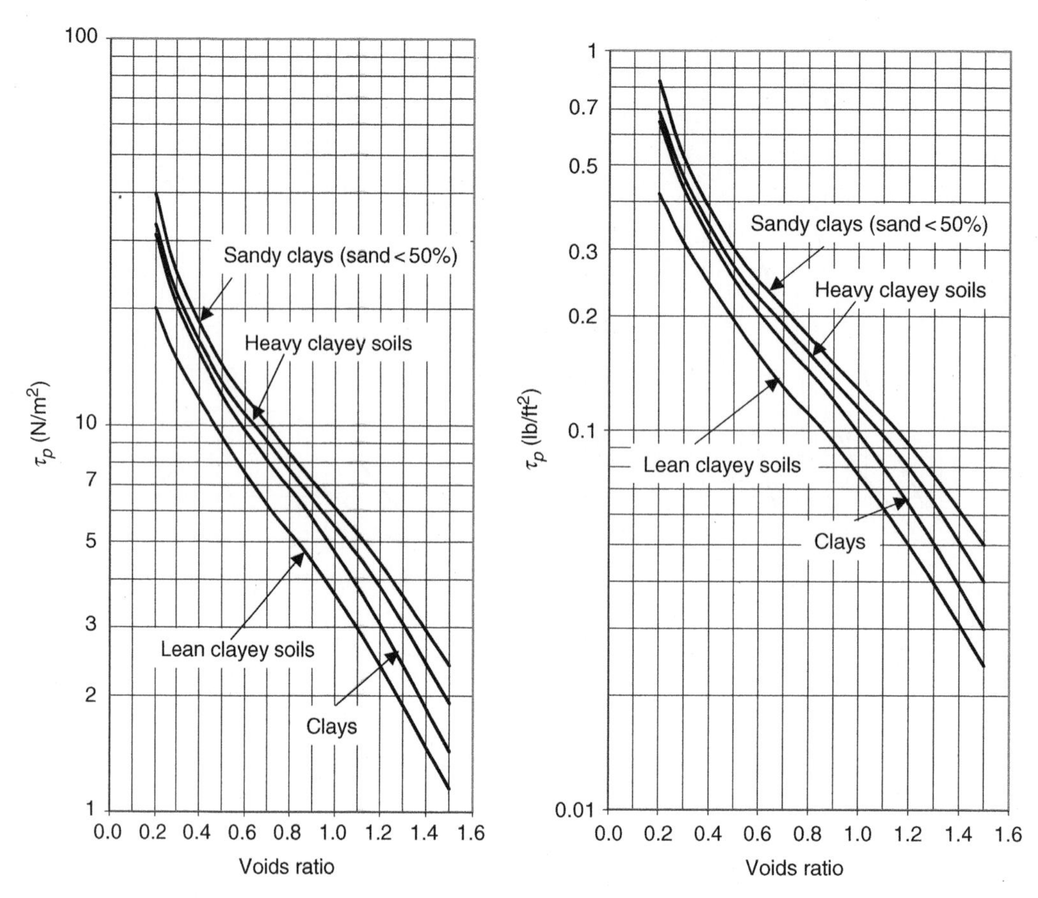

FIGURE 5.5 Permissible unit tractive force for cohesive soils as a function of voids ratio (source: Chow, 1959, with permission Estate of Ven Te Chow)

based on the information reported by Simon and Albertson (1960), and it represents only the average values. The average values should be used cautiously, since experiments show that significant deviations from the average values can occur.

The *permissible unit tractive force*, τ_p, is the maximum unit tractive force (combining the flow-induced shear force and gravitational forces acting on soil particles) that will not cause erosion. This can also be interpreted as the resistive force per unit area opposing the movement of soil particles. If τ_b exceeds the permissible tractive force, the channel bottom will be eroded. Likewise, if τ_s exceeds the permissible unit tractive force, the sides will erode. For cohesive soils, the voids ratio can be used to determine the permissible unit tractive force as shown in Figure 5.5. This figure was constructed using the information from a similar figure presented by Chow (1959). Alternatively, as suggested by Smerdon and Beaseley (Chen and Cotton, 1988), the plasticity index can be used to determine the permissible unit tractive force as shown in Figure 5.6. For non-cohesive soils, as suggested by Thibodeaux (Chen and Cotton, 1988), the permissible tractive force is a function of the mean diameter of the channel material as shown in Figure 5.7.

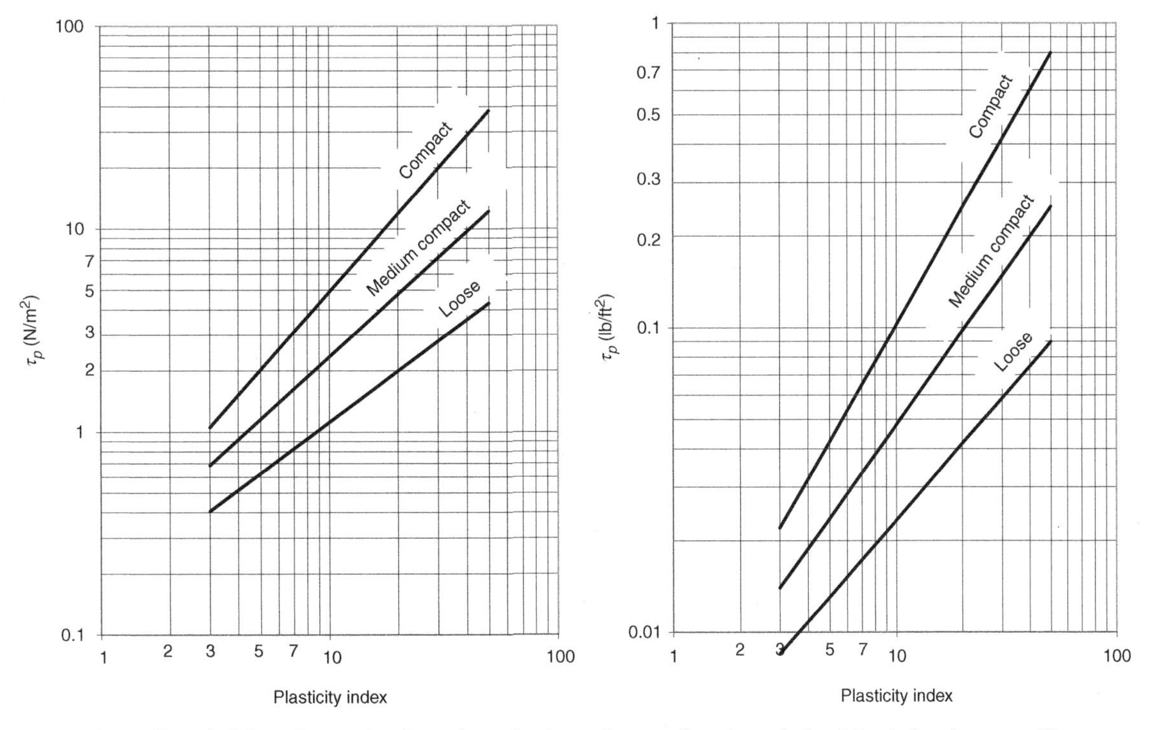

FIGURE 5.6 Permissible unit tractive force for cohesive soils as a function of plasticity index (source: Chen and Cotton, 1988)

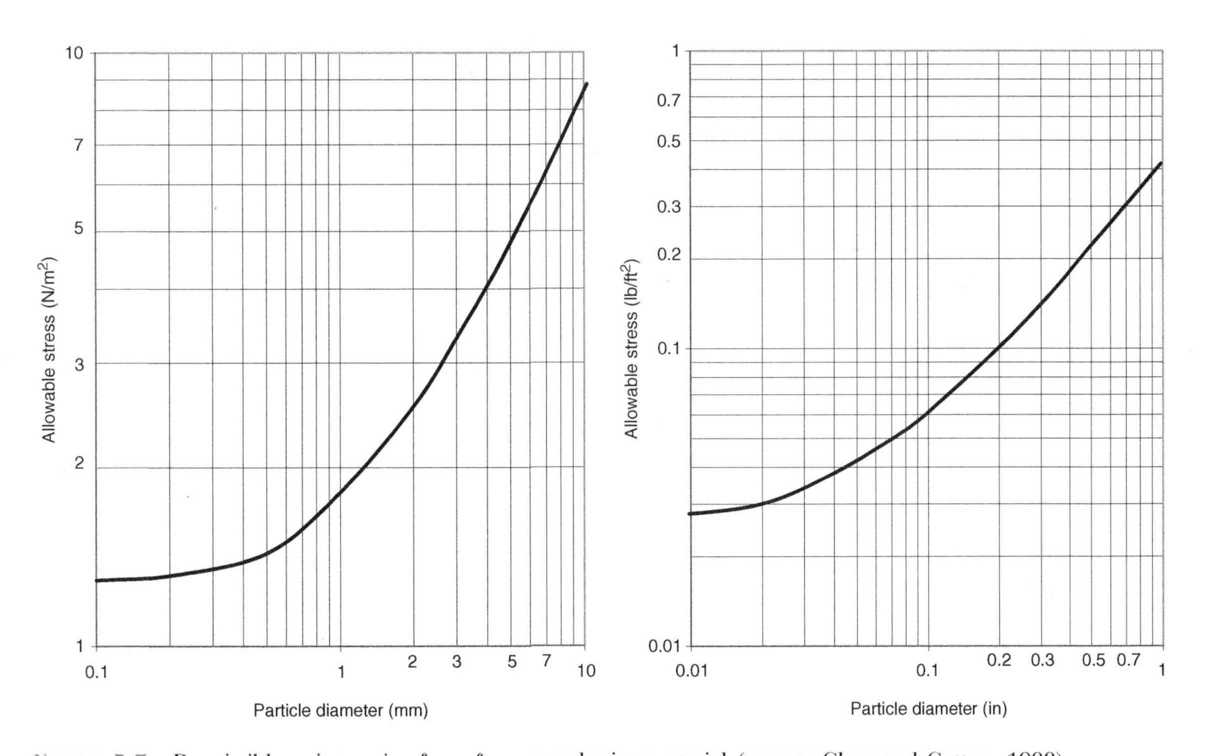

FIGURE 5.7 Permissible unit tractive force for non-cohesive material (source: Chen and Cotton, 1988)

TABLE 5.3 Reduction factors for sinuous channels

Degree of sinuousness	C_p
Straight	1.0
Slightly sinuous	0.90
Moderately sinuous	0.75
Very sinuous	0.60

The values of the permissible unit tractive force obtained from Figures 5.5 through 5.7 are for straight channels. For sinuous channels, these values should be multiplied by a reduction factor, C_p. Table 5.3 lists the reduction factors suggested by Lane (1955).

In the tractive force method, a channel cross-section is dimensioned so that neither the channel bottom nor the sides will be eroded under the design conditions. For cohesive soils the channel bottom is usually critical, whereas for non-cohesive soils the sides usually govern the design.

Denoting the permissible unit tractive force obtained from Figures 5.5, 5.6 or 5.7 by τ_p, channels in cohesive soils will be designed using $\tau_b \leq C_p\tau_p$, or

$$K_b\gamma y S_0 \leq C_p\tau_p \tag{5.10}$$

Therefore, the limiting flow depth becomes

$$y_{LIM} = \frac{C_p\tau_p}{K_b\gamma S_0} \tag{5.11}$$

For non-cohesive soils, the design is based on $\tau_s \leq C_p\tau_p$, or

$$\frac{K_s\gamma y S_0}{K} \leq C_p\tau_p \tag{5.12}$$

In this case, the limiting depth is

$$y_{LIM} = \frac{KC_p\tau_p}{K_s\gamma S_0} \tag{5.13}$$

The procedure to size a channel section for cohesive soils consists of the following steps, noting that the design discharge Q and the bottom slope S_0 are given:

1. For the channel material specified, select a Manning roughness factor, n, from Table 3.1; the side slope, m, from Table 5.1; and the straight channel permissible unit tractive force, τ_p, from Figure 5.5 or 5.6. Select a C_p value from Table 5.3, based on the sinuousness of the channel.

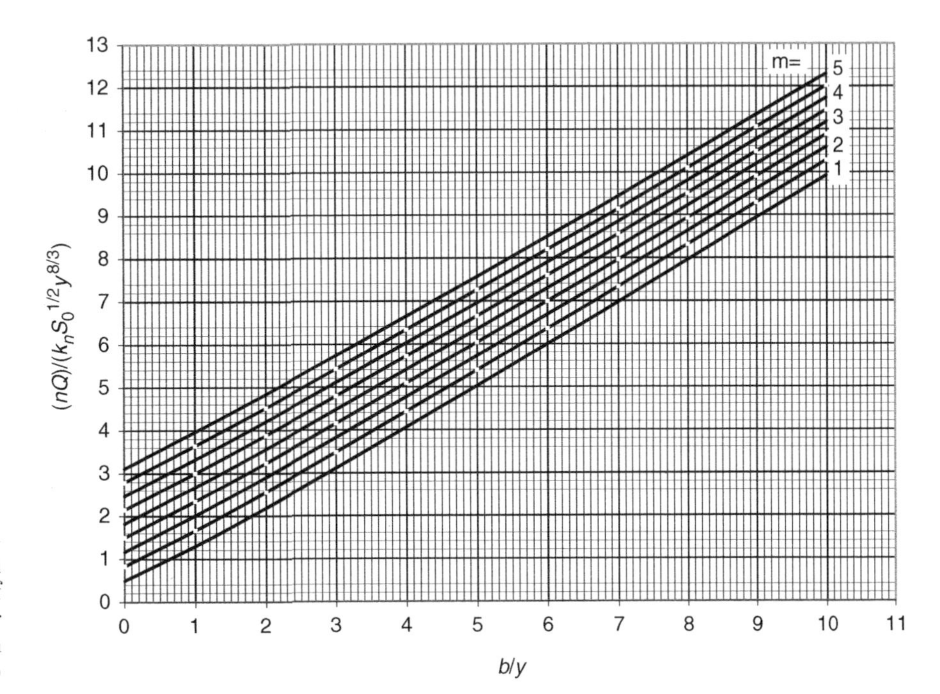

FIGURE 5.8 Graphical representation of Equation 5.14 (after Akan, 2001, with permission NKC)

2. Determine the limiting flow depth y_{LIM} by using Equation 5.11. Select a flow depth, y, equal to or less than y_{LIM}.
3. Determine the channel bottom width, b, by using Equation 5.14:

$$Q = \frac{k_n\sqrt{S_0}}{n}\frac{A^{5/3}}{P^{2/3}} = \frac{k_n\sqrt{S_0}}{n}\frac{[(b+my)y]^{5/3}}{\left(b+2y\sqrt{1+m^2}\right)^{2/3}} \tag{5.14}$$

However, solving this equation will require trial and error, since the equation is implicit in b. Also, while Equation 5.14 provides a mathematically exact value for b, we usually modify this value for practicality. It is therefore easier to use either Equation 5.15 or Figure 5.8 (Akan, 2001) to obtain an approximate value for b, and then modify it (increase it to a round number) for practicality.

$$b = 1.186y\left[\frac{nQ}{k_nS_0^{1/2}y^{8/3}} - \frac{m^{5/3}}{(2\sqrt{1+m^2})^{2/3}}\right]^{0.955} \tag{5.15}$$

4. Calculate the normal flow depth corresponding to the 'practical' channel width, then calculate the Froude number and verify that it is not close to the critical value of 1.0.
5. Determine K_s from Figure 5.3, and check the stability of the channel sides by ensuring that $(K_s\gamma yS_0)/K < C_p\tau_p$. Also, confirm the stability of the channel bottom, verifying that $(K_b\gamma yS_0) < C_p\tau_p$.
6. Determine a freeboard by using Equation 5.1.

For non-cohesive soils the procedure is similar, except that the forces on the channel sides govern the design. Given the design discharge Q and the bottom slope S_0, we can proceed as follows:

1. For the channel material specified, select a Manning roughness factor, n, from Table 3.1; side slope, m, from Table 5.1; the angle of repose, α_R, from Figure 5.4; K_s from Figure 5.3; and the straight channel permissible unit tractive force, τ_p, from Figure 5.7. Select a C_p value based on the sinuousness of the channel, using Table 5.3.
2. Determine the limiting flow depth y_{LIM} by using Equation 5.13. Select a flow depth, y, equal to or less than y_{LIM}.
3. Determine an approximate bottom width, b, by using Equation 5.15 or Figure 5.8. Modify this approximate width for practicality by increasing its magnitude to a round figure.
4. Calculate the normal flow depth corresponding to the 'practical' channel width, then calculate the Froude number and verify that it is not close to the critical value of 1.0.
5. Check the stability of the channel bottom by verifying that $(K_b\gamma y S_0) < C_p\tau_p$, and the stability of the channel sides by verifying that $(K_s\gamma y S_0)/K < C_p\tau_p$.
6. Determine a freeboard by using Equation 5.1.

EXAMPLE 5.2 A moderately sinuous channel will be excavated into stiff clay having a void ratio of 0.3. The channel will have a bottom slope of $S_0 = 0.0016$, and it will convey $Q = 9.5\ \text{m}^3/\text{s}$. Proportion the channel section.

By using Tables 3.1 and 5.1 as guides, we pick $n = 0.020$ and $m = 1.5$ (milder than the steepest recommended slope). Likewise, by using Figure 5.5 for a voids ratio of 0.3, we obtain $\tau_p = 20\ \text{N/m}^2$. Because the channel is moderately sinuous, $C_p = 0.75$ from Table 5.3. Now, using Equation 5.11 with $K_b = 1.0$,

$$y_{LIM} = \frac{C_p\tau_p}{K_b\gamma S_0} = \frac{(0.75)(20)}{(1)(9800)(0.0016)} = 0.96\ \text{m}$$

Let us pick $y = 0.96$ m and use Equation 5.15 to find an approximate b as

$$b = 1.186(0.96)\left[\frac{(0.020)(9.5)}{(1.0)(0.0016)^{1/2}(0.96)^{8/3}} - \frac{1.5^{5/3}}{(2\sqrt{1+1.5^2})^{2/3}}\right]^{0.955} = 4.75\ \text{m}$$

We could obtain a similar result by using Figure 5.8. Let us first evaluate the dimensionless parameter:

$$\frac{nQ}{k_n S_0^{1/2} y^{8/3}} = \frac{(0.02)(9.5)}{(1.0)(0.0016)^{1/2}(0.96)^{8/3}} = 5.30$$

With this value and $m = 1.5$, we obtain $b/y = 4.95$ from Figure 5.8. Therefore, $b = (0.96)(4.95) = 4.75$ m. For practicality, let us choose $b = 5.0$ m. Now, by using the methods discussed in Chapter 3, we can calculate the corresponding normal depth as being 0.93 m.

For $y = 0.93$ m, $b = 5.0$ m, and $m = 1.5$, the flow area becomes $A = (b + my)y = [5.0 + 1.5(0.93)]0.93 = 5.95\ \text{m}^2$, and the top width is $T = b + 2my = 5.0 + 2(1.5)0.93 = 7.79$ m. Therefore, $V = Q/A = 9.5/5.95 = 1.60$ m/s, $D = A/T = 5.95/7.79 = 0.76$ m, and $F_r = V/\sqrt{gD} = 1.60/\sqrt{(9.81)(0.76)} = 0.59$. This value is sufficiently below the critical value of 1.0.

To check the channel sides, we obtain $K_s = 0.77$ from Figure 5.3b for $m = 1.5$. Then $\tau_s = K_s \gamma y S_0/K = 0.77(9800)(0.93)(0.0016)/1.0 = 11.22\ \text{N/m}^2$, which is less than the allowable value of $C_p \tau_p = (0.75)(20) = 15\ \text{N/m}^2$, so the sides will not erode. Likewise, for the channel bottom, $K_b \gamma y S_0 = 1.0(9800)(0.93)(0.0016) = 14.58\ \text{N/m}^2$, which is less than $15\ \text{N/m}^2$. Thus the channel bottom is also stable.

Finally, from Equation. 5.1, with an interpolated value of $C = 0.53$ m, the freeboard is obtained as $F = \sqrt{0.53(0.93)} = 0.70$ m.

EXAMPLE 5.3 A straight trapezoidal channel will be excavated into cohesionless earth containing fine gravel having an average particle size of 0.3 inches. The particles are very rounded. The bottom slope is 0.0009, and the design discharge is 120 ft^3/s. Using $n = 0.020$ and $m = 3.0$, proportion the channel section.

The soil is cohesionless, and the sides will govern the design. From Figure 5.4, $\alpha_R = 31°$, and from Figure 5.7, $\tau_p = 0.14\ \text{lb/ft}^2$. Because the channel is straight, we use $C_p = 1.0$. Also, by using Figure 5.3 with $m = 3$, we obtain $K_s = 0.85$. From Equation 5.9,

$$K = \sqrt{1 - \frac{1}{(1 + m^2)\sin^2\alpha}} = \sqrt{1 - \frac{1}{(1 + 3^2)\sin^2 31°}} = 0.79$$

The limiting depth, y_{LIM}, is obtained by using Equation 5.13 as

$$y_{LIM} = \frac{KC_p\tau_p}{K_s\gamma S_0} = \frac{(0.79)(1.0)(0.14)}{(0.85)(62.4)(0.0009)} = 2.32\ \text{ft}$$

Let us pick $y = 2.30$ ft and use Equation 5.15 to obtain an approximate value of b as

$$b = 1.186(2.30)\left[\frac{(0.020)(120)}{(1.49)(0.0009)^{1/2}(2.30)^{8/3}} - \frac{3.0^{5/3}}{(2\sqrt{1+3.0^2})^{2/3}}\right]^{0.955} = 10.25\ \text{ft}$$

Alternatively, we could use Figure 5.8 to determine b. We would first evaluate the dimensionless term $(nQ)/(k_n S_0{}^{1/2} y^{8/3}) = (0.020)(120)/[1.49 (0.0009)^{1/2}(2.30)^{8/3}] = 5.82$. Then, with this value and $m = 3$, from Figure 5.8 we obtain $b/y = 4.45$. Thus $b = 4.45(2.30) = 10.24$ ft. For practicality, let us choose $b = 10.50$ ft. By using the methods discussed in Chapter 3, we can now obtain a normal depth of $y = 2.28$ ft.

For $y = 2.28$ ft, $b = 10.50$ ft, and $m = 3$, the flow area becomes $A = (b + my)y = [10.50 + 3(2.28)]2.28 = 39.53\ \text{ft}^2$, and the top width is $T = b + 2my = 10.50 + 2(3)2.28 = 24.18$ ft. Therefore, $V = Q/A = 120/39.53 = 3.04$ fps, $D = A/T = 39.53/24.18 = 1.64$ ft, and $F_r = V\sqrt{gD} = 3.04/\sqrt{(32.2)(1.64)} = 0.42$. This value is sufficiently below the critical value of 1.0.

We will now check the stability of the channel bottom. The permissible unit tractive force is $C_p \tau_p = 1.0(0.14) = 0.14\ \text{lb/ft}^2$. The maximum unit tractive force on the channel bottom is $\tau_b = K_b \gamma y S_0 = (1.0)(62.4)(2.28)0.0009 = 0.13\ \text{lb/ft}^2$, which is smaller than the permissible value. Thus the channel bottom is stable. Likewise, for the sides $K_s \gamma y S_0/K = (0.85)(62.4)(2.28)(0.0009)/0.79 = 0.14\ \text{lbs/ft}^2$, which does not exceed the permissible value. Thus, the sides are also stable.

Finally, by using Equation 5.1 with $C = 1.53$ ft, we obtain a freeboard $F = 1.87$ ft.

5.2.3 CHANNEL BENDS

Centrifugal forces result in a superelevation of the water surface where open channel flow occurs around a bend. A higher water surface occurs at the outside of the bend than at the inside of the bend. We can estimate the difference by using

$$\Delta y = \frac{V^2 T}{g R_c} \tag{5.16}$$

where Δy = difference in water surface elevation at the outside and inside the bend, V = average cross-sectional velocity, T = flow width at the water surface, g = gravitational acceleration, and R_c = mean radius of the bend as shown in Figure 5.9. We must ensure that the freeboard at a channel bend is adequate to contain the raised water surface.

Flow around a bend also causes higher shear stresses on the channel bottom and sides due to secondary currents. This increase can be accounted for by introducing a correction factor, C_p, as discussed in Section 5.2.2 for sinuous channels. However, many channels are formed of generally straight reaches and some bends. We can use $C_p = 1.0$ to design such channels and protect the bends and some distance downstream by lining. The protection length, L_p, downstream of the bend shown in Figure 5.9 may be estimated using

$$L_p = K_p \frac{R^{7/6}}{n_b} \tag{5.17}$$

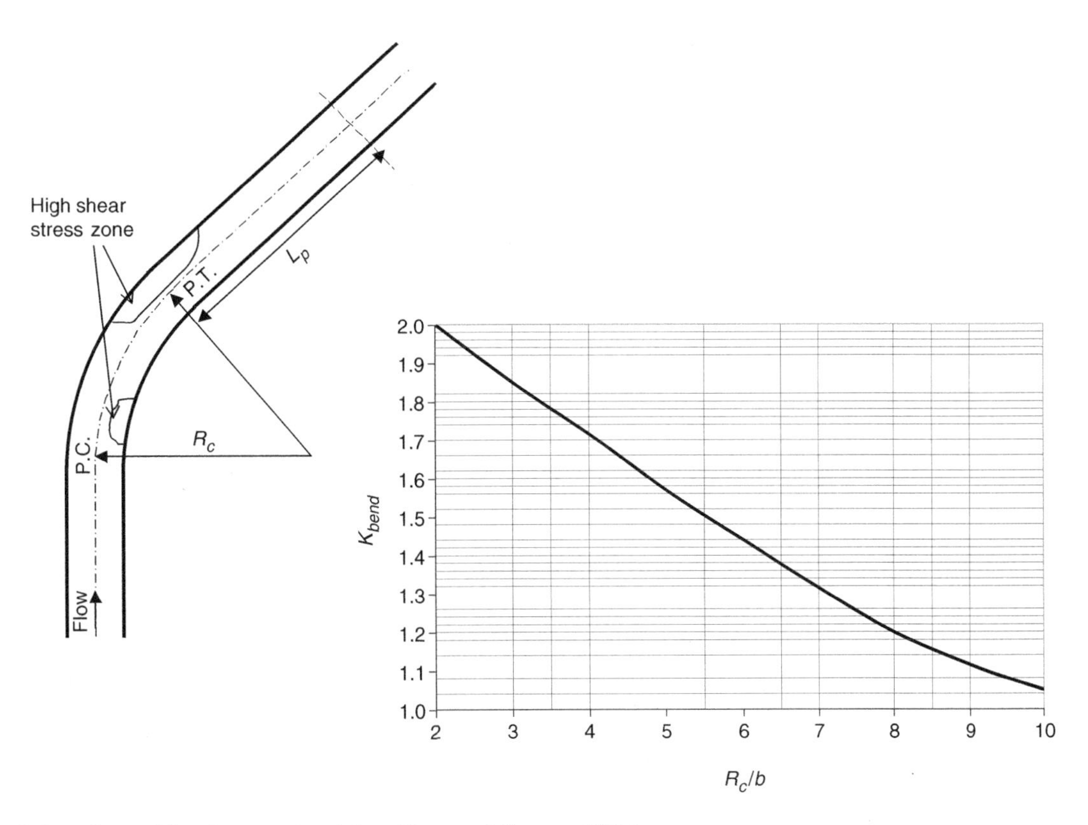

FIGURE 5.9 Channel bend protection (after Chen and Cotton, 1988)

where R = hydraulic radius; n_b = Manning roughness factor at the bend, and $K_p = 0.604\,\text{ft}^{-1/6}$ for US customary units or $0.736\,\text{m}^{-1/6}$ for metric units.

The increased shear stresses at the bend can be determined using

$$\tau_{bs} = K_{bend}\,\tau_s \tag{5.18}$$

and

$$\tau_{bb} = K_{bend}\,\tau_b \tag{5.19}$$

where τ_{bs} = side shear stress at a channel bend, τ_{bb} = bottom shear stress at a bend, τ_s and τ_b = the side and bottom shear stresses at an equivalent straight section, respectively, and K_{bend} = a dimensionless factor to be obtained from Figure 5.9. An unlined channel design must be checked to ensure that $\tau_p > \tau_{bs}$ and $\tau_p > \tau_{bb}$ where τ_p = permissible unit tractive force. If the bend shear stresses exceed the permissible unit tractive force, the bend and the protection length, L_p, downstream must be lined to avoid scouring of the channel.

EXAMPLE 5.4 Consider the trapezoidal channel designed in Example 5.3. The alignment of this channel will now include a bend that has a radius of $R_c = 50$ ft. Determine how the design needs to be modified at the bend.

From Example 5.3, $Q=120\,\text{ft}^3/\text{s}$, $b=10.50\,\text{ft}$, $m=3$, $y=2.28\,\text{ft}$, $A=39.53\,\text{ft}^2$, $\tau_p=0.14\,\text{lb/ft}^2$, $\tau_s=0.14\,\text{lb/ft}^2$, $\tau_b=0.13\,\text{lb/ft}^2$, $V=3.04\,\text{fps}$, $T=24.18\,\text{ft}$, and $F=1.87\,\text{ft}$. Using the expression given for wetted perimeter, P, in Table 1.1:

$$P = b + 2y\sqrt{1+m^2} = 10.50 + 2(2.28)\sqrt{1+3^2} = 24.92\,\text{ft}$$

and the hydraulic radius becomes $R=A/P=39.53/24.92=1.59\,\text{ft}$. Let us now check the superelevation due to the bend. From Equation 5.9,

$$\Delta y = \frac{V^2 T}{g R_c} = \frac{(3.04)^2(24.18)}{(32.2)(50)} = 0.14\,\text{ft}$$

This value is well below the freeboard of $F=1.87\,\text{ft}$ determined in Example 5.3, and therefore there is no need to increase the freeboard.

Next, the increased shear stresses due to the bend will be considered. For $R_c=50\,\text{ft}$ and $b=10.50\,\text{ft}$, $R_c/b=4.76$; and from Figure 5.9, $K_{bend}=1.58$. Then, from Equations 5.18 and 5.19, respectively,

$$\tau_{bs} = K_{bend}\,\tau_s = (1.58)(0.14) = 0.22\,\text{lb/ft}^2$$

and

$$\tau_{bb} = K_{bend}\,\tau_b = (1.58)(0.13) = 0.21\,\text{lb/ft}^2$$

Both the bottom and side tractive forces at the bend exceed the permissible unit tractive force of $\tau_p=0.14\,\text{lb/ft}^2$, therefore lining of the channel at the bend and along the protection length downstream will be required. Suppose the lining material is unfinished concrete that has $n_b=0.016$. Then, from Equation 5.17 with $K_p=0.604\,\text{ft}^{-1/6}$,

$$L_p = K_p\frac{R^{7/6}}{n_b} = 0.604\frac{(1.59)^{7/6}}{0.016} = 64.85\,\text{ft}$$

5.3 Design of Channels with Flexible Linings

The basic design principles for channels with most flexible linings are the same as those for unlined channels. However, the design procedures differ, particularly for grass-lined channels, for which the Manning roughness factor varies with the flow depth and the condition of the grass cover. Either the maximum permissible velocity or the tractive force approach may be used. However, the tractive force approach is discussed herein because it is more physically based. The channel bends will be treated using the same procedures as for the unlined channels.

5.3.1 DESIGN OF CHANNELS LINED WITH VEGETAL COVER

The basic criterion is that the shear forces exerted by the flow will not exceed the permissible unit tractive force. For channels lined with vegetal cover, the shear stresses on the channel bottom are more critical than those on the sides. The bottom shear force can be calculated using Equation 5.6, repeated here as

$$\tau_b = K_b \gamma y S_0 \tag{5.20}$$

where $K_b = 1.0$, τ_b = bottom shear stress caused by the flow, γ = specific weight of water, y = flow depth, and S_0 = bottom slope. The channel must be sized so that $\tau_b \leq C_p \tau_p$ where τ_p = permissible unit tractive force for straight channels and C_p = reduction factor for sinuousness.

The grass (or vegetal) covers are grouped into five classes (A, B, C, D, and E) regarding the degree of retardance, as shown in Table 3.3 in Chapter 3. The permissible unit tractive force, τ_p, depends on the retardance class. The values of τ_p for different classes are given in Table 5.4.

The Manning roughness factor also depends on the retardance class, the hydraulic radius, R, and the channel slope S_0. As reported by Chen and Cotton (1988), the Manning roughness factor, n, can be expressed as

$$n = \frac{(RK_v)^{1/6}}{C_n + 19.97 \log[(RK_v)^{1.4} S_0^{0.4}]} \tag{5.21}$$

where $K_v = 3.28\,\text{m}^{-1} = 1.0\,\text{ft}^{-1}$. The value of C_n depends on the retardance class as shown in Table 3.3.

It should be clear from Table 3.3 that a given type of vegetal cover can belong to different retardance classes depending on the season of the year and the height to which it is cut. For example, Bermuda grass belongs to retardance class B and has a relatively high retardance when it is about 12 inches tall with good stand. However, the same kind of grass will belong to class E and will have a very low retardance if it is cut to a 2.5-inch height. A channel lined with vegetation must be designed to function satisfactorily for all the retardance classes to which the selected vegetal cover may belong.

TABLE 5.4 Permissible unit tractive force for vegetal lining materials (Chen and Cotton1988)

	Permissible unit tractive force	
Lining type	**(lb/ft^2)**	**(N/m^2)**
Retardance class A	3.70	177.2
Retardance class B	2.10	100.5
Retardance class C	1.00	47.9
Retardance class D	0.60	28.7
Retardance class E	0.35	16.7

We should note that the lower retardance classes, such as D and E, are critical from the viewpoint of channel erosion. The high retardance classes, such as A and B, on the other hand, are critical from the viewpoint of channel conveyance. Therefore, it is logical to complete the design of a vegetal-lined channel in two stages. For example, the lowest and the highest retardance classes for the Bermuda grass are E and B, respectively. In the first phase of design, we size the channel for stability – that is, we determine the cross-sectional dimensions so that the class E permissible stress will not be exceeded when the design flow occurs. In the second phase, we review and modify the design for required conveyance capacity. For retardance class B, the design flow depth will be higher due to the increased retardance. Accordingly, we will increase the channel depth.

5.3.1.1 Phase 1: design for stability

In this phase, we use the lowest retardance class for the vegetal cover being considered. Given the design discharge, Q, and the bottom slope, S_0, the procedure is as follows:

1. Determine the maximum permissible unit tractive force τ_p from Table 5.4 and the sinuousness reduction factor C_p from Table 5.3, and select the side slope m using Table 5.1 as a guide.
2. Calculate the limiting depth, y_{LIM}, by using Equation 5.11. Select a flow depth, y, equal to or less than y_{LIM}.
3. Guess the bottom width, b.
4. Calculate the hydraulic radius $R=A/P$, where $A=(b+my)y$ and $P=b+2y\sqrt{1+m^2}$.
5. Determine the Manning roughness factor from Equation 5.21.
6. Knowing A, R, n, and S_0, calculate Q from the Manning formula (Equation 5.14). If the calculated Q is the same as the given design discharge, proceed to step (7). Otherwise, go back to step (3).
7. Check that the Froude number is sufficiently lower than the critical value of 1.0.

Note that steps (3) through (6) of this procedure require a lengthy trial-and-error solution. The use of Figure 5.10 to pick a value for b would facilitate this solution significantly. In this figure,

$$\beta = C_n + 19.97\log(K_v^{1.4}S_0^{0.4}y^{1.4}) \qquad (5.22)$$

Figure 5.10 was constructed using the procedure described by Akan and Hager (2001). It will yield precise values if $m=3$, slightly overestimated values if $m>3$, and slightly underestimated values if $m<3$.

EXAMPLE 5.5 A straight trapezoidal channel ($C_p=1.0$) lined with Bermuda grass will be sized to carry $Q=80$ cfs. The bottom slope is $S_0=0.002$, and the side slope is selected as $m=3.0$. Determine the bottom width, b, for the channel so that erosion will not occur when the retardance of the Bermuda grass is at its lowest.

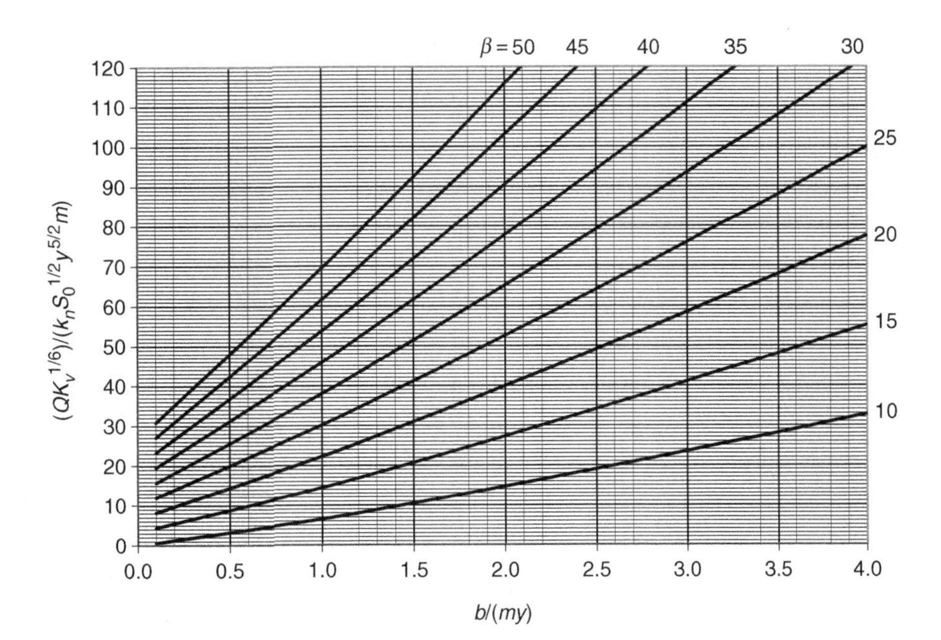

FIGURE 5.10 Chart to determine bottom width of grass-lined channels

From Table 3.3, the lowest retardance class for Bermuda grass is E; from Table 5.4, $\tau_p = 0.35\,\text{lb/ft}^2$ for class E. Also, $C_n = 37.7$ (from Table 3.3). By using Equation 5.11 with $K_b = 1.0$,

$$y_{LIM} = \frac{C_p \tau_p}{K_b \gamma S_0} = \frac{(1.0)(0.35)}{(1.0)(62.4)(0.002)} = 2.80\,\text{ft}$$

Suppose we select $y = 2.75$ ft. We will now try different values for b until the Manning formula yields $Q = 80$ cfs. Let us employ Figure 5.10 to pick the first trial value of b. By using Equation 5.22,

$$\begin{aligned}\beta &= C_n + 19.97 \log(K_v^{1.4} S_0^{0.4} y^{1.4}) \\ &= 37.7 + 19.97 \log[(1.0)^{1.4}(0.002)^{0.4}(2.75)^{1.4}] = 28.4\end{aligned}$$

Let us also evaluate the dimensionless parameter

$$\frac{QK_v^{1/6}}{k_n S_0^{1/2} y^{5/2} m} = \frac{80(1.0)^{1/6}}{(1.49)(0.002)^{1/2}(2.75)^{5/2}(3.0)} = 31.9$$

Then from Figure 5.10 we obtain $b/my = 0.80$, and therefore $b = 0.80(3)(2.75) = 6.6$ ft.

Let us now determine whether $b = 6.6$ ft satisfies the Manning formula:

$$P = b + 2y\sqrt{1 + m^2} = 6.6 + 2(2.75)\sqrt{1 + 3^2} = 24.0\,\text{ft}$$

$$A = (b + my)y = [6.6 + 3(2.75)](2.75) = 40.8\,\text{ft}^2$$

$$R = A/P = 40.8/24.0 = 1.70\,\text{ft}$$

Next, using $K_v = 1.0\,\text{ft}^{-1}$ for US customary units and $C_n = 37.7$ for class E, from Equation 5.21 we obtain

$$n = \frac{(RK_v)^{1/6}}{C_n + 19.97\log[(RK_v)^{1.4}S_0^{0.4}]}$$
$$= \frac{[(1.70)(1.0)]^{1/6}}{37.7 + 19.97\log\{[(1.70)(1.0)]^{1.4}(0.002)^{0.4}\}} = 0.0484$$

Next, from the Manning formula,

$$Q = \frac{1.49}{0.0484}(40.8)(1.70)^{2/3}(0.002)^{1/2} = 80.0\,\text{cfs}$$

The calculated Q is equal to the design discharge of 80 cfs; therefore, $b = 6.6$ ft is satisfactory. We can determine the corresponding Froude number as being 0.26 – an acceptable value.

5.3.1.2 Phase 2: modification for required conveyance

In this phase we use the highest retardance class for the vegetal cover used. Given the design discharge, Q, and the bottom slope, S_0, and knowing the bottom width b and size slope m from the first phase of design, the procedure is as follows:

1. Guess the flow depth, y.
2. Calculate the hydraulic radius $R = A/P$.
3. Determine the Manning roughness factor from Equation 5.21.
4. With known A, R, n, and S_0, calculate Q from the Manning formula (Equation 5.14). If the calculated Q is the same as the design discharge, proceed to step (5); otherwise, return to step (1).
5. Add an appropriate freeboard to the flow depth to determine the depth of the channel.

We can facilitate this trial-and-error procedure significantly by using Figure 3.6 (Chapter 3) to determine the first trial value of y. In this figure,

$$\alpha = C_n + 19.97\log\left(\frac{K_v^{1.4}b^{1.4}S_0^{0.4}}{m^{1.4}}\right) \qquad (5.23)$$

EXAMPLE 5.6 Modify the channel section sized in Example 5.5 so that it can accommodate the design discharge of 80 cfs when the retardance of Bermuda grass is highest.

From Example 5.5, $b = 6.6$ ft, $m = 3$, $S_0 = 0.002$, and $Q = 80$ cfs. From Table 3.3, the highest retardance class for Bermuda grass is B, for which $C_n = 23.0$. Let us use Figure 3.6 to determine the first trial value of y (y_n in the figure is the same as y here). By using Equation 5.23,

$$\alpha = C_n + 19.97 \log\left(\frac{K_v^{1.4} b^{1.4} S_0^{0.4}}{m^{1.4}}\right)$$

$$= 23.0 + 19.97 \log\left(\frac{(1.0)^{1.4}(6.6)^{1.4}(0.002)^{0.4}}{3^{1.4}}\right) = 11.0$$

Now, evaluate the dimensionless parameter

$$\frac{QK_v^{1.6} m^{3/2}}{k_n S_0^{1/2} b^{5/2}} = \frac{(80.0)(1.0)^{1/6}(3.0)^{3/2}}{(1.49)(0.002)^{1/2}(6.6)^{5/2}} = 55.7$$

From Figure 3.6, we obtain $my/b = 1.76$, or $y = (1.76)(6.6)/3.0 = 3.87$ ft. Let us now determine whether this depth satisfies the Manning formula. Let us first evaluate

$$A = (b + my)y = [6.6 + (3.87)]3.87 = 70.5 \text{ ft}^2$$

$$P = b + 2y\sqrt{1 + m^2} = 6.6 + 2(3.87)\sqrt{1 + 3^2} = 31.1 \text{ ft}$$

$$R = A/P = 70.5/31.1 = 2.27 \text{ ft}.$$

Then, by using Equation 5.21 with $K_v = 1.0 \text{ ft}^{-1}$ for customary US units and $C_n = 23.0$ for retardance class B,

$$n = \frac{(RK_v)^{1/6}}{C_n + 19.97 \log[(RK_v)^{1.4} S_0^{0.4}]}$$

$$= \frac{[(2.27)(1.0)]^{1/6}}{23.0 + 19.97 \log\{[(2.27)(1.0)]^{1.4}(0.002)^{0.4}\}} = 0.101$$

Next, from the Manning formula,

$$Q = \frac{1.49}{0.101}(70.5)(2.27)^{2/3}(0.002)^{1/2} = 80.3 \text{ cfs}$$

The calculated Q is close enough to 80 cfs, and therefore we will accept $y = 3.87$ ft. The corresponding Froude number can be determined as being 0.13 – an acceptable value.

5.3.2 DESIGN OF RIPRAP CHANNELS

The basic criterion for the design of riprap channels is that, under the design flow conditions, the maximum unit tractive force (or shear stress) on the channel bottom and sides must not exceed the permissible values. As recommended by

Anderson *et al.* (1970), the permissible unit tractive force for riprap material can be found by using

$$\tau_p = C_r d_{50} \tag{5.24}$$

where d_{50} = mean riprap size and $C_r = 4.0\ \text{lb/ft}^3 = 628.5\ \text{N/m}^3$.

The maximum unit tractive force on the channel bottom can be found using Equation 5.6, with $K_b = 1.0$, repeated here as

$$\tau_b = K_b \gamma y S_0 \tag{5.25}$$

The maximum unit tractive force on the side of the channel can be found using Equations 5.8 and 5.9, repeated here, respectively, as

$$\tau_s = \frac{K_s \gamma y S_0}{K} \tag{5.26}$$

and

$$K = \sqrt{1 - \frac{1}{(1+m^2)\sin^2 \alpha_R}} \tag{5.27}$$

As before, K_s can be obtained from Figure 5.3. For riprap material, Figure 5.11 should be used to obtain the angle of repose, α_R. This figure, constructed using the information provided by Anderson *et al.* (1970), also includes the recommended side slopes for riprap channels.

As discussed in Chapter 3, the Manning roughness factor in riprap channels can be calculated by using Equation 3.34, repeated here as

$$n = C_m (K_v d_{50})^{1/6} \tag{5.28}$$

where C_m = constant coefficient, K_v = unit conversion factor = $3.28\ \text{m}^{-1} = 1.0\ \text{ft}^{-1}$, and d_{50} = mean stone diameter. This equation was first suggested by Strickler (Henderson, 1966), with $C_m = 0.034$ for gravel-bed streams. Other suggested values for C_m are 0.039 (Hager, 2001) and 0.038 (Maynord, 1991). However, in shallow channels carrying a small discharge, the Manning roughness factor also depends on the ratio R/d_{50}. Based on the findings of Blodgett and McConaughy (1985), Chen and Cotton (1988) recommend the relationship

$$n = \frac{(K_v R)^{1/6}}{8.60 + 19.98 \log(R/d_{50})} \tag{5.29}$$

for $Q < 50$ cfs ($1.4\ \text{m}^3/\text{s}$).

If Equation 5.28 is adopted for the Manning roughness factor, the procedure to design a riprap channel is very similar to that discussed for the design of unlined non-cohesive channels in Section 5.2.2. However, Equation 5.24

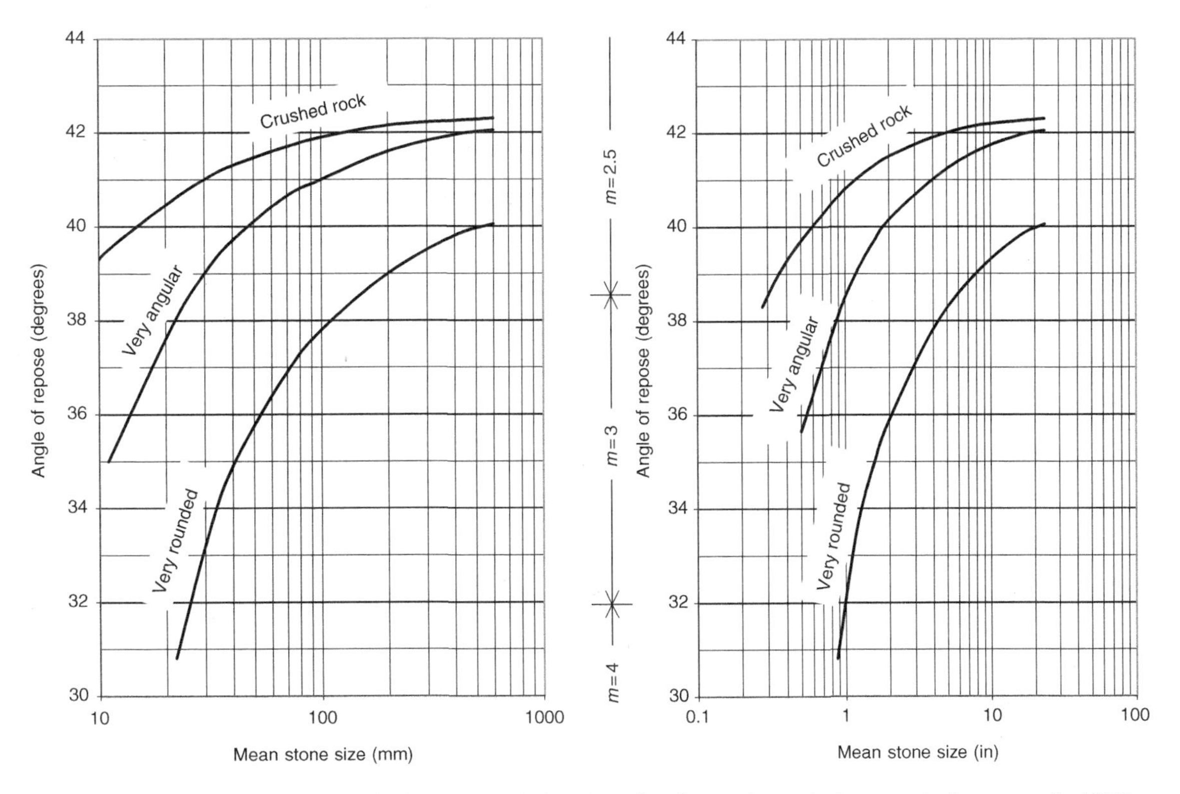

FIGURE 5.11 Angle of repose and recommended size slope for riprap channels (source: Anderson *et al.*, 1970)

should be used to determine τ_p for larger stones, and Figure 5.11 should be used to determine α_R. More specifically, the procedure is:

1. For the riprap size picked, calculate the Manning roughness factor, n, using Equation 5.28, and the straight channel permissible unit tractive force, τ_p by using Equation 5.24. Determine the angle of repose, α_R, and side slope, m, from Figure 5.11, and K_s from Figure 5.3. Calculate K using Equation 5.27. Select a C_p value from Table 5.3 based on the sinuousness of the channel.
2. Determine the limiting flow depth y_{LIM} by using Equation 5.13, repeated here as

$$y_{LIM} = \frac{KC_p\tau_p}{K_s\gamma S_0} \tag{5.30}$$

3. Select a flow depth, y, equal to or less than y_{LIM}.
4. Determine the channel bottom width, b, by using the Manning formula, rewritten as Equation 5.31:

$$Q = \frac{k_n\sqrt{S_0}}{n}\frac{A^{5/3}}{P^{2/3}} = \frac{k_n\sqrt{S_0}}{n}\frac{[(b+my)y]^{5/3}}{\left(b+y\sqrt{1+m^2}\right)^{2/3}} \tag{5.31}$$

However, solving this equation will require trial-and-error, since the equation is implicit in b. Also, while Equation 5.31 provides a mathematically exact value for b, we usually modify this value for practicality. It is therefore easier to use either Equation 5.32 or Figure 5.8 (Akan, 2001) to obtain an approximate value for b and then modify it (increase it to a round number) for practicality:

$$b = 1.186y\left[\frac{nQ}{k_n S_0^{1/2} y^{8/3}} - \frac{m^{5/3}}{(2\sqrt{1+m^2})^{2/3}}\right]^{0.955} \quad (5.32)$$

5. Determine the normal flow depth using the 'practical' channel width, b. Calculate the Froude number and verify that it is not close to the critical value of 1.0.
6. Check the stability of the channel bottom by verifying that $(K_b \gamma y S_0) < C_p \tau_p$ and of the channel sides by verifying that $(K_s \gamma y S_0/K) < C_p \tau_p$.
7. Determine a freeboard and modify the channel dimensions for practical purposes if needed.

If Equation 5.29 is adopted for the Manning roughness factor the procedure will be slightly different, since the roughness factor will depend on the flow depth and the hydraulic radius. In this case:

1. For the riprap size picked, calculate the straight channel permissible unit tractive force, τ_p by using Equation 5.24. Determine the angle of repose, α_R, and side slope, m, from Figure 5.11, and K_s from Figure 5.3. Calculate K using Equation 5.27. Select a C_p from Table 5.3 based on the sinuousness of the channel.
2. Determine the limiting flow depth y_{LIM} by using Equation 5.30.
3. Select a flow depth, y, equal to or less than y_{LIM}.
4. Pick a bottom width, b, calculate the hydraulic radius using the expression

$$R = \frac{(b+my)y}{b+2y\sqrt{1+m^2}} \quad (5.33)$$

and determine the corresponding Manning roughness factor using Equation (5.29).
5. Calculate the discharge using Equation 5.31. If the calculated discharge is the same as the design discharge given, the bottom width, b, picked in the previous step is acceptable. Otherwise, try another b. This lengthy trial-and-error procedure can be facilitated significantly by using Figure 5.12 to select the first trial value of b.
6. Calculate the Froude number and verify that it is not close to the critical value of 1.0.
7. Check the stability of the channel bottom by ensuring that $(K_b \gamma y S_0) < C_p \tau_p$.
8. Select a freeboard and modify the channel dimensions for practical purposes if needed.

It is important to note that the riprap gradation should follow a smooth size distribution. This will ensure that the interstices formed by larger stones are filled with smaller stones in an interlocking fashion. In general, riprap

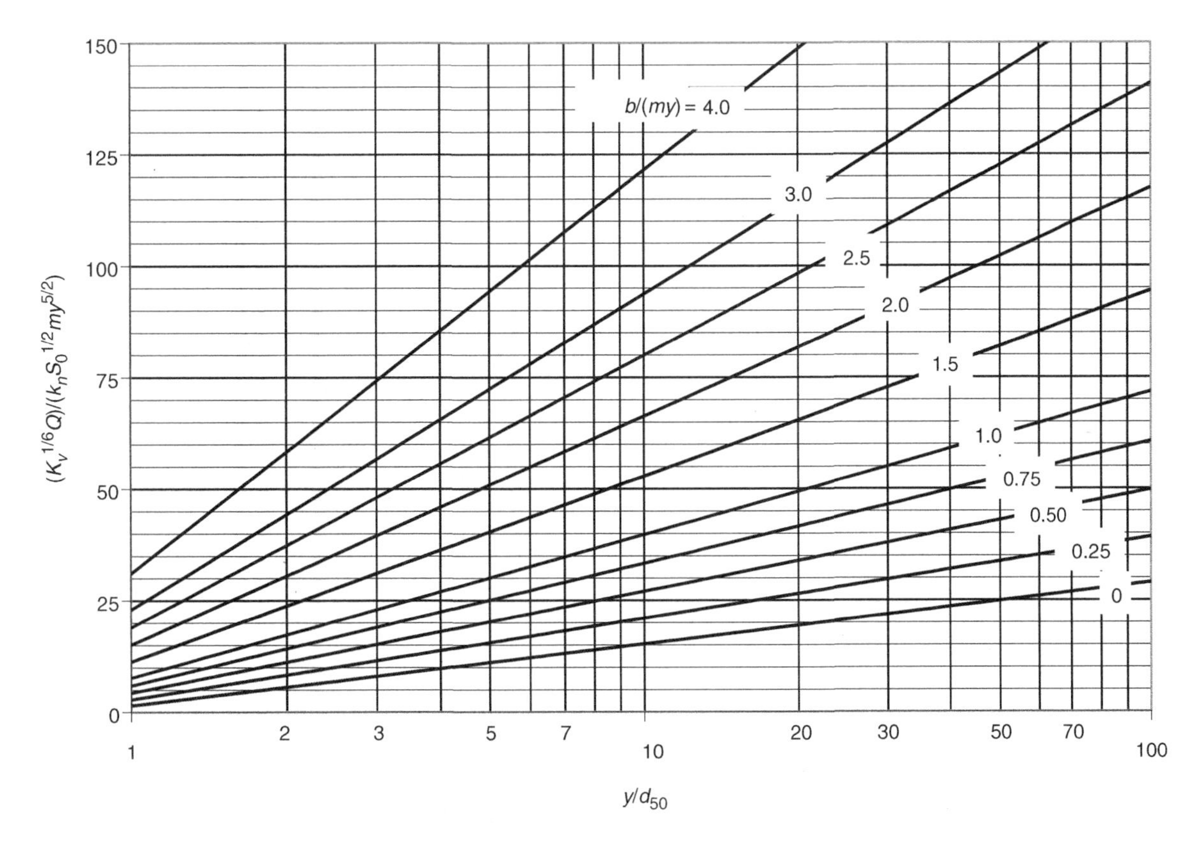

FIGURE 5.12 Design chart for riprap channels

constructed with angular stones is preferred. Round stones are acceptable as riprap if they are not placed on side slopes steeper than 3H : 1V, that is $m = 3$. An underlying filter made of either an engineering fabric or granular blanket may be placed when riprap is used. The thickness of the riprap lining should be equal to the diameter of the largest rock size in the gradation. For most gradations, the thickness will be between 1.5 and 3.0 times the mean riprap diameter (Chen and Cotton, 1988).

EXAMPLE 5.7 A riprap-lined straight channel will be designed to convey a peak discharge of 1000 cfs. The topography of the area where the channel is to be constructed and the channel alignment indicate that $S_0 = 0.004$. Proportion the channel section if the riprap chosen is very angular with $d_{50} = 0.4$ ft.

By using Equation 5.28 with $C_m = 0.039$ and $K_v = 1.0$/ft,

$$n = C_m(K_v d_{50})^{1/6} = 0.039[(1.0)(0.4)]^{1/6} = 0.0335$$

and from Equation 5.24 with $C_r = 4.0\,\text{lb/ft}^3$

$$\tau_p = C_r\, d_{50} = (4.0)(0.4) = 1.6\,\text{lb/ft}^2$$

Also, we obtain $\alpha_R = 41.2°$ and $m = 2.5$ from Figure 5.11, and $K_s = 0.83$ from Figure 5.3. Because the channel is straight, $C_p = 1.0$. Now, by using Equations 5.27 and 5.30,

$$K = \sqrt{1 - \frac{1}{(1+m^2)\sin^2 \alpha_R}} = \sqrt{1 - \frac{1}{(1+2.5^2)\sin^2 41.2}} = 0.83$$

$$y_{LIM} = \frac{KC_p\tau_p}{K_s\gamma S_0} = \frac{(0.83)(1.0)(1.6)}{(0.83)(62.4)(0.004)} = 6.41 \text{ ft}$$

Any design depth smaller than 6.41 ft is acceptable. Suppose we pick $y = 6.00$ ft. Now, let us use Figure 5.8 to determine an approximate b. First, we will evaluate

$$\frac{nQ}{k_n S_0^{1/2} y^{8/3}} = \frac{(0.0335)(1000)}{(1.49)(0.004)^{1/2}(6.00)^{8/3}} = 3.00$$

Then from Figure 5.8 with $m = 2.5$, we obtain $b/y = 1.73$ or $b = (1.73)(6.00) = 10.40$ ft. Let us modify this to $b = 11.0$ ft for practicality. Now, by using the methods discussed in Chapter 3, we can calculate the normal depth for this width as $y = 5.91$ ft. The corresponding Froude number is found as being $F_r = 0.59$ – an acceptable value.

Let us now check the stability of the channel bottom. With $K_b = 1.0$, we obtain $\tau_b = (K_b\ \gamma\ y\ S_0) = (1.0)(62.4)(5.91)(0.004) = 1.48$ lb/ft^2. This is smaller than the permissible value $C_p\tau_p = (1.0)(1.6) = 1.6$ lb/ft^2, therefore the bottom will be stable. For the sides of the channel, $(K_s\gamma y S_0/K) = (0.83)(62.4)(5.91)(0.004)/0.83 = 1.48$ lb/ft$^2 < 1.6$ lb/ft^2, and the sides will also be stable.

EXAMPLE 5.8 A straight channel conveying 800 cfs has a slope of $S_0 = 0.005$, bottom width of $b = 8.0$ ft, and a side slope of $m = 3.0$. Determine a very angular riprap size for this channel.

We will not follow the design procedure given in the foregoing section, because in this problem we already know the bottom width of the channel. Instead, we will try different riprap sizes to determine the proper size. Table 5.5 summarizes the calculations.

TABLE 5.5 Summary of results for Example 5.8

d_{50} (ft)	n	τ_p (lbs/ft^2)	y_n (ft)	α_R (deg.)	K	K_s	y_{LIM} (ft)
0.25	0.031	1.00	5.02	41.0	0.88	0.85	3.32
0.33	0.032	1.32	5.09	41.2	0.88	0.85	4.38
0.50	0.035	2.00	5.30	41.4	0.88	0.85	6.63
0.75	0.037	3.00	5.43	41.7	0.88	0.85	9.95
1.00	0.039	4.00	5.56	41.8	0.88	0.85	13.27

Listed in column 1 are the different stone sizes tried. The Manning roughness factors in column 2 are obtained using Equation 5.28 with $C_m = 0.039$, and the allowable shear stress values in column 3 are obtained by using Equation 5.24. The normal depths in column 4 are determined by solving the Manning formula as discussed in Chapter 3. The angle of repose values in column 5 are obtained from Figure 5.11, the K values in column 6 are calculated by using Equation 5.27, and the K_s values in column 7 are chosen from Figure 5.3. The limiting flow depths in column 8 are calculated by using Equation 5.30 with $C_p = 1$ for straight channels.

In Table 5.5, $y_n > y_{LIM}$ for the mean stone sizes of 0.25 ft and 0.33 ft. Therefore these stone sizes are not acceptable, since erosion would occur under the normal flow conditions. The mean stone sizes of 0.50 ft and larger will be stable, but $d_{50} = 0.50$ ft is probably the most economical. Also, note that $m = 2.5$ is suggested for $d_{50} = 0.50$ ft in Figure 5.11. The existing side slope, $m = 3.0$, of the channel is milder and therefore acceptable.

EXAMPLE 5.9 A straight roadside channel to carry 45 cfs will be riprap lined with very angular stone. The mean stone size is 2.0 in = 0.167 ft. Proportion the channel section if the bottom slope is $S_0 = 0.005$.

Because $Q < 50$ cfs, we will use Equation 5.29 to calculate the Manning roughness factor and follow the procedure developed for this equation. From Figure 5.11 we obtain $\alpha_R = 40°$ and $m = 2.5$, and from Figure 5.3 we obtain $K_s = 0.83$. Then, by using Equation 5.27,

$$K = \sqrt{1 - \frac{1}{(1 + m^2)\sin^2 \alpha_R}} = \sqrt{1 - \frac{1}{(1 + 2.5^2)\sin^2 40}} = 0.82$$

Likewise, by using Equation 5.24,

$$\tau_p = (4.0)(2.0/12.0) = 0.66 \text{ lb/ft}^2$$

Now, by using Equation 5.30 with $C_p = 1.0$ for straight channels,

$$y_{LIM} = \frac{K C_p \tau_p}{K_s \gamma S_0} = \frac{(0.82)(1.0)(0.66)}{(0.83)(62.4)(0.005)} = 2.09 \text{ ft}$$

Let us pick $y = 2.0$ ft. We will use Figure 5.12 to facilitate the solution. We will first evaluate the term

$$\frac{K_v^{1/6} Q}{k_n S_0^{1/2} m y^{5/2}} = \frac{(1.0)^{1/6}(45.0)}{(1.49)\sqrt{0.005}(2.5)(2.0)^{5/2}} = 30.2$$

Then, with $(y/d_{50}) = (2.0)/(0.167) = 12.0$, Figure 5.12 yields $b/(my) = 0.55$. Therefore, $b = (0.55)(2.5)(2.0) = 2.75$ ft

Let us now verify whether this design satisfies the Manning formula. We can calculate

$$A = (b + my)y = [2.75 + 2.5(2.0)]2.0 = 15.5\,\text{ft}$$

$$P = b + 2y\sqrt{1+m^2} = 2.75 + 2(2.0)\sqrt{1+2.5^2} = 13.5\,\text{ft}$$

$$R = AP = 15.5/13.5 = 1.15$$

Now, by using Equation 5.29,

$$n = \frac{(K_v R)^{1/6}}{8.60 + 19.98\log(R/d_{50})} = \frac{[(1.0)(1.15)]^{1/6}}{8.60 + 19.98\log(1.15/0.167)} = 0.040$$

Finally, substituting into the Manning formula,

$$Q = \frac{1.49}{n} A R^{2/3} S_0^{1/2} = \frac{1.49}{0.04}(15.5)(1.15)^{2/3}(0.005)^{1/2} = 44.8$$

This is very close to the design discharge, $Q = 45$ cfs, and therefore the suggested dimensions are acceptable. Let us now check the stability of the channel bottom. With $K_b = 1.0$, we obtain $\tau_b = (K_b \gamma y S_0) = (1.0)(62.4)(2.00)(0.005) = 0.62\,\text{lb/ft}^2$. This is smaller than the allowable value of $C_p \tau_p = (1.0)(0.66) = 0.66\,\text{lb/ft}^2$, and therefore the channel bottom will be stable. We can also show that $F_r = 0.46$ – an acceptable value. We may increase the bottom width to $b = 3.0$ ft for practicality. As a result of this modification, the flow depth, y, would decrease and the channel sides and the bottom would remain stable.

5.3.3 TEMPORARY FLEXIBLE LININGS

Temporary flexible linings include woven paper net, jute net, fiberglass roving, curled wood mat, synthetic mat, and straw with net. They provide only temporary protection against erosion while allowing vegetation to establish in a channel already sized and constructed. Therefore, design of a temporary flexible lining involves simply determining a lining material that has a higher permissible unit tractive force than the maximum unit tractive force caused on the channel bottom by the design discharge. The sides of the channel section are not as critical as the bottom.

The permissible unit tractive forces for different types of temporary lining are given in Table 5.6. Also included in the same table are the Manning roughness factors for different depth ranges. Given the channel dimensions and the design discharge, the procedure to determine a temporary lining material for straight channels is as follows:

1. Select a lining, and estimate the range of the flow depth.
2. Obtain n from Table 5.6.

TABLE 5.6 Permissible unit tractive force and Manning roughness factor for temporary lining materials (Chen and Cotton 1988)

	Permissible unit tractive force		Manning roughness n		
			Depth range		
Lining material	(lb/ft^2)	(N/m^2)	0–0.5 ft (0–15 cm)	0.5–2.0 ft (15–60 cm)	>2.0 ft (>60 cm)
Woven paper net	0.15	7.2	0.016	0.015	0.015
Jute net	0.45	21.6	0.028	0.022	0.019
Single fiberglass roving	0.60	28.7	0.028	0.021	0.019
Double fiberglass roving	0.85	40.7	0.028	0.021	0.019
Straw with net	1.45	69.4	0.065	0.033	0.025
Curled wood mat	1.55	74.2	0.066	0.035	0.028
Synthetic mat	2.00	95.7	0.036	0.025	0.021

3. Determine the flow depth, y, from the Manning formula (Equation 5.14), using one of the procedures discussed in Chapter 3.
4. If the calculated depth is within the range assumed in step (1), proceed to the next step. Otherwise, return to step (1).
5. Determine τ_p from Table 5.6, and calculate τ_b using Equation 5.20. If $\tau_p > \tau_b$, the type of lining selected is acceptable. Otherwise, return to step (1).

At channel bends the bottom shear stress can be determined using Equation 5.19, repeated here as

$$\tau_{bb} = K_{bend}\,\tau_b \tag{5.34}$$

where K_{bend} is a dimensionless factor to be obtained from Figure 5.9. Because of the increased unit tractive force, the channel bends may require a stronger lining material than the straight portions of a channel.

EXAMPLE 5.10 Determine a temporary lining for a straight channel that is trapezoidal in cross-section with $b = 4.0$ ft, $m = 3$, and $S_0 = 0.01$. The design discharge is 45 cfs.

Let us select jute net as an initial lining alternative. Assuming a depth range of 0.5 ft to 2.0 ft, from Table 5.6, $n = 0.022$. Equation 5.14 is written as

$$45 = \frac{1.49\sqrt{0.01}}{0.022}\,\frac{[(4.0+3y)y]^{5/3}}{[(4.0+2y\sqrt{1+3^2})]^{2/3}}$$

and solved by trial and error to obtain $y = 1.12$ ft. This depth is within the assumed range, and therefore the selected roughness coefficient is acceptable. From Table 5.6, for jute net $\tau_p = 0.45$ lb/ft^2. Also, using Equation 5.20, $\tau_b = (1.0)62.4(1.12)0.01 = 0.70$ lb/ft^2. Because $\tau_b > \tau_p$, jute net cannot be used as the lining material for this channel. For the second trial, we should choose a lining material that has a τ_p greater than the calculated τ_b in the previous trial. Let us now select double fiberglass roving, for which $\tau_p = 0.85$ lb/ft^2. For this material, $n = 0.021$ is obtained from Table 5.6 for the depth range of 0.5 ft to 2.0 ft. Substituting into Equation 5.14 we determine the flow depth as 1.10 ft, which is within the range 0.5 ft to 2.0 ft. Then, from Equation 5.20, we obtain $\tau_b = 0.69\ \text{lb/ft}^2$. Because $\tau_p > \tau_b$, double fiberglass roving is acceptable.

EXAMPLE 5.11 The channel considered in Example 5.10 has a 45° bend with a centerline radius of $R_c = 20$ ft. Determine the channel lining required at the bend.

In Example 5.10, it was found that double fiberglass roving was acceptable for lining the straight reaches of the channel. The normal flow depth was found to be 1.10 ft. From Figure 5.9, with $R_c/b = 20/4 = 5$, we obtain $K_{bend} = 1.56$. Then, using Equation 5.34, $\tau_{bb} = K_{bend}\tau_b 1.59(0.69) = 1.07\ \text{lb/ft}^2$. This exceeds the permissible unit tractive force, $\tau_p = 0.89$, of double fiberglass roving. Therefore, a stronger material is needed to line the channel bend. Let us try straw with net as the bend lining material. For this material, $\tau_p = 1.45\ \text{lb/ft}^2$ (Table 5.6). Assuming a depth range of 0.5 ft to 2.0 ft, $n = 0.033$ from Table 5.6. Then, approximating the flow at the bend by normal flow and solving Equation 5.14 by trial and error for y, we obtain $y = 1.37$ ft. This is within the assumed range, and the selected n value is valid. Then, using Equation 5.34, $\tau_{bb} = K_{bend}\tau_b = K_{bend}K_b\gamma y S_0 = 1.56(1.0) \times (62.4)1.37(0.01) = 1.33\ \text{lb/ft}^2$. Because $\tau_{bb} < \tau_p$, straw with net is an acceptable lining material for the bend.

Straw with net will extend through the bend and for a distance, L_p, downstream. The downstream distance will be found using Equation 5.17. However, first we need to determine the hydraulic radius. Noting that

$$A = (b + my)y = [4.0 + (3)(1.37)](1.37) = 11.1\ \text{ft}^2$$

and

$$P = b + 2y\sqrt{1 + m^2} = 4.0 + 2(1.37)\sqrt{1 + 3^2} = 12.7\ \text{ft}$$

we obtain $R = A/P = 11.1/12.7 = 0.87$ ft. Then, with $R = 0.87$ ft, $n_b =$ bend roughness coefficient $= 0.033$, and $K_p = 0.604$ for US customary units, we obtain $L_p = 0.604(0.87)^{7/6}/0.033 = 15.5$ ft.

5.4 DESIGN OF RIGID BOUNDARY CHANNELS

Channels lined with materials such as concrete, asphaltic concrete, soil cement, and grouted riprap are considered to have rigid boundaries. These channels are non-erodible due to the high shear strength of the lining material. For the most part, there are not any design constraints on the maximum velocity or tractive force from an erosion standpoint. On the contrary, it is desirable to maintain flow velocities at a higher rate than a minimum permissible velocity below which siltation and sedimentation occur. The minimum permissible velocity in this regard is about 2.0 fps = 0.60 m/s under the design discharge conditions, particularly for channels in which the design discharge occurs frequently.

If rigid channel lining materials become deformed or displaced, or deteriorate, the channel flow may tend to work itself behind the lining material. This condition often leads to failure of the lining, sometimes with serious consequences. For this reason, rigid channel linings should be selected carefully, and used only where appropriate.

Three different approaches are discussed herein to proportion a rigid boundary channel section. These approaches differ only in the way the bottom width of the channel is selected. In the first approach, the channel width is determined from experience curves. In the second approach, the selected channel maximizes the channel conveyance for a fixed flow area. In the third approach, the lining cost is minimized.

5.4.1 EXPERIENCE CURVE APPROACH

The experience curve given in Figure 5.13 shows the average relationship of the bottom width and canal capacity for lined trapezoidal sections as recommended by the US Bureau of Reclamation. A similar curve was previously presented by Chow (1959). The experience curve can be used as a guide to select the bottom width of the channel. For a given design discharge Q and

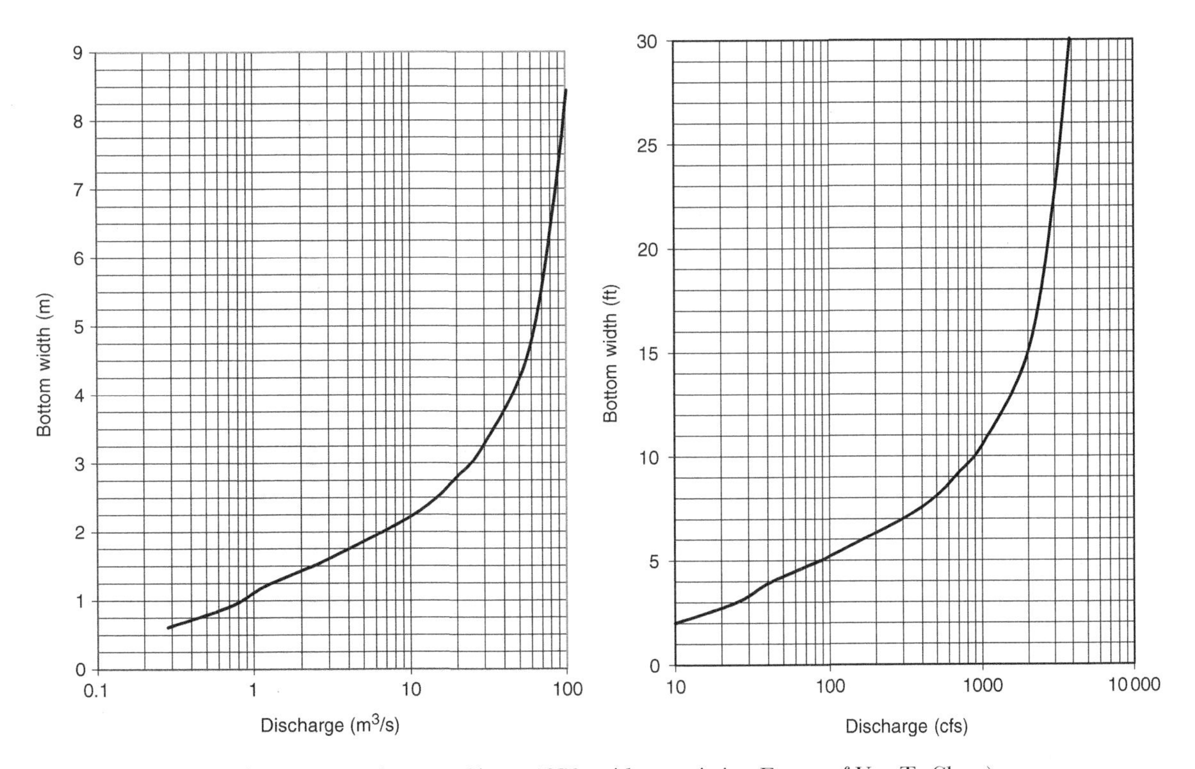

FIGURE 5.13 Experience curves (source: Chow, 1959, with permission Estate of Ven Te Chow)

longitudinal bottom slope S_0, a trapezoidal section can be sized using the following procedure:

1. Select m and determine n for the specified lining material.
2. Select a bottom width, b, using Figure 5.13 as a guide.
3. Substitute all the known quantities into the Manning formula repeated here for a trapezoidal section as

$$Q = \frac{k_n\sqrt{S_0}}{n}\frac{A^{5/3}}{P^{2/3}} = \frac{k_n\sqrt{S_0}}{n}\frac{[(b+my)y]^{5/3}}{\left(b+2y\sqrt{1+m^2}\right)^{2/3}} \tag{5.35}$$

   and solve for y by trial and error.
4. Determine the Froude number and check that it is sufficiently lower than the critical value of 1.0.
5. Determine the height of lining and the freeboard from Figure 5.1.

Example 5.12 A lined, trapezoidal channel is to be sized to carry a design discharge of 350 cfs. The lining material is asphalt. The longitudinal slope of the channel is $S_0 = 0.0016$, and the side slope is $m = 2.0$. Determine the appropriate channel proportions.

For asphalt, $n = 0.017$ from (Table 3.1). From the experience curve, Figure 5.13, we obtain $b = 8.0$ ft, approximately. Then, using $k_n = 1.49$ for customary US units, and substituting $Q = 350$ cfs, $S_0 = 0.0016$, $m = 2.0$, and $n = 0.0017$, Equation 5.35 becomes

$$350 = \frac{1.49\sqrt{0.0016}}{0.017}\frac{(8y+2y^2)^{5/3}}{\left[8+2y\sqrt{1+2^2}\right]^{2/3}}$$

Solving for y by trial and error, we obtain $y = 3.70$ ft. For this depth,

$$A = (b+my)y = [8+2(3.70)]3.70 = 57.07\ \text{ft}^2$$

$$T = b + 2my = 8 + 2(2)3.70 = 22.80\ \text{ft}$$

$$D = A/T = 57.07/22.80 = 2.50\ \text{ft}$$

$$V = Q/A = 350/57.07 = 6.13\ \text{fps}$$

$$F_r = V/\sqrt{gD} = 6.13/\sqrt{(32.2)(2.50)} = 0.68$$

The calculated Froude number is sufficiently below the critical value of 1.0. The design is acceptable.

Finally, using Figure 5.1, we determine that the vertical distance from the free surface to the top of the lining should be 1.1 ft, and the freeboard from the free surface to the top of the bank should be 2.70 ft.

5.4.2 BEST HYDRAULIC SECTION APPROACH

An inspection of the Manning formula, Equation 5.35, reveals that, everything else remaining constant, the discharge carried under the normal flow condition will increase with decreasing wetted perimeter, P. Thus, for a given flow area, the channel section having the shortest wetted perimeter will have the maximum conveyance capacity. Such a channel section is called the *best hydraulic section*. Although the best hydraulic section is not necessarily the most economic section, this concept can be used to guide the sizing of rigid boundary channel sections. As shown by Chow (1959), the best trapezoidal section with fixed side slopes m has a flow depth to bottom width ratio of

$$\frac{b}{y} = 2\left(\sqrt{1+m^2} - m\right) \tag{5.36}$$

The procedure to size a trapezoidal section using the best hydraulic section approach is as follows:

1. Select m and determine n for the specified lining material.
2. Evaluate the ratio, b/y, using Equation 5.36.
3. Rearrange the Manning formula as

$$y = \frac{\left[(b/y) + 2\sqrt{1+m^2}\right]^{1/4}}{\left[(b/y) + m\right]^{5/8}} \left(\frac{Qn}{k_n\sqrt{S_0}}\right)^{3/8} \tag{5.37}$$

   and solve for y explicitly knowing all the terms on the right-hand side. Then find b using Equation 5.36. Modify b for practicality if needed.
4. Verify that the Froude number is sufficiently below the critical value of 1.0.
5. Determine the height of lining and the freeboard using Figure 5.1.

EXAMPLE 5.13 A lined channel of trapezoidal section will be sized using the best hydraulic section approach. The channel bottom slope is $S_0 = 0.0016$, the side slope is $m = 2.0$, and the design discharge is $Q = 15\ \text{m}^3/\text{s}$. The lining material is asphalt. Proportion the channel dimensions.

From Table 3.1, $n = 0.017$ for asphalt. Substituting $m = 2$ in Equation 5.36, we find

$$\frac{b}{y} = 2\left(\sqrt{1+2^2} - 2\right) = 0.47$$

Next, using Equation 5.37 with $k_n = 1.0$ for the metric unit system,

$$y = \frac{\left[(0.47) + 2\sqrt{1+2^2}\right]^{1/4}}{\left[(0.47) + 2\right]^{5/8}} \left[\frac{(15.0)(0.017)}{1.0\sqrt{0.0016}}\right]^{3/8} = 1.70\ \text{m}$$

Then $b = 0.47(1.70) = 0.80$ m. For this section,

$$A = (b + my)y = [0.80 + 2(1.70)]1.70 = 7.11\,\text{m}^2$$

$$T = b + 2my = 0.80 + 2(2)1.70 = 7.60\,\text{m}$$

$$D = A/T = 7.11/7.60 = 0.94\,\text{m}$$

$$V = Q/A = 15/7.11 = 2.11\,\text{m/s}$$

$$F_r = V/\sqrt{gD} = 2.11/\sqrt{(9.81)(0.94)} = 0.69$$

This is sufficiently lower than the critical value of 1.0.

Finally, From Figure 5.1, the height of lining above the free surface is 0.37 m. Also, the freeboard above the free surface is 0.90 m.

5.4.3 MINIMUM LINING COST APPROACH

The minimum lining cost procedure for the design of trapezoidal lined channels was developed by Trout (1982), and was presented previously by French (1985). Given the design discharge, the channel longitudinal slope, and the side slope, the channel section is proportioned such that the lining cost will be minimized. To use this procedure, we should know the cost of the base lining and side lining materials per unit area, and the cost of the corner materials per unit length.

Defining

$$K_1 = 20(m^2+1) - \left[1 + 4\left(\frac{U_B}{U_S}\right)\right] 4m\sqrt{m^2+1} \tag{5.38}$$

and

$$K_2 = 6\left(1 - \frac{U_B}{U_S}\right)\sqrt{m^2+1} - 10m\left(\frac{U_B}{U_S}\right) \tag{5.39}$$

where U_B = cost of base lining material per unit area of the specified thickness and U_S = cost of side lining material per unit area for the specified thickness, the ratio of b/y minimizing the lining cost is

$$\frac{b}{y} = \frac{2K_1}{-K_2 + [K_2^2 + 20(U_B/U_S)K_1]^{1/2}} \tag{5.40}$$

The procedure for designing a channel section using the minimum lining cost approach is very similar to that given for the best hydraulic section approach. The only difference is that Equation 5.40 should be used in place of Equation 5.37 to find the ratio b/y.

After the channel has been proportioned, the total material cost of the channel per unit length is computed as

$$U = C_B + C_S = b\,U_B + U_C + 2\,U_S(y + F)\sqrt{m^2 + 1} \tag{5.41}$$

where U = total material cost of the channel lining per unit length, C_B = material cost for the channel base per unit length, C_S = material cost of the sides per unit channel length, b = bottom width of the channel, U_C = the combined cost of corner materials per unit channel length, y = flow depth, F = distance from water surface to top of lining, and m = side slope of both sides of the channel.

We should note that the lining cost is only one of many cost components. To determine the total cost we need to consider other components such as the cost of land, excavation and construction costs, and permit fees.

EXAMPLE 5.14 A trapezoidal channel is to be sized to carry 15 m³/s using the minimum lining cost approach. For this channel, $S_0 = 0.0016$, $n = 0.015$, and $m = 1.0$. The unit cost of the base lining material is \$50 per square meter, and that of side lining material is \$40 per square meter. The combined cost of the corner materials is \$10 per meter. Proportion the channel section and determine the total cost of the lining.

From the problem statement, $U_B = 50$ and $U_S = 40$. Using Equations 5.38 and 5.39,

$$K_1 = 20(1^2 + 1) - \left[1 + 4\left(\frac{50}{40}\right)\right]4(1)\sqrt{1^2 + 1} = 6.06$$

$$K_2 = 6\left(1 - \frac{50}{40}\right)\sqrt{1^2 + 1} - 10(1)\left(\frac{50}{40}\right) = -14.62$$

Next, From Equation 5.40,

$$\frac{b}{y} = \frac{2(6.06)}{-(-14.62) + \left[(-14.62)^2 + 20(50/40)(6.06)\right]^{1/2}} = 0.36$$

Now, using Equation 5.37,

$$y = \frac{\left(0.36 + 2\sqrt{1 + 1^2}\right)^{1/4}}{(0.36 + 1)^{5/8}}\left[\frac{15(0.015)}{1.0\sqrt{0.0016}}\right]^{3/8} = 2.11\text{ m}$$

Therefore $b = 0.36(2.11) = 0.76$ m. For this section,

$$A = (b + my)y = [0.76 + (1)(2.11)]2.11 = 6.06\text{ m}^2$$

$$T = b + 2my = 0.76 + 2(1)2.11 = 4.98\text{ m}$$

$$D = A/T = 6.06/4.98 = 1.22\text{ m}$$

$$V = Q/A = 15/6.06 = 2.48\,\text{m/s}$$
$$F_r = V/\sqrt{gD} = 2.48/\sqrt{(9.81)(1.22)} = 0.72$$

This is sufficiently below the critical value of 1.0 and is acceptable. Finally, from Figure 5.1, the required height of lining above the water surface is 0.37 m. Also from Figure 5.1, the freeboard measured from the water surface to the top of the bank is 0.89 m. With the unit costs, bottom width, normal depth, freeboard, and side slopes determined, the total material cost of the lining is computed from Equation 5.41 as

$$U = 0.76(50) + 10 + 2(40)(2.11 + 0.89)\sqrt{1^2 + 1} = \$387.75/\text{m}$$

5.5 Channel Design for Non-Uniform Flow

Open channels are usually designed for normal flow conditions, and the procedures presented in this chapter are all based on the normal flow depth. However, modifications to the design may be needed where the flow depth deviates from the normal depth. For example, if an earthen channel carrying subcritical flow terminates at a free fall, protection of the downstream portion of the channel may be needed due to the increased velocities and shear stresses. Note that, for non-uniform flow, we should use the friction slope, S_f, rather than the bottom slope, S_0, in Equations 5.6 and 5.8 to calculate unit tractive force. In other words, for non-uniform flow:

$$\tau_b = K_b \gamma y S_f \tag{5.42}$$

and

$$\tau_s = \frac{K_s \gamma y S_f}{K} \tag{5.43}$$

Also, when channels are part of a channel system, the mutual interaction of the connected channels may pull the flow conditions away from the normal conditions in some channels. For example, an M1 curve in a channel caused by a milder downstream channel will result in flow depths higher than the normal depth. This can potentially cause flooding if the channel is sized for normal flow conditions without adequate freeboard. To avoid problems of this sort, we should perform the gradually-varied flow calculations for the whole system once the individual channels are sized.

Example 5.15 Consider the proposed channel system shown in Figure 5.14. All the channels are to be lined with concrete. The proposed individual channel

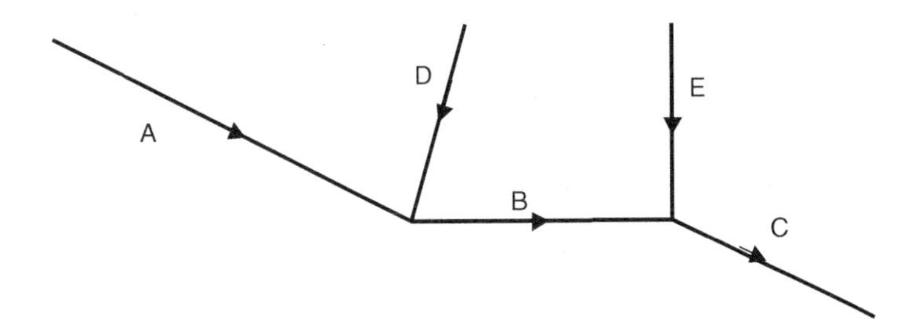

FIGURE 5.14 Channel system for Example 5.15

TABLE 5.7 Data for Example 5.15

Channel	Bottom width (ft)	Manning, n	Discharge (cfs)	Side slope, m	Length (ft)	Slope	Height of lining (ft)	Depth of channel (ft)
A	6	0.016	200	2	2000	0.002	3.75	5.50
B	8	0.016	400	2	1000	0.001	5.00	5.50
C	10	0.016	600	3	1000	0.002	4.50	5.50
D	6	0.016	200	2	1000	0.001	4.00	5.50
E	6	0.016	200	2	1000	0.002	3.75	5.50

characteristics are given in Table 5.7. In the table, the height of lining is measured from the channel invert to the top of the lining. The depth of channel is measured from the channel invert to the bank shoulder. Suppose the design criteria require that the distance from the water surface to the top of the lining be at least 0.50 ft and the distance from the water surface to the bank shoulder be at least 1.0 ft.

1. Consider each channel individually and determine whether these channels would be adequate if normal flow were to occur in each channel. Check whether the design criteria are satisfied based on the normal flow conditions.
2. Analyze the channel system as a whole to determine whether it is adequately designed. Assume that channel C terminates at a free overflow. Use a depth increment of about 0.10 ft for gradually-varied flow calculations.

In part (1) of this example, the normal flow depths are calculated using the procedures described in Chapter 3. The results are presented in Table 5.8, along with the minimum required height of lining and the channel depth. Comparison of the minimum required values with the proposed values shows that the proposed design is acceptable under normal flow conditions.

In part (2), the gradually-varied flow calculations are performed using the direct step method (see Chapter 4) with a depth increment of $\Delta y = 0.10$ ft. We first calculate the water surface profile in channel C, and then proceed in the upstream direction. Equation 4.30 is adopted as the junction equation (that is, the water surface is assumed to be continuous at the junctions). The flow depths calculated at the downstream and upstream ends of the channels are given in Table 5.8. The larger of the downstream and upstream depths

TABLE 5.8 Results of Example 5.15

	Normal flow			Gradually varied flow				Initially proposed	
Channel	Flow depth (ft)	Min. required height of lining (ft)	Min. required channel depth (ft)	Flow depth at downstream end (ft)	Flow depth at upstream end (ft)	Min. required height of lining (ft)	Min. required channel depth (ft)	Height of lining (ft)	Depth of channel (ft)
A	2.83	3.33	3.83	4.23	2.83	4.73	5.23	3.75	5.50
B	4.31	4.81	4.81	3.79	4.23	4.73	5.23	5.00	5.50
C	3.79	4.29	4.79	3.44	3.79	4.29	4.79	4.50	5.50
D	3.35	3.85	4.35	4.23	3.66	4.73	5.23	4.00	5.50
E	2.83	3.33	3.83	3.79	2.83	4.29	4.79	3.75	5.50

govern the minimum required lining height and the channel depth. A review of the results in Table 5.8 reveals that the proposed channel depths are adequate for all the channels. However, the lining height should be increased in channels A, D, and E.

PROBLEMS

P.5.1 An unlined earthen channel will carry $Q = 74$ cfs over a bottom slope of $S_0 = 0.0009$. Proportion the channel section if the maximum permissible velocity is $V_{MAX} = 2.80$ fps, the recommended sideslope is $m = 2$, and the Manning roughness factor is $n = 0.021$.

P.5.2 An unlined earthen channel will carry $Q = 20\ m^3/s$ over a bottom slope of $S_0 = 0.0009$. Proportion the channel section if the maximum permissible velocity is $V_{MAX} = 1.50$ m/s, the recommended side slope is $m = 2$, and the Manning roughness factor is $n = 0.021$.

P.5.3 A straight, unlined clay channel is to be sized to carry $Q = 173$ cfs over a bottom slope of $S_0 = 0.0025$. The voids ratio of the channel material is 0.30. Proportion the channel section using $n = 0.021$ and $m = 1.5$.

P.5.4 A straight, unlined clay channel is to be sized to carry $Q = 4.80\ m^3/s$ over a bottom slope of $S_0 = 0.0025$. The voids ratio of the channel material is 0.30. Proportion the channel section using $n = 0.021$ and $m = 1.5$.

P.5.5 A straight, unlined channel will be excavated into a very angular cohesionless soil with a mean particle diameter of 0.07 in. The channel is to carry $Q = 40$ cfs over a slope of $S_0 = 0.0003$. Using $n = 0.021$ and $m = 2.5$, proportion the channel section.

P.5.6 A straight, unlined channel will be excavated into a very angular cohesionless soil with a mean particle diameter of 5 mm. The channel is to carry $Q = 1.0\ m^3/s$ over a slope of $S_0 = 0.0009$. Using $n = 0.021$ and $m = 2.5$, proportion the channel section.

P.5.7 The alignment of the channel sized in Problem P.5.5 will include a bend that has a radius of $R_c = 30$ ft. Determine whether the design needs to be modified at the bend, and if so, how.

P.5.8 The alignment of the channel sized in Problem P.5.6 will include a bend that has a radius of $R_c = 8.0$ m. Determine whether the design needs to be modified at the bend and, if so, how.

P.5.9 A straight roadside channel will be riprap lined with very rounded stone. The mean stone size is 2.0 in. Proportion the channel section if the design discharge is $Q = 40$ cfs and the channel bottom slope is $S_0 = 0.006$.

P.5.10 A straight roadside channel will be riprap lined with very rounded stone. The mean stone size is 50 mm. Proportion the channel section if the design discharge is $Q = 1.40\ m^3/s$ and the channel bottom slope is $S_0 = 0.006$.

P.5.11 A very sinuous segment of an earthen channel carrying $Q = 1200$ cfs frequently erodes. The channel cross-section can be approximated as a trapezoid with a bottom width of $b = 10$ ft and side slopes of $m = 3.0$. The bottom slope is $S_0 = 0.003$. The channel will be lined with very angular riprap to prevent the erosion problem. Select a proper riprap size. To determine the Manning roughness factor, use Equation 5.28 with cm = 0.039.

P.5.12 A very sinuous segment of an earthen channel carrying $Q = 35\ \text{m}^3/\text{s}$ frequently erodes. The channel cross-section can be approximated as a trapezoid with a bottom width of $b = 3$ m and side slopes of $m = 3.0$. The bottom slope is $S_0 = 0.003$. The channel will be lined with very angular riprap to prevent the erosion problem. Select a proper riprap size. To determine the Manning roughness factor, use Equation 5.28 with cm = 0.039.

P.5.13 A straight trapezoidal channel lined with grass–legume mixture will be sized to carry $Q = 200$ cfs. The channel will have a bottom slope of $S_0 = 0.002$ and side slopes of $m = 3$. Proportion the channel section.

P.5.14 A straight trapezoidal channel lined with grass–legume mixture will be sized to carry $Q = 7\ \text{m}^3/\text{s}$. The channel will have a bottom slope of $S_0 = 0.002$ and side slopes of $m = 3$. Proportion the channel section.

P.5.15 A lined channel is being considered as an alternative to the channel sized in Problem P.5.5. The lining material is concrete with $n = 0.014$. It is suggested that the steepest side slopes feasible be used. Proportion the channel section.

P.5.16 A lined channel is being considered as an alternative to the channel sized in Problem P.5.6. The lining material is concrete with $n = 0.014$. It is suggested that the steepest side slopes feasible be used. Proportion the channel section.

P.5.17 Suppose the earthen channel sized in Example 5.3 is very long but terminates at a free overfall. Determine the flow depth and the maximum unit tractive force on the channel bed and sides 20 ft upstream of the brink. Would erosion occur at this location if the channel were not protected by lining? Use the direct step method with space increments of about 0.05 ft, and assume the Manning formula is applicable.

P.5.18 Suppose the earthen channel sized in Example 5.4 is very long but terminates at a free overfall. Determine the flow depth and the maximum unit tractive force on the channel bed and sides 10 m upstream of the brink. Would erosion occur at this location if the channel were not protected by lining? Use the direct step method with space increments of about 0.01 m, and assume the Manning formula is applicable.

REFERENCES

Akan, A. O. (2001). Tractive force channel design aid. *Canadian Journal of Civil Engineering*, **28(5)**, 865–867.

Akan, A. O. and Hager, W. W. (2001). Design aid for grass-lined channels. *Journal of Hydraulic Engineering, ASCE*, **127(3)**, 236–237.

Anderson, A. T., Paintal, G. S. and Davenport, J. T. (1970). *Tentative Design Procedure for Riprap Lined Channels*. NCHRP Report 108, National Cooperative Highway Research Program, National Research Council, Washington, DC.

Bankston, J. D. and Baker, F. E. (1995). *Open Channel Flow in Aquaculture*. Publication No. 374, Southern Regional Aquaculture Center, Texas A&M University, College Station, TX.

Blodgett, J. C. and McConaughy, C. E. (1985). *Evaluation of Design Practices for Riprap Protection of Channels near Highway Structures*. US Geological Survey, prepared in cooperation with the Federal Highway Administration Preliminary Draft, Sacramento, CA.

Chaudhry, M. H. (1993). *Open-Channel Flow*. Prentice Hall, Englewood Cliffs, NJ.

Chen, Y. H. and Cotton, G. K. (1988). *Design of Roadside Channels with Flexible Linings*. Hydraulic Engineering Circular No. 15, Publication No. FHWA-IP-87-7, US Department of Transportation, Federal Highway Administration, McLean, VA.

Chow, V. T. (1959). *Open-Channel Hydraulics*. McGraw-Hill Book Co., New York, NY.

French, R. H. (1985). *Open-Channel Hydraulics*. McGraw-Hill Book Co., New York, NY.

Hager, W. W. (2001). *Wastewater Hydraulics: Theory and Practice*. Springer-Verlag, New York, NY.

Henderson, F. M. (1966). *Open Channel Flow*. Prentice Hall, Upper Saddle River, NJ.

Lane, E. W. (1955). Stable channel design. *Transactions, ASCE*, **120,** 1234–1260.

Maynord, S. T. (1991). Flow resistance of riprap. *Journal of Hydraulic Engineering, ASCE*, **117(6),** 687–695.

Simon, D. B. and Albertson, M. L. (1960). Uniform water conveyance channels in alluvial material. *Journal of the Hydraulics Division, ASCE*, **86(HY5),** 33–71.

Trout, T. J. (1982). Channel design to minimize lining material cost. *Journal of the Irrigation and Drainage Division, ASCE*, **108(IR4),** 242–249.

US Army Corps of Engineers (1991). Hydraulic design of flood control channels. *Engineer Manual*, EM 1110-2-1601, Department of the Army, Washington, DC.

6 Hydraulic structures

Hydraulic structures are used to control and manage the flow of water in natural and built systems. They include flow measurement structures such as weirs, conveyance structures such as culverts, and flood control structures such as dams. In this chapter, we will consider several types of hydraulic structures associated with open-channel flow.

6.1 Flow Measurement Structures

Measurement of flow in open channels is essential for better management of limited supplies of water. Accurate measurement practices help provide equitable distribution of water between competing demands, and conserve the water supplies by minimizing waste due to excess delivery. Most flow measurement structures are emplaced in a channel. They are used to determine the discharge indirectly from measurements of the flow depth.

6.1.1 SHARP-CRESTED WEIRS

A sharp-crested weir is an overflow structure consisting of a thin, vertical plate with a sharp-edged crest mounted perpendicular to the flow direction, as shown in Figure 6.1. The flow accelerates as it approaches the weir, resulting in a drop (*drawdown*) in the water surface. The water does not contact or cling to the downstream weir plate, but springs clear. The falling sheet of water springing from the weir plate is called the *nappe*. The section cut with a sharp upstream corner into the thin plate is called the *weir notch* or the *overflow section*. The depth measurements are taken at an upstream section not affected by the drawdown. This section is called the *approach section*, and is located at a distance about two to three times the head over the weir.

Downstream of a sharp-crested weir, *free flow* occurs when the weir allows free access of air under the nappe. The weir will be *submerged* if downstream water rises near or above the crest elevation. Submerged weir conditions should be avoided for accurate flow measurement.

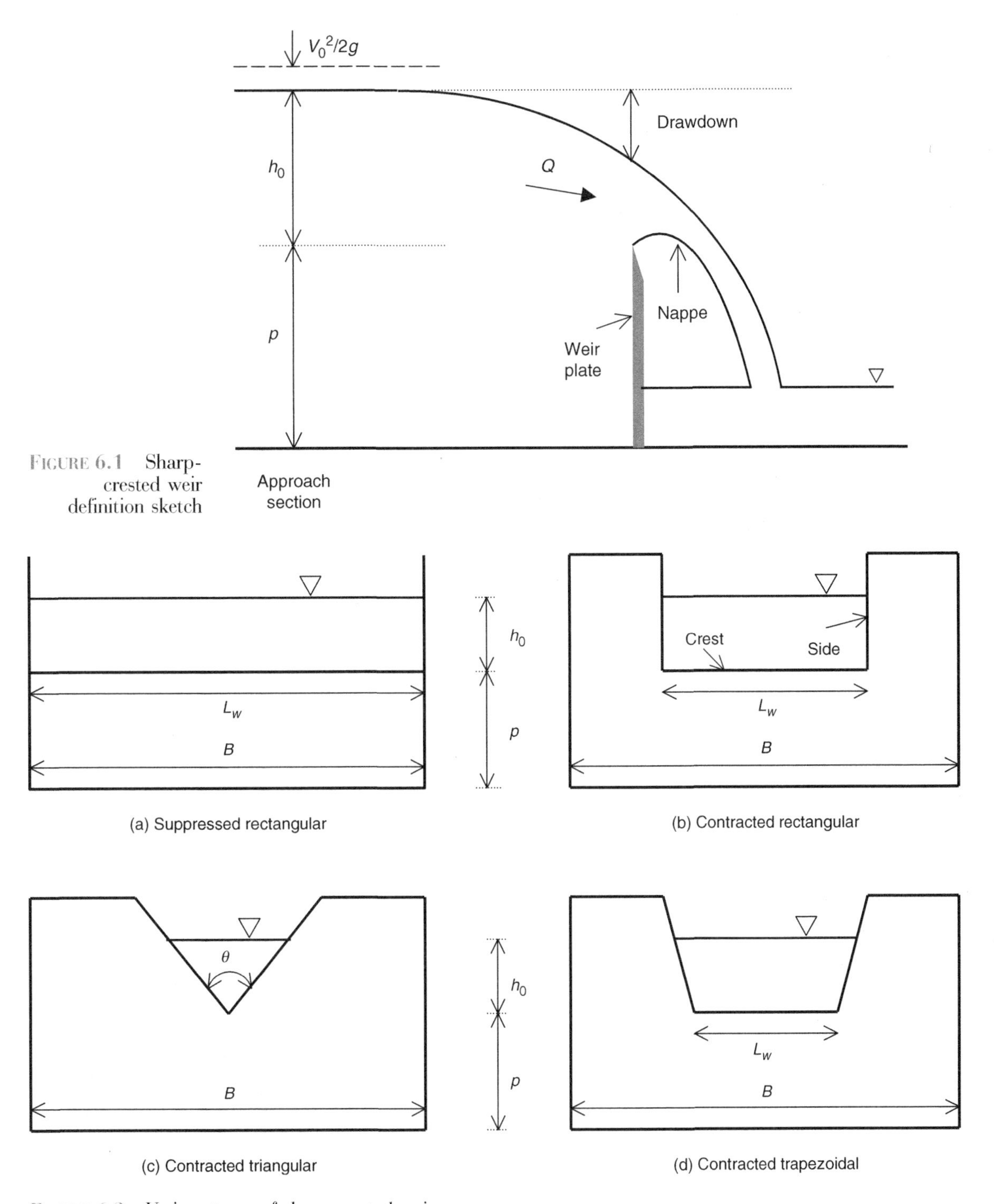

FIGURE 6.1 Sharp-crested weir definition sketch

FIGURE 6.2 Various types of sharp-crested weirs

Sharp-crested weirs are commonly named by the shape of their blade overflow opening. Figure 6.2 shows various rectangular, trapezoidal, and triangular weirs. The triangular weir is also called a *V-notch* weir. The bottom edge of the notch in the vertical plate is called the *crest*, and the side edges (which are vertical or flare up and outward) are the *sides* of the weir. In the case of a V-notch weir, the point of the triangle is the *crest*. For a suppressed rectangular weir, the sidewalls of the flow channel serve also as the sides of the weir. In this case, the flow approaching the weir plate is contracted vertically due to the weir crest and the drawdown. The contracted weirs, on the other hand, cause side contraction as well as vertical contraction, since the overflow section is narrower than the flow channel. If the bed and the walls of the flow channel are sufficiently far from the weir crest and the sides that the channel boundaries do not affect the contraction of the nappe, the weir is called *fully contracted*.

6.1.1.1 Rectangular sharp-crested weirs

We can use the energy equation to obtain a relationship between the approach flow characteristics and the discharge over a weir. For example, with reference to Figures 6.1 and 6.2a: neglecting the head-loss between the approach section and the weir, assuming the pressure is atmospheric within the flow section over the crest, and ignoring the drawdown, we obtain

$$Q = \frac{2}{3} L_w \sqrt{2g} \left[\left(1 + \frac{V_0^2}{2gh_0}\right)^{3/2} - \left(\frac{V_0^2}{2gh_0}\right)^{3/2} \right] h_0^{3/2} \tag{6.1}$$

for a suppressed rectangular weir (Sturm, 2001). However, this equation is not practical despite all the assumptions involved. Instead, a simpler equation in the form of

$$Q = \frac{2}{3} \sqrt{2g} C_d L_w h_0^{3/2} \tag{6.2}$$

is often adopted, where C_d = discharge coefficient, which accounts for the approach flow velocity head, the head-loss, and the effect of the drawdown (or vertical contraction). This coefficient is determined through experimental studies. Here, we will further simplify Equation 6.2 by introducing $k_w = (2/3) C_d$ = weir discharge coefficient. Then, we can rewrite Equation 6.2 as

$$Q = k_w \sqrt{2g} L_w h_0^{3/2} \tag{6.3}$$

Based on the experimental studies of Kindsvater and Carter (1957), we can write Equation 6.3 in a more general form for all types of rectangular sharp crested weirs as

$$Q = k_w \sqrt{2g} L_{ew} h_{e0}^{3/2} \tag{6.4}$$

where h_{e0} = effective head over the crest, and L_{ew} = effective crest length. The effective head is evaluated as

$$h_{e0} = h_0 + h_k \tag{6.5}$$

where $h_k = 0.001\,\text{m} = 0.003\,\text{ft}$. The effective crest length is determined as

$$L_{ew} = L_w + L_k \tag{6.6}$$

where the length correction, L_k, depends on the crest-length to channel-width ratio (L_w/B) as shown in Figure 6.3. The weir discharge coefficient depends on the (L_w/B) ratio as well as the (h_0/p) ratio as shown in Figure 6.4. Both Figures 6.3 and 6.4 are constructed based on the experimental findings of Kindsvater and Carter (1957). Note that in Figure 6.4, p = crest height above the bottom of the approach channel as shown in Figure 6.2. Also, B in Figures 6.3 and 6.4 represents the channel width for a rectangular approach channel. For other cross-sectional shapes of the approach channel, the average width of the approach flow section should be used.

The US Bureau of Reclamation (2001) suggests the following limits on the weir dimensions: the crest length, L_w, should be at least 0.5 ft (0.15 m); the crest height, p, should be at least 4 in (0.10 m); the head measured at the approach section, h_0, should be at least 0.20 ft (0.06 m); the (h_0/p) ratio should be less than 2.4; and the downstream water level should be at least 2 in (0.05 m) below the crest.

EXAMPLE 6.1 A trapezoidal irrigation canal has a bottom width of $b = 1.2\,\text{m}$, side slopes of $m = 2$ (2H : 1V), and a longitudinal bottom slope of $S_0 = 0.0005$. A rectangular sharp-crested weir placed in this channel has a crest height of $p = 0.70\,\text{m}$, and a crest length of $L_w = 0.80\,\text{m}$. The water surface elevation at the approach section is $h_0 = 0.77\,\text{m}$ above the weir crest. Determine the discharge in the canal.

The flow depth at the approach section is $y = 0.70 + 0.77 = 1.47\,\text{m}$. The top width of the flow at this section is $T = b + 2my = 1.2 + 2(2.0)(1.47) = 7.08\,\text{m}$. Thus the average channel width becomes $B = (b+T)/2 = (1.20 + 7.08)/2 = 4.14\,\text{m}$. We will use Equation 6.4 to determine the discharge. With $L_w/B = 0.80/4.14 = 0.19$ and $h_0/p = 0.77/0.70 = 1.1$, we obtain $L_k = 0.0024\,\text{m}$ and $k_w = 0.392$ from Figures 6.3 and 6.4, respectively. Also, the head correction is $h_k = 0.001\,\text{m}$. Then $L_{ew} = 0.80 + 0.0024 = 0.8024\,\text{m}$ and $h_{e0} = 0.77 + 0.001 = 0.771\,\text{m}$ from Equations 6.6 and 6.5, respectively. Substituting into Equation 6.4, we obtain

$$Q = k_w\sqrt{2g}L_{ew}h_{e0}^{3/2} = 0.392\sqrt{2(9.81)}(0.8024)(0.771)^{3/2} = 0.94\,\text{m/s}^3$$

Obviously, in this case the corrections on the crest length and the head are negligible, and we could have ignored them. Also, assuming that the Manning roughness factor is $n = 0.020$ for this channel, we can find that the normal flow depth is 0.64 m by using the methods discussed in Chapter 3. Therefore the weir will cause the flow depth to rise above the normal depth in the channel, resulting in an M1 profile (see Chapter 4) upstream of the approach section.

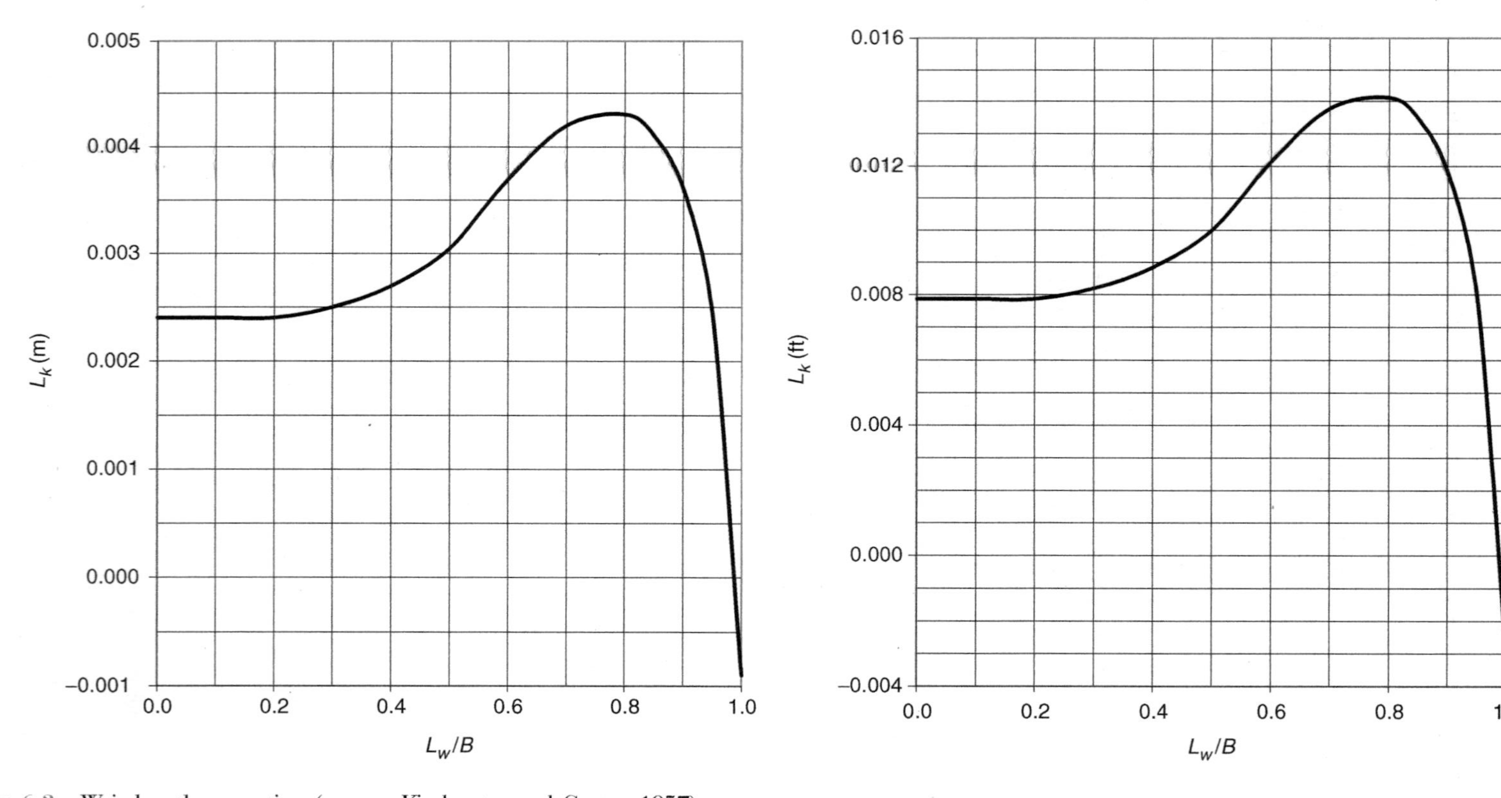

FIGURE 6.3 Weir length correction (source: Kindsvater and Carter, 1957)

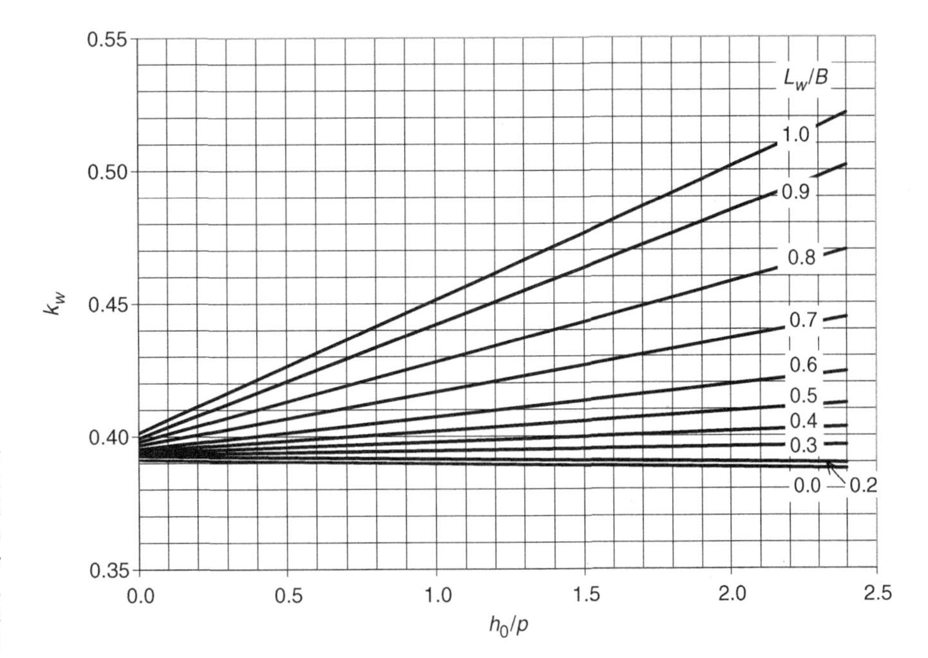

FIGURE 6.4 Weir discharge coefficient for sharp-crested rectangular weirs (source: Kindsvater and Carter, 1957, with permission ASCE)

6.1.1.2 Sharp-crested V-notch weirs

For a sharp-crested V-notch (or triangular) weir, the discharge–head relationship can be expressed as (Bos, 1989)

$$Q = C_e \frac{8}{15}\sqrt{2g}\tan\frac{\theta}{2}h_{e0}^{5/2} \tag{6.7}$$

where θ = notch angle (see Figure 6.2), C_e = discharge coefficient, and h_{e0} = effective head as defined by Equation 6.5. We will simplify this relationship by defining $k_w = (8/15)C_e$, where k_w = weir discharge coefficient for a sharp-crested V-notch weir. In terms of k_w, Equation 6.7 becomes

$$Q = k_w\sqrt{2g}\tan\frac{\theta}{2}h_{e0}^{5/2} \tag{6.8}$$

The correction, h_k, for the head is given in Figure 6.5 for fully contracted sharp-crested V-notch weirs, and the weir discharge coefficient, k_w, is given in Figure 6.6. Both figures are constructed by using the information presented by Kulin and Compton (1975) and the US Bureau of Reclamation (2001). As reported by Bos (1989), a V-notch is considered to be fully contracted if the (h_0/p) ratio is less than or equal to 0.4 and the (h_0/B) ratio is less than or equal to 0.2. Also, the water surface downstream from the weir should be at least 0.2 ft (0.06 m) below the notch, and h_0 should be greater than 0.2 ft (0.06 m) and smaller than 1.25 ft (0.38 m) (US Bureau of Reclamation, 2001).

EXAMPLE 6.2 A 45° V-notch weir is installed in a 2-ft wide rectangular laboratory flume. The crest height is 1.0 ft, and the water surface elevation at the approach section is 0.38 ft above the crest. Determine the discharge in the flume.

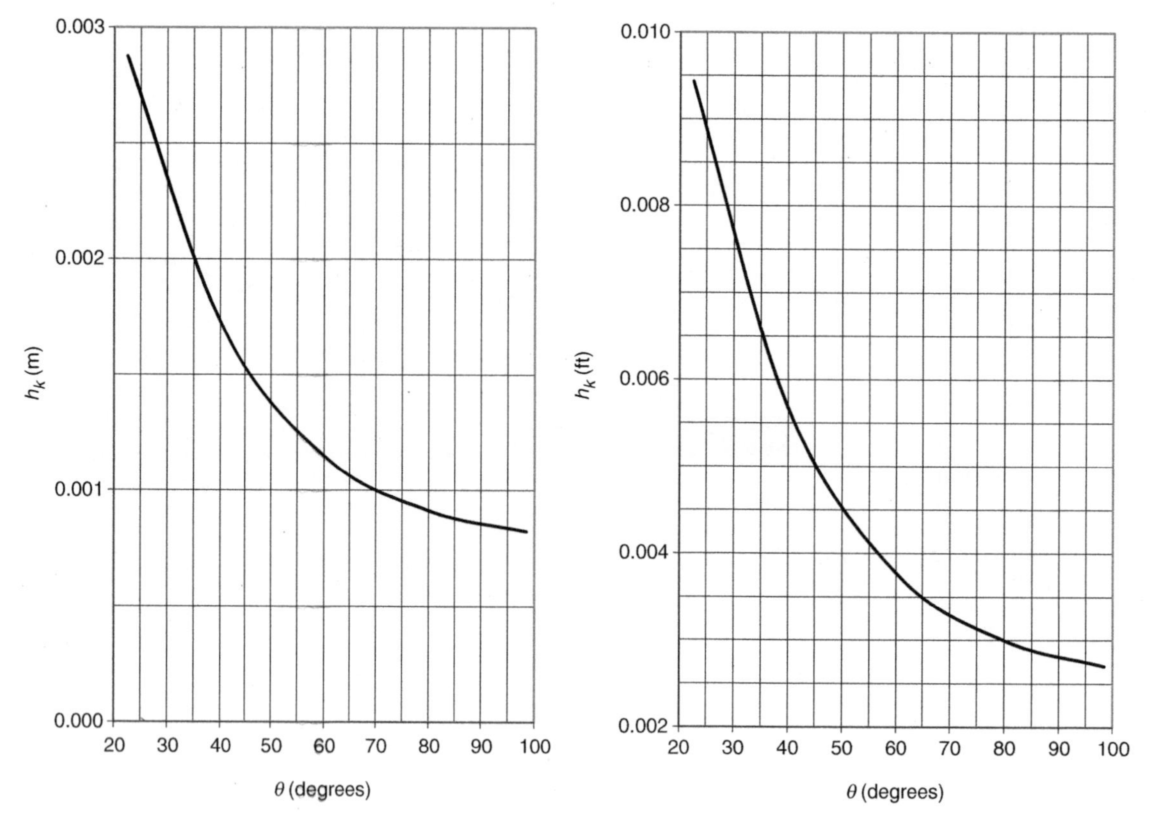

FIGURE 6.5 Head correction for sharp-crested V-notch weirs (source: Kulin and Compton, 1975; US Bureau of Reclamation, 2001)

From the problem statement, we have $B = 2.0$ ft, $p = 1.0$ ft, $\theta = 45°$, and $h_0 = 0.38$ ft. For this case, $h_0/p = 0.38/1.0 = 0.38 < 0.40$, and $h_0/B = 0.38/2.0 = 0.19 < 0.20$. Therefore, the V-notch is fully contracted. For $\theta = 45°$, we have $h_k = 0.005$ ft and $k_w = 0.309$ from Figures 6.5 and 6.6, respectively. Substituting into Equation 6.8,

$$Q = k_w\sqrt{2g}\tan\frac{\theta}{2}h_{e0}^{5/2} = (0.309)(\sqrt{2(32.2)}\left(\tan\frac{45°}{2}\right)(0.38 + 0.005)^{5/2} = 0.095 \text{ cfs}$$

6.1.1.3 Cipoletti weirs

A Cipolletti weir is trapezoidal in shape, with weir sides sloping 1 horizontal over 4 vertical (1H : 4V). The discharge–head relationship for a fully contracted Cipoletti weir can be expressed as

$$Q = k_w\sqrt{2g}L_w h_0^{3/2} \tag{6.9}$$

with $k_w = 0.42$ if the velocity head of the approach flow is negligible. As reported by Bos (1989), for a fully contracted Cipotelli weir neither the (h_0/p) nor the (h_0/B) ratio should exceed 0.50. The height of the weir crest, p, should be at least

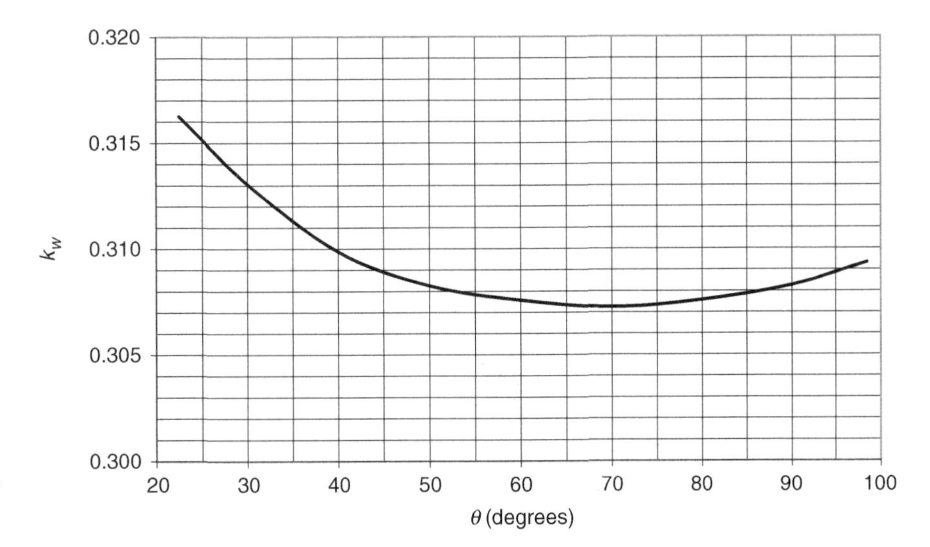

FIGURE 6.6
Discharge coefficient for fully contracted sharp-crested V-notches (source: Kulin and Compton, 1975; US Bureau of Reclamation, 2001)

1.0 ft (0.30 m), and the distance from the sides of the trapezoidal weir to the sides of the channel should be at least $2 \times h_0$ with a minimum of 1.0 ft (0.30 m). The measured head, h_0, over the weir crest should be between 0.20 ft (0.06 m) and 2.0 ft (0.6 m). The tailwater should be at least 0.20 ft (0.06 m) below the weir crest.

6.1.2 BROAD-CRESTED WEIRS

Broad-crested weirs have a horizontal crest with a finite length, L_b, in the flow direction, as shown in Figure 6.7. A weir is classified as broad-crested if $12.5 > (L_b/h_0) > 3.0$ (Sturm, 2001). Streamlines become straight and parallel over a broad-crested weir, with the critical depth occurring at some point over the crest. Various cross-sectional shapes, such as parabolic and triangular, are possible for broad-crested weirs. However, we will limit our discussion to rectangular broad-crested weirs.

We can write the energy equation between the approach section and the critical flow section as

$$E_0 - h_L = h_0 + \frac{V_0^2}{2g} - h_L = y_c + \frac{V_c^2}{2g} \tag{6.10}$$

where h_L = head loss, y_c = critical depth, and V_c = critical flow velocity. From Chapter 2, we recall that for a rectangular section $V_c = q/y_c$, and $y_c = (q^2/g)^{1/3}$, where q = discharge per unit width. Substituting these into Equation 6.10, noting that $q = Q/L_w$ for this case, and rearranging the equation we obtain

$$Q = \frac{2}{3}\left(\frac{2}{3}g\right)^{1/2} L_w \left(h_0 + \frac{V_0^2}{2g} - h_L\right)^{3/2} \tag{6.11}$$

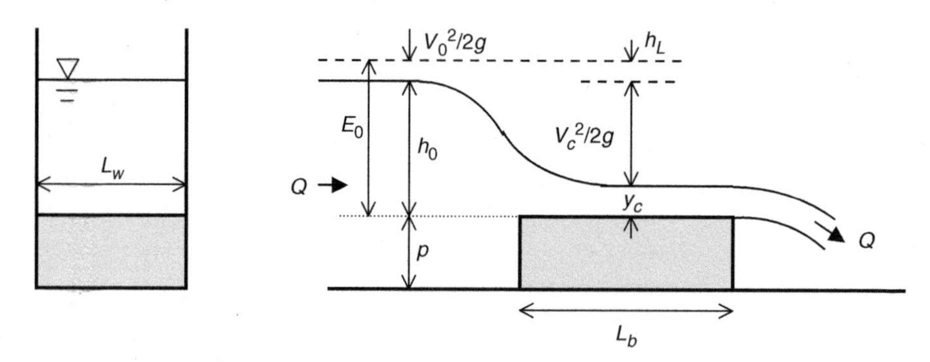

FIGURE 6.7
Broad-crested weir

In a field installation we measure the flow depth, h_0, and therefore an equation expressing Q in terms of h_0 is practical. For this purpose, traditionally, Equation 6.11 is simplified as (Bos, 1989)

$$Q = C_d C_v \frac{2}{3}\left(\frac{2}{3}g\right)^{1/2} L_w h_0^{3/2} \tag{6.12}$$

The coefficient C_d accounts for the head loss, and is expressed as

$$C_d = 0.93 + 0.10\frac{E_0}{L_b} \tag{6.13}$$

The coefficient C_v accounts for the approach velocity head, and is expressed as

$$C_v = \left(\frac{E_0}{h_0}\right)^{3/2} \tag{6.14}$$

Here we will further simplify Equation 6.12 to the form of

$$Q = k_w\sqrt{2g}L_w h_0^{3/2} \tag{6.15}$$

where k_w = broad-crested weir discharge coefficient, expressed as

$$k_w = k_d C_v \tag{6.16}$$

with $k_d = (2/3)(1/3)^{1/2}C_d$. Substituting Equation 6.13 for C_d,

$$k_d = 0.358 + 0.038\frac{E_0}{L_b} \tag{6.17}$$

Equations 6.14–6.17 are used to determine the discharge over a broad-crested weir. However, because E_0 is not measured, an iterative scheme is needed to solve these equations.

EXAMPLE 6.3 A broad-crested weir has a crest length of $L_b = 0.75$ m, crest width of $L_w = 1.0$ m, and crest height of $p = 0.30$ m. The water surface at the approach section is 0.20 m above the crest – that is, $h_0 = 0.20$ m. Determine the discharge.

We will first neglect the velocity head of the approach flow – in other words, we will assume that $E_0 = h_0$. With this assumption, $C_v = 1.0$ from Equation 6.14, and by using Equation 6.17

$$k_d = 0.358 + 0.038 \frac{0.20}{0.75} = 0.368$$

Then, from Equation 6.16, $k_w = (0.368)(1.0) = 0.368$. Substituting this into Equation 6.15,

$$Q = k_w \sqrt{2g} L_w h_0^{3/2} = (0.368)\sqrt{2(9.81)}(1.0)(0.20)^{3/2} = 0.146 \, \text{m}^3/\text{s}$$

We will now refine the solution by taking the approach velocity head into account based on the calculated discharge. The total depth at the approach section is $0.30 + 0.20 = 0.50$ m, thus, $V_0 = 0.146/[(0.50)(1.0)] = 0.292$ m/s. The corresponding velocity head becomes $V_0^2/(2g) = (0.292)^2/[2(9.81)] = 0.004$ m. Thus, $E_0 = 0.20 + 0.004 = 0.204$ m. We can now recalculate C_v and k_d, using Equations 6.14 and 6.17, respectively, as

$$C_v = \left(\frac{E_0}{h_0}\right)^{3/2} = \left(\frac{0.204}{0.20}\right)^{3/2} = 1.03$$

and

$$k_d = 0.358 + 0.038 \frac{E_0}{L_b} = 0.358 + 0.038 \frac{0.204}{0.75} = 0.368$$

Then, by using Equation 6.16, $k_w = (0.368)(1.03) = 0.379$. Substituting this into Equation 6.15, we obtain $Q = 0.15 \, \text{m}^3/\text{s}$. We can now update the velocity head again using this discharge, and repeat the calculations. The next set of calculations results in $Q = 0.15 \, \text{m}^3/\text{s}$. This is the same as the result of the previous iteration, and is accepted as the final result.

6.1.3 FLUMES

Flumes are open-channel flow segments built with contracted sidewalls and/or raised bottoms. Among the various types of flumes available as flow measuring devices summarized by US Bureau of Reclamation (2001), the Parshall flume is employed most widely. A schematic of a Parshall flume is shown in Figure 6.8, with the dimensions given in Table 6.1 for various sizes. Flumes with throat widths of less than 8 ft have a rounded entrance with a 25% floor slope.

The flow passes through the critical depth at the throat section when the downstream depth is shallow. This condition is known as free flow. A unique water surface profile develops within the flume for each discharge under the free flow conditions, and it is adequate to take one depth measurement, h_0, to determine the discharge. However, high downstream depths cause submerged

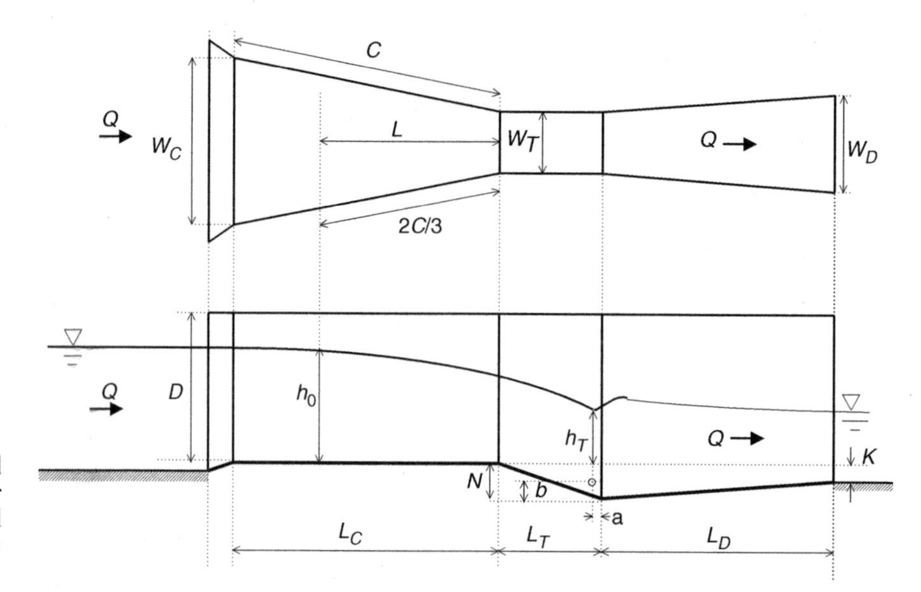

FIGURE 6.8 Parshall Flume (after Kilpatrick and Schneider, 1983)

TABLE 6.1 Standard Parshall Flume Dimensions (After Kilpatrick and Schneider 1983)

Widths			Axial lengths			Vertical dimensions			Gage points				Free flow capacity	
W_T (ft)	W_C (ft)	W_D (ft)	L_C (ft)	L_T (ft)	L_D (ft)	D (ft)	N (ft)	K (ft)	C (ft)	L (ft)	a (ft)	b (ft)	Min. (cfs)	Max. (cfs)
1.0	2.77	2.00	4.41	2.0	3.0	3.0	0.75	0.25	4.50	3.00	0.167	0.25	0.11	16.1
1.5	3.36	2.50	4.66	2.0	3.0	3.0	0.75	0.25	4.75	3.17	0.167	0.25	0.15	24.6
2.0	3.96	3.00	4.91	2.0	3.0	3.0	0.75	0.25	5.00	3.33	0.167	0.25	0.42	33.1
3.0	5.16	4.00	5.40	2.0	3.0	3.0	0.75	0.25	5.50	3.67	0.167	0.25	0.61	50.4
4.0	6.35	5.00	5.88	2.0	3.0	3.0	0.75	0.25	6.00	4.00	0.167	0.25	1.30	67.9
5.0	7.55	6.00	6.38	2.0	3.0	3.0	0.75	0.25	6.50	4.33	0.167	0.25	1.60	85.6
6.0	8.75	7.00	6.86	2.0	3.0	3.0	0.75	0.25	7.00	4.67	0.167	0.25	2.60	103.5
7.0	9.95	8.00	7.35	2.0	3.0	3.0	0.75	0.25	7.50	5.00	0.167	0.25	3.00	121.4
8.0	11.15	9.00	7.84	2.0	3.0	3.0	0.75	0.25	8.00	5.33	0.167	0.25	3.50	139.5
10.0	15.60	12.00	14.00	3.0	6.0	4.0	1.12	0.50	9.00	6.00			6.0	300.0
12.0	18.40	14.67	16.0	3.0	8.0	5.0	1.12	0.50	10.00	6.67			8.0	520.0
15.0	25.00	18.33	25.00	4.0	10.0	6.0	1.50	0.75	11.50	7.67			8.0	900.0
20.0	30.00	24.00	25.00	6.0	12.0	7.0	2.25	1.00	14.00	9.33			10.0	1340.0
25.0	35.00	29.33	25.00	6.0	13.0	7.0	2.25	1.00	16.50	11.00			15.0	1660.0
30.0	40.40	34.67	26.00	6.0	14.0	7.0	2.25	1.00	19.00	12.67			15.0	1990.0
40.0	50.80	45.33	27.00	6.0	16.0	7.0	2.25	1.00	24.00	16.00			20.0	2640.0
50.0	60.80	56.67	27.00	6.0	20.0	7.0	2.25	1.00	29.00	19.33			25.0	3280.0

flow conditions. In such a case a second depth measurement, h_T, is needed to determine the discharge. The percentage of submergence for Parshall flumes is defined as $100(h_T/h_0)$. For flumes having a throat width of 1–8 ft, the submergence should exceed 70% to affect the discharge measurement in the flume. For flumes with larger throat widths, the threshold submergence is 80% (Kilpatrick and Schneider, 1983).

The head–discharge relationship under the free flow conditions can be approximately expressed as (Davis, 1963):

$$Y_0 + \frac{Q_0^2}{2Y_0^2(1+0.4X_0)^2} = 1.351Q_0^{0.645} \tag{6.18}$$

where

$$Y_0 = \frac{h_0}{W_T} \tag{6.19}$$

$$X_0 = \frac{L}{W_T} \tag{6.20}$$

$$Q_0 = \frac{Q_f}{W_T^{5/2} g^{1/2}} \tag{6.21}$$

and Q_f = free flow discharge. The solution of this equation requires a trial-and-error method. For flumes with throat widths not exceeding 6 ft, we can replace Equation 6.18 with a simpler expression (Dodge, 1963):

$$Q_0 = \frac{Y_0^{1.5504}}{1.3096X_0^{0.0766}} \tag{6.22}$$

For submerged conditions, the discharge is calculated by using

$$Q_s = Q_f - k_s Q_c \tag{6.23}$$

where Q_s = submerged flow discharge, k_s = discharge correction factor, and Q_c = discharge correction unadjusted to flume size (Kilpatrick and Schneider, 1983). Figures 6.9 and 6.10 can be used to determine k_s and Q_c, depending on the throat size. In these figures, the percentage of submergence is $100h_T/h_0$.

EXAMPLE 6.4 A standard Parshall flume has a throat width of W_T = 4.0 ft. Determine the free flow discharge corresponding to h_0 = 2.4 ft.

For W_T = 4.0 ft, from Table 6.1 we obtain L = 4.0 ft. Then, by using Equations 6.19, 6.20, and 6.22,

$$Y_0 = \frac{h_0}{W_T} = \frac{2.4}{4.0} = 0.6$$

$$X_0 = \frac{L}{W_T} = \frac{4.0}{4.0} = 1.0$$

$$Q_0 = \frac{Y_0^{1.5504}}{1.3096X_0^{0.0766}} = \frac{(0.6)^{1.5504}}{1.3096(1.0)^{0.0766}} = 0.3459$$

Finally, by rearranging Equation 6.21 and evaluating Q_f,

$$Q_f = Q_0 W_T^{5/2} g^{1/2} = (0.3459)(4.0)^{5/2}(32.2)^{1/2} = 62.8\,\text{cfs}$$

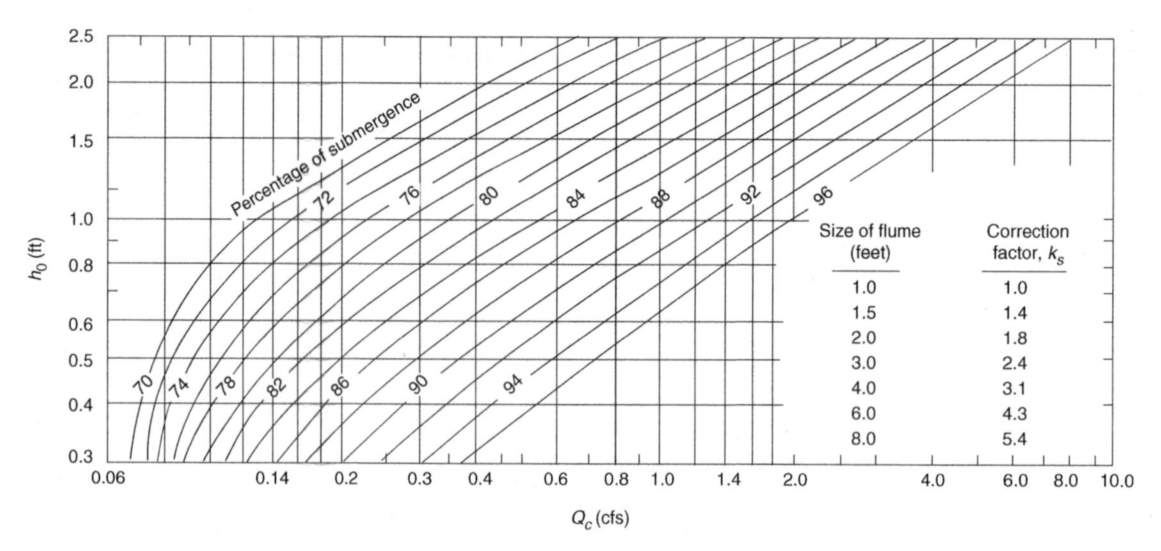

FIGURE 6.9 Submerged discharge correction for throat lengths of 1–8 ft (after Kilpatrick and Schneider, 1983)

EXAMPLE 6.5 Suppose the downstream depth is $h_T = 1.82$ ft in the Parshall flume considered in Example 6.4. Determine the discharge.

The percentage of submergence is 100(1.82/2.4) = 76%. From Figure 6.9, with $h_0 = 2.4$ ft and $W_T = 4.0$ ft, we obtain $Q_c = 1.15$ cfs and $k_s = 3.1$. Then, by using Equation 6.23,

$$Q_s = Q_f - k_s Q_c = 62.8 - (3.1)(1.15) = 59.2 \text{ cfs}.$$

6.2 CULVERTS

Culverts are short drainage conduits that convey stormwater through highway and railway embankments. They are also used as outlet structures for detention basins. Most culverts are circular, rectangular (box), or elliptical in cross-section. Other commonly used shapes include arch and pipe-arch culverts. Most culverts are made of concrete, corrugated aluminum, and corrugated steel. Concrete culverts may be reinforced. Some are lined with another material, such as asphalt, to prevent corrosion and reduce flow resistance.

The inlet configuration plays an important role in the hydraulic performance of culverts. A variety of prefabricated and constructed-in-place inlet installations are commonly used. These include projecting culvert barrels, concrete headwalls, end sections, and culvert ends mitered to conform to the fill slope. Figure 6.11 depicts various standard inlet types.

A variety of flow types can occur in a culvert, depending on the upstream and downstream conditions, the inlet geometry, and the conduit characteristics. A culvert may flow full, partially full (in subcritical or supercritical flow

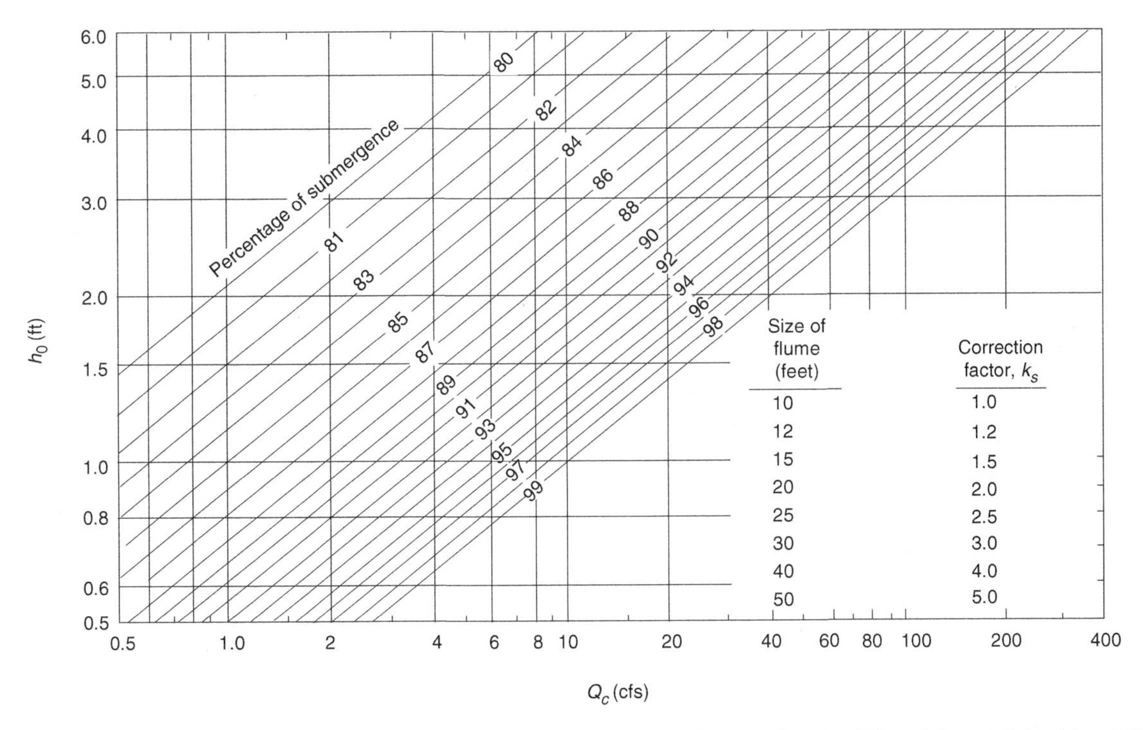

FIGURE 6.10 Submerged discharge correction for throat lengths of 10–50 ft (after Kilpatrick and Schneider, 1983)

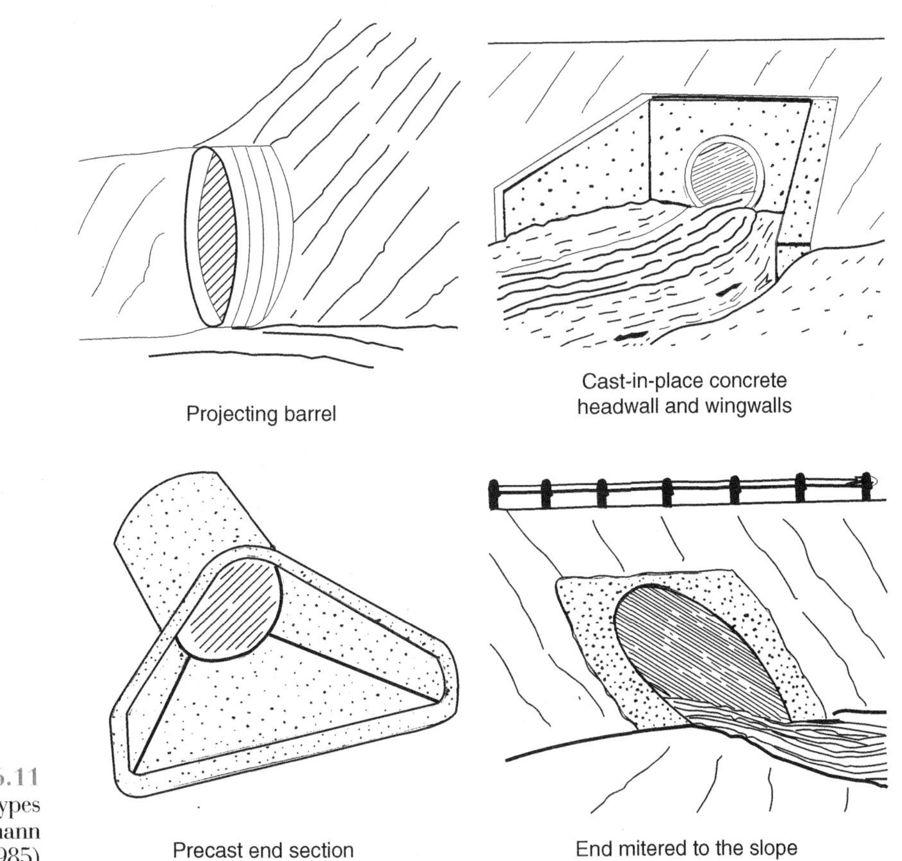

FIGURE 6.11 Standard inlet types (after Normann *et al.*, 1985)

conditions), or a combination of both. Partially full flow can be subcritical or supercritical. Flow conditions may change over time for any given culvert. Various US Geological Survey (Bodhaine, 1976) and Federal Highway Administration (Normann *et al.*, 1985) publications are available in the literature on culvert hydraulics. The Federal Highway Administration (FHWA) (Normann *et al.*, 1985) procedures are adapted herein for the most part. However, the equations are modified into a form that can be employed using any consistent unit system.

Flow in a culvert can be controlled either by the inlet (upstream) or by the outlet (downstream). Inlet control occurs when the conveyance capacity of the culvert barrel is higher than the inlet will accept; otherwise, outlet control flow occurs. We discuss the hydraulics of inlet and outlet control in the following sections.

6.2.1 INLET CONTROL FLOW

Inlet control flow generally occurs in steep, smooth culverts. The culvert will flow partially full under supercritical conditions, as shown in Figures 6.12a and 6.12c. However, if the downstream end of the culvert is submerged a hydraulic jump can form, after which the culvert will flow full as in Figure 6.12b.

The hydraulic behavior of the inlet is similar to that of a weir if the inlet is unsubmerged. If the inlet is submerged, it will perform similarly to an orifice.

According to the FHWA (Normann *et al.*, 1985), the inlet will be considered unsubmerged if

$$\frac{Q}{AD^{0.5}\,g^{0.5}} \leq 0.62 \tag{6.24}$$

where Q = discharge, A = cross-sectional area of the culvert, D = interior height of the culvert, and g = gravitational acceleration. Two forms of equations are available for unsubmerged inlets. The form I equation is

$$\frac{HW}{D} = \frac{y_c}{D} + \frac{V_c^2}{2gD} + K_I\left(\frac{Q}{AD^{0.5}\,g^{0.5}}\right)^{M_I} + k_s S \tag{6.25}$$

where HW = headwater depth above the upstream invert of the culvert, y_c = critical depth, V_c = velocity at critical depth, $k_s = 0.7$ for mitered inlets and -0.5 for non-mitered inlets, S = culvert barrel slope, and K_I, M_I = empirical constants. The values of K_I and M_I are given in Table 6.2 for various inlet configurations.

For circular culverts, the critical depth, y_c, can be determined by using Equation 2.6 or Figure 2.2. Then, from the geometry of a circular pipe,

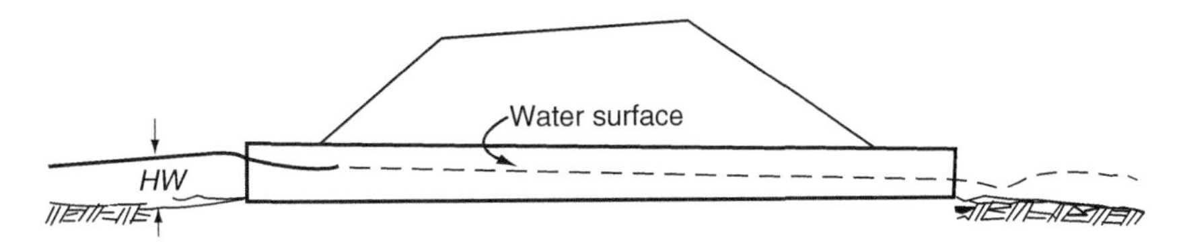

(a) Outlet unsubmerged

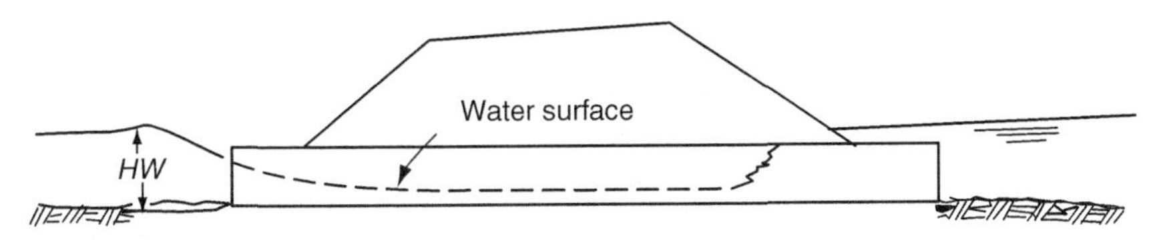

(b) Outlet submerged
inlet unsubmerged

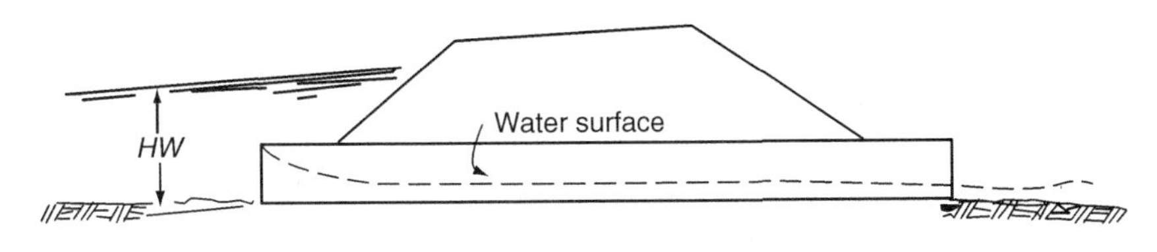

(c) Inlet submerged

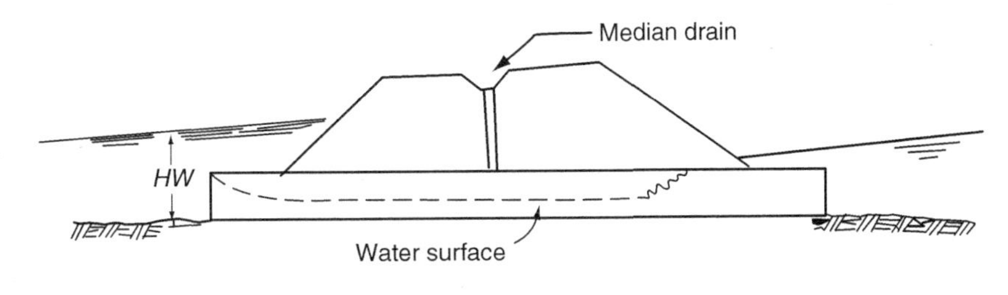

(d) Outlet submerged

FIGURE 6.12 Types of inlet control flow (after Normann *et al.*, 1985)

the corresponding area is found as $A_c = (2\theta - \sin 2\theta)D^2/8$ (see Table 1.1), and the velocity, V_c, can be determined by using

$$V_c = \frac{8Q}{(2\theta - \sin 2\theta)D^2} \tag{6.26}$$

where θ (in radians) is

$$\theta = \pi - \arccos\left(\frac{2y_c}{D} - 1\right) \tag{6.27}$$

As we recall from Chapter 2, for rectangular channels or box culverts,

$$y_c = \left(\frac{Q^2}{gb^2}\right)^{1/3} \tag{6.28}$$

TABLE 6.2 Culvert Inlet Control Flow Coefficients (Adapted From Normann *et al.*, 1985)

Shape and material	Inlet edge description	K_I	M_I	K_{II}	M_{II}	c	Y
Circular concrete	Square edge with headwall	0.3155	2.0			1.2816	0.67
Circular concrete	Groove end with headwall	0.2512	2.0			0.9402	0.74
Circular concrete	Groove end projecting	0.1449	2.0			1.0207	0.69
Circular corrugated metal	Headwall	0.2512	2.0			1.2204	0.69
Circular corrugated metal	Mitered to slope	0.2113	1.33			1.4909	0.75
Circular corrugated metal	Projecting	0.4596	1.50			1.7807	0.54
Circular	Beveled ring, 45° bevels	0.1381	2.50			0.9660	0.74
Circular	Beveled ring, 33.7° bevels	0.1381	2.50			0.7825	0.83
Rectangular box	30–75° Wingwall flares	0.1475	1.00			1.1173	0.81
Rectangular box	90° and 15° Wingwall flares	0.2243	0.75			1.2880	0.80
Rectangular box	0° Wingwall flare	0.2243	0.75			1.3621	0.82
Corrugated metal box	90° Headwall	0.2673	2.00			1.2204	0.69
Corrugated metal box	Thick wall projecting	0.3025	1.75			1.3492	0.64
Corrugated metal box	Thin wall projecting	0.4596	1.50			1.5971	0.57
Horizontal ellipse concrete	Square edge with headwall	0.3220	2.0			1.2816	0.67
Horizontal ellipse concrete	Groove end with headwall	0.1381	2.5			0.9402	0.74
Horizontal ellipse concrete	Groove end projecting	0.1449	2.0			1.0207	0.69
Vertical ellipse concrete	Square edge with headwall	0.3220	2.0			1.2816	0.67
Vertical ellipse concrete	Groove end with headwall	0.1381	2.5			0.9402	0.74
Vertical ellipse concrete	Groove end projecting	0.3060	2.0			1.0207	0.69
Rectangular box	45° Wingwall flare $d = 0.043D$			1.623	0.667	0.9950	0.80
Rectangular box	18–33.7° Wingwall flare $d = 0.083D$			1.547	0.667	0.8018	0.83
Rectangular box	90° Headwall with ¾″ chamfers			1.639	0.667	1.2075	0.79
Rectangular box	90° Headwall with 45° bevels			1.576	0.667	1.0111	0.82
Rectangular box	90° Headwall with 33.7° bevels			1.547	0.667	0.8114	0.865
Rectangular box	¾″ Chamfers; 45° skewed headwall			1.662	0.667	1.2944	0.73
Rectangular box	¾″ Chamfers; 30° skewed headwall			1.697	0.667	1.3685	0.705
Rectangular box	¾″ Chamfers; 15° skewed headwall			1.735	0.667	1.4506	0.73
Rectangular box	45° Bevels; 10–45° skewed headwall			1.585	0.667	1.0525	0.75
Rectangular box with ¾″ chamfers	45° Non-offset wingwall flares			1.582	0.667	1.0916	0.803
Rectangular box with ¾″ chamfers	18.4° Non-offset wingwall flares			1.569	0.667	1.1624	0.806
Rectangular box with ¾″ chamfers	18.4° Non-offset wingwall flares with 30° skewed barrel			1.576	0.667	1.2429	0.71
Rectangular box with top bevels	45° Wingwall flares – offset			1.582	0.667	0.9724	0.835
Rectangular box with top bevels	33.7° Wingwall flares – offset			1.576	0.667	0.8144	0.881
Rectangular box with top bevels	18.4° Wingwall flares –offset			1.569	0.667	0.7309	0.887
Circular	Smooth tapered inlet throat			1.699	0.667	0.6311	0.89
Circular	Rough tapered inlet throat			1.652	0.667	0.9306	0.90
Rectangular	Tapered inlet throat			1.512	0.667	0.5764	0.97
Rectangular concrete	Side tapered – less favorable edges			1.783	0.667	1.5005	0.85
Rectangular concrete	Side tapered – more favorable edges			1.783	0.667	1.2172	0.87
Rectangular concrete	Slope tapered – less favorable edges			1.592	0.667	1.5005	0.65
Rectangular concrete	Slope tapered – more favorable edges			1.592	0.667	1.2172	0.71

where b = width of the box culvert. Also, $V_c^2/2g = 0.5y_c$. Thus, Equation 6.25 can be rewritten for box culverts as

$$\frac{HW}{D} = \frac{3}{2D}\left(\frac{Q^2}{gb^2}\right)^{1/3} + K_I\left(\frac{Q}{AD^{0.5}g^{0.5}}\right)^{M_I} + k_s S \tag{6.29}$$

The form II equation for unsubmerged inlets is

$$\frac{HW}{D} = K_{II}\left(\frac{Q}{AD^{0.5}g^{0.5}}\right)^{M_{II}} \tag{6.30}$$

where K_{II} and M_{II} are empirical constants given in Table 6.2. Both form I and form II equations are acceptable for practical purposes, and the choice between the two is governed by the availability of the empirical coefficients (Table 6.2) for the type of the culvert being considered.

The inlet will be submerged if

$$\frac{Q}{AD^{0.5}g^{0.5}} \geq 0.70 \tag{6.31}$$

The flow equation for submerged inlets is

$$\frac{HW}{D} = c\left(\frac{Q}{AD^{0.5}g^{0.5}}\right)^2 + Y + k_s S \tag{6.32}$$

where S = slope, c and Y are empirical constants given in Table 6.2, and

$$\begin{aligned} k_s &= 0.7 \quad \text{for inlets mitered to embankment slope} \\ k_s &= -0.5 \quad \text{for inlets not mitered to embankment slope} \end{aligned} \tag{6.33}$$

A transition from unsubmerged to submerged condition occurs for $0.62 < (Q/AD^{0.5}g^{0.5}) < 0.70$. A linear interpolation between the submerged and unsubmerged inlet equations can be used for the transition zone.

There are several nomographs presented by the FHWA (Normann *et al.*, 1985) for quick calculations of culvert flow. Figures 6.13 and 6.14 are included as examples of concrete pipe culverts and box culverts, respectively.

EXAMPLE 6.6 A circular concrete culvert has a diameter D = 3 ft, a slope of S = 0.025, and a square edge inlet with a headwall. The inlet is mitered to embankment slope. Determine the headwater depth, HW, when the culvert conveys Q = 30 cfs under inlet control conditions.

To determine whether the inlet is submerged, we first calculate

$$\frac{Q}{AD^{0.5}g^{0.5}} = \frac{30}{(\pi(3)^2/4)\,(3)^{0.5}(32.2)^{0.5}} = 0.43 < 0.62$$

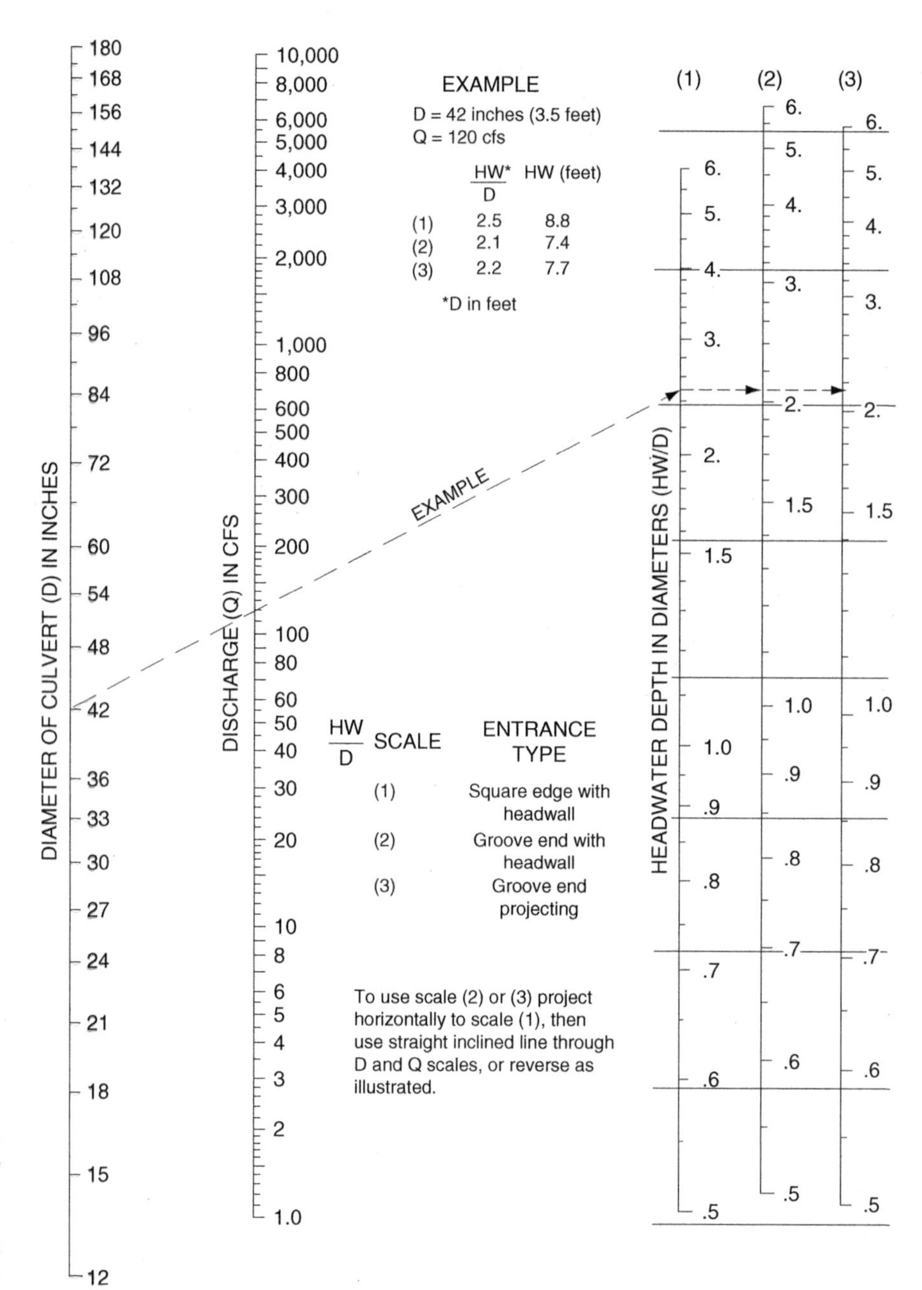

FIGURE 6.13 Headwater depth for concrete pipe culverts with inlet control (after Normann *et al.*, 1985)

Therefore, the inlet is unsubmerged. Table 6.2 lists $K_I = 0.3155$ and $M_I = 2.0$, while no values are provided for K_{II} and M_{II} for this inlet configuration. We will use the form I equation (Equation 6.25) to determine the headwater depth. Let us first determine the critical depth y_c. With reference to Figure 2.2, and denoting the culvert diameter by D,

$$\frac{Q}{g^{0.5}D^{2.5}} = \frac{30.0}{32.2^{0.5}3.0^{2.5}} = 0.34$$

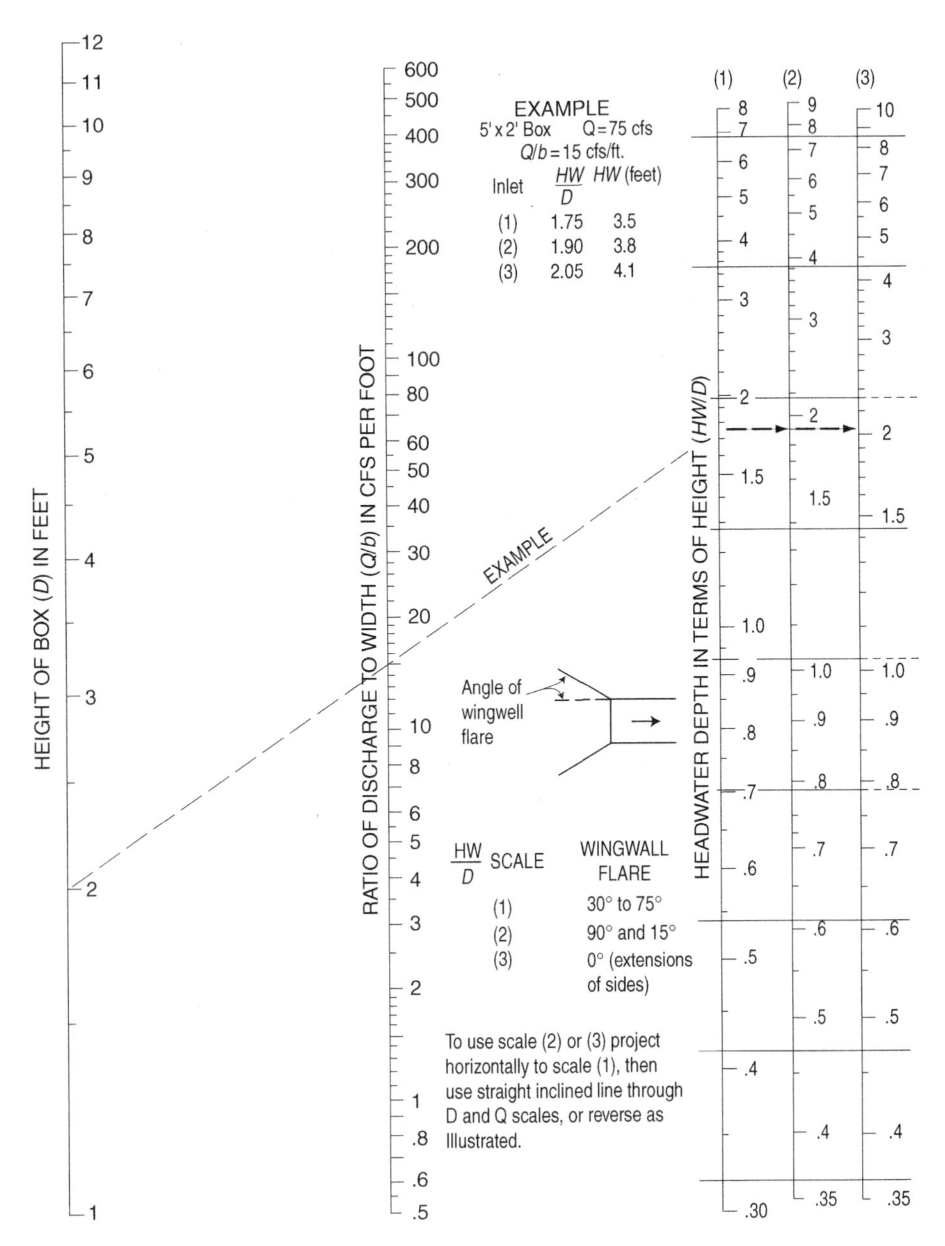

FIGURE 6.14 Headwater depth for box culverts with inlet control (after Normann *et al.*, 1985)

Then, from Figure 2.2, $y_c/D = 0.59$. Thus $y_c = 0.59(3.0) = 1.77$ ft. Next by using Equation 6.27,

$$\theta = (3.14) - \arc\cos\left[\frac{2(1.77)}{3.0} - 1\right] = 1.75 \text{ rad} = 100^\circ$$

With $2\theta = 2(1.75) = 3.50$ rad., from Equation 6.26

$$V_c = \frac{8(30)}{(3.50 - \sin 3.5)(3)^2} = 6.92 \text{ fps}$$

Now, by using Equation 6.25 with $k_s = 0.7$ for a mitered inlet,

$$\frac{HW}{D} = \frac{1.77}{3.0} + \frac{6.92^2}{2(32.2)(3.0)} + 0.3155(0.43)^{2.0} + (0.7)(0.025) = 0.91$$

and therefore $HW = 0.91\ (3.0) = 2.73$ ft

EXAMPLE 6.7 A 100-ft long circular culvert has a diameter $D = 4$ ft and a bottom slope $S = 0.02$. The culvert has a smooth tapered inlet, not mitered to the embankment slope. Determine the headwater depth when the culvert carries 120 cfs under inlet control conditions.

To determine whether the inlet is submerged we first calculate

$$\frac{Q}{AD^{0.5}g^{0.5}} = \frac{120}{(\pi(4)^2/4)(4)^{0.5}(32.2)^{0.5}} = 0.84 > 0.70$$

Therefore, the inlet is submerged, and we will use Equation 6.32. From Table 6.2, we obtain $c = 0.6311$ and $Y = 0.89$ for a circular culvert with a smooth tapered inlet throat. Also, $k_s = -0.5$ since the end is not mitered (Equation 6.33). Then, by using Equation 6.32,

$$\frac{HW}{D} = 0.6311(0.84)^2 + 0.89 - 0.5(0.02) = 1.33$$

and therefore $HW = (1.33)(4.0) = 5.32$ ft.

6.2.2 OUTLET CONTROL FLOW

A culvert may flow full or partially full under the outlet control conditions. When partially full, outlet control culvert flow is subcritical. Several outlet control flow types are depicted in Figure 6.15. Conditions a, d, and e shown in this figure are most common.

6.2.2.1 Full-flow conditions

Neglecting the difference between the velocity heads of the flow approaching a culvert and that downstream of a culvert, the energy equation for a culvert flowing full is written as

$$HW = TW - SL + \left(1 + k_e + \frac{2gn^2L}{k_n^2 R^{4/3}}\right)\frac{Q^2}{2gA^2} \tag{6.34}$$

where TW = tailwater depth measured from the downstream invert of the culvert, S = culvert slope, L = culvert length, g = gravitational acceleration, n = Manning roughness factor, R = hydraulic radius, A = cross-sectional area,

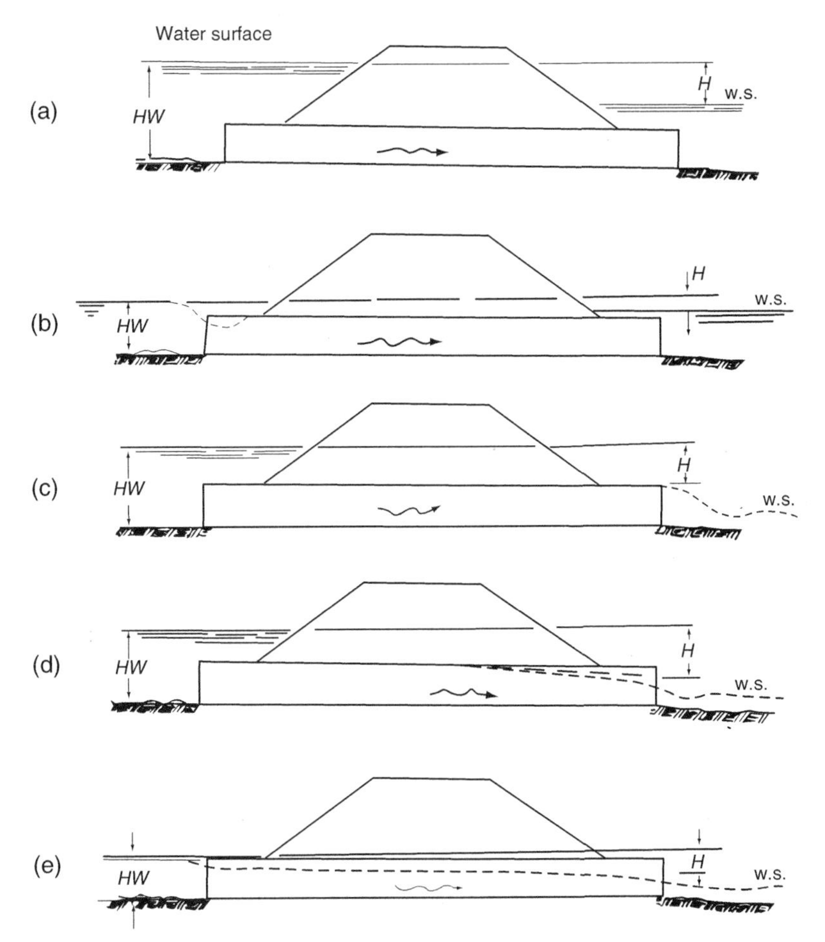

FIGURE 6.15 Types of outlet control flow (after Normann *et al.*, 1985)

$k_n = 1.0\,\text{m}^{1/3}/\text{s} = 1.49\,\text{ft}^{1/3}/\text{s}$, and k_e = entrance loss coefficient given in Table 6.3 as reported by Normann *et al.* (1985).

Equation 6.34 includes friction losses as well as entrance and exit losses. Figures 6.16 and 6.17 present nomographs for full flow in concrete pipe culverts and concrete box culverts, respectively. Full-flow nomographs for other types of culverts are also available in the literature (Normann *et al.*, 1985).

EXAMPLE 6.8 A reinforced concrete rectangular box culvert has the following properties: $D = 1.0\,\text{m}$, $b = 1.0\,\text{m}$, $L = 40\,\text{m}$, $n = 0.012$ and $S = 0.002$. The inlet is square-edged on three edges and has a headwall parallel to the embankment, and the outlet is submerged with $TW = 1.3\,\text{m}$. Determine the headwater depth, HW, when the culvert is flowing full at $Q = 3.0\,\text{m}^3/\text{s}$.

From Table 6.3, we obtain $k_e = 0.5$. Also, for a box culvert, $A = bD = (1.0)(1.0) = 1.0\,\text{m}^2$ and $R = bD/(2b + 2D) = (1.0)(1.0)/[2(1.0) + 2(1.0)] = 0.25\,\text{m}$ under full-flow conditions. Therefore, by using Equation 6.34,

$$HW = 1.3 - (0.002)(40) + \left[1 + 0.5 + \frac{2(9.81)(0.012)^2(40)}{(1.0)^2(0.25)^{4/3}}\right]\frac{(3.0)^2}{2(9.81)(1.0)^2} = 2.24\,\text{m}$$

TABLE 6.3 Entrance Loss Coefficients (After Normann *et al.*, 1985)

Type of structure and design of entrance	Coefficient k_e
Pipe, concrete:	
Projecting from fill socket end (groove-end)	0.2
Projecting from fill, sq. cut end	0.5
Headwall or headwall and wingwalls:	
• Socket end of pipe (groove-end)	0.2
• Square-edge	0.5
• Rounded (radius = 1/12*D*)	0.2
Mitered to conform to fill slope	0.7
End-section conforming to fill slope	0.5
Beveled edges, 33.7° or 45° bevels	0.2
Side- or slope-tapered inlet	0.2
Pipe or pipe-arch, corrugated metal:	
Projecting from fill (no headwall)	0.9
Headwall or headwall and wingwalls square-edge	0.5
Mitered to conform to fill slope, paved or unpaved slope	0.7
End-section conforming to fill slope	0.5
Beveled edges, 33.7° or 45° bevels	0.2
Side- or slope-tapered inlet	0.2
Box, reinforced concrete:	
Headwall parallel to embankment (no wingwalls):	
• Square-edged on three edges	0.5
• Rounded on three edges to radius of 1/12 barrel dimension, or beveled edges on three sides	0.2
Wingwalls at 30–75° to barrel:	
• Square-edged at crown	0.4
• Crown edge rounded to radius of 1/12 barrel dimension or beveled top edge	0.2
Wingwalls at 10–25° to barrel:	
• Square-edged at crown	0.5
Wingwalls parallel (extension of sides):	
• Square-edged at crown	0.7
Side- or slope-tapered inlet	0.2

6.2.2.2 Partly full-flow conditions

For partly full flow in culverts controlled by the outlet, an accurate relationship between discharge and headwater elevation can be obtained by using the gradually-varied flow calculations discussed in Chapter 4. In these calculations, the downstream depth is set equal to the higher of the tailwater depth (TW) and the critical depth (y_c). If the calculated water surface profile intersects the top of the barrel, full-flow equations are used between that point and the upstream end of the culvert. The head loss (h_{Lf}) for the full-flow segment is calculated as

$$h_{Lf} = \left(1 + k_e + \frac{2gn^2 L_f}{k_n^2 R^{4/3}}\right) \frac{Q^2}{2gA^2} \tag{6.35}$$

where L_f = length of the full-flow segment.

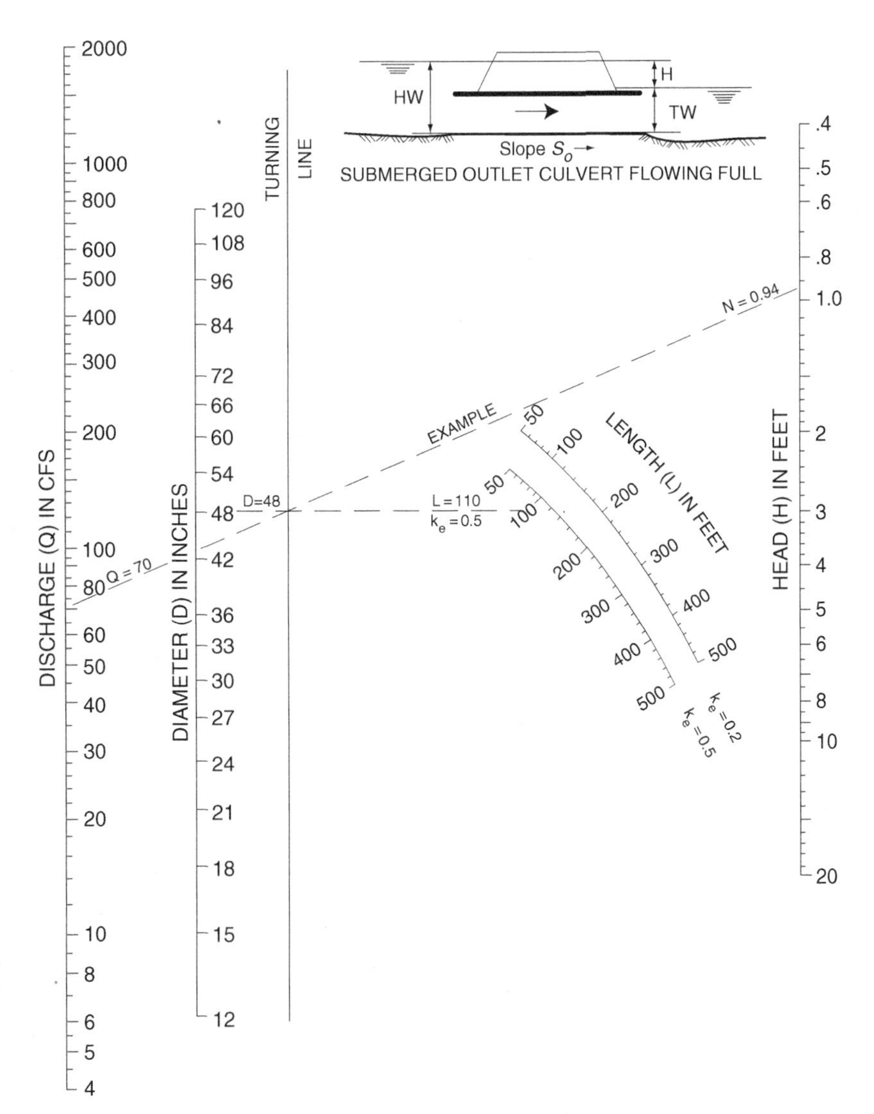

FIGURE 6.16 Head loss in concrete pipe culverts flowing full with $n = 0.012$ (after Normann *et al.*, 1985)

To avoid tedious gradually-varied flow calculations, the FHWA (Normann *et al.*, 1985) developed an approximate method for partly-full outlet control culvert flow. In this method, the headwater elevation is calculated using

$$HW = H_D - SL + \left(1 + k_e + \frac{2g\,n^2 L}{k_n^2 R^{4/3}}\right)\frac{Q^2}{2gA^2} \tag{6.36}$$

in which R and A are calculated assuming the culvert is full. Also, H_D is set equal to the tailwater depth, TW, if $TW > (y_c + D)/2$ where y_c = critical depth and D = interior height of the culvert. Otherwise, $H_D = (y_c + D)/2$.

Equation 6.36 is deemed satisfactory when the culvert flows full over at least part of its length, as shown in Figure 6.15d. The approximate method becomes less accurate if free-surface flow occurs over the entire length of the culvert, in which case the results are acceptable only if $HW > (0.75D)$. For lower headwater elevations, gradually-varied flow calculations are required.

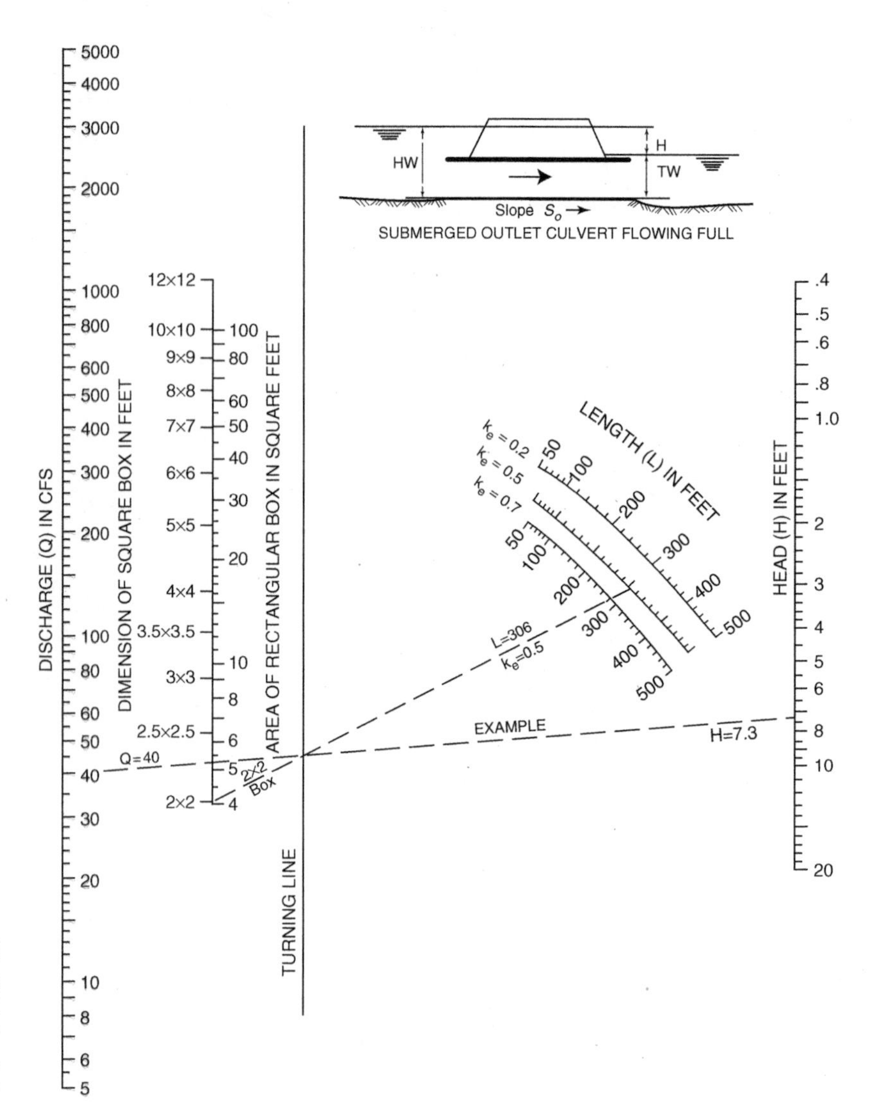

FIGURE 6.17 Head loss in concrete box culverts flowing full with $n = 0.012$ (after Normann *et al.*, 1985)

EXAMPLE 6.9 Determine the headwater elevation for the culvert of Example 6.8 if the tailwater is 0.6 m and the discharge is $Q = 2.0\,\text{m}^3/\text{s}$.

We first calculate the critical depth, by using Equation 6.28, as

$$y_c = \left[\frac{2.0^2}{(9.81)(1.0)^2}\right]^{1/3} = 0.74\,\text{m}$$

Then $(y_c + D)/2 = (0.74 + 1.0)/2 = 0.87\,\text{m}$. This is higher than $TW = 0.6\,\text{m}$, and therefore $H_D = 0.87\,\text{m}$. Then, by using Equation 6.36,

$$HW = 0.87 - (0.002)(40) + \left[1 + 0.5 + \frac{(2)(9.81)(0.012)^2(40)}{(1)^2(0.25)^{4/3}}\right]\frac{2.0^2}{2(9.81)(1.0)^2} = 1.24\,\text{m}$$

Noting that $1.24\,\text{m} > (0.75)(1.0)\,\text{m}$, the calculated result is acceptable.

6.2.3 SIZING OF CULVERTS

As discussed in the preceding sections, the equations describing the flow in a culvert depend on the flow condition (inlet or outlet control) and the nature of the flow (full or partly full). The flow is most likely to be governed by outlet control if the culvert slope is mild. For mild slopes, full flow will occur if $TW > D$; otherwise, the flow will be partly full. The flow is most likely to be governed by inlet control if the culvert slope is steep. An exception is that full flow may occur if $TW > D$.

Stormwater drainage culverts placed under highway and railway embankments are sized to accommodate a design discharge without overtopping the embankment. In a typical situation the design discharge and the tailwater elevation are known, and the culvert is sized to prevent the headwater elevation from exceeding an allowable value. The FHWA (Normann *et al.*, 1985) suggests a 'minimum performance' approach to sizing culverts as drainage structures. In this approach, no attempt is made to determine whether inlet control or outlet control flow will actually occur under the design flow conditions; instead, both flow conditions are checked and the one resulting in a more conservative design is picked. In other words, a culvert size is selected so that for the design discharge the calculated headwater elevation will not exceed the maximum allowable value under either inlet control or outlet control conditions.

6.3 OVERFLOW SPILLWAYS

Spillways are hydraulic structures provided for storage and detention dams in order to release surplus water or floodwater from a reservoir and convey it to a downstream river or channel. The terms *service spillway* and *primary spillway* refer to principal spillways used to regulate flow from reservoirs and pass frequent floods. *Auxiliary* or *secondary spillways* operate infrequently during large floods exceeding the capacity of the principal spillways. The *emergency spillways* are provided for additional safety, should emergencies (such as enforced shutdown of outlet works of a dam or malfunctioning of the spillway gates) occur. Emergency spillways also act like auxiliary spillways if the design flood is exceeded. There are various types of spillways, including *ogee spillways*, *chute spillways*, *side-channel spillways*, and *morning glory spillways*. The US Bureau of Reclamation (1987) provides a thorough discussion of most spillway types. In this section we will discuss *ogee spillways*, also called *overflow spillways*.

As reported by the US Bureau of Reclamation (1987), the ogee spillways are shaped such that the upper curve of the spillway body conforms to the profile of the lower nappe of a ventilated sheet falling from a sharp-crested weir. Flow over the crest adheres to the face of the profile by preventing access of air to the underside of the sheet. At the design discharge, the flow glides over the crest with no interference from the boundary. The profile below the upper curve of the ogee is continued tangent along a slope. A reverse curve at the bottom

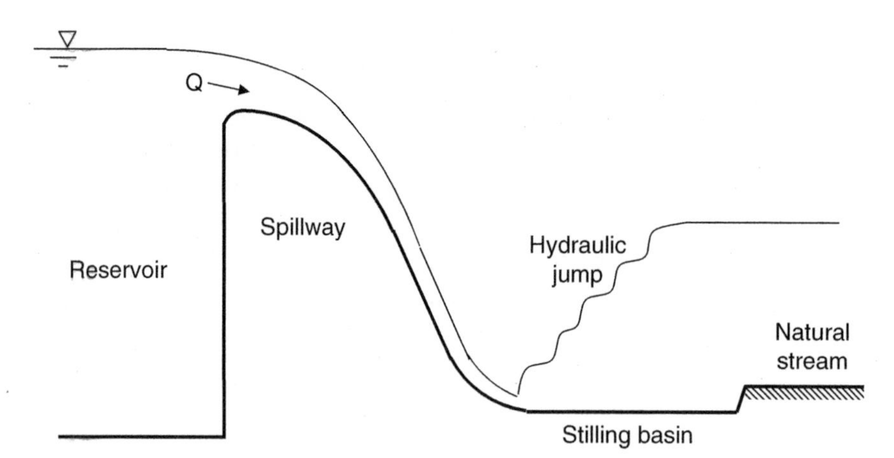

FIGURE 6.18 Ogee spillway

turns the flow to the apron of a *stilling basin* downstream, as shown in Figure 6.18. The face of the spillway can be vertical (as shown in the figure) or inclined.

The flow upstream of the spillway is subcritical at a very low velocity. The flow accelerates as it approaches the spillway crest, becoming critical as water passes over the crest. Below the crest, down the chute, supercritical flow occurs, gaining very high velocities as the potential energy is converted to kinetic energy. At or near the terminus of the chute, a hydraulic jump is forced for flow to change from the supercritical to the subcritical state before joining the natural stream channel downstream. However, the hydraulic jump may be partially or fully drowned by a high tailwater. Even higher tailwater elevations can submerge the spillway crest and affect the discharge.

Some spillways are gated, and are called *controlled* spillways. Also, bridges are often provided across the spillways for pedestrians and vehicular traffic. Piers may be used to support the bridge, and on controlled spillways the piers are used to support the crest gates. If present, the abutments, piers, and gates affect the flow over the spillway.

6.3.1 SHAPE FOR UNCONTROLLED OGEE CREST

An ogee crest is shaped to approximate the profile of the under-nappe of a jet flowing over a sharp-crested weir. The shape of such a profile depends on the head, the inclination of the upstream face of the overflow section, and the elevation of the spillway crest above the floor (US Bureau of Reclamation, 1987). The form shown in Figure 6.19 adequately represents this shape for most conditions. In the figure, the profile is defined as it relates to two axes placed at the apex of the crest. That portion upstream from the origin is defined as a compound circular curve, while the portion downstream is defined by the equation

$$\frac{y}{H_0} = -K\left(\frac{x}{H_0}\right)^n \tag{6.37}$$

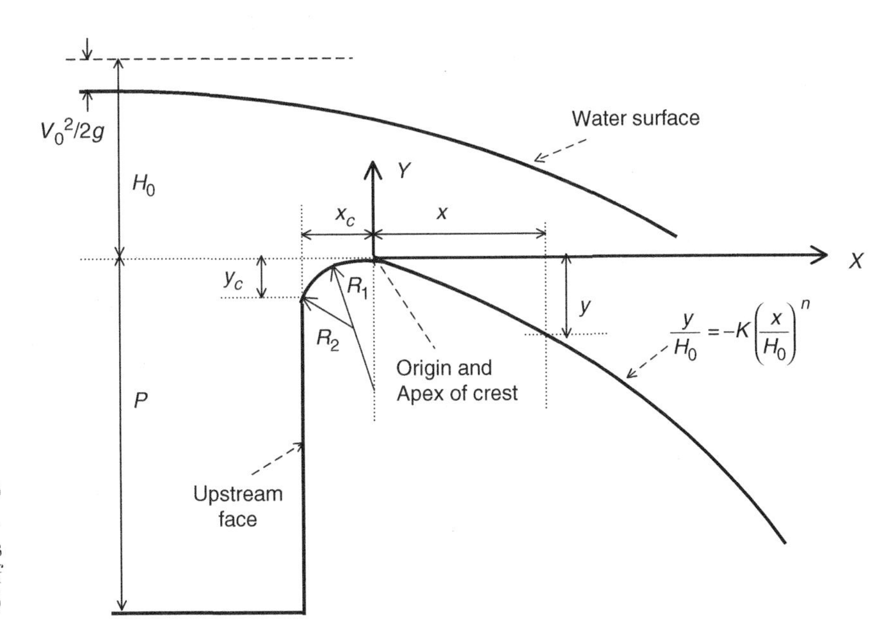

FIGURE 6.19 Elements of nappe-shaped crest profiles (after US Bureau of Reclamation, 1987)

where H_0 is the design head as shown in Figure 6.19, and K and n are constants whose values depend on the inclination of the upstream face and the velocity head of the approach flow. The US Bureau of Reclamation (1987) has provided various charts to determine K, n, R_1, R_2, y_C, and x_C. If the upstream face is vertical and the approach velocity head is negligible, we can use $K = 0.50$, $n = 1.872$, $R_1 = 0.53\,H_0$, $R_2 = 0.235\,H_0$, $y_C = 0.127\,H_0$, and $x_C = 0.283\,H_0$.

6.3.2 DISCHARGE OVER AN UNCONTROLLED OGEE CREST

The spillway crest is shaped for a design head, H_0. For this head, the discharge, Q, over the crest is given as

$$Q = k_{w0}\sqrt{2g}L_e H_0^{3/2} \tag{6.38}$$

where k_{w0} = discharge coefficient for design head, and L_e = effective crest length. Figure 6.20 displays the values of k_{w0} as a function of the crest height to design head ratio (P/H_0).

A spillway will often operate under heads different from the design head. In that event, the discharge is given as

$$Q = k_w\sqrt{2g}L_e H_e^{3/2} \tag{6.39}$$

where k_w = discharge coefficient and H_e = existing head including the velocity head. We can use Figure 6.21 in conjunction with Figure 6.20 to determine the discharge coefficient for spillways with a vertical face. Figure 6.21 also assumes that the flow over the crest is not affected by the downstream apron elevation or the tailwater depth. Additional charts are provided by the US Bureau of

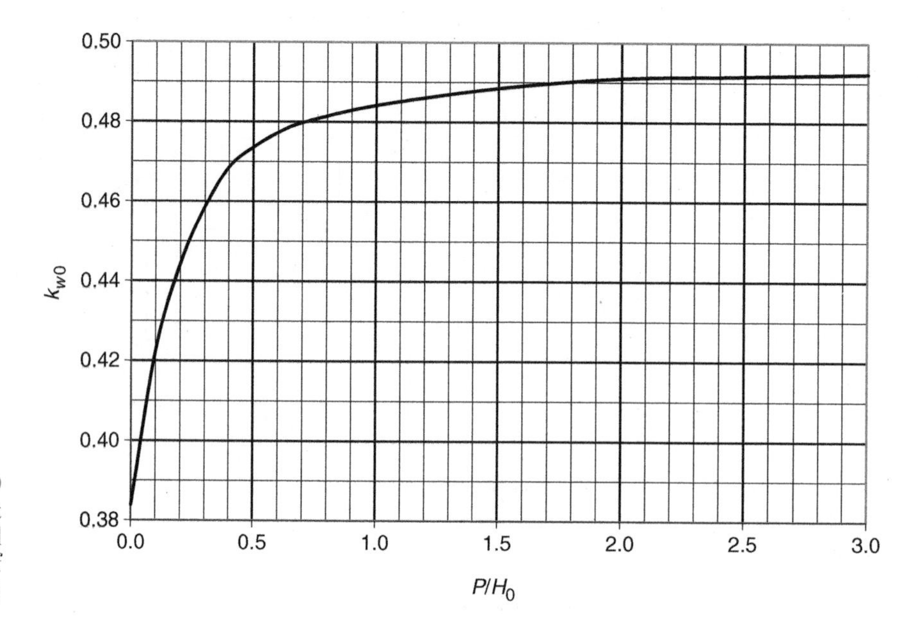

FIGURE 6.20 Discharge coefficient for design head (after US Bureau of Reclamation, 1987)

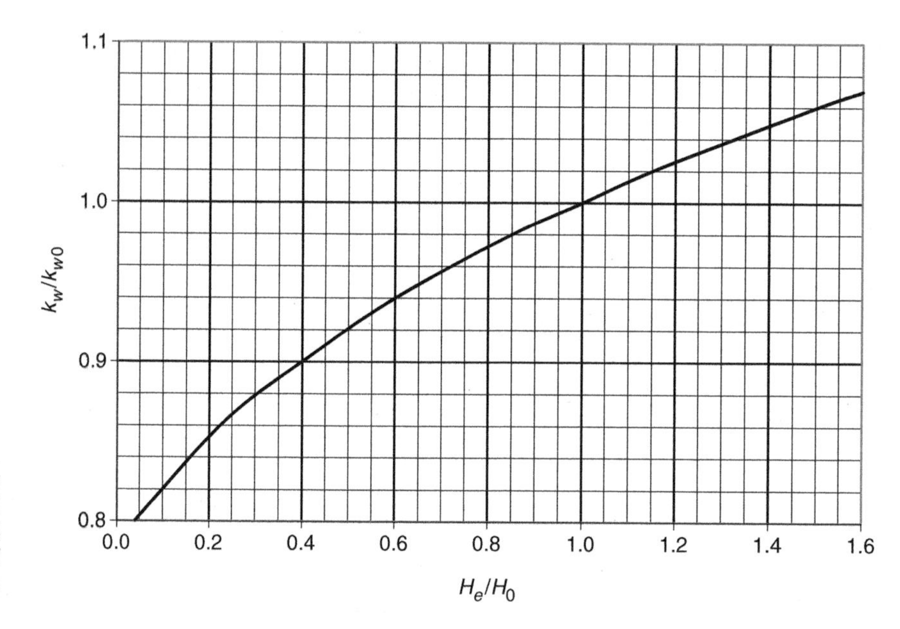

FIGURE 6.21 Spillway discharge coefficient for vertical faces (after US Bureau of Reclamation, 1987)

Reclamation (1987) to account for such effects if needed, and for inclined spillway faces. In most situations the approach velocity head is negligible, and therefore H_e is equal to the water surface elevation in the reservoir above the spillway crest.

The effective crest length, L_e, is less than the net crest length L where abutments and piers are present and are shaped to cause side contraction of the overflow. The effective crest length is calculated by using

$$L_e = L - 2(NK_p + K_a)H_e \tag{6.40}$$

where L_e = effective crest length, L = net crest length, N = number of piers, K_p = pier contraction coefficient, and K_a = abutment contraction coefficient. Detailed charts are presented by the US Army Corps of Engineers (1990) to determine the pier contraction coefficient. Here, we will use the approximate values of $K_p = 0.0$ for pointed-nose piers, $K_p = 0.01$ for round-nosed piers, and $K_p = 0.02$ for square-nosed piers with rounded corners. Likewise, the approximate abutment contraction coefficients are $K_a = 0.2$ for square abutments with headwall at 90° to direction of flow, and $K_a = 0.10$ for rounded abutments with headwall at 90° to direction of flow.

EXAMPLE 6.10 An uncontrolled overflow ogee crest for a spillway is to be designed that will discharge 3000 cfs at a design head of 5.0 ft. The upstream face of the crest is vertical. A bridge is to be provided over the crest, with bridge spans not to exceed 20 ft. The piers are 1.5 ft wide with rounded noses. The abutments are rounded with a headwall at 90° to the direction of flow. The vertical distance between the spillway crest and the floor of the reservoir is 9.0 ft. Determine the length of the spillway crest.

From the problem statement, we know that $Q = 3000$ cfs, $P = 9.0$ ft, $H_0 = 5.0$ ft, $K_p = 0.01$, and $K_a = 0.10$. With $P/H_0 = 9.0/5.0 = 1.8$, we obtain $k_{w0} = 0.49$ from Figure 6.20. Then, rearranging Equation 6.38 to determine L_e,

$$L_e = \frac{Q}{k_w\sqrt{2g}H_0^{3/2}} = \frac{3000}{(0.49)\sqrt{2(32.2)}(5.0)^{3/2}} = 68.2 \text{ ft}$$

We will need three bridge piers, since the bridge spans are not to exceed 20 ft. Now we can write Equation 6.40 for the design head and rearrange it to determine L as

$$L = L_e + 2(NK_p + K_a)H_0 = 68.2 + 2[3(0.01) + 0.10](5.0) = 69.5 \text{ ft}$$

Thus, the net crest length, not including the piers, is 69.5 ft. Noting that each of the three piers is 1.5 ft wide, the total crest length will be $69.5 + 3(1.5) = 74.0$ ft.

EXAMPLE 6.11 Suppose that in Example 6.10 the maximum expected head is 6.0 ft above the weir crest – in other words, the weir crest is shaped for a design head smaller than the maximum expected head. This is not unusual, since such a design is more economical. Obtain a discharge–head relationship for this spillway for heads varying from 1 ft to 6 ft.

The calculations are summarized in Table 6.4. Listed in column (1) are the heads we pick. Entries in column (2) are obtained by dividing those in column (1) by the design head, 5.0 ft. The entries in column (3) are obtained from Figure 6.21. We multiply these by 0.49, the value of k_{w0} from Example 6.10, to obtain the entries in column (4). We use Equation 6.40 to calculate the effective length values in column (5). Finally, we determine the discharge values in column (6) by using Equation 6.39.

TABLE 6.4 Calculations for Example 6.11

H_e (ft)	H_e/H_0	k_w/k_{w0}	k_w	L_e (ft)	Q (cfs)
1.0	0.2	0.85	0.42	69.2	233
2.0	0.4	0.90	0.44	69.0	689
3.0	0.6	0.94	0.46	68.7	1318
4.0	0.8	0.97	0.48	68.5	2111
5.0	1.0	1.00	0.49	68.2	3000
6.0	1.2	1.025	0.50	67.9	4004

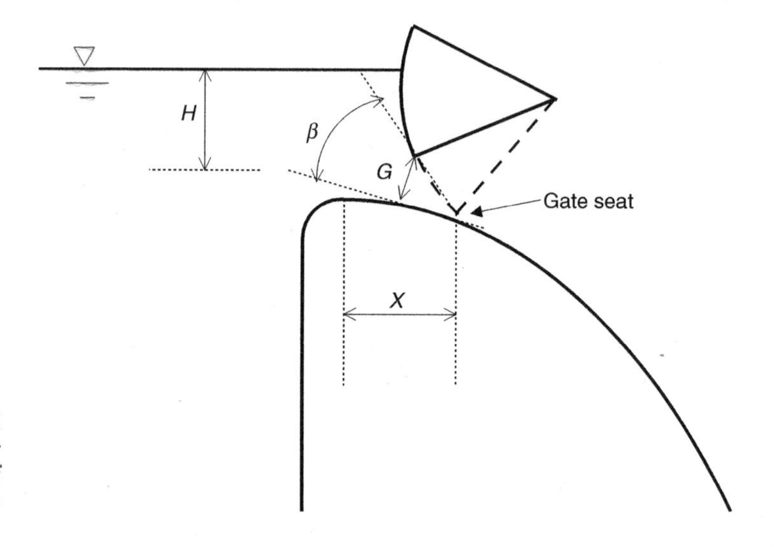

FIGURE 6.22 Gated spillway crest (after US Army Corps of Engineers, 1987)

6.3.3 DISCHARGE OVER GATE-CONTROLLED OGEE CRESTS

Controlled ogee spillways include crest gates that serve as a movable damming surface to adjust the flow over the spillway. *Tainter gates* are commonly used for this purpose. These gates rotate around an axis to adjust the gate opening to control the flow. Figure 6.22 shows a schematic of a gated ogee spillway, and Figure 6.23 displays a downstream view of a typical tainter gate.

The discharge over a gated ogee crest at partial gate openings is similar to flow through an orifice, and we can compute it as

$$Q = k_G G B_G \sqrt{2gH} \tag{6.41}$$

where k_G = discharge coefficient, G = shortest distance from the gate lip to the crest curve (see Figure 6.22), B_G = width of the gate opening, and H = head to the center of the gate opening, including the approach velocity head if not negligible.

Figure 6.24, constructed from information presented by the US Corps of Engineers (1987), provides average values of the discharge coefficient for tainter

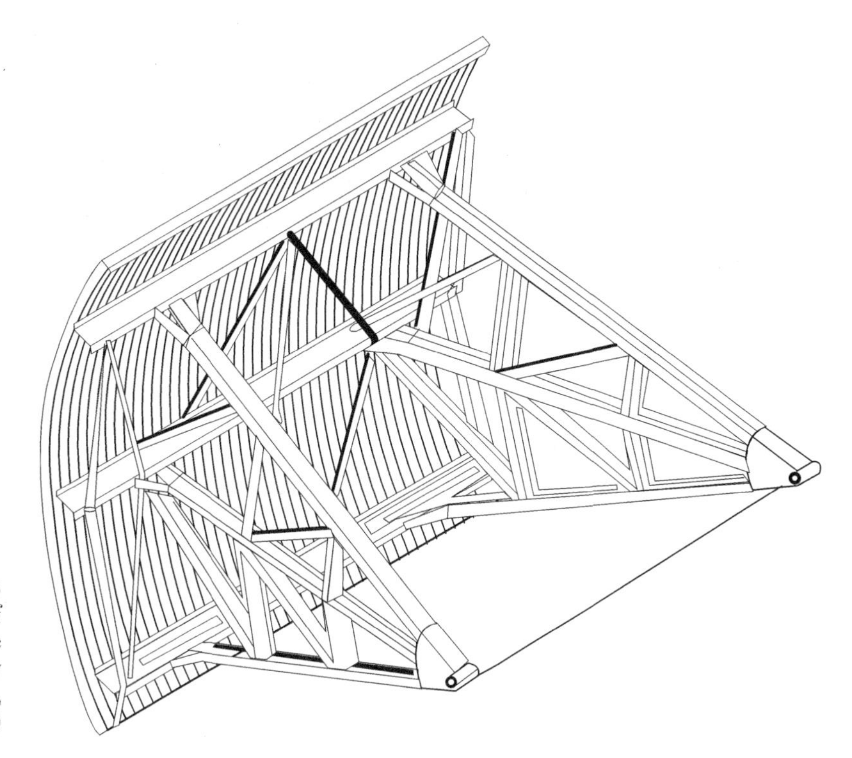

FIGURE 6.23
Downstream view of a typical tainter gate (after US Army Corps of Engineers, 2000)

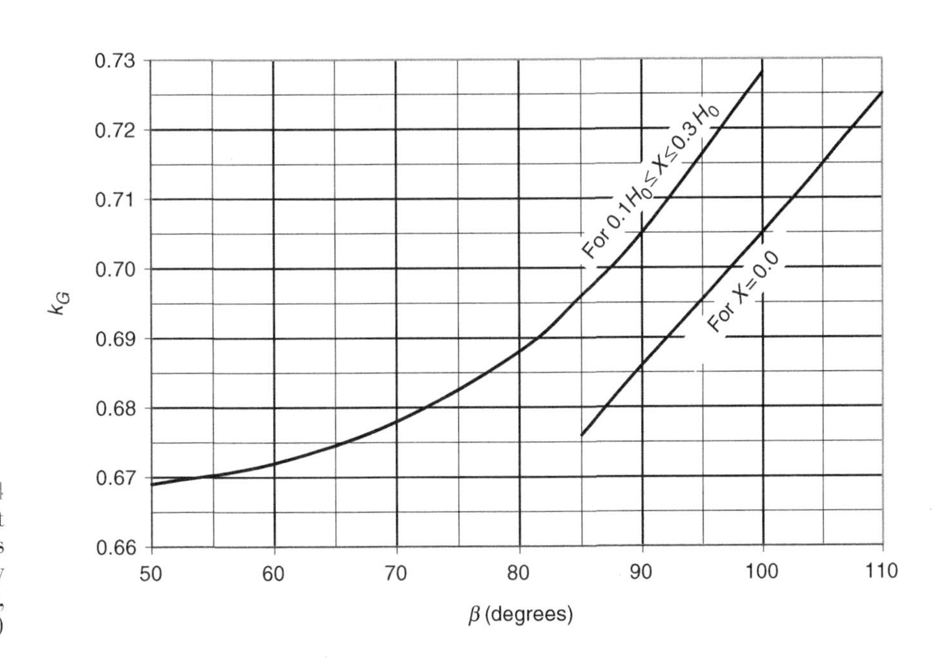

FIGURE 6.24
Discharge coefficient for gated spillways (after US Army Corps of Engineers, 1987)

gates under non-submerged flow conditions. The data used to develop this figure are based on tests with three or more bays in operation. Discharge coefficients for single-bay operation are not available. In Figure 6.24, β = angle formed by the tangent to the gate lip and the tangent to the crest curve at the nearest point of the crest curve (see Figure 6.22), X = horizontal distance between the gate seat and the apex of the crest, and H_0 = design head used to shape the crest.

EXAMPLE 6.12 Tainter gates are used to control the flow over a spillway of which the crest was shaped for $H_0 = 37.0$ ft. Each gate has a width of $B_G = 42.0$ ft, and the gate seat is 4.0 ft downstream of the apex ($X = 4$ ft). Determine the discharge under each gate for $H = 10$, 25, and 35 ft when $G = 3.96$ ft and $\beta = 67°$.

We will use Figure 6.24 to obtain the discharge coefficient k_G. From the problem statement, $X/H_0 = 4.0/37.0 = 0.11$ or $X = 0.11 H_0$. Then, with $\beta = 67°$, we obtain $k_G = 0.676$ from the figure.

Now we can use Equation 6.41 to calculate the discharge. For $H = 10$ ft,

$$Q = k_G G B_G \sqrt{2gH} = (0.676)(3.96)(42.0)\sqrt{2(32.2)(10)} = 2853 \text{ cfs}$$

Similarly, we obtain $Q = 4511$ cfs and 5338 cfs for $H = 25$ ft and 35 ft, respectively.

6.4 STILLING BASINS

As the water flows over the spillway crest and down the spillway body, it gains very high velocities as the potential energy is converted to kinetic energy. At the toe of the spillway the flow is supercritical, and it has high enough energy to cause erosion in the streambeds and banks downstream. *Stilling basins* are used for the flow to dissipate part of this energy before it is conveyed to the downstream river channel. The energy dissipation occurs through a hydraulic jump in the stilling basin. The floor elevation, length, and width of a stilling basin should be designed to ensure a stable jump that is contained within the basin.

6.4.1 POSITION OF HYDRAULIC JUMP

The position of a hydraulic jump below a spillway depends on the spillway head and height, the discharge, the tailwater depth, and the width of the stilling basin. Figure 6.25 depicts various possibilities for the water surface profile downstream of a spillway. In case A the hydraulic jump occurs at the spillway toe, while in case B it occurs some distance downstream. Case C represents a drowned jump.

We can determine the flow depth, y_1, at the toe of the spillway by writing the energy equation between this section and a section just upstream of the spillway crest. Neglecting the energy loss between the two sections, we can write

$$Z_u + P + H_e = y_1 + \frac{V_1^2}{2g} \tag{6.42}$$

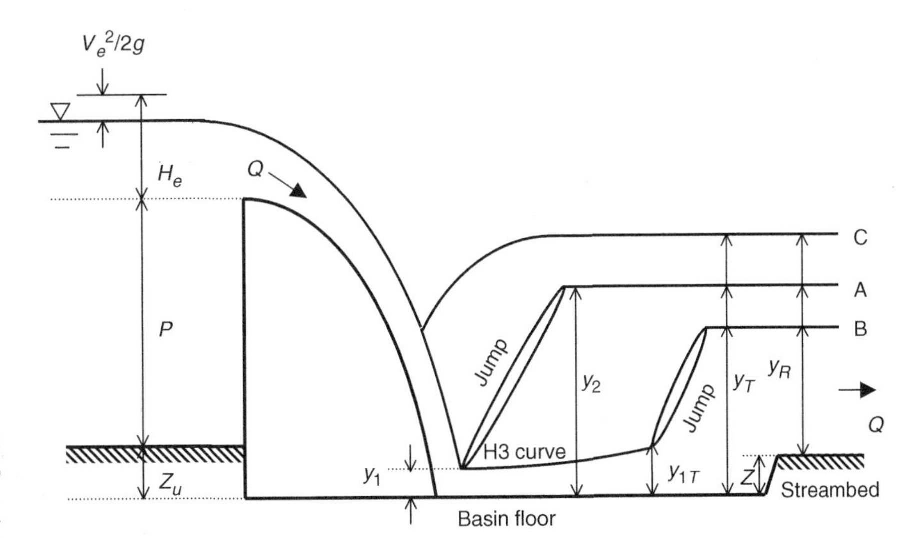

FIGURE 6.25 Hydraulic jump positions downstream of a spillway

Most stilling basins are rectangular in cross-section with a constant width, b_B. Noting that for a rectangular basin $V_1 = Q/(y_1 b_B)$, we can rewrite Equation 6.42 as

$$Z_u + P + H_e = y_1 + \frac{Q^2}{2gy_1^2 b_B^2} \tag{6.43}$$

We can solve this equation for y_1 by trial and error. We recall from Chapter 2 that two positive values of y_1, one subcritical and one supercritical, would satisfy this equation. We are interested here in the supercritical depth.

In Figure 6.25, y_2 is the conjugate depth – in other words, if a hydraulic jump occurred right at the toe of the spillway, the flow depth after the jump would be y_2. As we recall from Chapter 2, the two depths y_1 and y_2 should satisfy the hydraulic jump equation (Equation 2.26), rewritten here as

$$y_2 = \frac{y_1}{2}\left(\sqrt{1 + 8F_{r1}^2} - 1\right) \tag{6.44}$$

where F_{r1} = Froude number at Section 1. For a rectangular section with a constant width, b_B, we can evaluate the Froude number as

$$F_r = \frac{V}{\sqrt{gy}} = \frac{Q}{b_B\sqrt{gy^3}} \tag{6.45}$$

It should be clear from Equations 6.43 and 6.44 that both y_1 and y_2 depend on the spillway height, the head, and the discharge over the spillway. However, we have not yet demonstrated that the hydraulic jump actually occurs right at the toe of the spillway. Indeed, in many cases it does not, as we will discuss below.

In Figure 6.25, y_R represents the flow depth in the downstream river channel. From the continuity principle, the discharge Q in the river must be the same as the discharge over the spillway. However, the flow depth y_R depends on the cross-sectional properties of the channel, the Manning roughness factor, and the longitudinal slope. Ideally, y_R should be determined by the use of gradually varied flow calculations (see Chapter 4) if detailed information is available on the river channel further downstream. Otherwise, assuming the flow is normal in the channel (Chapter 3), we can use the Manning formula to determine an approximate value for y_R.

The elevation of the floor of the stilling basin is not necessarily the same as the natural bed elevation of the downstream river. Suppose the stilling basin is at a lower elevation than the river bed, and the elevation difference is Z, as shown in Figure 6.25. Then, in the same figure, we can define y_T = tailwater depth $= y_R + Z$.

We can now discuss the various profiles shown in Figure 6.25. If $y_2 = y_T$, then a hydraulic jump will occur right at the toe of the spillway as in profile A. If $y_2 > y_T$, the jump will not occur at the toe. In this event, an H3 curve (Chapter 4) will form downstream of the toe, along which the flow depth will increase until it reaches y_{1T} as represented by profile B in Figure 6.25. Note that the tailwater depth y_T is conjugate to y_{1T}, and the two are related through the hydraulic jump equation written as

$$y_{1T} = \frac{y_T}{2}\left(\sqrt{1 + 8F_{rT}^2} - 1\right) \qquad (6.46)$$

where F_{rT} is the Froude number corresponding to the tailwater depth y_T. The distance between the toe of the spillway and the hydraulic jump can be determined by applying the gradually-varied flow calculations between the depths y_1 and y_{1T}.

If $y_2 < y_T$, the jump will be forced upstream and drowned over the spillway body as shown in Figure 6.25 by profile C. A drowned jump does not dissipate a significant amount of energy and is not desired in a stilling basin. However, condition B is not desirable either, since it would require a longer and more expensive stilling basin to contain the jump. Profile A is the ideal condition. We can achieve this condition by adjusting the elevation of the stilling basin so that $y_2 = y_T = y_R + Z$. However, we should also note that this adjustment is possible for a single value of the discharge (or a single head like the design head) over the spillway. It is necessary to check the conditions for other possible discharges, as well to ensure that the jump will not move out of the stilling basin into the downstream river.

EXAMPLE 6.13 The crest of the spillway shown in Figure 6.26 is shaped for a design head of 12 ft with an effective crest length of 20 ft. As shown in the figure, the crest elevation is 131 ft, and the elevation of the reservoir floor is 101 ft. Thus the height of the spillway over the reservoir floor is 131 − 101 = 30 ft. A hydraulic jump forms over a horizontal apron, which is 20 ft wide. The apron elevation is

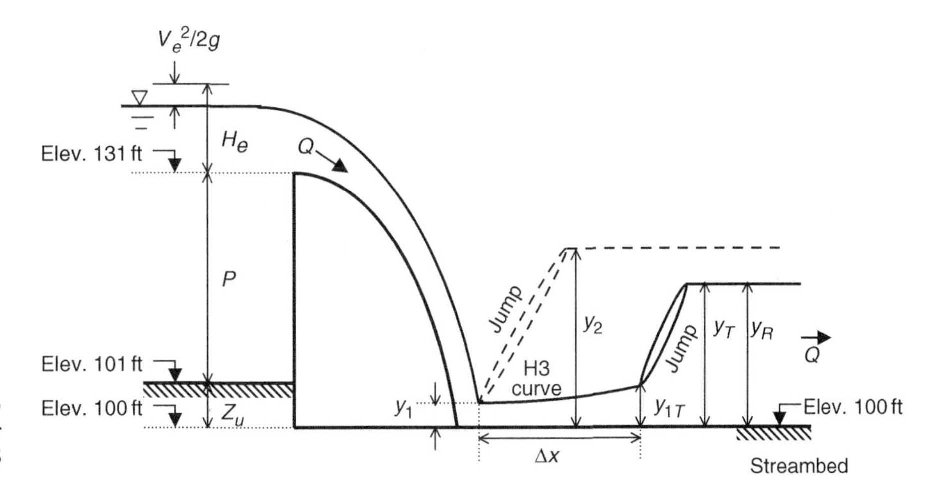

FIGURE 6.26 Definition sketch for Example 6.13

100 ft. It is 101 − 100 = 1.0 ft below the reservoir floor, but is at the same elevation as the natural streambed downstream. The natural stream can be approximated by a trapezoidal channel that has a bottom width of $b = 20$ ft, side slopes of $m = 1.5$, a Manning roughness factor of $n = 0.022$, and a longitudinal bottom slope of $S_0 = 0.0001$.

Determine the position of the hydraulic jump with respect to the spillway toe for the design head condition.

From the problem statement, we have $P = 30$ ft, $Z_u = 1.0$ ft, $H_0 = 12$ ft, $L_e = 20$ ft and $b_B = 20$ ft. We will first calculate the discharge for the design head. For $P/H_0 = 30/12 = 2.5$, we obtain $k_{w0} = 0.491$ from Figure 6.20. Then, by using Equation 6.38,

$$Q = k_{w0}\sqrt{2g}L_eH_0^{3/2} = (0.491)\sqrt{2(32.2)}(20.0)(12.0)^{3/2} = 3276\text{ cfs}$$

We will now calculate the flow depth at the spillway toe. For this, we write Equation 6.43 for the design head as

$$1.0 + 30.0 + 12.0 = y_1 + \frac{(3276)^2}{2(32.2)y_1^2(20.0)^2}$$

By trial and error, we obtain $y_1 = 3.24$ ft. Next we will calculate the conjugate depth (sequent depth) y_2. By using Equation 6.45,

$$F_{r1} = \frac{Q}{b_B\sqrt{gy_1^3}} = \frac{3276}{(20.0)\sqrt{32.2(3.24)^3}} = 4.96$$

Then, from Equation 6.44,

$$y_2 = \frac{y_1}{2}\left(\sqrt{1 + 8F_{r1}^2} - 1\right) = \frac{3.24}{2}\left(\sqrt{1 + 8(4.96)^2} - 1\right) = 21.13\text{ ft}$$

We will now determine the flow depth in the natural stream, assuming the flow in the stream is normal at depth y_R. As we recall from Chapter 3, for a trapezoidal section the Manning formula (Equation 3.26) can be written as

$$Q = \frac{k_n}{n} AR^{2/3} S_0^{1/2} = \frac{k_n}{n}(b+my)y\left(\frac{(b+my)y}{b+2y\sqrt{1+m^2}}\right)^{2/3} S_0^{1/2}$$

where $k_n = 1\,\text{m}^{1/3}/\text{s} = 1.49\,\text{ft}^{1/3}/\text{s}$, y = flow depth, n = Manning roughness factor, A = flow area, b = bottom width, m = side slope, R = hydraulic radius, and S_0 = longitudinal bottom slope. Substituting the known values into the equation,

$$3276 = \frac{1.49}{0.022}(20.0+1.5y_R)y_R\left(\frac{(20.0+1.5y_R)y_R}{20.0+2y_R\sqrt{1+(1.5)^2}}\right)^{2/3}(0.0001)^{1/2}$$

As discussed in Chapter 3, we can solve this equation by trial and error or by using Figure 3.4. In this case, the solution is obtained as $y_R = 19.87$ ft. Because the apron elevation is the same as the streambed elevation ($Z = 0$ in Figure 6.25), we have $y_T = y_R = 19.87$ ft.

Comparing the calculated values of $y_2 = 21.13$ ft and $y_T = 19.87$ ft, we can see that y_2 is larger. Therefore, the jump will occur at some distance, Δx, downstream of the toe, as shown by the solid lines in Figure 6.26. Over this distance Δx the depth will increase to y_{1T} following an H3 profile. We can use Equations 6.45 and 6.46 to calculate y_{1T} as

$$F_{rT} = \frac{Q}{b_B\sqrt{gy_T^3}} = \frac{3276}{(20)\sqrt{32.2(19.87)^3}} = 0.326$$

$$y_{1T} = \frac{y_T}{2}\left(\sqrt{1+8F_{rT}^2}-1\right) = \frac{19.87}{2}\left(\sqrt{1+8(0.326)^2}-1\right) = 3.58\text{ ft}$$

We can determine the distance Δx by gradually-varied flow calculations. For this, we need the velocity and the friction slope at depths y_1 and y_{1T}. The velocity at depth y_1 is $V_1 = 3276/[(20.0)(3.24)] = 50.6$ fps, and that at depth y_{1T} is $V_{1T} = 3276/[(20.0)(3.58)] = 45.8$ fps. The friction slope is calculated by using Equation 4.11, rewritten here as

$$S_f = \frac{n^2}{k_n^2}\frac{V^2}{R^{4/3}}$$

Noting that for a rectangular channel of width b_B the hydraulic radius is $R = (yb_B)/(b_B + 2y)$, and using $n = 0.016$ for a concrete apron, the friction slope corresponding to $y_1 = 3.24$ ft is calculated as

$$S_{f1} = \frac{(0.016)^2}{(1.49)^2}\frac{(50.6)^2}{\{(20)(3.24)/[20.+2(3.24)]\}^{4/3}} = 0.0897$$

TABLE 6.5 Discharge and Depth Calculations for Example 6.14

H_e (ft)	H_e/H_0	k_w/k_{w0}	k_w	Q (cfs)	Z_u+P+H_e (ft)	y_1 (ft)	F_{r1}	y_2 (ft)	y_R (ft)	y_T (ft)
7.2	0.6	0.940	0.461	1431	38.2	1.47	7.07	13.99	13.36	13.36
9.6	0.8	0.972	0.477	2278	40.6	2.29	5.78	17.64	16.74	16.74
12.0	1.0	1.000	0.491	3276	43.0	3.24	4.96	21.13	19.87	19.87
14.4	1.2	1.025	0.503	4414	45.4	4.29	4.38	24.50	22.81	22.81

Likewise, we obtain $S_{f1T}=0.0666$. Thus, the average friction slope between the two sections is $S_{fm}=0.0781$. We can now use Equation 4.13 to determine the distance Δx. For this case, Equation 4.13 becomes

$$\Delta x = \frac{(y_{1T}+(V_{1T}^2/2g)) - (y_1+(V_1^2/2g))}{S_o - S_{fm}}$$

$$= \frac{(3.58+((45.8)^2/2(32.2)) - (3.24+((50.6)^2/2(32.2)))}{0.0-0.0781} = 87.6\,\text{ft}$$

Note that we could use smaller depth increments to calculate the distance to the jump more accurately (Chapter 4).

EXAMPLE 6.14 The maximum head expected to occur over the spillway considered in Example 6.13 is 14.4 ft. Determine the position of the jump for spillway heads of 7.2, 9.6, 12.0, and 14.4 ft.

We will first calculate the spillway discharge using Equation 6.39 for the various heads specified in the problem statement. These calculations are summarized in columns (1) through (5) in Table 6.5. Note that the entries in column (3) are obtained from Figure 6.21, and that $k_{w0}=0.491$ as determined in Example 6.13.

Next we calculate the flow depth at the spillway toe using Equation 6.42, as summarized in columns (6) and (7) of Table 6.5. The entries in column (6) are calculated, noting $P=30.0$ ft and $Z_u=1.0$ ft. Equation 6.43, with $b_B=20.0$ ft, is solved by trial and error to obtain the y_1 values in column (7). The conjugate depths, y_2, are calculated by using Equations 6.45 and 6.44. The results are shown in columns (8) and (9) of Table 6.5.

We now determine the tailwater depths. The values of y_R in column (10) of Table 6.5 represent the normal depths in the natural stream corresponding to the Q values listed in column (5). The Manning formula is used to determine the normal depths as described in Example 6.13. Because the natural streambed and the apron are at the same elevation, the y_T values listed in column (11) are the same as y_R values in column (10).

A comparison of the y_2 and y_T values calculated and listed in Table 6.5 indicates that y_2 is larger for all the spillway heads considered. Therefore, the jump will

TABLE 6.6 Calculations for Jump Position in Example 6.14

H_e (ft)	Q (cfs)	y_1 (ft)	y_T (ft)	F_{rT}	y_{1T} (ft)	V_1 (fps)	V_{1T} (fps)	S_{f1}	S_{f1T}	S_{fm}	Δx (ft)
7.2	1431	1.47	13.36	0.26	1.59	48.63	44.96	0.1957	0.1527	0.1742	30.0
9.6	2278	2.29	16.74	0.29	2.50	49.67	45.54	0.1239	0.0948	0.1094	53.9
12.0	3276	3.24	19.87	0.33	3.58	50.60	45.81	0.0897	0.0665	0.0781	87.6
14.4	4414	4.29	22.81	0.36	4.80	51.45	45.95	0.0705	0.0507	0.0606	128.8

form some distance downstream from the spillway toe. Table 6.6 summarizes the calculations for the jump position. The entries in columns (1) to (4) have already been determined. The other entries are calculated following the procedure described in Example 6.13. A review of the results clearly shows that the position of the hydraulic jump varies with the head over the spillway.

6.4.2 HYDRAULIC JUMP CHARACTERISTICS

A comprehensive series of tests conducted by the Bureau of Reclamation have indicated that the form and characteristics of hydraulic jumps are related to the Froude number just upstream of the jump (US Bureau of Reclamation, 1987). Various forms of the hydraulic jump phenomena corresponding to different ranges of the Froude number are illustrated in Figure 6.27.

As reported by the US Bureau of Reclamation (1987), for Froude numbers from 1.0 to about 1.7 the incoming flow is only slightly below critical depth, and the change from this low stage to the high stage flow is gradual with a slightly ruffled water surface. A series of small rollers begins to develop on the surface as the Froude number approaches 1.7, and these become more intense with increasing values of the Froude number. However, other than the surface roller phenomenon, relatively smooth flows prevail throughout the Froude number range up to about 2.5. The form of hydraulic jumps for the range of Froude numbers from 1.7 to 2.5 is shown as form A in Figure 6.27.

For Froude numbers between 2.5 and 4.5, an oscillating form of jump occurs. This oscillating flow causes undesirable surface waves that carry far downstream. The form of hydraulic jumps for this range of Froude numbers is designated as form B in Figure 6.27.

For Froude numbers between 4.5 and 9.0, a stable and well-balanced jump occurs. Turbulence is confined to the main body of the jump, and the water surface downstream is comparatively smooth. Form C in Figure 6.27 represents the hydraulic jumps for this range.

As the Froude number increases above 9.0, the turbulence within the jump and the surface rollers becomes increasingly active. This results in a rough water surface with strong water waves downstream from the jump. The form of

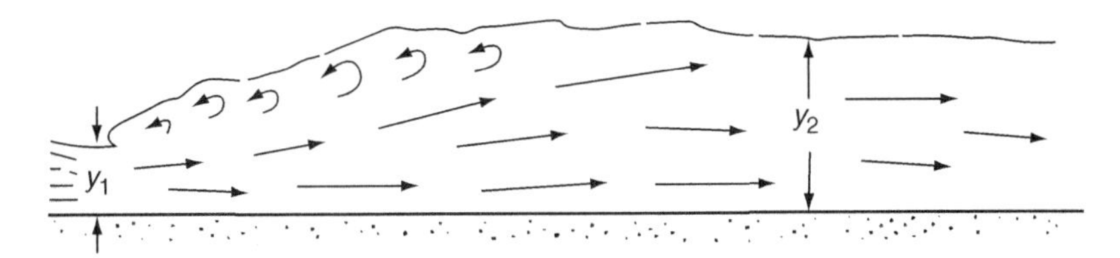

Form A: F_{r1} between 1.7 and 2.5

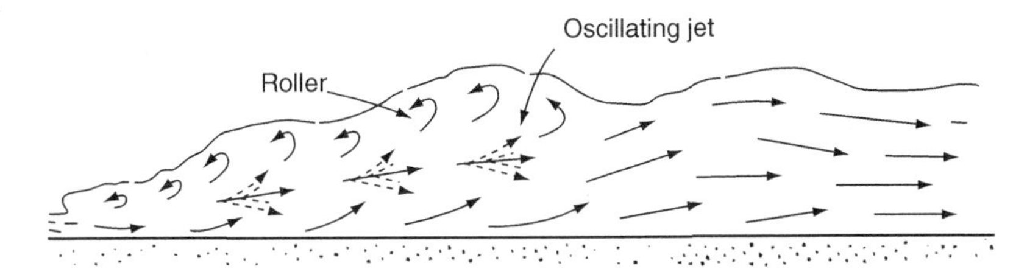

Form B: F_{r1} between 2.5 and 4.5

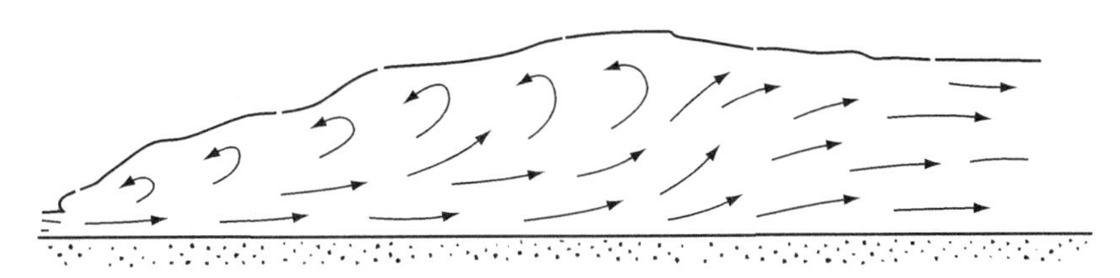

Form C: F_{r1} between 4.5 and 9.0

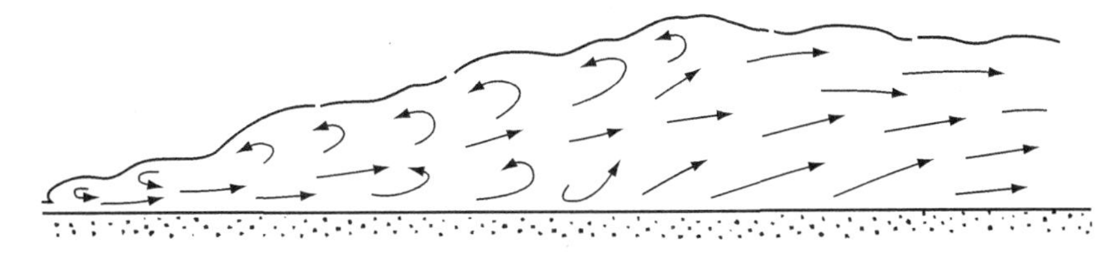

Form D: F_{r1} greater than 9.0

FIGURE 6.27 Forms of hydraulic jump related to Froude number (after US Bureau of Reclamation, 1987)

hydraulic jumps for this range of Froude numbers is designated as form D in Figure 6.27.

6.4.3 STANDARD STILLING BASIN DESIGNS

Various standard stilling basin designs have been developed, based on experience, observations, and model studies. Figure 6.28 displays three standard types of stilling basins developed by the US Bureau of Reclamation (1987). All three types include *chute blocks* at the entrance to produce a shorter jump. These blocks also tend to stabilize the jump. The *end sills*, dentated or solid, placed at the end of the stilling basin further reduce the length of the jump. Baffle piers placed in intermediate positions across the stilling basin increase the energy dissipation by impact action. The selection of the type of stilling basin is governed by the upstream Froude number.

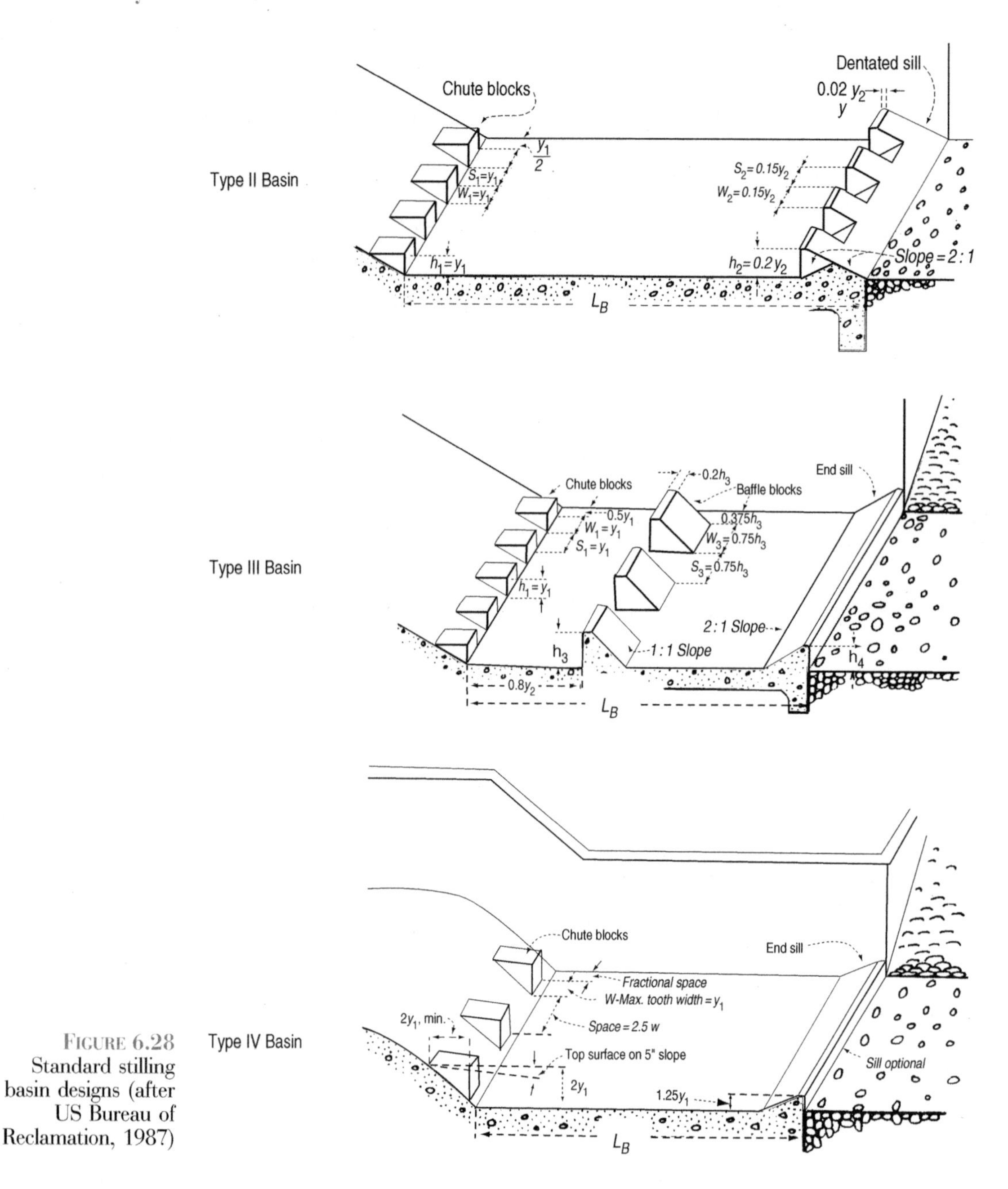

FIGURE 6.28 Standard stilling basin designs (after US Bureau of Reclamation, 1987)

For Froude numbers of less than 2.5, no baffles or other dissipating devices are used. However, the apron lengths beyond the point where the depth starts to change should be not less than about five times y_2.

For Froude numbers between 2.5 and 4.5, type IV basins are recommended. Also, auxiliary wave dampeners or wave suppressors must sometimes be used to provide smooth surface flow downstream. Because of the tendency of the jump to sweep out, and as an aid in suppressing wave action, the water depths in the basin should be about 10% greater than the computed conjugate depth. In other

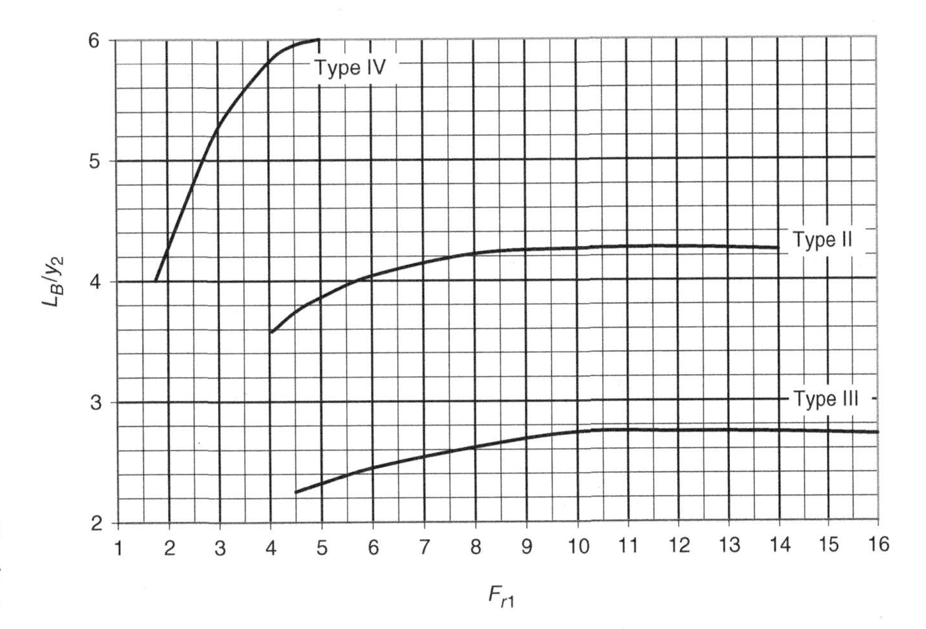

FIGURE 6.29 Stilling basin length (source: US Bureau of Reclamation, 1987)

words, these basins should be designed such that $y_T = 1.10y_2$. The length of the basin can be obtained from Figure 6.29.

For Froude numbers higher than 4.5, type III basins can be adopted where incoming velocities, V_1, do not exceed 60 ft/s (20 m/s). The type III basin uses chute blocks, impact baffle blocks, and an end sill to shorten the jump length and to dissipate the high-velocity flow within the shortened basin length. The height of the chute blocks is $h_1 = y_1$, as shown in Figure 6.28. The height of the baffle blocks can be expressed as $h_3 = y_1[1.30 + 0.164(F_{r1} - 4.0)]$ and the height of the end sill is $h_4 = y_1[1.25 + 0.056(F_{r1} - 4.0)]$, as suggested by Roberson *et al.* (1997). The basin length is obtained from Figure 6.29.

For Froude numbers higher than 4.5 where incoming velocities exceed 60 ft/s, or where impact baffle blocks are not used, the type II basin may be adopted. Because the dissipation is accomplished primarily by hydraulic jump action, the basin length will be greater than that indicated for the type III basin. However, the chute blocks and dentated end sill will still effectively reduce the length. The tailwater depth in the basin should be about 5% greater than the computed conjugate depth – that is, $y_T = 1.05y_2$. The basin length can be obtained from Figure 6.29.

The tailwater considerations govern the selection of the floor elevation for the stilling basins. The final design should ensure that the hydraulic jump will not be swept out into the natural stream channel for any discharge that may occur over the spillway. This can be achieved by comparing the conjugate and tailwater depths calculated for the possible range of discharges. The conjugate depths should be smaller than but close to the tailwater depths for an efficient design. Otherwise, if the conjugate depth is higher than the tailwater depth, the jump will sweep out. If the conjugate depth is much smaller, the jump will be completely drowned.

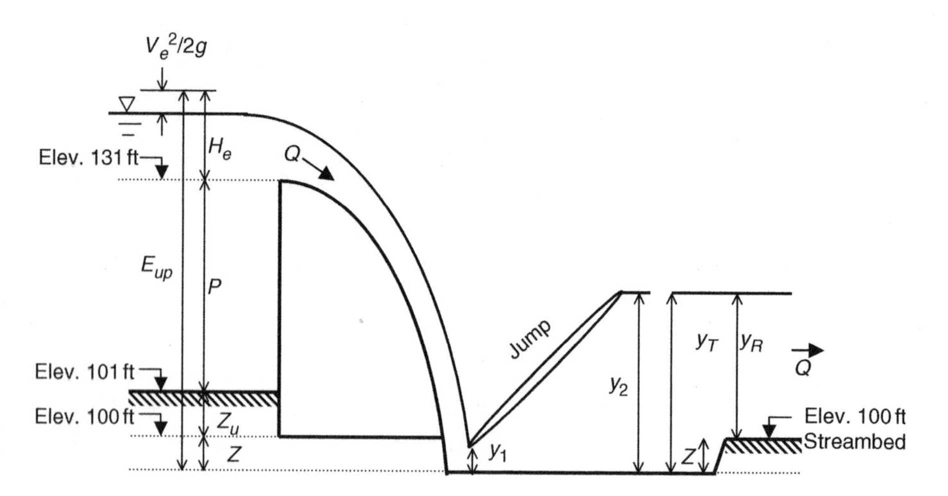

FIGURE 6.30 Definition sketch for Example 6.15

TABLE 6.7 Example 6.15: Calculations for $Z = 1.5$ ft

H_e (ft)	Q (cfs)	y_R (ft)	E_{up} (ft)	y_1 (ft)	V_1 (fps)	F_{r1}	y_2 (ft)	$y_2 - Z$ (ft)
7.2	1431	13.36	39.7	1.44	49.64	7.28	14.15	12.65
9.6	2278	16.74	42.1	2.25	50.66	5.95	17.84	16.34
12.0	3276	19.87	44.5	3.18	51.59	5.10	21.38	19.88
14.4	4414	22.81	46.9	4.21	52.43	4.50	24.79	23.29

EXAMPLE 6.15 Design a stilling basin for the spillway considered in Examples 6.13 and 6.14. More specifically, select the type of stilling basin, and determine the floor elevation and the basin length.

In Example 6.14, the apron and the natural streambed are at the same elevation and the conjugate depth y_2 is consistently higher than the tailwater depth, y_T. As a result, the hydraulic jump occurs some distance downstream of the toe. In a stilling basin, it is desirable for the jump to form at the toe. This can be achieved by adjusting the floor elevation of the stilling basin to match the conjugate and tailwater depths. In this case we will lower the stilling basin floor elevation by Z, as shown in Figure 6.30.

We do not know the magnitude of Z, and we will determine it by trial and error. Some of the results obtained in Example 6.13 and 6.14 are useful for this purpose, and these are summarized in the first three columns of Table 6.7. Note that the entries in these columns do not depend on Z.

Let us now try $Z = 1.5$ ft. With the floor elevation lowered by Z, Equation 6.43 can be written as (see Figure 6.30)

$$E_{up} = Z + Z_u + P + H_e = y_1 + \frac{Q^2}{2gy_1^2 b_B^2} \tag{6.47}$$

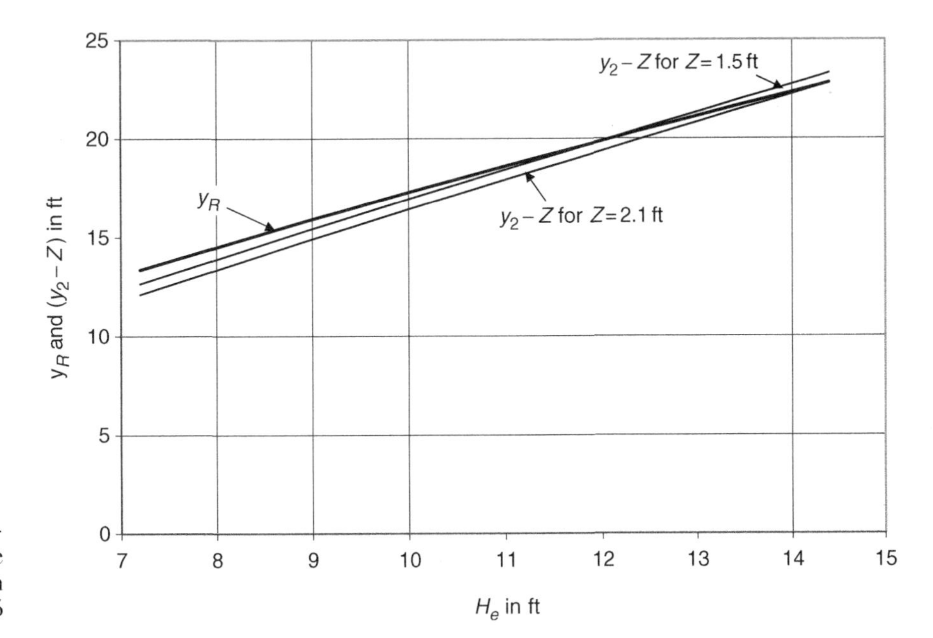

FIGURE 6.31 Tailwater and conjugate depths calculated in Example 6.15

Noting that $Z_u = 1.0$ ft, $P = 30$ ft, and $b_B = 20$ ft, we solve Equation 6.47 for y_1 by trial and error for the H_e and Q values listed in Table 6.7. For example, at the design head $H_e = H_0 = 12$ ft and $Q = 3276$ cfs, we have $E_{up} = 1.5 + 1.0 + 30 + 12 = 44.5$ ft as shown in column (4) of Table 6.7. Solving Equation 6.47, we obtain $y_1 = 3.18$ ft and thus $V_I = Q/(b_B y_I) = 51.59$ fps. Then, from Equations 6.45 and 6.44, we obtain $F_{r1} = 5.10$ and $y_2 = 21.38$ ft, respectively. Similar calculations are performed in the same manner for the various heads, and the results are summarized in Table 6.7.

A review of Table 6.7 will reveal that, for the design head as well as the other heads considered, $F_{r1} > 4.5$ and $V_1 < 60$ fps. Therefore, stilling basin type III is recommended. For a type III basin, it is required that the conjugate depth should be equal to or slightly less than the tailwater depth to ensure that the jump will be contained within the stilling basin. In other words, with reference to Figure 6.30, we need $y_2 \leq y_T$ or $(y_2 - Z) \leq y_R$. A plot of y_R and $(y_2 - Z)$ versus H_e displayed in Figure 6.31 for $Z = 1.5$ ft reveals that, for heads higher than the design head, $(y_2 - Z) > y_R$. This is not acceptable, since it may cause the jump to sweep out into the natural channel downstream. Therefore, a larger value of Z should be picked.

Next we try $Z = 2.1$ ft. Table 6.8 summarizes the results obtained with $Z = 2.1$ ft. Again a type III basin is recommended, since $F_{r1} > 4.5$ and $V_1 < 60$ fps for all the flow conditions considered. Also, a plot of y_R and $(y_2 - Z)$ versus H_e for $Z = 2.1$ ft in Figure 6.31 shows that $(y_2 - Z) = y_R$ for the maximum head over the spillway and $(y_2 - Z) \leq y_R$ for smaller heads. Therefore, $Z = 2.10$ ft is adequate to contain the jump within the stilling basin.

We can now determine the length of the stilling basin. The Froude number corresponding to the maximum head is 4.55. From Figure 6.29, for $F_{r1} = 4.55$,

TABLE 6.8 Example 6.15: Calculations for $Z = 2.1$ ft

H_e (ft)	Q (cfs)	y_R (ft)	E_{up} (ft)	y_1 (ft)	V_1 (fps)	F_{r1}	y_2 (ft)	$y_2 - Z$ (ft)
7.2	1431	13.36	40.3	1.43	50.03	7.37	14.21	12.11
9.6	2278	16.74	42.7	2.23	51.05	6.02	17.92	15.82
12.0	3276	19.87	45.1	3.15	51.98	5.16	21.47	19.37
14.4	4414	22.81	47.5	4.18	52.82	4.55	24.90	22.80

we obtain $L_B/y_2 = 2.3$. Therefore, the length of the stilling basin should be $L_B = 2.3(24.90) = 57.3$ ft. The chute blocks and the baffle blocks can be sized following Figure 6.28.

6.5 CHANNEL TRANSITIONS

Channel transitions are used where changes in the channel cross-sectional geometry are necessary. Transitions from trapezoidal channels to rectangular flumes and *vice versa* are most common. The hydraulic characteristics of flow at channel transitions depend on whether the flow is subcritical or supercritical.

6.5.1 CHANNEL TRANSITIONS FOR SUBCRITICAL FLOW

Figure 6.32 depicts the most common types of transitions connecting trapezoidal and rectangular channels under subcritical flow conditions. Each transition type can be used for both expansions and contractions. Where two rectangular channels are connected, an abrupt cross-sectional change with a wall built perpendicular to the flow direction, a straight-line transition, or a quadrant transition can be used. The transitions should be designed to minimize the flow disturbances and the head loss.

6.5.1.1 Energy loss at transitions

The energy head loss due to friction is usually negligible in transition structures. The head loss occurs mainly due to the changes in the channel cross-sectional geometry. This loss is often expressed in terms of the change in the velocity head as

$$h_T = C_c \Delta h_v \tag{6.48}$$

for contractions and

$$h_T = C_e \Delta h_v \tag{6.49}$$

where h_T = head loss due to the transition, Δh_v = difference in velocity head across the transition, C_c = contraction coefficient, and C_e = expansion coefficient.

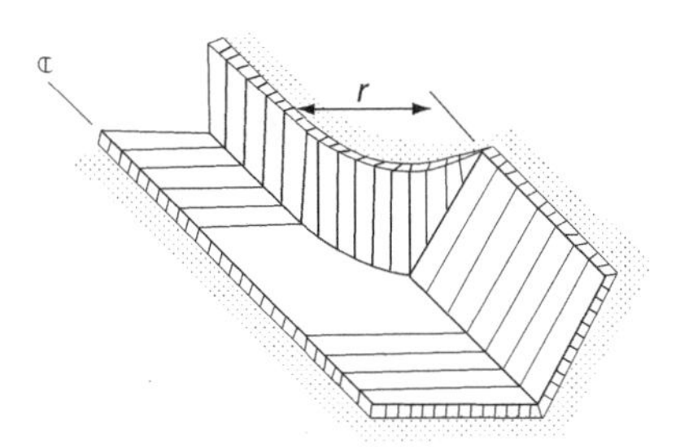

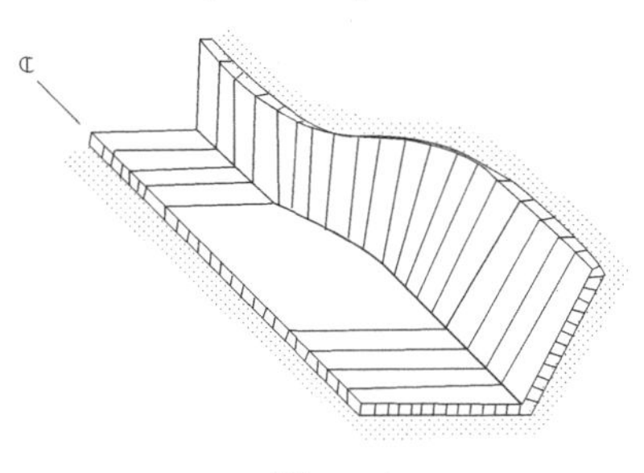

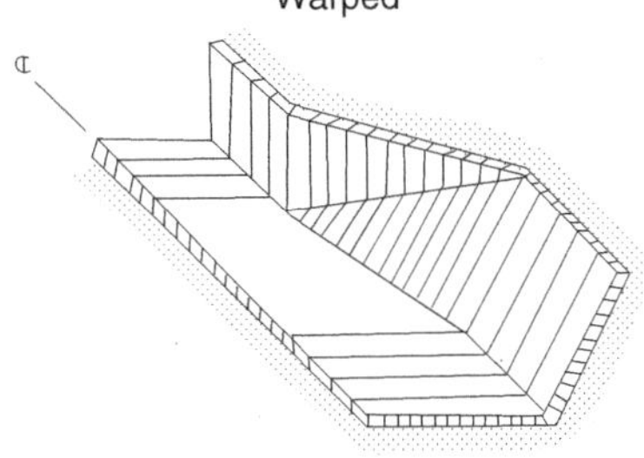

FIGURE 6.32 Common types of transition (after US Army Corps of Engineers, 1991)

TABLE 6.9 Transition Loss Coefficients (after US Army Corps of Engineers, 1991)

Transition type	C_c	C_e	**Source**
Warped	0.10	0.20	Chow (1959), Brater and King (1976)
Cylindrical quadrant	0.15	0.20	Chow (1959)
Wedge	0.30	0.50	US Bureau of Reclamation (1967)
Straight line	0.30	0.50	Chow (1959)
Square end	0.30	0.75	Chow (1959)

The suggested values of the contraction and expansion coefficients are given in Table 6.9 for various transition types.

6.5.1.2 Water surface profile at transitions

Under subcritical flow conditions, the flow depth and velocity at the downstream end of the transition (like y_E and V_E at section E in Figure 6.33) depend on the characteristics of the downstream channel. They are determined

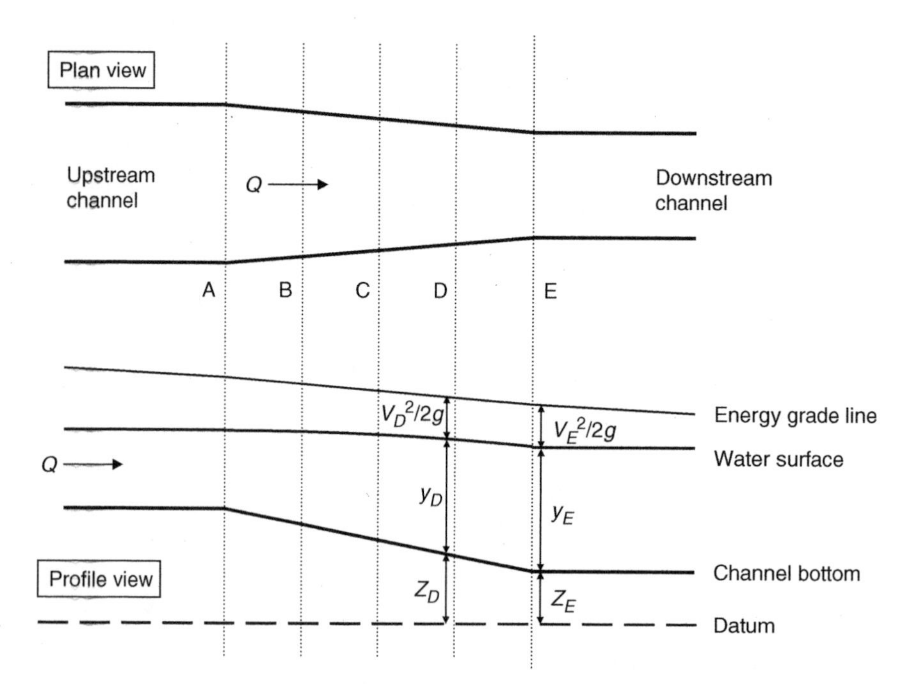

FIGURE 6.33 Definition sketch for a contraction

from the gradually-varied flow calculations for the downstream channel (see Chapter 4). We can then apply the energy equation to determine the flow depths at selected locations within the transition structure. For example, to determine the flow depth at section D in Figure 6.33, we write the energy equation as

$$Z_D + y_D + \frac{V_D^2}{2g} - (h_T)_{DE} = Z_E + y_E + \frac{V_E^2}{2g} \tag{6.50}$$

where $(h_T)_{DE}$ = transition loss between D and E. This loss depends on the type of the transition. For a contraction as in Figure 6.33,

$$(h_T)_{DE} = C_c\left(\frac{V_E^2}{2g} - \frac{V_D^2}{2g}\right) \tag{6.51}$$

If the channel section expands from D to E in the flow direction, we would use

$$(h_T)_{DE} = C_e\left(\frac{V_D^2}{2g} - \frac{V_E^2}{2g}\right) \tag{6.52}$$

For the case of contraction, substituting Equation 6.51 into 6.50 and rearranging we obtain

$$y_D + (1 + C_c)\frac{V_D^2}{2g} = (Z_E - Z_D) + y_E + (1 + C_c)\frac{V_E^2}{2g} \tag{6.53}$$

For the case of expansion, Equations 6.50 and 6.52 will yield

$$y_D + (1 - C_e)\frac{V_D^2}{2g} = (Z_E - Z_D) + y_E + (1 - C_e)\frac{V_E^2}{2g} \tag{6.54}$$

In Equations 6.53 and 6.54, all the terms on the right-hand side are known. On the left-hand side, given the discharge, the velocity, V_D, can be expressed in terms of the flow depth, y_D. Therefore, the only unknown is y_D, and this can be determined by trial and error.

We can use the same approach to calculate the flow depths in sequence at the further upstream sections. For example, to calculate the flow depth at section C in Figure 6.33, the energy equation is written between sections C and D and solved for y_C. Alternatively, to determine the flow depth at C, we can write the energy equation between C and E and solve it for y_C. Because of the way the transition loss is expressed, both approaches will lead to the same results (see Example 6.16).

If we are interested only in the flow depth at the upstream end of the transition, as in section A in Figure 6.33, we can simply write the energy equation between the upstream and downstream ends of the transition and solve it for the unknown depth. For the case of contraction, as in Figure 6.33, the energy equation between the end sections becomes

$$y_A + (1 + C_c)\frac{V_A^2}{2g} = (Z_E - Z_A) + y_E + (1 + C_c)\frac{V_E^2}{2g} \tag{6.55}$$

For an expanding transition, that is if the flow is expanding from section A to E,

$$y_A + (1 - C_e)\frac{V_A^2}{2g} = (Z_E - Z_A) + y_E + (1 - C_e)\frac{V_E^2}{2g} \tag{6.56}$$

EXAMPLE 6.16 Suppose the flow conditions at section E are known in Figure 6.33. We are to determine the flow depth at C.

There are two ways to achieve this. In the first approach, we first calculate the flow conditions at D using the energy equation between D and E, and then we write the energy equation between C and E to solve for the depth at C. In the second approach, we can write the energy equation between C and E and solve for the depth at C directly without any intermediate calculations for section D. Show that both approaches will yield the same result.

Initially, let us formulate the first approach. Equation 6.53 represents the energy equation between D and E. We can write a similar equation for C and D as

$$y_C + (1 + C_c)\frac{V_C^2}{2g} = (Z_D - Z_C) + y_D + (1 + C_c)\frac{V_D^2}{2g} \tag{6.57}$$

If we equate the sum of the left-hand sides of Equations 6.53 and 6.57 with the sum of the right-hand sides,

$$y_D + (1 + C_c)\frac{V_D^2}{2g} + y_C + (1 + C_c)\frac{V_C^2}{2g} = (Z_E - Z_D) + y_E$$
$$+ (1 + C_c)\frac{V_E^2}{2g} + (Z_D - Z_C) + y_D + (1 + C_c)\frac{V_D^2}{2g}$$

Dropping the similar terms on both sides, we obtain

$$y_C + (1+C_c)\frac{V_C^2}{2g} = (Z_E - Z_C) + y_E + (1+C_c)\frac{V_E^2}{2g} \tag{6.58}$$

This is also the energy equation between C and E – i.e. the equation we would use in the second approach. Therefore, the two approaches are equivalent, and they will produce the same result.

EXAMPLE 6.17 A straight-line transition connects a 10-ft wide rectangular channel carrying a discharge of 120 cfs to a 5-ft wide rectangular flume, as shown in Figure 6.33. The bottom elevation also decreases linearly in the flow direction as shown in the figure. Suppose the length of the transition is 20 ft, and the total drop in the bottom elevation is 1.20 ft. The sections D, C, B, and A are, respectively 5, 10, 15, and 20 ft from section E. The flow depth at section E is 4 ft. Determine the depth at D, C, B, and A.

From the problem statement, the width of the flume at section E is $b_E = 5.0$ ft and the depth $y_E = 4.0$ ft. Therefore, $V_E = 120/(5.0)(4.0) = 6.0$ fps and $V_E{}^2/2g = (6.0)^2/(2.0)(32.2) = 0.56$ ft. At section D, the channel width $b_D = 5.0 + (5.0/20.0)(10.0 - 5.0) = 6.25$ ft. Also, $Z_E - Z_D = -(5.0/20.0)(1.20) = -0.3$ ft. The velocity head at section D can be expressed as

$$\frac{V_D^2}{2g} = \frac{1}{2g}\left(\frac{Q}{b_D y_D}\right)^2 = \frac{1}{2(32.2)}\left(\frac{120}{6.25 y_D}\right)^2 = \frac{5.72}{y_D^2}$$

Also, from Table 6.9, $C_c = 0.30$ for a straight-line transition. Substituting all the known terms in Equation 6.53,

$$y_D + (1+0.30)\frac{5.72}{y_D^2} = -0.30 + 4.0 + (1+0.30)(0.56)$$

or

$$y_D + \frac{7.44}{y_D^2} = 4.43$$

We can solve this expression by trial and error to obtain $y_D = 3.95$ ft.

We can determine the flow depths at C, B, and A in the same manner. Noting that $b_C = 7.5$ and $(Z_E - Z_C) = -0.60$ ft, the velocity head at C is expressed as

$$\frac{V_C^2}{2g} = \frac{1}{2(32.2)}\left(\frac{120}{7.5 y_C}\right)^2 = \frac{3.98}{y_C^2}$$

and the energy equation becomes

$$y_C + (1+0.30)\frac{3.98}{y_C^2} = -0.60 + 4.0 + (1+0.30)(0.56)$$

or

$$y_C + \frac{5.17}{y_C^2} = 4.13$$

By trial and error, we obtain $y_C = 3.77$ ft.

Noting that $b_B = 8.75$ ft and $(Z_E - Z_B) = -0.90$ ft, and using the same procedure, we obtain $y_B = 3.52$ ft. Likewise, for section A, $b_A = 10.0$ ft and $(Z_E - Z_A) = -1.20$ ft, and the solution is obtained as being $y_A = 3.25$ ft.

6.5.1.3 Design of channel transitions for subcritical flow

Channel transitions should be designed to minimize the flow disturbances and the head loss resulting from the cross-sectional geometry. However, the design procedure is arbitrary for the most part, since broadly accepted relationships are not available to proportion a channel transition structure. Limited model studies in the past led to a few guidelines. Generally, a transition structure resulting in a smooth water surface is considered satisfactory. This can be achieved with continuous and gradual changes in the bed profile, the channel width, and the channel side slopes if the transition is from a rectangular to a trapezoidal section or *vice versa*.

The cylindrical quadrant-type transition shown in Figure 6.32 is used for expansions from rectangular to trapezoidal sections, and for contractions from trapezoidal to rectangular sections. The radius of the quadrant will be half the difference between the top widths of the channels connected.

For wedge-type transitions, Roberson *et al.* (1997) recommend that the expansion angle be 22.5° for expanding transitions, and the contraction angle be 27.5° for contracting transitions. The expansion and the contraction angles, θ, can be defined as

$$\theta = \arctan \frac{\Delta T}{2L_T} \tag{6.59}$$

where ΔT = difference in the top widths of the channels connected and L_T = transition length. Winkel recommends that θ be limited to 6° (US Army Corps of Engineers, 1991).

For warped transitions, Morris and Wiggert (1972) suggest that, for an expanding transition, $L_T \geq 2.25\,\Delta T$, which is equivalent to $\theta \leq 12.5°$. Roberson *et al.* (1997) recommend that $\theta = 12.5°$ both for expanding and contracting warped transitions. Based on experimental studies, for warped transitions joining a rectangular flume with a trapezoidal channel having a side slope of m_C (that is 1 vertical over m_C horizontal), Vittal and Chiranjeevi (1983) obtained the expression

$$L_T = 2.35\Delta b + 1.65 m_C y_C \tag{6.60}$$

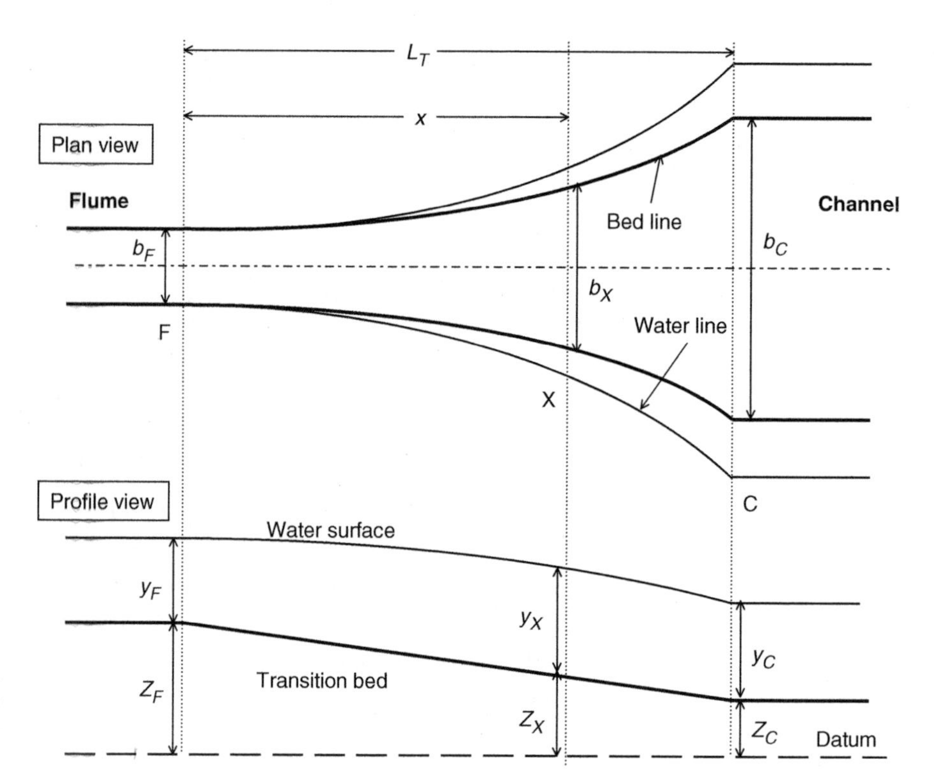

FIGURE 6.34 Transition between rectangular flume and trapezoidal channel

where $\Delta b = |b_C - b_F|$ = difference in bottom widths of the flume and the channel, b_C = bottom width of the channel, b_F = width of the rectangular flume, and y_C = flow depth in the channel (see Figure 6.34).

They also recommend the expressions

$$m_X = m_C - m_C\left(1 - \frac{x}{L_T}\right)^{1/2} \tag{6.61}$$

and

$$b_X = b_F + (b_C - b_F)\frac{x}{L_T}\left[1 - \left(1 - \frac{x}{L_T}\right)^{e}\right] \tag{6.62}$$

for flow sections within the transition where x = distance from the flume, b_X = bottom width, m_X = side slope, and $e = 0.80\text{–}0.26 m_C^{1/2}$. If the flow is from the flume towards the trapezoidal channel in Figure 6.34, that is if the transition is expanding, the energy equation written between the end sections F and C will be similar to Equation 6.56:

$$y_F + (1 - C_e)\frac{V_F^2}{2g} = (Z_C - Z_F) + y_C + (1 - C_e)\frac{V_C^2}{2g} \tag{6.63}$$

Similarly, noting that section X is located at a distance x from the flume, the energy equation between the sections X and C becomes

$$y_X + (1 - C_e)\frac{V_X^2}{2g} = (Z_C - Z_X) + y_C + (1 - C_e)\frac{V_C^2}{2g} \tag{6.64}$$

If the flow is from the trapezoidal channel towards the rectangular flume in Figure 6.34, that is if the transition is contracting, Equations 6.61 and 6.62 can still be used. However, the energy equation between the end sections C and F will be similar to Equation 6.55, rewritten here as

$$y_C + (1 + C_c)\frac{V_C^2}{2g} = (Z_F - Z_C) + y_F + (1 + C_c)\frac{V_F^2}{2g} \tag{6.65}$$

Likewise, the energy equation written between sections X and F will be

$$y_X + (1 + C_c)\frac{V_X^2}{2g} = (Z_C - Z_X) + y_C + (1 + C_c)\frac{V_C^2}{2g} \tag{6.66}$$

If the bed elevations of the flume and the channel are fixed, we can simply connect the two by a straight line (as in Figure 6.34) to determine the transition bed elevations. Then, we can calculate the water surface profile by solving Equation 6.64 for expansions and Equation 6.66 for contractions to determine the flow depth y_X at selected values of x. This procedure is similar to that of Example 6.17. However, if the bed elevations are not fixed, we can select the flow depths (corresponding to a smooth water surface) and calculate the bed elevations by using Equation 6.64 or 6.66 as appropriate.

EXAMPLE 6.18 A warped expansion transition is to be designed to connect a rectangular flume to a trapezoidal channel as shown in Figure 6.34. The channel has a bottom width of 10 ft and side slopes of 1 vertical over 2 horizontal, and it carries 120 cfs at a depth of 3.5 ft. The invert elevation of the channel at section C is 50 ft. The rectangular flume is 5.7 ft wide, and is desired to keep the flow depth constant at 3.5 ft throughout the transition. Design the transition.

Let us first determine the invert elevation of the flume at section F. From the problem statement, the givens are $Q = 120$ cfs, $b_C = 10$ ft, $y_C = 3.5$ ft, $m_C = 2$, $Z_C = 50$ ft, and $b_F = 5.7$ ft. For the channel (or section C in Figure 6.34), we can determine that

$$A_C = (b_C + m_C y_C)y_C = [10.0 + 2(3.5)](3.5) = 59.5 \text{ ft}^2$$

$$V_C = \frac{Q}{A_C} = \frac{120}{59.5} = 2.02 \text{ fps}$$

$$\frac{V_C^2}{2g} = \frac{(2.02)^2}{2(32.2)} = 0.063 \text{ ft.}$$

Similarly, for the rectangular flume we have $A_F = (5.7)(3.5) = 19.95 \text{ ft}^2$, $V_F = 120/19.95 = 6.02$ fps, and $V_F^2/2g = (6.02)^2/2(32.2) = 0.562$ ft. From Table 6.9, for a warped expansion $C_e = 0.20$. Substituting the known quantities into Equation 6.63,

$$3.5 + (1 - 0.2)(0.562) = (50.0 - Z_F) + 3.5 + (1 - 0.2)(0.063)$$

TABLE 6.10 Summary of Results for Example 6.18

x (ft)	m_X	b_X (ft)	y_X (ft)	A_X (ft^2)	V_X (fps)	$V_X^2/2g$ (ft)	Z_X (ft)
19.25	1.29	7.93	3.5	43.60	2.75	0.12	49.96
16.50	1.00	7.15	3.5	37.29	3.22	0.16	49.92
13.75	0.78	6.63	3.5	32.70	3.67	0.21	49.88
11.00	0.59	6.26	3.5	29.07	4.13	0.26	49.84
8.25	0.42	6.00	3.5	26.12	4.59	0.33	49.79
5.50	0.27	5.83	3.5	23.67	5.07	0.40	49.73
2.75	0.13	5.73	3.5	21.64	5.55	0.48	49.67

Solving this equation for Z_F, we obtain $Z_F = 49.60$ ft.

We use Equation 6.60 to determine the length of the transition. Substituting the known quantities in Equation 6.60, we obtain

$$L_T = 2.35\Delta b + 1.65 m_C y_C = 2.35(10.0 - 5.7) + 1.65(2)(3.5) = 21.65 \text{ ft}.$$

Let us use $L_T = 22.0$ ft. We can now determine the side slopes and the bottom width of the transition at selected locations by using Equations 6.61 and 6.62, respectively. The calculated values are shown in columns (2) and (3) of Table 6.10. Then, we can determine the flow area, $A_X = (b_X + m_X y_X) y_X$, and the velocity $V_X = Q/A_X$ as summarized in Table 6.10. Finally, substituting all the known quantities into Equation 6.64, we determine the bed elevation, Z_X. The results are tabulated in column (8) of Table 6.10.

6.5.2 CHANNEL TRANSITIONS FOR SUPERCRITICAL FLOW

The design of channel transitions for supercritical flow is complicated by the presence of *standing wave fronts* (or *oblique wave fronts*) caused by the changes in the channel geometry. If the transition is not designed properly, these standing waves will be carried into the downstream channel and will cause a rough and irregular water surface. Ippen and Dawson (1951) investigated the problem of oblique wave fronts and developed a set of equations that can be used to design supercritical contractions.

6.5.2.1 Standing wave fronts in supercritical flow

In Section 4.3, we saw that the gravity waves caused by a disturbance in supercritical flow propagate downstream (see Figure 4.4d). A standing wave front develops along a straight line drawn tangent to the edges of the gravity waves from the point where the disturbance is first created. This wave front makes an angle, β, with the flow direction, evaluated as

$$\sin\beta = \frac{c}{V} = \frac{\sqrt{gD}}{V} = \frac{1}{F_r} \qquad (6.67)$$

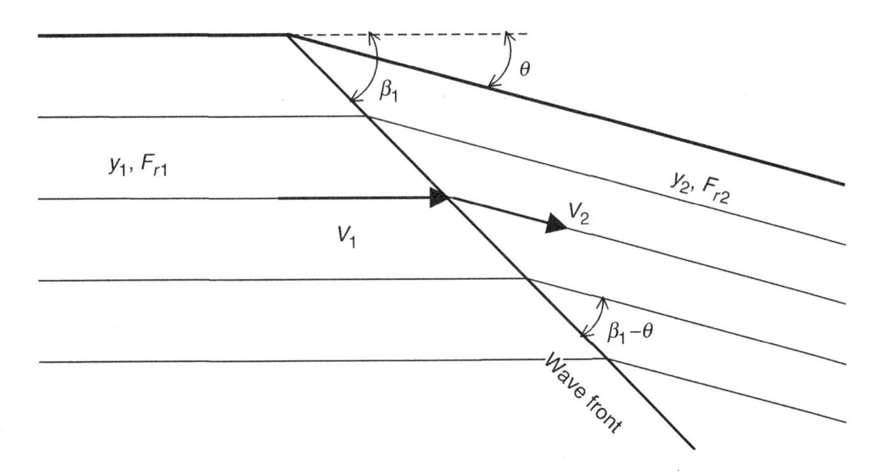

FIGURE 6.35
Standing wave caused by change in wall direction

where D = hydraulic depth, c = wave celerity, and F_r = Froude number. For a rectangular channel, $D = y$.

An inward change in the channel wall direction by an angle θ (as shown in Figure 6.35) also causes a flow disturbance in supercritical flow and gives rise to a standing wave front. Let the flow depth, the velocity, and the Froude number be denoted respectively by y_1, V_1, and F_{r1} upstream of the wave front, and by y_2, V_2, and F_{r2} downstream. The angle between the wave front and the upstream flow lines is β_1. Therefore, the component of V_1 normal to the wave front is $V_1 \sin\beta_1$, and the component parallel to the wave front is $V_1 \cos\beta_1$. Likewise, the angle between the wave front and the flow lines downstream is $(\beta_1 - \theta)$. The component of V_2 normal to the wave front is then $V_2 \sin(\beta_1 - \theta)$, and the component parallel to the wave front is $V_1 \cos(\beta_I - \theta)$.

The continuity equation for the discharge normal to the wave front (and written per unit flow width) yields the relationship

$$y_1 V_1 \sin\beta_1 = y_2 V_2 \sin(\beta_1 - \theta) \tag{6.68}$$

Likewise, the momentum equation in the direction normal to the wave front becomes

$$\frac{y_1^2}{2} + \frac{y_1 V_1^2 (\sin\beta_1)^2}{g} = \frac{y_2^2}{2} + \frac{y_2 V_2^2 [\sin(\beta_1 - \theta)]^2}{g} \tag{6.69}$$

The velocity components parallel to the wave front should be equal, since there are no forces along the front to cause a change. Thus

$$V_1 \cos\beta_1 = V_2 \cos(\beta_1 - \theta) \tag{6.70}$$

Dividing Equation 6.68 by Equation 6.70 and rearranging, we obtain

$$\frac{y_2}{y_1} = \frac{\tan\beta_1}{\tan(\beta_1 - \theta)} \tag{6.71}$$

Likewise, solving Equation 6.68 for V_2, substituting this into Equation 6.69, and rearranging:

$$\sin\beta_1 = \frac{1}{F_{r1}}\left[\frac{y_2}{2y_1}\left(\frac{y_2}{y_1}+1\right)\right]^{1/2} \tag{6.72}$$

or

$$\frac{y_2}{y_1} = \frac{1}{2}\left[\sqrt{1+8F_{r1}^2(\sin\beta_1)^2}-1\right] \tag{6.73}$$

The similarity between Equation 6.73 and the hydraulic jump equation (Equation 2.26) is noteworthy. For $\beta_1 = 90°$, Equation 6.73 reduces to Equation 2.26. For this reason, we sometimes refer to a standing wave front as *oblique jump*. Substituting Equation 6.71 into Equation 6.72,

$$\sin\beta_1 = \frac{1}{F_{r1}}\left[\frac{\tan\beta_1}{2\tan(\beta_1-\theta)}\left(\frac{\tan\beta_1}{\tan(\beta_1-\theta)}+1\right)\right]^{1/2} \tag{6.74}$$

Also, we can manipulate Equation 6.68 to obtain

$$\frac{F_{r2}}{F_{r1}} = \frac{\sin\beta_1}{\sin(\beta_1-\theta)}\left(\frac{y_1}{y_2}\right)^{3/2} \tag{6.75}$$

Given the flow conditions upstream of the wave front, and the angle θ, we can determine β_1 by using Equation 6.74. Then we can employ Equations 6.71 and 6.75, respectively, to calculate y_2 and F_{r2}. However, Equation 6.74 is implicit in β_1 and requires a trial-and-error solution. Alternatively, we can use Figure 6.36, which provides a graphical solution to Equation 6.74.

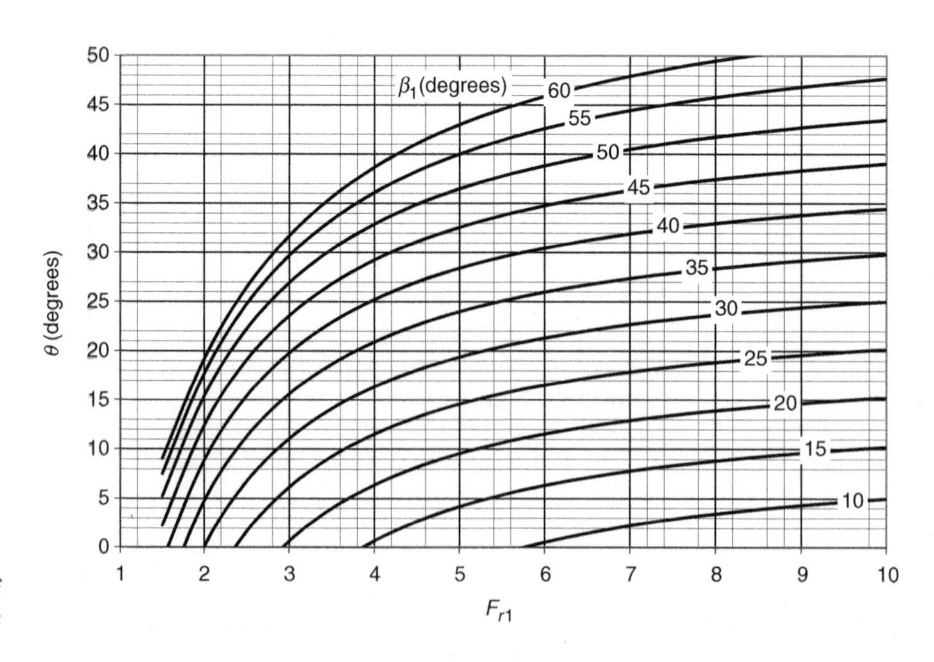

FIGURE 6.36 Graphical solutions of Equation 6.74

EXAMPLE 6.19 One of the sidewalls of a rectangular channel is deflected inward by an angle of $\theta = 25°$, causing a standing wave front. Upstream of the wave front, the flow depth is $y_1 = 0.28$ m and the velocity is $V_1 = 9.12$ m/s. Determine the flow depth and velocity downstream.

To solve this problem, we will first calculate the Froude number upstream of the wave front. For a rectangular channel,

$$F_{r1} = \frac{V_1}{\sqrt{gy_1}} = \frac{9.12}{\sqrt{(9.81)(0.28)}} = 5.50$$

Then, with $F_{r1} = 5.5$ and $\theta = 25°$, we obtain $\beta_1 = 35°$ from Figure 6.36. Next, by using Equation 6.71,

$$y_2 = \frac{\tan\beta_1}{\tan(\beta_1 - \theta)} y_1 = \frac{\tan 35°}{\tan(35° - 25°)} 0.28 = 1.11\,\text{m}$$

Then, employing Equation 6.75,

$$F_{r2} = F_{r1}\frac{\sin\beta_1}{\sin(\beta_1 - \theta)}\left(\frac{y_1}{y_2}\right)^{3/2} = 5.50\frac{\sin 35°}{\sin(35° - 25°)}\left(\frac{0.28}{1.11}\right)^{3/2} = 2.30$$

Finally, from the definition of Froude number,

$$V_2 = F_{r2}\sqrt{gy_2} = 2.30\sqrt{(9.81)(1.11)} = 7.59\,\text{m/s}$$

6.5.2.2 Rectangular contractions for supercritical flow

Let the width of a channel be reduced from b_1 to b_3 through a straight-walled contraction as shown in Figure 6.37. The contraction angle in this case is

$$\theta = \arctan\frac{b_1 - b_3}{2L} \tag{6.76}$$

The standing waves (AB and A′B in Figure 6.37) formed due to the change in the wall direction on both sides of the wall will make an angle β_1 with the initial flow direction. We can use Equation 6.74 (or Figure 6.36) to determine this angle, and Equations 6.71 and 6.75 to determine the depth y_2 and the Froude number F_{r2} downstream. The centerline of the channel shown in Figure 6.37 is also a separation streamline due to the symmetry. In other words there is no flow across the centerline. Therefore, another change in the flow direction caused at the centerline will give rise to the wave fronts BC and BC′ as shown in the figure. By analogy to Equations 6.74, 6.71, and 6.75, we can write the expressions

$$\sin\beta_2 = \frac{1}{F_{r2}}\left[\frac{\tan\beta_2}{2\tan(\beta_2 - \theta)}\left(\frac{\tan\beta_2}{\tan(\beta_2 - \theta)} + 1\right)\right]^{1/2} \tag{6.77}$$

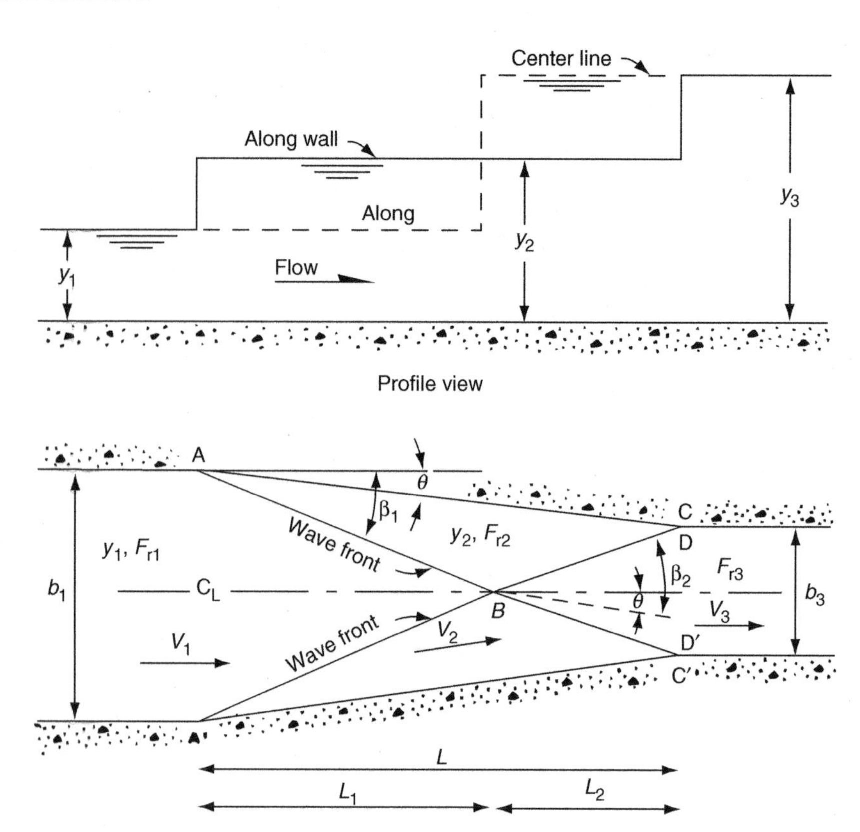

FIGURE 6.37 Rectangular contraction supercritical flow (after US Army Corps of Engineers, 1991)

$$\frac{y_3}{y_2} = \frac{\tan \beta_2}{\tan(\beta_2 - \theta)} \tag{6.78}$$

$$\frac{F_{r3}}{F_{r2}} = \frac{\sin \beta_2}{\sin(\beta_2 - \theta)} \left(\frac{y_2}{y_3}\right)^{3/2} \tag{6.79}$$

to determine the angle β_2, the flow depth y_3 and the Froude number F_{r3} downstream of the wave fronts BC and BC′. The standing waves will not be carried into the downstream channel if the second set of standing waves end precisely at the end of the transition, as shown in Figure 6.37. In other words, points C and C′, where the second set of standing wave fronts meet the two side walls, coincide with the points D and D′, where the channel width is reduced exactly to b_3. Therefore,

$$L = L_1 + L_2 = \frac{b_1}{2 \tan \beta_1} + \frac{b_3}{2 \tan(\beta_2 - \theta)} \tag{6.80}$$

EXAMPLE 6.20 The width of a rectangular channel carrying 40 cfs at depth 0.35 ft will be reduced from 10 ft to 5 ft through a straight-walled transition.

Determine the length of the transition, and the flow conditions downstream of the transition.

The givens in this problem are $b_1 = 10$ ft, $b_3 = 5$ ft, $Q = 40$ cfs, and $y_1 = 0.35$ ft. Therefore, $V_1 = 40/[(10)(0.35)] = 11.43$ fps and

$$F_{r1} = \frac{V_1}{\sqrt{gy_1}} = \frac{11.43}{\sqrt{(32.2)(0.35)}} = 3.40$$

We will determine the transition length L by trial and error. Let us try $L = 8.0$ ft. Then, by using Equation 6.76,

$$\theta = \arctan\frac{b_1 - b_3}{2L} = \arctan\frac{10.0 - 5.0}{2(8.0)} = 17.35^\circ$$

Next we will determine the angle β_1. We could use Figure 6.36 to find this angle, but for more precise results we solve Equation 6.74 by trial and error and obtain $\beta_1 = 34.14^\circ$. Now, by using Equations 6.71 and 6.75,

$$y_2 = \frac{\tan\beta_1}{\tan(\beta_1 - \theta)} y_1 = \frac{\tan 34.14^\circ}{\tan(34.14^\circ - 17.35^\circ)} 0.35 = 0.79 \text{ ft}$$

$$F_{r2} = F_{r1}\frac{\sin\beta_1}{\sin(\beta_1 - \theta)}\left(\frac{y_1}{y_2}\right)^{3/2} = 3.40\frac{\sin 34.14^\circ}{\sin(34.14^\circ - 17.35^\circ)}\left(\frac{0.35}{0.79}\right)^{3/2} = 1.96$$

Employing Equations 6.77, 6.78, and 6.79 in the same manner, we obtain $\beta_2 = 55.87^\circ$, $y_3 = 1.46$ ft, and $F_{r3} = 1.033$. Note that by using F_{r2} in lieu of F_{r1} and β_2 in lieu β_1, we could use Figure 6.36 to determine β_2. However, we need to solve Equation 6.77 numerically by trial and error for more precise results. Finally, by using Equation 6.80,

$$L = \frac{b_1}{2\tan\beta_1} + \frac{b_3}{2\tan(\beta_2 - \theta)} = \frac{10.0}{2\tan(34.14^\circ)} + \frac{5.0}{2\tan(55.87^\circ - 17.35^\circ)} = 10.5 \text{ ft}$$

This is different from the assumed length of 8.0 ft, and therefore 8.0 ft is not acceptable. We will repeat the same procedure with different trial values of L until the assumed and calculated values are equal. The calculations are summarized in Table 6.11. The trial values for L are in column (1). The entries in columns (2), (3), (4), and (5) are calculated by using Equations 6.76, 6.74, 6.71, and 6.75, respectively. Likewise, the entries in columns (6), (7), and (8) are obtained by using Equations 6.77, 6.78, and 6.79, respectively. Equation 6.80 is used to determine the entries in column (11). The calculations are repeated with different trial values of L until the calculated L in column (11) is the same as the assumed L in column (1).

A review of Table 6.11 will show that the transition length for this contraction structure is 16.8 ft. The resulting flow depth downstream of the transition is $y_3 = 0.80$, and the Froude number is $F_{r3} = 1.98$. It is important to note that

TABLE 6.11 Summary Calculations for Example 6.20

L (ft)	$\theta(°)$	$\beta_1(°)$	y_2 (ft)	F_{r2}	$\beta_2(°)$	y_3 (ft)	F_{r3}	L_1 (ft)	L_2 (ft)	L (ft)
8.00	17.35	34.14	0.79	1.96	55.87	1.46	1.03	7.37	3.14	10.51
10.50	13.39	29.81	0.68	2.21	42.07	1.12	1.45	8.73	4.57	13.30
13.30	10.65	29.96	0.61	2.40	35.64	0.94	1.73	9.83	5.36	15.19
15.20	9.34	25.63	0.58	2.50	32.94	0.85	1.88	10.42	5.72	16.14
16.2	8.77	25.06	0.56	2.54	31.80	0.82	1.94	10.69	5.88	16.57
16.80	8.46	24.76	0.55	2.56	31.22	0.80	1.98	10.84	5.96	16.80

$F_{r3} = 1.98$ is not near unity (1.0) in this case. Transitions designs resulting in Froude numbers close to 1.0 are unacceptable because of the possibility of choking.

6.5.2.3 Rectangular expansions for supercritical flow

Designing an expansion structure for supercritical flow is challenging due to the possibility of flow separation from the boundaries in addition to local standing waves. Rouse *et al.* (1951) developed a series of design curves based on experimental and analytical studies, and this study is the basis for our discussion here. The equations we present below approximate the design curves of Rouse *et al.* (1951), and are also similar to those reported by the US Army Corps of Engineers (1991).

Figure 6.38 depicts the schematic of a typical transition for rectangular channels where the channel width expands from b_1 at section PC to b_2 at section PT. Each sidewall of the transition forms a relatively short convex curve upstream between sections PC and PRC, and a longer concave curve downstream between PRC and PT. The equations describing the transition width are expressed in terms of the dimensionless parameters

$$x^* = \frac{x}{b_1 F_{r1}} \tag{6.81}$$

$$b^* = \frac{b}{b_1} \tag{6.82}$$

$$r = \frac{b_2}{b_1} \tag{6.83}$$

where x = distance measured from the upstream section PC and b = width of the transition structure at x. Obviously, at section PC, $x^*_{PC} = 0$ and $b^*_{PC} = 1$.

We can determine the distance to section PRC by using

$$x^*_{PRC} = \left[\frac{22}{15}(r - 1)\right]^{2/3} \tag{6.84}$$

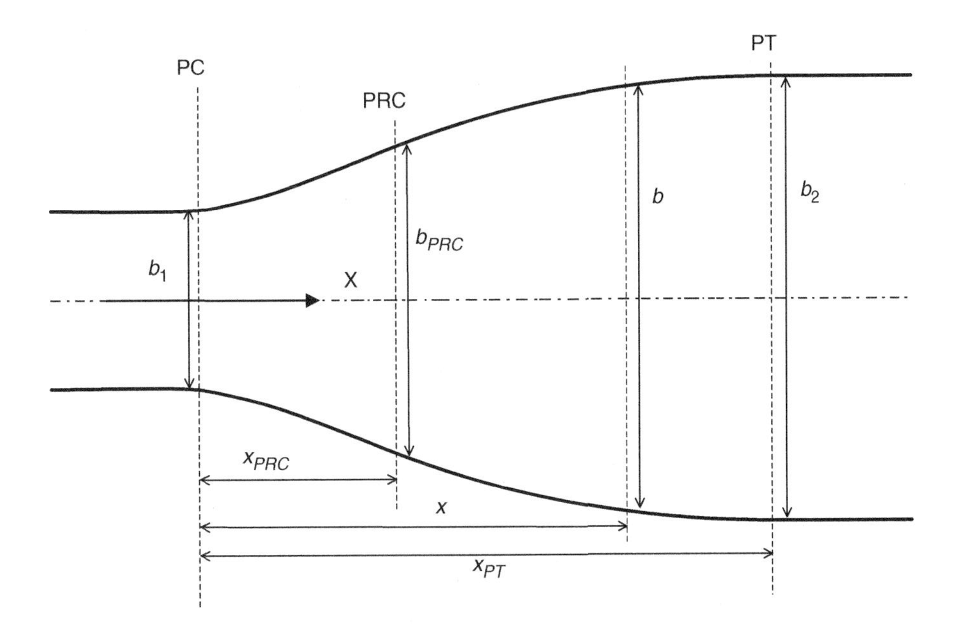

FIGURE 6.38 Supercritical rectangular expansion (after US Army Corps of Engineers, 1991)

and the width at section PRC is found by using

$$b^*_{PRC} = \frac{11}{30}r + \frac{19}{30} \tag{6.85}$$

Between the sections PC and PRC, the transition width is determined by using

$$b^* = \frac{1}{4}(x^*)^{3/2} + 1 \tag{6.86}$$

At the channel section PT we have $b^*_{PT} = r$. The distance between the channel sections PC and PT is the length of the transition structure, and this is determined by using

$$x^*_{PT} = \frac{13}{4}r - \frac{9}{4} \tag{6.87}$$

Between the sections PRC and PT, the transition width is calculated by using

$$b^* = r - 2s(x^*_{PT} - x^*)^t \tag{6.88}$$

where

$$t = \frac{(x^*_{PT} - x^*_{PRC})}{r - b^*_{PRC}}\frac{3}{8}(x^*_{PRC})^{1/2} \tag{6.89}$$

and

$$s = \frac{r - b^*_{PRC}}{2(x^*_{PT} - x^*_{PRC})^t} \tag{6.90}$$

EXAMPLE 6.21 The width of a rectangular channel carrying 50 cfs at depth 1.10 ft will be expanded from 4 ft to 10 ft. Design the transition structure.

The givens in this problem are $b_1 = 4$ ft, $b_2 = 10$ ft, $Q = 50$ cfs, $y_1 = 1.10$ ft, and $r = 10/4 = 2.5$. Therefore, at section PC we have $V_1 = 50/[(4)(1.10)] = 11.36$ fps and

$$F_{r1} = \frac{V_1}{\sqrt{gy_1}} = \frac{11.36}{\sqrt{(32.2)(1.10)}} = 1.909$$

We can now determine the locations of sections PRC and PT. By using Equation 6.84,

$$x^*_{PRC} = \left[\frac{22}{15}(r-1)\right]^{2/3} = \left[\frac{22}{15}(2.5-1)\right]^{2/3} = 1.692$$

Then, $x_{PRC} = (x^*_{PRC})(b_1)(F_{r1}) = (1.692)(4)(1.909) = 12.92$ ft. Likewise, by using Equation 6.87,

$$x^*_{PT} = \frac{13}{4}r - \frac{9}{4} = \frac{13}{4}(2.5) - \frac{9}{4} = 5.875$$

and $x_{PT} = (5.875)(4.0)(1.909) = 44.87$ ft. Therefore, the length of the transition structure will be 44.87 ft. Now we can calculate the width at various sections. For section PRC, we can use Equation 6.85

$$b^*_{PRC} = \frac{11}{30}r + \frac{19}{30} = \frac{11}{30}(2.5) + \frac{19}{30} = 1.55$$

and $b_{PRC} = (b^*_{PRC})(b_1) = (1.55)(4.0) = 6.20$ ft. For sections between PC and PRC, that is for $0 < x < 12.92$ ft, we use Equation 6.86 to determine the width. For example, for $x = 4$ ft, $x^* = 4.0/[(4.0)(1.909)] = 0.524$ (see Equation 6.81) and, by using Equation 6.86,

$$b^* = \frac{1}{4}(x^*)^{3/2} + 1 = \frac{1}{4}(0.524)^{3/2} + 1 = 1.095$$

Then $b = (1.095)(4.0) = 4.38$ ft. Equation 6.88, along with Equations 6.89 and 6.90, is used to determine the width at the sections between PRC and PT, that is for 12.92 ft $< x <$ 44.87 ft. Let us first evaluate t and s, by using Equations 6.89 and 6.90, as

$$t = \frac{(x^*_{PT} - x^*_{PRC})}{r - b^*_{PRC}}\frac{3}{8}(x^*_{PRC})^{1/2} = \frac{(5.875 - 1.692)}{2.5 - 1.55}\frac{3}{8}(1.692)^{1/2} = 2.148$$

and

$$s = \frac{r - b^*_{PRC}}{2(x^*_{PT} - x^*_{PRC})^t} = \frac{2.5 - 1.55}{2(5.875 - 1.692)^{2.148}} = 0.022$$

TABLE 6.12 Example 6.21

Section	x (ft)	x^*	b^*	b (ft)
PR	0.00	0.00	1.00	4.00
	4.00	0.52	1.09	4.38
	8.00	1.05	1.27	5.07
PRC	12.92	1.69	1.55	6.20
	20.00	2.619	1.945	7.78
	25.00	3.273	2.157	8.63
	30.00	3.928	2.316	9.26
	35.00	4.583	2.424	9.70
	40.00	5.237	2.483	9.93
PT	44.87	5.875	2.500	10.00

Now, for a section $x = 30$ ft from section PC, we have $x^* = 30/[(4.0)(1.909)] = 3.928$. Then, by using Equation 6.88,

$$b^* = r - 2s(x^*_{PT} - x^*)^t = 2.5 - 2(0.022)(5.875 - 3.928)^{2.148} = 2.316$$

and $b = (2.316)(4.0) = 9.27$ ft. Table 6.12 summarizes the transition width calculations at various other locations.

PROBLEMS

P.6.1 Another rectangular sharp-crested weir placed at the downstream end of the canal considered in Example 6.1 also has a crest height of $p = 0.70$ m and a crest length of $L_w = 0.80$ m. The water surface at the approach section of this weir is $h_o = 0.72$ m above the crest. Determine the rate of loss of water due to seepage between the two weirs.

P.6.2 A trapezoidal irrigation canal has a bottom width of $b = 4$ ft, side slopes of $m = 2$, and a longitudinal bottom slope of $S_0 = 0.0005$. A rectangular sharp-crested weir placed in this canal has a crest height of $p = 2.30$ ft and a crest length of $L_w = 2.60$ ft. Another weir of the same characteristics is placed in the canal some distance downstream. The water surface elevation at the approach section of the upstream weir is $h_0 = 2.50$ ft above the crest. For the downstream weir, $h_0 = 2.35$ ft. Determine the rate of loss of water due to seepage in the canal between the two weirs.

P.6.3 A rectangular weir with $L_b = 0.80$ m, $L_w = 1.0$ m, and $p = 0.40$ m is placed in a rectangular channel that is 1.0 m wide. Determine the minimum and maximum discharge that can be measured by this weir using the broad-crested weir equations.

P.6.4 A rectangular weir with $L_b = 2.50$ ft, $L_w = 3.0$ ft, and $p = 1.20$ ft is placed in a rectangular channel that is 3.0 ft wide. Determine the minimum and maximum discharge that can be measured by this weir using the broad-crested weir equations.

P.6.5 A circular concrete culvert has a diameter $D=4$ ft, bottom slope $S=0.01$ and length $L=100$ ft. The culvert inlet is grooved with a headwall and is not mitered. Determine the maximum discharge this culvert can convey under inlet control conditions if the headwater depth, HW, is not to exceed 6.0 ft.

P.6.6 A circular concrete culvert has a diameter $D=1.5$ m, bottom slope $S=0.01$ and length $L=30$ m. The culvert inlet is grooved with a headwall and is not mitered. Determine the maximum discharge this culvert can convey under inlet control conditions if the headwater depth, HW, is not to exceed 2.3 m.

P.6.7 A 4 ft by 4 ft square culvert has 30° wingwall flares and the inlet is not mitered. The culvert has a slope of $S=0.01$. The water marks observed after a storm event indicated that the headwater rose to a depth of $HW=3.2$ ft. Determine the maximum discharge occurred during the storm event.

P.6.8 A 1.5 m by 1.5 m square culvert has 30° wingwall flares and the inlet is not mitered. The culvert has a slope of $S=0.01$. The water marks observed after a storm event indicated that the headwater rose to a depth of $HW=0.8$ m. Determine the maximum discharge occurred during the storm event.

P.6.9 A rectangular box culvert has a 90° headwall with 33.7° bevels, and the inlet is mitered. The width of the culvert is $b=4$ ft and the height is $D=5$ ft. The culvert is 100 ft long with a slope of $S=0.01$. Prepare a plot of HW versus Q for this culvert if it is to operate under inlet control.

P.6.10 A rectangular box culvert has a 90° headwall with 33.7° bevels, and the inlet is mitered. The width of the culvert is $b=1$ m and the height is $D=1.5$ m. The culvert is 30 m long with a slope of $S=0.01$. Prepare a plot of HW versus Q for this culvert if it is to operate under inlet control.

P.6.11 A 100-ft long horizontal concrete pipe culvert ($n=0.012$) is to be sized to carry 38 cfs. The tailwater depth is $TW=3.5$ ft. The inlet will be square-edged, and the headwater depth, HW, is not to exceed 4.5 ft. Select a suitable culvert diameter.

P.6.12 A 35-meter long horizontal pipe culvert ($n=0.012$) has a groove-end inlet. The culvert diameter is $D=1.0$ m and tailwater depth is $TW=1.2$ m. Can this culvert convey $Q=2\text{ m}^3/\text{s}$ if the headwater depth, HW, is not to exceed 2.0 m.

P.6.13 A 100-ft long concrete box culvert ($n=0.012$) will be laid on a slope $S=0.001$. The inlet will have a headwall parallel to the embankment and the entrance will be beveled on three sides. What should be the width, b, of the culvert if $D=3$ ft, $TW=3.2$ ft, $Q=60$ cfs, and $HW\leq 8$ ft.

P.6.14 A 40-m long concrete box culvert ($n=0.012$) will be laid on a slope $S=0.001$. The inlet will have a headwall parallel to the embankment and the entrance will be beveled on three sides. What should be the width, b, of the culvert if $D=1$ m, $TW=1.1$ m, $Q=1.8\text{ m}^3/\text{s}$, and $HW\leq 3.0$ m.

P.6.15 For the culvert sized in Problem P.6.13 prepare a plot of HW versus Q for Q varying between 10 cfs and 60 cfs.

P.6.16 A corrugated metal pipe culvert has a length of $L=120$ ft, slope of $S=0.001$, roughness factor of $n=0.024$, and diameter of $D=3$ ft. The inlet is projecting from fill with no headwall. The tailwater depth is $TW = 1.0$ ft. The flow is controlled by the outlet. Determine the headwater depth, HW, for $Q = 40$, 60, and 80 cfs. Can this culvert carry 80 cfs if the headwater depth, HW, is not to exceed 8 ft?

P.6.17 A corrugated metal pipe culvert has a length of $L = 40$ m, slope of $S = 0.001$, roughness factor of $n = 0.024$, and diameter of $D = 1$ m. The inlet is projecting from fill with no headwall. The tailwater depth is $TW = 0.3$ m. The flow is controlled by the outlet. Determine the headwater depth, HW, for $Q = 1.0$, 1.5, and 2.0 m^3/s. Can this culvert carry 2 m^3/s if the headwater depth, HW, is not to exceed 2.7 m?

P.6.18 An uncontrolled overflow ogee crest for a spillway is to discharge 80 m^3/s at a design head of 1.5 m. The crest is 3 m above the reservoir bottom. A bridge to be provided over the crest will be supported by 0.5-m wide piers with round noses. The bridge spans are not to exceed 6 m. The abutments are rounded with a headwall perpendicular to the flow direction.

(a) Determine the total length of the weir crest
(b) Obtain a discharge–head relationship for $H_e = 0.25$, 0.50, 0.75, 1.0, 1.25, and 1.50 m.

P.6.19 What would be the total length of the weir crest in Example 6.10 if the piers were 2 ft wide.

P.6.20 Suppose in Figure 6.25, $Z = 0$, $Z_u = 0$, $P = 3$ m, $H_o = 1.5$ m, and $L_e = b_B = 10$ m. The apron downstream of the spillway is horizontal. Determine the flow depth, y_1, at the toe of the spillway for $H_e = 0.5$, 1.0, and 1.5 m.

P.6.21 Suppose in Figure 6.25, $Z = 0$, $Z_u = 0$, $P = 90$ ft, $H_0 = 10$ ft and $L_e = b_B = 200$ ft. The apron downstream of the spillway is horizontal. Determine the flow depth, y_1, at the toe of the spillway for $H_e = 5.0$, 7.5, and 10.0 ft.

P.6.22 Suppose the downstream channel in Problem P.6.20 can be approximated by a rectangular channel with $b = 10$ ft, $n = 0.02$ and $S_0 = 0.001$. Determine the location of the hydraulic jump for $H_e = 0.5$, 1.0, and 1.5 m. Assume $b_B = L_e$ and $n_B = 0.016$.

P.6.23 Suppose the downstream channel in Problem P.6.21 can be approximated by a rectangular channel with $b = 200$ ft, $n = 0.025$, and $S_0 = 0.0003$. Determine the locations of the hydraulic jump for $H_e = 5.0$, 7.5, and 10.0 ft. Assume $b_B = L_e$ and $n_B = 0.016$.

P.6.24 Suppose the discharge in Example 6.17 is increased to 150 cfs, and the corresponding flow depth at section E is 4.6 ft. Determine the flow depth at sections D, C, B, and A.

P.6.25 A straight line transition connects a 4.0 m wide rectangular channel carrying 4 m^3/s to a 2.0 m wide flume as shown in Figure 6.33. The transition is 8 m long and the bottom elevation decreases linearly from 10.4 m at the channel to 10.0 m at the flume. The flow depth at the flume is 1.8 m. Determine the flow depth in the transition structure 2, 4, and 6 m from the flume.

P.6.26 The width of a rectangular channel carrying 2.2 m^3/s at a depth 0.15 m will be reduced from 4 m to 2 m through a straight wall transition. Determine the length of the transition and the flow depth at the downstream end of the transition.

P.6.27 The width of a rectangular channel carrying 80 cfs at a depth 0.50 ft will be reduced from 12 ft to 6 ft through a straight wall transition. Determine the length of the transition and the flow depth at the downstream end of the transition.

P.6.28 The width of a rectangular channel carrying 150 cfs at a depth 1.5 ft will be expanded from 6 ft to 12 ft. Design the transition structure.

P.6.29 The width of a rectangular channel carrying 4.2 m^3/s at a depth 0.5 m will be expanded from 2 m to 4 m. Design the transition structure.

REFERENCES

Bodhaine, G. L. (1976). Measurement of peak discharge at culverts by indirect methods. In: *Techniques of the Water Resources Investigations of the US Geological Survey*, Book 3, Chapter 3. Government Printing Office, Washington, DC.

Bos, M. G. (1989). *Discharge Measurement Structures*. International Institute for Land Reclamation and Improvement, Wageningen, The Netherlands.

Brater, E. F. and King, H. W. (1976). *Handbook of Hydraulics*. McGraw-Hill Book Co., New York, NY.

Chow, V. T. (1959). *Open-Channel Hydraulics*. McGraw-Hill Book Co., New York, NY.

Davis, S. (1963). Unification of Parshall flume data. *Transactions, ASCE*, **128,** 339–421.

Dodge, R. A. (1963). Discussion of unifications of Parshall flume data by S. Davis. *Transactions, ASCE*, **128,** 339–421.

Ippen, A. T. and Dawson, J. H. (1951). Design of channel contractions. *Transactions, ASCE*, **116,** 326–346.

Kilpatrick, F. A. and Schneider, V. R. (1983). Use of flumes in measuring discharge. In: *Techniques of Water Resources Investigations of the US Geological Survey*, Book 3, Chapter A14. US Government Printing Office, Washington, DC.

Kindsvater, C. E. and Carter, R. W. C. (1957). Discharge characteristics of rectangular thin plate weirs. *Journal of the Hydraulics Division, ASCE*, **83(HY6),** 1–36.

Kulin, G. and Compton, P. R. (1975). *A Guide to Methods and Standards for the Measurement of Water Flow*. Special Publication 421, Institute for Basic Standards, National Bureau of Standards, Washington, DC.

Morris, H. M. and Wiggert, J. M. (1972). *Applied Hydraulics in Engineering*. John Wiley and Sons, New York, NY.

Normann, J. M., Houghtalen, R. J. and Johnston, W. J. (1985). *Hydraulic Design of Highway Culverts*. Federal Highway Administration, Hydraulic Design Series No. 5, McLean, VA.

Roberson, J. A., Cassidy, J. J. and Chaudhry, M. H. (1997). *Hydraulic Engineering*. John Wiley and Sons, Inc., New York, NY.

Rouse, H., Bhootha, B. V. and Hsu, E. Y. (1951). Design of channel expansions. *Transactions, ASCE*, **116,** 347–363.

Sturm, T. W. (2001). *Open Channel Hydraulics*. McGraw-Hill Book Co., New York, NY.

US Army Corps of Engineers (1987). *Hydraulic Design Criteria*. Coastal and Hydraulics Laboratory, Vicksburg, MS.

US Army Corps of Engineers (1990). *Hydraulic Design of Spillways*. Engineer Manual, EM 1110-2-1603, Department of the Army, Washington, DC.

US Army Corps of Engineers (1991). *Hydraulic Design of Flood Control Channels*. Engineer Manual, EM 1110-2-1601, Department of the Army, Washington, DC.

US Army Corps of Engineers (2000). *Design of Spillway Tainter Gates*. Engineer Manual, EM 1110-2-2702, Department of the Army, Washington, DC.

US Bureau of Reclamation (1967). General design information for structures. In: *Canals and Related Structures*, Design Standards No. 3, Chapter 2. US Government Printing Office, Washington, DC.

US Bureau of Reclamation (1987). *Design of Small Dams*. US Government Printing Office, Washington, DC.

US Bureau of Reclamation (2001). *Water Measurement Manual*. Technical Publication, Water Resources Research Laboratory, US Government Printing Office, Washington, DC.

Vittal, N. and Chiranjeevi, V. V. (1983). Open-channel transitions: rational method of design. *Journal of Hydraulic Engineering, ASCE*, **109(1),** 99–115.

7 Bridge hydraulics

Bridges over streams and rivers pose challenging hydraulics problems. The flow constrictions caused by bridge abutments and piers give rise to additional energy losses due to the contraction of the flow lines at the upstream side and expansion of the flow lines on the downstream side of the bridge. Unlike the smooth and gradual channel contraction problems studied in Chapters 2 and 4, the flow constriction due to a bridge is abrupt. Therefore the resulting energy losses can be significant, and they need to be taken into account when the energy method is employed to calculate the water surface profiles through the bridge. If the momentum approach is used, then the external forces, such as the drag forces due to the bridge piers, need to be considered.

Another problem concerning hydraulic engineers is scour at bridges. By scour, we refer to erosion caused by water on the soil surrounding the bridge abutments and piers. Excessive scour during floods is a common cause of bridge failure. Therefore, it is important to estimate the scour depths accurately when designing new bridges or for evaluating the vulnerability of existing bridges.

7.1 Modeling Bridge Sections

The major portion of the discussion in the subsequent sections follows the procedures adopted in the Hydraulic Engineering Center River Analysis System (HEC-RAS) of the US Army Corps of Engineers (2002).

7.1.1 CROSS-SECTION LOCATIONS

As we recall from Chapter 4, in the standard step method for gradually-varied flow calculations, the flow depths are determined at selected sections. The flow sections in the vicinity of a bridge need to be carefully located, since the effect of a bridge on water surface profiles extends beyond the bridge section. Figure 7.1 displays a plan view of the basic cross-section layout suggested in the HEC-RAS model. Note that this layout includes only the cross-sections needed for bridge calculations; obviously, additional cross-sections both upstream of Section 4 and

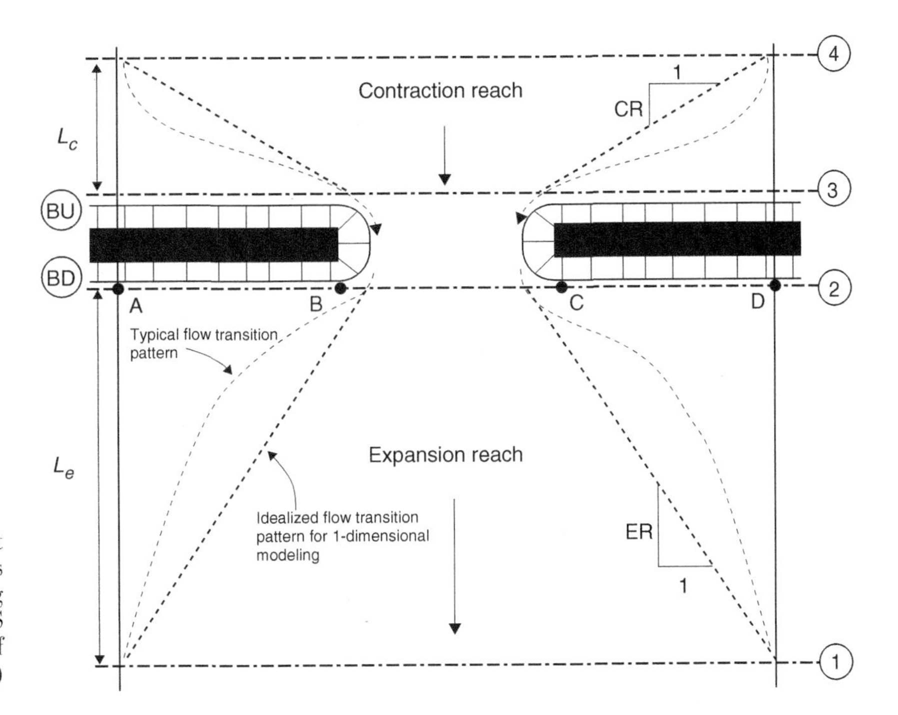

FIGURE 7.1 Layout of cross-sections for modeling bridges (after US Army Corps of Engineers, 2002)

TABLE 7.1 Ranges of expansion ratios (after US Army Corps of Engineers, 2002)

b/B	S_0 (ft/min)	$n_{ob}/n_c = 1$	$n_{ob}/n_c = 2$	$n_{ob}/n_c = 3$
0.10	1	1.4–3.6	1.3–3.0	1.2–2.1
	5	1.0–2.5	0.8–2.0	0.8–2.0
	10	1.0–2.2	0.8–2.0	0.8–2.0
0.25	1	1.6–3.0	1.4–2.5	1.2–2.0
	5	1.5–2.5	1.3–2.0	1.3–2.0
	10	1.5–2.0	1.3–2.0	1.3–2.0
0.50	1	1.4–2.6	1.3–1.9	1.2–1.4
	5	1.3–2.1	1.2–1.6	1.0–1.4
	10	1.3–2.0	1.2–1.5	1.0–1.4

downstream of Section 1 are used for water surface calculations in the further upstream and further downstream segments of the river.

Section 1 is located sufficiently downstream from the bridge so that the flow is fully expanded. The expansion distance depends on the degree and shape of the constriction, the flow rate, and the velocity. The ranges of expansion ratios provided in Table 7.1 can be used as a guide to determine this distance. To obtain L_e, we multiply the average of the distances A to B and C to D in Figure 7.1 by the expansion ratio obtained from Table 7.1.

In this Table 7.1, b/B is the ratio of the bridge opening width to the total flood plain width, S_0 is the longitudinal bottom slope, and n_c and n_{ob} are, respectively, the Manning roughness factors for the main channel and for the overbank.

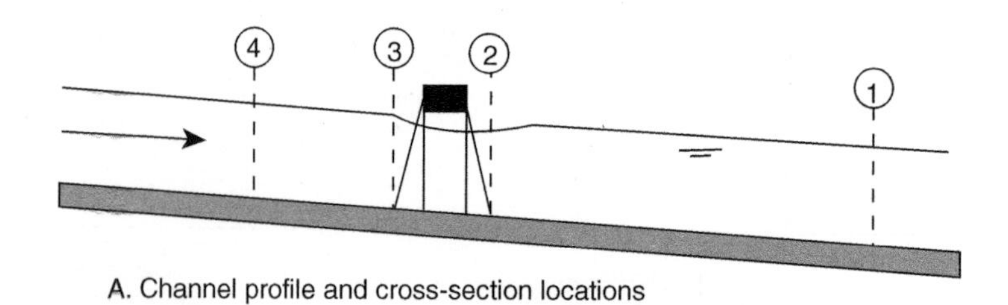

A. Channel profile and cross-section locations

B. Bridge cross-section on natural ground

C. Portion of cross-sections 2 & 3 that is ineffective for low flow

FIGURE 7.2
Cross-sections near bridges (after US Army Corps of Engineers, 2002)

Cross-section 2 is a short distance downstream of the bridge. This section is usually located at the downstream toe of the roadway embankment, as shown in Figure 7.2. Likewise, cross-section 3 is a short distance upstream of the bridge placed at the upstream toe of the embankment. Because the flow lines are contracted near the bridge, both sections 2 and 3 include ineffective areas that should be excluded from the flow area. The shaded areas in Figure 7.2C represent the ineffective areas. We can assume a 1:1 contraction and expansion rate of flow lines in the close vicinity of the bridge. In other words, if the cross-section is 10 ft from the bridge face, the ineffective flow areas should be placed 10 ft away from each side of the bridge opening.

Cross-section 4 is sufficiently upstream of the bridge so that the flow lines are approximately parallel. The distance between sections 3 and 4 (the contraction length) depends on the degree and shape of the constriction, the flow rate, and the velocity. The contraction length, L_c, is generally shorter than the expansion length, L_e. The recommended contraction lengths vary between 1 and 1.5 times the average of distances A to B and C to D in Figure 7.1.

Cross-section BU is placed just inside the bridge structure at the upstream end. Likewise, the cross-section BD is placed inside the bridge structure at the downstream end. The bridge deck, abutments, and bridge piers determine the characteristics of these cross-sections.

7.1.2 LOW-FLOW TYPES AT BRIDGE SITES

Low flow exists when the flow passes through the bridge opening in the form of open-channel flow. In other words, the water surface is below the highest point on the low chord of the bridge opening. Low flow can further be classified into classes A, B, and C, depending on whether the flow is subcritical or supercritical, and whether choking occurs. Class A flow remains completely subcritical through the bridge, and class C flow remains completely supercritical. Class B occurs when the flow is choked due to the bridge constriction. In this case, the water surface passes through the critical depth in the bridge section. A hydraulic jump may occur upstream of the bridge or downstream, depending on whether the flow approaching the bridge is subcritical or supercritical (see Sections 4.6.4.1 and 4.6.4.2). If a bridge is properly designed, class B flow should not occur.

7.1.3 LOW-FLOW CALCULATIONS AT BRIDGE SITES

We will focus on class A flow in this section, since it is the most common type. Among the different methods available, we will include the energy method, the momentum method, and the Yarnell method. In all three methods, for class A flow (subcritical), the calculations start from the downstream end and proceed in the upstream direction. The energy and the momentum methods are applicable to class C flow (supercritical) as well. However, for class C flow the calculations start from the upstream end and proceed in the downstream direction.

7.1.3.1 Flow choking at bridge sections

The low-flow calculations for class A flow include verification that the flow will not choke. This can be achieved using either the energy equations or the momentum equations.

Suppose the narrowest bridge section is BR. The energy equation between section 3 (see Figure 7.1) and BR and that between sections BR and 2 can be written respectively as

$$z_{b3} + y_3 + \alpha_3 \frac{V_3^2}{2g} - \Delta H_{3BR} = z_{bBR} + E_{bBR} \qquad (7.1)$$

and

$$z_{bBR} + E_{bBR} - \Delta H_{BR2} = z_{b2} + y_2 + \alpha_2 \frac{V_2^2}{2g} \qquad (7.2)$$

where the subscripts 2, 3, and BR indicate the flow sections, z_b = elevation of the channel bottom, α = energy correction factor, V = cross-sectional average velocity, g = gravitational acceleration, E = specific energy, ΔH_{3BR} = energy head loss between sections 3 and BR, and ΔH_{BR2} = energy head loss between sections BR and 2. Let us now denote the specific energy corresponding to critical flow at section BR by E_{crit}. Then, as we recall from Chapter 2, the flow will not choke if

$$z_{bBR} + E_{bBR} > z_{bBR} + E_{crit}. \tag{7.3}$$

On the basis of Equations 7.1 and 7.2, this can be written as

$$z_{b3} + y_3 + \alpha_3 \frac{V_3^2}{2g} - \Delta H_{3BR} > z_{bBR} + E_{crit} \tag{7.4}$$

or

$$z_{b2} + y_2 + \alpha_2 \frac{V_2^2}{2g} + \Delta H_{BR2} > z_{bBR} + E_{crit} \tag{7.5}$$

Either Equation 7.4 or Equation 7.5 can be used to verify that flow will not choke. However, for class A flow, which is subcritical, water surface calculations are performed from downstream towards upstream – in other words, the conditions at section 2, rather than 3, would be known at the time the choking needs to be checked. Therefore, Equation 7.5 is more convenient to use. Also, dropping the term ΔH_{BR2} from the left-hand side would be more conservative, and it would simplify the inequality as

$$z_{b2} + y_2 + \alpha_2 \frac{V_2^2}{2g} > z_{bBR} + E_{crit} \tag{7.6}$$

Therefore, in the energy approach we can use this inequality to verify that choking will not occur.

We can also use the momentum approach to check choking. As we recall from Chapter 2, choking will not occur if

$$M_{BR} > M_{crit} \tag{7.7}$$

where M_{BR} = specific momentum at section BR (the most constricted bridge section) and M_{crit} = specific momentum corresponding to critical flow at section BR. Again, for subcritical flow, conditions at section 2 (rather than 3) will be known at the time choking needs to be checked. The momentum equation between the sections BR and 2 can be written as

$$M_{BR} - \frac{F_f}{\gamma} - \frac{F_p}{\gamma} + \frac{W_x}{\gamma} = M_2 \tag{7.8}$$

where F_f = friction force between sections BR and 2, F_p = forces exerted by piers on the flow, and W_x = component in the flow direction of weight of water present between sections BR and 2. Solving Equation 7.8 for M_{BR} and substituting into Equation 7.7,

$$M_2 + \frac{F_f}{\gamma} + \frac{F_p}{\gamma} - \frac{W_x}{\gamma} > M_{crit} \tag{7.9}$$

for flow not to choke. The US Army Corps of Engineers (2002) suggests an approximation to Equation 7.9 in the form

$$M_2 > M_{crit} \tag{7.10}$$

This inequality is adequate for the most part to verify that choking will not occur. However, if the two sides of the inequality turn out to be close, then Equation 7.9 should be used for greater precision. Evaluation of the terms F_f, F_p, and W is discussed later in this chapter.

7.1.3.2 Energy method for low-flow calculations

The energy method is very similar to the standard step method used for gradually-varied flow calculations in natural channels. The method is based on Equation 4.34, repeated here as

$$z_{bU} + y_U + \alpha_U \frac{V_U^2}{2g} - \frac{1}{2}(\Delta X)S_{fU} = z_{bD} + y_D + \alpha_D \frac{V_D^2}{2g} + \frac{1}{2}(\Delta X)S_{fD} + h_e \tag{7.11}$$

where U and D, respectively, denote the upstream and downstream sections, and z_b = elevation of channel bottom above a horizontal datum, y = flow depth, g = gravitational acceleration, ΔX = distance between the upstream and downstream sections, S_f = friction slope, α = energy correction coefficient, V = cross-sectional average velocity, and h_e = eddy loss.

For a compound channel section, as in Figure 3.10, the friction slope is evaluated by using Equation 3.41, rewritten here as

$$S_f = \left(\frac{Q}{\sum K_i}\right)^2 \tag{7.12}$$

where i = index referring to the i-th subsection of the compound channel section, and K = conveyance, calculated as (see Equation 3.40)

$$K_i = \frac{k_n}{n_i} A_i R_i^{2/3} = \frac{k_n}{n_i} \frac{A_i^{5/3}}{P_i^{2/3}} \tag{7.13}$$

The discharge in the i-th subsection is

$$Q_i = K_i S_f^{1/2} \tag{7.14}$$

The energy coefficient is evaluated by using Equation 1.21, written here as

$$\alpha = \frac{\sum V_i^3 A_i}{V^3 \sum A_i} = \frac{\left(\sum A_i\right)^2}{\left(\sum K_i\right)^3} \sum \frac{K_i^3}{A_i^2} \tag{7.15}$$

The eddy loss is evaluated by using

$$h_e = k_e \left| \alpha_U \frac{V_U^2}{2g} - \alpha_D \frac{V_D^2}{2g} \right| \tag{7.16}$$

where $k_e = 0.3$ is suggested for contracting flow and $k_e = 0.5$ for expanding flow.

In a typical situation, the conditions at section 1 (see Figure 7.1) will be known from the gradually-varied flow calculations further downstream. Then we can perform the calculations for the standard step method in sequence from section 1 to 2, 2 to BD, BD to BU, BU to 3, and 3 to 4. Note that at sections 2 and 3, the edges of the ineffective areas are usually not included in the wetted perimeter (see Figure 7.2C). The bridge sections BU and BD are treated just like a compound channel section. However, the area of the bridge below the water surface is subtracted from the total area, and the wetted perimeter is increased where the water is in contact with the bridge structure. Suggested values for the eddy coefficient are $k_e = 0.5$ between sections BD and 2 and between sections 2 and 1 (expansion), and $k_e = 0.3$ between sections BU and 3 and between sections 3 and 4.

Example 7.1 Suppose a bridge is located between sections 1 and 4 of the reach considered in Example 4.16 (see Figure 4.30). We are to calculate the water surface profile through the bridge constriction.

Let us revisit the channel reach considered in Example 4.16. Recall that the channel is straight, and the sections marked in Figure 4.30 are 600 m apart. The cross-section of the channel in this reach can be approximated as shown in Figure 4.31. The Manning roughness factor is 0.025 for the main channel, and 0.05 for the left overbank and the right overbank areas. The bed elevation, z_b, at section 0.7 is 64 m. The channel has a longitudinal bottom slope of 0.0005. Therefore, the bed elevations at sections 0.8, 1, 4, and 4.3 are 64.3 m, 64.6 m, 64.9 m, and 65.2 m, respectively. In Example 4.16, we calculated the water surface profile for $Q = 250\ \text{m}^3/\text{s}$, given the water surface elevation at section 0.7 is 66.3 m.

Suppose the information available on the bridge can be summarized as follows. The downstream toe of the roadway embankment is 330 m from section 1 and the downstream edge of the bridge structure is 333 m from section 1. Suppose the width of the roadway embankment is 14 m. Now, referring to Figures 7.1 and 7.2, we can place section 2 at 330 m from section 1, section BD at 333 m from section 1, section BU at $333 + 14 = 347$ m from section 1, and section 3 at a distance $347 + 3 = 350$ m from section 1. Therefore, the distance between sections 3 and 4 is $600 - 350 = 250$ m. Let the bed elevation at section 2 be 64.76 m, and that at section 3 be 64.77 m. Figure 7.3 depicts a schematic representation of the cross-sectional geometry (not to scale) of these sections. At the bridge section, there are two 2-m wide abutments placed on the left

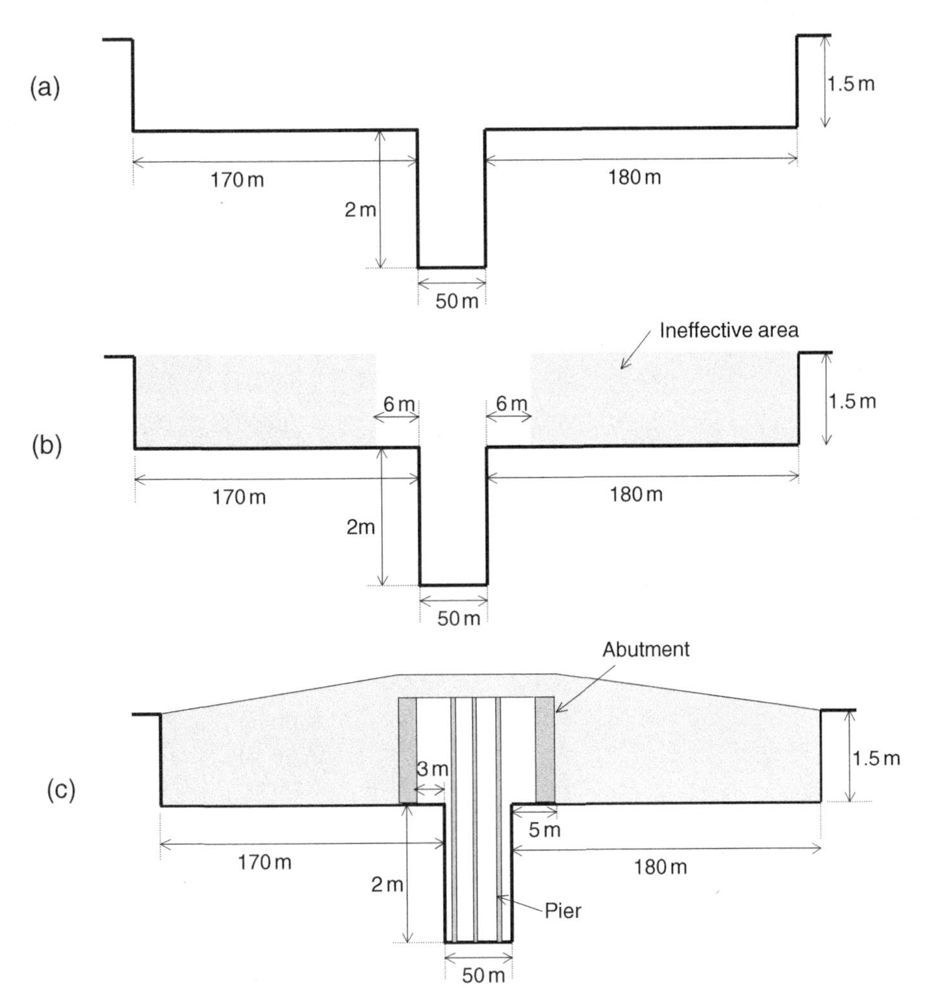

FIGURE 7.3 Schematics of channel sections for Example 7.1: (a) sections 1 and 4; (b) sections 2 and 3; (c) sections BU and BD

and right overbanks at a distance 3 m from the edges of the main channel. The road embankment covers the area on the outer sides of the abutments on both the left and the right overbanks. There are also three bridge piers in the main channel. Each pier is 1 m wide, extends the length of the bridge constriction, and has semicircular ends. Both sections 2 and 3 have ineffective areas, with edges of the ineffective areas placed 3 m from the edges of the bridge opening – that is, 6 m from the edges of the main channel on both sides. (Recall that the distance between sections 2 and BD and that between sections BU and 1 is 3 m.)

Half the obstruction length at the bridge constriction is $(167+177)/2 = 172$ m, the ratio of the bridge opening width to flood plain width is $(56)/(400) = 0.14$, and $n_{ob}/n_c = (0.05)/(0.025) = 2$. Also, the longitudinal slope is 0.0005 (approximately 2.6 ft/mile). The distance between sections 2 and 1 is $L_e = 330$ ft, which corresponds to an expansion ratio of $(330)/(172) = 1.9$. A review of Table 7.1 indicates that this expansion ratio is within the acceptable range for the given situation. Therefore, section 1 is properly located as is. Likewise, the distance between sections 3 and 4 is 250 m, which corresponds to a contraction

TABLE 7.2 Summary of results for Example 7.1

			Elements			
Section		Whole section		Left overbank	Main channel	Right overbank
1	h (m)	67.11	y (m)	0.51	2.51	0.51
	z_b (m)	64.6	A (m^2)	86.24	125.37	91.32
	y (m)	2.51	P (m)	170.51	54.0	180.51
	V (m/s)	0.83	K (m^3/s)	1095.0	8792.8	1159.6
	α	2.97	Q (m^3/s)	24.78	199.98	26.24
	S_f	0.000512	V (m/s)	0.29	1.59	0.29
2	h (m)	67.28	y (m)	0.51	2.51	0.51
	z_b (m)	64.77	A (m^2)	3.06	125.51	3.06
	y (m)	2.51	P (m)	6.0	54.0	6.0
	V (m/s)	1.90	K (m^3/s)	39.1	8809.1	39.1
	α	1.07	Q (m^3/s)	1.10	247.80	1.10
	S_f	0.000791	V (m/s)	0.36	1.97	0.36
BD	h (m)	67.27	y (m)	0.50	2.50	0.50
	z_b (m)	64.77	A (m^2)	1.50	117.4	1.50
	y (m)	2.50	P (m)	3.50	66.0	3.50
	V (m/s)	2.08	K (m^3/s)	17.0	6895.3	17.0
	α	1.04	Q (m^3/s)	0.61	248.78	0.61
	S_f	0.001302	V (m/s)	0.41	2.12	0.41
BU	h (m)	67.27	y (m)	0.50	2.50	0.50
	z_b (m)	64.77	A (m^2)	1.50	117.4	1.50
	y (m)	2.50	P (m)	3.50	66.0	3.50
	V (m/s)	2.08	K (m^3/s)	17.0	6895.3	17.0
	α	1.04	Q (m^3/s)	0.61	248.78	0.61
	S_f	0.001302	V (m/s)	0.41	2.12	0.41
3	h (m)	67.33	y (m)	0.56	2.56	0.56
	z_b (m)	64.77	A (m^2)	3.36	127.97	3.36
	y (m)	2.56	P (m)	6.0	54.0	6.0
	V (m/s)	1.86	K (m^3/s)	45.6	9098.8	45.6
	α	1.08	Q (m^3/s)	1.24	247.52	1.24
	S_f	0.000740	V (m/s)	0.37	1.93	0.37
4	h (m)	67.63	y (m)	0.73	2.73	0.73
	z_b (m)	64.90	A (m^2)	124.01	136.46	131.30
	y (m)	2.73	P (m)	170.73	54.0	180.73
	V (m/s)	0.64	K (m^3/s)	2004.0	10127.2	2122.3
	α	3.01	Q (m^3/s)	35.15	177.63	37.22
	S_f	0.000308	V (m/s)	0.28	1.30	0.28

ratio of (250)/(172) = 1.45. This is within the acceptable range of 1 to 1.5, and therefore section 4 is also properly located.

In Example 4.16, we determined the flow condition at section 1 as summarized in Table 7.2. We will now calculate the water surface profile upstream of section 1 with the bridge in place.

First, we will use the standard step method to determine the water surface elevation at section 2. This elevation should satisfy Equation 7.11 with section 1 as the downstream and section 2 as the upstream section. By using a trial-and-error procedure as in Example 4.16 (or using a computer program), we can determine the conditions at 2. The results obtained for section 2 are given in Table 7.2. A review of the tabulated values for section 2 is instructive. Particularly, we should note that the wetted perimeter for each overbank segment is only 6 m, and it does not include the vertical edge of the ineffective area. Let the subscripts *lob, mc,* and *rob* represent the left overbank, main channel, and right overbank at channel section 2. Then, for $y_2 = 2.51$ m, we can determine that $y_{mc} = 2.51$ m, $A_{mc} = (2.51)(50.0) = 125.5$, and $P_{mc} = 50.0 + 2.0 + 2.0 = 54.0$ m. Likewise, for the overbank areas, $y_{lob} = y_{rob} = 0.51$ m, $A_{lob} = A_{rob} = (0.51)(6.0) = 3.06\text{ m}^2$, and $P_{lob} = P_{rob} = 6.0$ m. Assuming $n_{lob} = n_{rob} = 0.05$ and $n_{mc} = 0.025$, we can use Equation 7.13 to obtain $K_{lob} = K_{rob} = 39.1\text{ m}^3$ and $K_{mc} = 8809.1\text{ m}^3$. Then, by using Equation 7.12, $S_f = 0.000791$. Next, by using Equation 7.14, we obtain $Q_{lob} = Q_{rob} = 1.10\text{ m}^3/\text{s}$ and $Q_{mc} = 247.8\text{ m}^3/\text{s}$. Then, $V_{lob} = V_{rob} = 1.10/3.06 = 0.36$ m/s and $V_{mc} = 247.8/125.5 = 1.97$ m/s. The total flow area of the compound channel is $A = 3.06 + 125.5 + 3.06 = 131.6\text{ m}^2$, and the cross-sectional average velocity is $V = 250/131.6 = 1.90$ m/s. Also, from Equation 7.15, $\alpha = 1.07$.

We can verify that the solution obtained for section 2 satisfies Equation 7.11. The left-hand side of the equation is evaluated as

$$
\begin{aligned}
& z_{bU} + y_U + \alpha_U \frac{V_U^2}{2g} - \frac{1}{2}(\Delta X) S_{fU} \\
& \quad = 64.77 + 2.51 + 1.07 \frac{1.90^2}{2(9.81)} - \frac{1}{2}(330)(0.000791) = 67.35\text{ m}
\end{aligned}
$$

and the right-hand side becomes

$$
\begin{aligned}
& z_{bD} + y_D + \alpha_D \frac{V_D^2}{2g} + \frac{1}{2}(\Delta X) S_{fD} + h_e \\
& \quad = 64.60 + 2.51 + 2.97 \frac{0.83^2}{2(9.81)} + \frac{1}{2}(330)(0.000512) \\
& \quad + (0.5)\left| 1.07 \frac{1.90^2}{2(9.81)} - 2.97 \frac{0.83^2}{2(9.81)} \right| = 67.35\text{ m}
\end{aligned}
$$

The calculated value of the left-hand side is equal to that of the right-hand side. Therefore, Equation 7.11 is satisfied. Note that because the flow is expanding between sections 2 and 1, we use $k_e = 0.5$.

The next step is to verify that flow will not choke due to the bridge. We will use the condition given in Equation 7.6 for this. Recall that the subscript BR in Equation 7.6 refers to the most constricted bridge section. In this problem both the bottom elevation and the width of the bridge remain constant throughout the length of the bridge, so any bridge section can be used. We will now calculate the critical depth at the bridge section. More than one critical depth is possible in compound sections, but the one in the main channel usually governs. Referring to Figure 7.3c, and noting that each pier is 1 m wide, the flow width in the rectangular main channel is $50-(3)(1)=47$ m. Therefore, the discharge per unit width becomes $q=(250.0)/(47)=5.32\,\text{m}^2/\text{s}$. Now, by using Equation 2.3 for rectangular sections,

$$y_c = \sqrt[3]{\frac{q^2}{g}} = \sqrt[3]{\frac{(5.32)^2}{9.81}} = 1.42\,\text{m}$$

Note that, because $1.42\,\text{m} < 2.0\,\text{m}$, the flow will be in the main channel only under the critical flow condition (see Figure 7.3). The corresponding flow area is $(1.42)(47.0)=66.74\,\text{m}^2$. Assuming that $\alpha=1.0$ for the rectangular section, the specific energy corresponding to critical flow in the bridge section is

$$E_{crit} = 1.42 + 1.00\frac{250.0^2}{(66.74)^2}\frac{1}{2(9.81)} = 2.14\,\text{m}$$

Then the right-hand side of the inequality in Equation 7.6 becomes $64.77+2.14=66.91$ m. We can now use the flow variables at section 2, already calculated and summarized in Table 7.2, to calculate the left-hand side of the inequality as

$$z_{b2} + y_2 + \alpha_2\frac{V_2^2}{2g} = 64.77 + 2.51 + 1.07\frac{1.90^2}{2(9.81)} = 67.48\,\text{m}$$

Because $67.48 > 66.91$, the condition of Equation 7.6 is satisfied, and choking will not occur. The flow will remain subcritical. We can therefore proceed to calculate the water surface elevation in section BD and the further sections upstream using the standard step method. The results are summarized in Table 7.2, in which $h=z_b+y$ is water surface elevation.

7.1.3.3 Momentum method for low-flow calculations

In this approach the subcritical gradually varied flow calculations are performed for all the sections downstream of the bridge including section 2. After verifying that the flow will not choke, we apply the momentum equation between sections 2 and BD to find the flow condition at section BD. Likewise, we apply the momentum equation between sections BD and BU to determine the condition at BU, and between sections BU and 3 to determine the condition at Section 3. The gradually-varied flow calculations are then performed to determine the water surface profile further upstream.

We can rewrite the momentum equation (Equation 2.18) for sections 2 (downstream) and BD (upstream) as

$$A_{BD}Y_{BD} + \beta_{BD}\frac{Q^2}{gA_{BD}} = A_2Y_2 + \beta_2\frac{Q^2}{gA_2} - \frac{F_{bBD}}{\gamma} + \frac{F_f}{\gamma} - \frac{W_x}{\gamma} \qquad (7.17)$$

where A = flow area, Y = vertical distance from water surface to center of gravity of the flow area, Q = discharge, β = momentum correction factor, g = gravitational acceleration, F_f = external force opposing the flow due to friction, F_b = external force in the flow direction due to the force exerted by the obstructed area in section BD, and W_x = force due to the weight of water between sections 2 and BD in the direction of flow. The subscripts *BD* and 2 stand for the sections BD and 2, respectively.

Part of the area at section BD is obstructed due to the piers, the abutments, and the road embankment. If a bridge section is made of two overbanks and a main channel, then the total external force due to the obstructed area will be equal to the sum of external forces due to the obstructed areas in the main channel and the two overbanks. For example, in Figure 7.3b, representing section 2, the distance between the edge of the ineffective area and that of the main channel is 6 m. In Figure 7.3c, representing section BD, the distance between the inner edge of the abutment and the edge of the main channel is 3 m. Therefore, a 3-m wide area on each of the left and right overbanks is blocked at the bridge section. In addition, three piers, 1 m each, block an area in the main channel that is 3 m wide.

The flow exerts a force on the obstructed area, and the obstructed area exerts a force on the flow of the same magnitude but in the opposite direction. We will approximate this force by the hydrostatic pressure force, and we express it per unit weight of water as

$$\frac{F_{bBD}}{\gamma} = (A_p Y_p)_{BD} = (A_{plob} Y_{plob} + A_{pmc} Y_{pmc} + A_{prob} Y_{prob})_{BD} \tag{7.18}$$

where A_p = obstructed area due to the piers, abutments, and part of the road embankment at section BD (relative to section 2), Y_p = vertical distance from the water surface to the center of gravity of the obstructed area, and the subscripts *lob, mc,* and *rob*, respectively, stand for the left overbank, the main channel, and the right overbank.

The friction force between sections BD and 2 per unit weight of water can be approximated by

$$\frac{F_f}{\gamma} = \frac{\Delta X}{2}(A_{BD} S_{fBD} + A_2 S_{f2}) \tag{7.19}$$

where Δx = distance between sections 2 and BD, and S_f = friction slope. Likewise, the weight component in the flow direction per unit weight of water can be approximated as

$$\frac{W_x}{\gamma} = \frac{(\Delta X) S_0}{2}(A_{BD} + A_2) \tag{7.20}$$

TABLE 7.3 Typical drag coefficients for various pier shapes (after US Army Corps of Engineers, 2002)

Pier shape	C_D
Circular pier	1.20
Elongated piers with semi-circular ends	1.33
Elliptical piers with 2 : 1 length to width	0.60
Elliptical piers with 4 : 1 length to width	0.32
Elliptical piers with 8 : 1 length to width	0.29
Square-nosed piers	2.00
Triangular-nosed with 30° angle	1.00
Triangular-nosed with 60° angle	1.39
Triangular-nosed with 90° angle	1.60
Triangular-nosed with 120° angle	1.72

where S_0 = bottom slope. Substituting Equations 7.18 to 7.20 into Equation 7.17 and rearranging, we obtain

$$A_{BD}Y_{BD} + \beta_{BD}\frac{Q^2}{gA_{BD}} + A_{pBD}Y_{pBD} - \frac{\Delta X(S_{fBD} - S_0)A_{BD}}{2}$$
$$= A_2Y_2 + \beta_2\frac{Q^2}{gA_2} + \frac{\Delta X(S_{f2} - S_0)A_2}{2} \quad (7.21)$$

Likewise, we can write the momentum equation between sections BU and BD as

$$A_{BU}Y_{BU} + \beta_{BU}\frac{Q^2}{gA_{BU}} - \frac{\Delta X(S_{fBU} - S_0)A_{BU}}{2}$$
$$= A_{BD}Y_{BD} + \beta_{BD}\frac{Q^2}{gA_{BD}} + \frac{\Delta X(S_{fBD} - S_0)A_{BD}}{2} \quad (7.22)$$

Finally, the momentum equation between sections BU and 3 becomes

$$A_3Y_3 + \beta_3\frac{Q^2}{gA_3} - \frac{\Delta X(S_{f3} - S_0)A_3}{2} - \frac{C_D}{2}\frac{A_{pier}Q^2}{gA_3^2}$$
$$= A_{BU}Y_{BU} + \beta_{BU}\frac{Q^2}{gA_{BU}} + (A_pY_p)_{BU} + \frac{\Delta X(S_{fBU} - S_0)A_{BU}}{2} \quad (7.23)$$

where A_{pier} = area obstructed by the piers at section BU, and C_D = drag coefficient for flow going around the piers. Table 7.3 presents the recommended drag coefficients for various pier shapes derived from experimental data of Lindsey (US Army Corps of Engineers, 2002). Note that the fourth term on the left-hand side of Equation 7.23 is an additional external force (dynamic force) opposing the flow. This force is exerted by the piers on the flow, and is equal in magnitude but opposite in direction to the drag force exerted by the flow on the piers.

In Equations 7.17 to 7.23, the momentum coefficient for a compound channel formed of a left overbank (*lob*), right overbank (*rob*), and a main channel (*mc*) can be evaluated as

$$\beta = \frac{V_{lob}^2 A_{lob} + V_{mc}^2 A_{mc} + V_{rob}^2 A_{rob}}{V^2 A} \tag{7.24}$$

in which the variables in the denominator represent the whole compound section. Likewise, we can determine Y by using

$$Y = \frac{Y_{lob} A_{lob} + Y_{mc} A_{mc} + Y_{rob} A_{rob}}{A_{lob} + A_{mc} + A_{rob}} \tag{7.25}$$

EXAMPLE 7.2 Reconsider the bridge problem of Example 7.1 and calculate the flow profile through the bridge using the momentum method.

In the momentum method, all the sections downstream of the bridge, including section 2, are calculated using the standard step method. Therefore, the results obtained for section 2 in Example 7.1 are still valid and are summarized in Table 7.2. Before we proceed to section BD, we need to verify that the flow will not choke. We will use the condition of Equation 7.7 for this purpose. Recall that the subscript *BR* in Equation 7.7 refers to the most constricted bridge section. In this problem both the bottom elevation and the width of the bridge remain constant throughout the length of the bridge, so any bridge section can be used.

We defined the specific momentum, M, in Chapter 2 as

$$M = \beta \frac{Q^2}{gA} + YA$$

where β = momentum correction factor, and Y = distance from the water surface to centroid of the flow area. Referring to Table 7.2 and using Equation 7.25, we can calculate Y for section 2 as

$$Y = \frac{Y_{lob} A_{lob} + Y_{mc} A_{mc} + Y_{rob} A_{rob}}{A_{lob} + A_{mc} + A_{rob}}$$

$$= \frac{(0.51/2)(3.06) + (2.51/2)(125.51) + (0.51/2)(3.06)}{3.06 + 125.51 + 3.06} = 1.21\,\text{m}$$

where the subscript *lob* stands for left overbank, *mc* stands for main channel, and *rob* stands for right overbank.

The momentum correction factor is calculated by using Equation 7.24 as

$$\beta = \frac{V_{lob}^2 A_{lob} + V_{mc}^2 A_{mc} + V_{rob}^2 A_{rob}}{V^2 A}$$

$$= \frac{0.36^2(3.06) + 1.97^2(125.51) + 0.36^2(3.06)}{1.90^2(3.06 + 125.51 + 3.06)} = 1.03$$

Noting that $A_2 = 3.06 + 125.51 + 3.06 = 131.6\,\text{m}^2$ at section 2, the specific momentum is

$$M_2 = \left(1.03 \frac{250.0^2}{9.81(131.6)} + (1.21)(131.6)\right) = 209.1\,\text{m}^3$$

Next we will calculate M_{crit}, the specific momentum for critical flow at the bridge section. The critical depth and the corresponding flow area at the bridge section were found in Example 7.1 as being 1.42 m and 66.74 m^2, respectively. Assuming that $\beta = 1.0$ for the rectangular section, the specific momentum corresponding to critical flow in the bridge section becomes

$$M_{crit} = \left(1.00 \frac{250.0^2}{9.81(66.74)} + \frac{1.42}{2}(66.74)\right) = 142.8\,\text{m}^3$$

Because $M_2 > M_{crit}$, choking will not occur, and the flow will remain subcritical. We can therefore proceed to calculate the water surface elevation in sections BD, BU, and 3 using the momentum method.

As mentioned previously in the momentum method, all the sections downstream of the bridge, including section 2, are calculated using the standard step method. Therefore, the results obtained for section 2 in Example 7.1 are still valid. Then, using the results from Example 7.1 (as tabulated in Table 7.2) and noting that $Y_2 = 1.21$ m and $\beta_2 = 1.03$ (already calculated above when the condition of choking was checked), we can evaluate the right-hand side of Equation 7.21 as

$$A_2 Y_2 + \beta_2 g A_2 \frac{Q^2}{g A_2} + \frac{\Delta X (S_{f2} - S_0) A_2}{2}$$

$$= (131.6)(1.21) + 1.03 \frac{250.0^2}{9.81(131.6)} + \frac{3.0(0.000791 - 0)(131.6)}{2}$$

$$= 209.3\,\text{m}^3$$

Note that S_0 is set equal to zero, because the channel bottom elevation at section BD is assumed to be the same as at section 2. Now we will determine the flow depth at section BD, for which the left-hand side of Equation 7.21 is 209.3 m^3. By trial and error, we determine that $y_{BD} = 2.48$ m. Note that, for this depth, $y_{mc} = 2.48$ m, $A_{mc} = (2.48)(50 - 3) = 116.56\,\text{m}^2$ subtracting the width of three piers from the main channel width, and $P_{mc} = (50 - 3) + 2 + 2 + 3$

(2.48 + 2.48) = 65.88 including an increase of 2.48 + 2.48 = 4.96 m in the wetted perimeter due to the each of three piers. Likewise, for the overbank areas, $y_{lob} = y_{rob} = 0.48$ m, $A_{lob} = A_{rob} = (0.48)(3.0) = 1.44\ \text{m}^2$, and $P_{lob} = P_{rob} = 3.0 + 0.48 = 3.48$ m. Assuming $n_{lob} = n_{rob} = 0.05$ and $n_{mc} = 0.025$, we can use Equation 7.13 to obtain $K_{lob} = K_{rob} = 15.99\ \text{m}^3$ and $K_{mc} = 6820.35\ \text{m}^3$. Then, by using Equation 7.12, $S_f = 0.001331$. Next, by using Equation 7.14, we obtain $Q_{lob} = Q_{rob} = 0.58\ \text{m}^3$ and $Q_{mc} = 248.83\ \text{m}^3$. Then, $V_{lob} = V_{rob} = 0.58/1.44 = 0.40\ \text{m}^3/\text{s}$ and $V_{mc} = 248.83/116.56 = 2.13\ \text{m}^3/\text{s}$. The total flow area of the compound channel is $A = 1.44 + 116.56 + 1.44 = 119.44\ \text{m}^2$, and the cross-sectional average velocity is 250/119.44 = 2.09 m/s. Also, from Equation 7.24, $\beta = 1.02$, and from Equation 7.25, $Y = 1.215$ m.

Now we can evaluate the left-hand side of Equation 7.21 as

$$A_{BD}Y_{BD} + \beta_{BD}\frac{Q^2}{gA_{BD}} + A_{pBD}Y_{pBD} - \frac{\Delta X(S_{fBD} - S_0)A_{BD}}{2}$$

$$= (119.44)(1.215) + (1.02)\frac{250^2}{9.81(119.44)} + 2\frac{(3.0)(0.48)(0.48)}{2}$$

$$+ 3\frac{(1.0)(2.48)(2.48)}{2} - \frac{(3.0)(0.001331 - 0)(119.44)}{2} = 209.2\ \text{m}^3$$

This is practically equal to the right-hand side of the equation. Therefore, the flow depth of 2.48 m at section BD satisfies Equation 7.21.

We will next calculate the flow depth at section BU, the section just inside the bridge constriction at the upstream end. For this we will use Equation 7.22. Because we already know the condition at section BD, we can evaluate the right-hand side of the equation with $\Delta X = 14$ m as

$$A_{BD}Y_{BD} + \beta_{BD}\frac{Q^2}{gA_{BD}} + \frac{\Delta X(S_{fBD} - S_0)A_{BD}}{2} = (119.44)(1.215)$$

$$+ 1.02\frac{250^2}{9.81(119.44)} + \frac{14.0(0.001331 - 0)119.44}{2} = 200.6\ \text{m}^3$$

The left-hand side of Equation 7.22 should be equal to 200.6 m^3 for the correct value of the depth at section BU. By trial and error, we find this depth as being 2.50 m. Note that for $y_{BU} = 2.50$ m, we can determine that $y_{mc} = 2.50$ m, $A_{mc} = (2.50)(50 - 3) = 117.50\ \text{m}^2$, and $P_{mc} = (50 - 3) + 2 + 2 + 3(2.50 + 2.50)) = 66$ m. Likewise, for the overbank areas, $y_{lob} = y_{rob} = 0.50$ m, $A_{lob} = A_{rob} = (0.50)(3.0) = 1.50\ \text{m}^2$, and $P_{lob} = P_{rob} = 3.0 + 0.50 = 3.50$ m. Assuming $n_{lob} = n_{rob} = 0.05$ and $n_{mc} = 0.025$, we can use Equation 7.13 to obtain $K_{lob} = K_{rob} = 17.05\ \text{m}^3$ and $K_{mc} = 6903.89\ \text{m}^3$. Then, by using Equation 7.12, $S_f = 0.001298$. Next, by using Equation 7.14, we obtain $Q_{lob} = Q_{rob} = 0.61\ \text{m}^3$ and $Q_{mc} = 248.77\ \text{m}^3$. Then, $V_{lob} = V_{rob} = 0.61/1.50 = 0.41\ \text{m}^3/\text{s}$ and $V_{mc} = 248.77/117.50 = 2.12\ \text{m}^3/\text{s}$. The total flow area of the compound channel is $A = 1.5 + 117.5 + 1.5 = 120.5\ \text{m}^2$, and the cross-sectional average velocity is 250/120.5 = 2.07 m/s. Also, from Equation 7.24,

$\beta = 1.02$, and from Equation 7.25, $Y = 1.22$ m. Then the left-hand side of Equation 7.22 becomes

$$A_{BU}Y_{BU} + \beta_{BU}\frac{Q^2}{gA_{BU}} - \frac{\Delta X(S_{fBU} - S_0)A_{BU}}{2}$$
$$= (120.5)(1.22) + 1.02\frac{250^2}{9.81(120.5)} - \frac{14(0.001298 - 0)(120.5)}{2} = 199.9\,\text{m}^3$$

The left-hand side is very close to being equal to the right-hand side, and therefore the flow depth of 2.50 m at section BU is accepted.

We will now determine the flow depth at section 3 by using Equation 7.23. With the known values at section BU, and applying Equation 7.18 to Section BU to determine $(A_pY_p)_{BU}$, we can evaluate the right-hand side of Equation 7.23 as

$$A_{BU}Y_{BU} + \beta_{BU}\frac{Q^2}{gA_{BU}} + (A_pY_p)_{BU} + \frac{\Delta X(S_{fBU} - S_0)A_{BU}}{2}$$
$$= (120.5)(1.22) + 1.02\frac{250^2}{9.81(120.5)} + 2\frac{(3.0)(0.5)(0.5)}{2}$$
$$+ 3\frac{(1.0)(2.5)(2.5)}{2} + \frac{3.0(0.001298 - 0)(120.5)}{2} = 211.3\,\text{m}^3$$

The correct value of the flow depth at section 3 is the value that makes the left-hand side of Equation 7.23 equal to the right-hand side. By trial and error, we find this depth as being 2.55 m. For this depth, $y_{mc} = 2.55$ m, $A_{mc} = (2.55)(50.0) = 127.5$, and $P_{mc} = 50 + 2 + 2 = 54.0$ m. Likewise, for the overbank areas, $y_{lob} = y_{rob} = 0.55$ m, $A_{lob} = A_{rob} = (0.55)(6.0) = 3.30\,\text{m}^2$, and $P_{lob} = P_{rob} = 6.0$ m excluding the edges of the ineffective area. With $n_{lob} = n_{rob} = 0.05$ and $n_{mc} = 0.025$, we can use Equation 7.13 to obtain $K_{lob} = K_{rob} = 44.3\,\text{m}^3$ and $K_{mc} = 9043.3\,\text{m}^3$. Then, by using Equation 7.12, $S_f = 0.000749$. Next, by using Equation 7.14, we obtain $Q_{lob} = Q_{rob} = 1.21\,\text{m}^3/\text{s}$ and $Q_{mc} = 247.57\,\text{m}^3/\text{s}$. Then, $V_{lob} = V_{rob} = 1.21/3.3 = 0.37$ m/s and $V_{mc} = 247.57/127.5 = 1.94$ m/s. The total flow area of the compound channel is $A = 3.30 + 127.5 + 3.3 = 134.1\,\text{m}^2$, and the cross-sectional average velocity is $250/134.1 = 1.86$ m/s. Also, from Equation 7.24, $\beta = 1.04$, and from Equation 7.25, $Y = 1.22$ m. For piers with semicircular ends, we obtain $C_D = 1.33$ from Table 7.3, and the area obstructed by the three piers is $A_{pier} = 3(1.0)(2.55) = 7.65\,\text{m}^2$. Therefore, the left-hand side of Equation 7.23 becomes

$$A_3Y_3 + \beta_3\frac{Q^2}{gA_3} - \frac{\Delta X(S_{f3} - S_0)A_3}{2} - \frac{C_D}{2}\frac{A_{pier}Q^2}{gA_3^2}$$
$$= (134.1)(1.22) + 1.04\frac{250^2}{9.81(134.1)} - \frac{3.0(0.000749 - 0)134.1}{2} - \frac{1.33}{2}\frac{(7.65)250^2}{9.81(134.1)^2}$$
$$= 211.1\,\text{m}^3$$

This is very close to the calculated value of the right-hand side of the equation. Therefore, it is verified that the depth at section 3 is 2.55 m.

The depths at sections 4 and 4.3 are calculated by using the standard step method, discussed previously in Chapter 4 and in Example 7.1. The resulting depths are 2.73 m at section 4 and 2.62 m at section 4.3.

7.1.3.4 Yarnell equation for low-flow calculations

The Yarnell (US Army Corps of Engineers, 2002) equation is an empirical equation based on laboratory experiments. Given the water surface elevation just downstream of the bridge (section 2 in Figure 7.1), and the shape of the piers, the Yarnell equation estimates the water surface elevation at just upstream of the bridge (section 3 in Figure 7.1) as

$$h_3 = h_2 + 2K_Y\left(K_Y + 10\frac{\alpha_2 V_2^2}{2gy_2} - 0.6\right)(r_o + 15r_o^4)\frac{V_2^2}{2g} \tag{7.26}$$

where the subscripts 2 and 3 refer to channel sections 2 and 3. Also in Equation 7.26, $h = z_b + y =$ water surface elevation, $z_b =$ bottom elevation, $y =$ flow depth, $K_Y =$ Yarnell's pier shape coefficient, $V =$ average cross-sectional velocity, $g =$ gravitational acceleration, and $r_o =$ ratio of obstructed area by the piers to the total unobstructed area in section 2. Table 7.4 presents the Yarnell's pier coefficient for various pier shapes.

The equation is sensitive to the pier shape, the area obstructed by the piers, and the velocity. However, it does not directly account for the shape of the bridge opening, shape of the abutments, or width of the bridge. Therefore, it should be used when the energy losses at a bridge section are caused mainly by the piers. Also, the equation is applicable only to class A flow – that is, subcritical flow throughout. We can use either the energy or the momentum approach to verify that the flow will not choke, as in the preceding sections.

EXAMPLE 7.3 Revisit the bridge problem discussed in Example 7.1 and determine the flow depth at section 3 by using the Yarnell equation.

TABLE 7.4 Yarnell's pier coefficient K_Y for various pier shapes (after US Army Corps of Engineers, 2002)

Pier shape	K_Y
Semicircular nose and tail	0.90
Twin-cylinder piers with connecting diaphragm	0.95
Twin-cylinders without diaphragm	1.05
90° Triangular nose and tail	1.05
Square nose and tail	1.25
Ten pile trestle bent	2.50

In Example 7.1, we showed that the flow does not choke due to the bridge constriction. We also determined that $y_2 = 2.51$ m, $h_2 = 67.28$ m, and $V_2 = 1.90$ m/s. Also, a review of Table 7.2 will reveal that the flow area at section 2 is $3.06 + 125.5 + 3.06 = 131.6\,\text{m}^2$. Noting that at the bridge section there are three piers each 1 m in width, the area obstructed by the piers is equal to $3(1.0)(2.51) = 7.53\,\text{m}^2$. Therefore, $r_o = 7.53/131.6 = 0.057$. Also, from Table 7.4, $K_Y = 0.90$ for piers with semicircular ends. Substituting the known values into Equation 7.26,

$$h_3 = h_2 + 2K_Y\left(K_Y + 10\frac{\alpha_2 V_2^2}{2gy_2} - 0.6\right)(r_o + 15r_o^4)\frac{V_2^2}{2g}$$

$$= 67.28 + 2(0.9)\left[0.9 + 10\frac{(1.07)1.90^2}{2(9.81)(2.51)} - 0.6\right]\left[0.057 + 15(0.057)^4\right]\frac{1.90^2}{2(9.81)}$$

$$= 67.30\,\text{m}$$

Because the channel bottom elevation at section 3 is 64.77 m, we can determine the flow depth as $y_2 = 67.30 - 64.77 = 2.53$ m.

If desired, we can calculate the flow depths at further upstream channel sections by using the standard step method. In this problem, the flow depths at sections 4 and 4.3 are found to be 2.70 m and 2.61 m, respectively.

7.1.4 HIGH-FLOW CALCULATIONS AT BRIDGE SITES

High flows are defined as flows where the water surface elevation exceeds the maximum low chord of the bridge deck. Three general types of high flow can occur, depending on the water surface elevation, the crest elevation of the roadway embankment, and the low and high chord elevations of the bridge deck.

7.1.4.1 Sluice-gate type flow

Sluice-gate type flow occurs when the flow comes into contact with the side of the bridge at the upstream side but is below the low chord at the downstream side, as shown schematically in Figure 7.4. In this figure the subscripts 2 and 3 refer to the channel sections just downstream and upstream of the bridge, respectively, as described in Section 7.1.1 and Figure 7.1.

The hydraulic behavior of the flow in this case is similar to that of flow under a sluice gate, and we use the equation (Bradley, 1978)

$$Q = C_d A_{BU}\sqrt{2g}\left(y_3 + \frac{\alpha_3 V_3^2}{2g} - \frac{Z}{2}\right)^{1/2} \qquad (7.27)$$

where Q = discharge, A_{BU} = net area of the bridge opening at section BU (upstream end of the bridge), g = gravitational acceleration, α = energy correction factor, V = cross-sectional average velocity, Z = vertical distance from

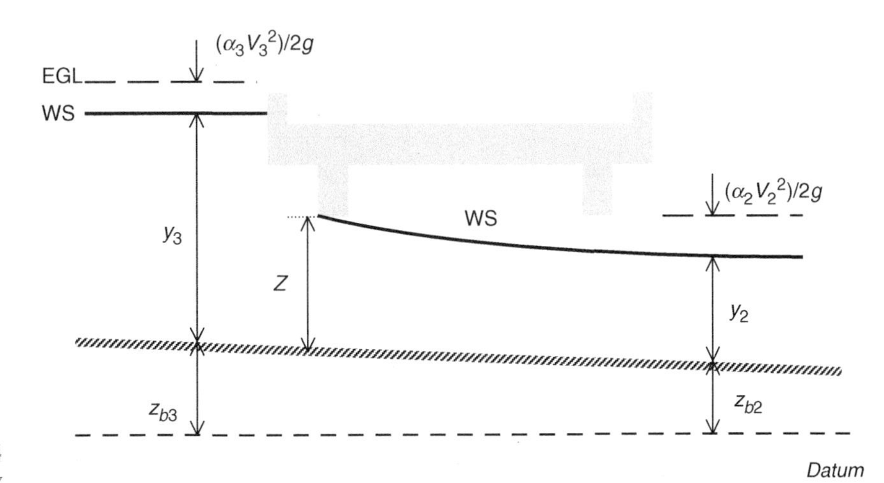

FIGURE 7.4
Sluice-gate type flow

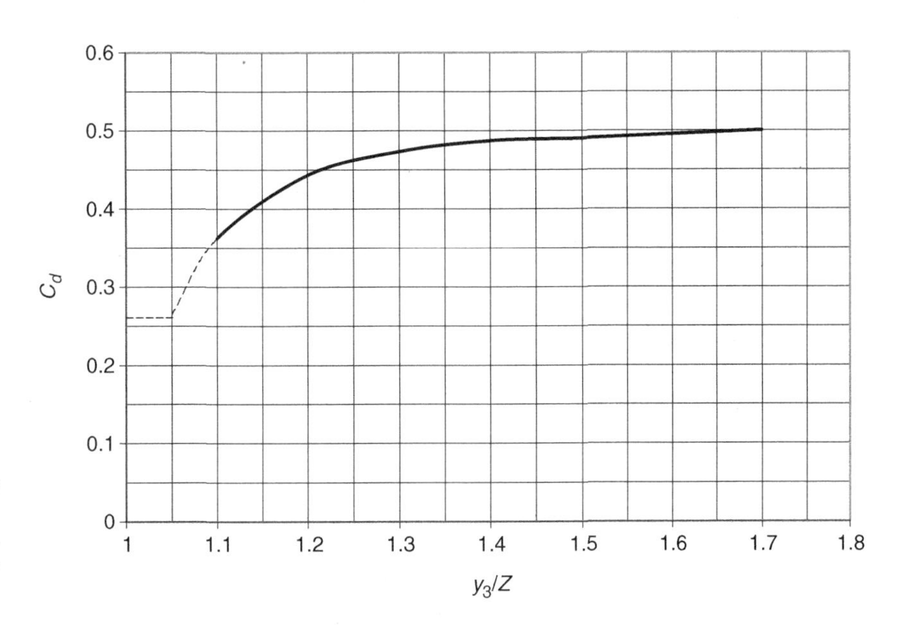

FIGURE 7.5
Coefficient of discharge for sluice-gate type flow (after US Army Corps of Engineers, 2002)

maximum bridge low chord to the river bed at section BU, and C_d = coefficient of discharge. As shown in Figure 7.5, the coefficient of discharge depends on the ratio y_3/Z. The values of C_d for $y_3/Z < 1.1$ are uncertain, since Equation 7.27 is not applicable for this range.

Recall that parts of channel sections 2 and 3 were designated as ineffective areas and were excluded in low-flow calculations. In the case of sluice-gate type flow, low flow still exists in channel Section 2. Therefore, ineffective areas should still be excluded in the gradually-varied flow calculations downstream of the bridge. However, high flow occurs at channel section 3, and this channel section no longer contains any ineffective areas. Therefore, the entire wetted area is used as the active flow area in the calculations for section 3.

EXAMPLE 7.4 Figure 7.6 displays a schematic (not to scale) of the cross-section of a bridge. The horizontal dimensions are the same at the downstream

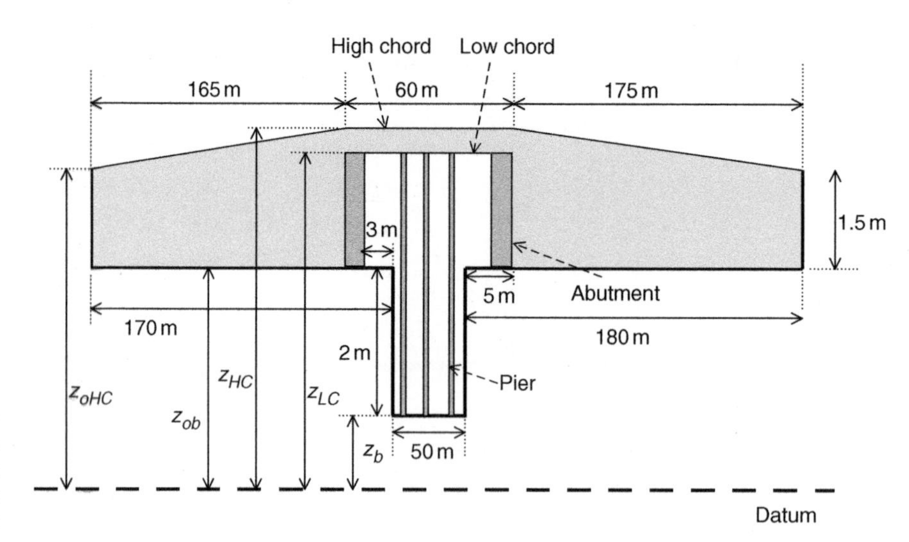

FIGURE 7.6 Schematic representation of example bridge section

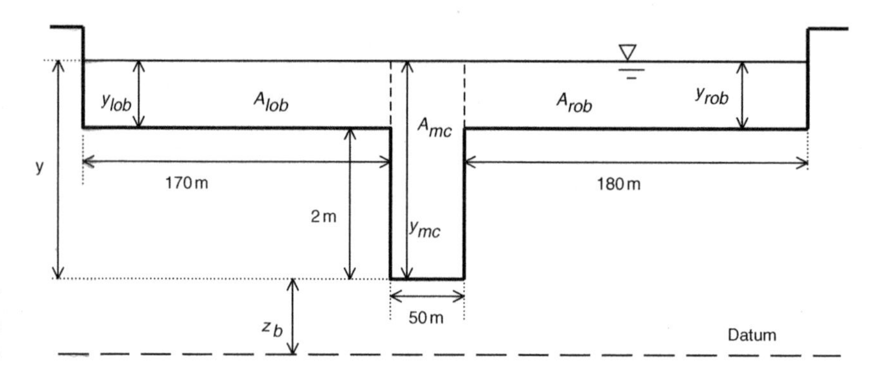

FIGURE 7.7 Example channel section

and upstream ends of the bridge, but the elevations are slightly different. At the downstream end, $z_b = 64.77$ m, $z_{LC} = 67.50$, $z_{HC} = 68.50$ m, $z_{ob} = 66.77$ m, and $z_{oHC} = 68.27$ m. At the upstream end, $z_b = 64.78$ m, $z_{LC} = 67.50$ m, $z_{HC} = 68.50$ m, $z_{ob} = 66.78$ m, and $z_{oHC} = 68.28$ m. Each of the three piers is 1 m wide. Figure 7.7 represents the cross-sectional dimensions of the channel reaches both upstream and downstream of the bridge. At section 3, just upstream of the bridge, $z_{b3} = 64.78$ m. At section 2, just downstream of the bridge, $z_{b2} = 64.77$ m.

Suppose for a discharge of 315 m^3/s the flow surface is below the low chord at section 2 but is higher than the low chord at section 3. Calculate the water surface elevation at section 3.

Because the flow described in the problem statement is of the sluice-gate type, we will use Equation 7.27. First, $Z = 67.50 - 64.78 = 2.72$ m. Next, the net area of bridge opening at the upstream end over each overbank is $(3.0)(67.50 - 66.78) = 2.16$ m^2. The net bridge opening in the main channel is $\{50.0 - 3(1.0)\}(67.50 - 64.78) = 127.84$ m^2. Therefore, $A_{BU} = 2.16 + 127.84 + 2.16 = 132.16$ m^2. We will now determine the flow depth y_3 at section 3 that satisfies Equation 7.27. This will require a trial-and-error procedure.

TABLE 7.5 Trial and error calculations for Example 7.4

y_3 (m)	Y_3/Z	C_d	α_3	V_3 (m/s)	Right-hand side (m³/s)
3.00	1.10	0.360	2.86	0.63	275
3.05	1.12	0.370	2.85	0.61	286
3.08	1.13	0.390	2.82	0.59	304
3.12	1.15	0.410	2.80	0.57	323
3.10	1.14	0.402	2.81	0.58	315

Let us guess that $y_3 = 3.0$ m. Then, referring to Figure 7.7, we have $y_{mc} = 3.00$ m, $A_{mc} = (3.0)(50.0) = 150\ \text{m}^2$, and $P_{mc} = (50.0) + 2.0 + 2.0 = 54.0$ m. Likewise, for the left overbank, $y_{lob} = 1.0$ m, $A_{lob} = (1.0)(170.0) = 170\ \text{m}^2$, and $P_{lob} = 1.0 + 170.0 = 171.0$ m. For the right overbank, $y_{rob} = 1.0$ m, $A_{rob} = (1.0)(180.0) = 180.0\ \text{m}^2$, and $P_{rob} = 1.0 + 180.0 = 181.0$ m. Assuming that $n_{lob} = n_{rob} = 0.05$ and $n_{mc} = 0.025$, we can use Equation 7.13 to obtain $K_{lob} = 3386.7\ \text{m}^3$, $K_{rob} = 3586.7\ \text{m}^3$, and $K_{mc} = 11\,856.3\ \text{m}^3$. Then, by using Equation 7.12, $S_f = 0.00028$. Next, by using Equation 7.14, we obtain $Q_{lob} = 56.7\ \text{m}^3/\text{s}$, $Q_{rob} = 60.0\ \text{m}^3/\text{s}$, and $Q_{mc} = 198.3\ \text{m}^3/\text{s}$. Then, $V_{lob} = 56.7/170 = 0.33$ m/s, $V_{rob} = 60.0/180 = 0.33$ m/s, and $V_{mc} = 198.3/150.0 = 1.32$ m/s. The total flow area of the compound channel is $A = 170.0 + 150.0 + 180.0 = 500.0\ \text{m}^2$, and the cross-sectional average velocity is $V = 315.0/500.0 = 0.63$ m/s. Also, by using Equation 7.15, we calculate that $\alpha = 2.86$. Finally, with $y_3/Z = 3.0/2.72 = 1.10$, we obtain $C_d = 0.36$ from Figure 7.5. Substituting these into the right-hand side of Equation 7.27,

$$C_d A_{BU}\sqrt{2g}\left(y_3 - \frac{Z}{2} + \frac{\alpha_3 V_3^2}{2g}\right)^{1/2}$$

$$= (0.36)(132.16)\sqrt{2(9.81}\left(3.0 - \frac{2.72}{2} + \frac{(2.86)0.63^2}{2(9.81)}\right)^{1/2} = 275\ \text{m}^3/\text{s}$$

This is different from the given discharge of 315 m³/s, and thus we need to try different values for y_3. Table 7.5 summarizes the results obtained for the different flow depths tried. For the correct answer, $y_3 = 3.10$ m, the right-hand side of the equation is equal to the given discharge of 315 m³/s.

7.1.4.2 Orifice-type flow

Orifice-type flow occurs when both the upstream and downstream sides of the bridge are submerged, as shown schematically in Figure 7.8. For this case, we use the equation (US Corps of Engineers, 2002)

$$Q = C_o A_{br}\sqrt{2g\left(z_{b3} + y_3 + \frac{\alpha_3 V_3^2}{2g} - z_{b2} - y_2\right)} \tag{7.28}$$

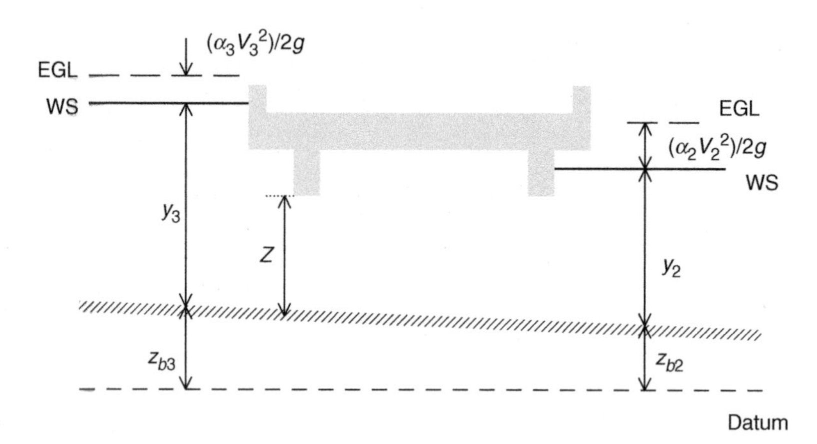

FIGURE 7.8 Orifice type flow

where z_b = channel bed elevation, C_o = discharge coefficient for fully submerged orifice flow, and A_{br} = net area of the bridge opening. A typical value for the discharge coefficient is $C_o = 0.8$. Also, if the net area of the bridge opening is different at upstream and downstream ends, we may use the average value.

To facilitate the water surface calculations, we can rearrange Equation 7.28 as

$$z_{b3} + y_3 + \frac{\alpha_3 V_3^2}{2g} = \frac{1}{2g}\left(\frac{Q}{C_o A_{br}}\right)^2 + z_{b2} + y_2 \tag{7.29}$$

Note that high-flow conditions exist at both channel sections 2 and 3, and neither contains any ineffective areas.

EXAMPLE 7.5 Suppose the discharge in the channel reach considered in Example 7.4 is 375 m^3/s and the corresponding flow depth at Section 2 is 2.94 m. Determine the flow depth at section 3.

Recall that Figure 7.6 displays a schematic (not to scale) of the cross-section of the bridge considered in Example 7.4. The horizontal dimensions are the same at the downstream and upstream ends of the bridge, but the elevations are slightly different. At the downstream end, $z_{b2} = 64.77$ m, $z_{LC} = 67.50$, $z_{HC} = 68.50$ m, $z_{ob} = 66.77$ m, and $z_{oHC} = 68.27$ m. At the upstream end, $z_{b3} = 64.78$ m, $z_{LC} = 67.50$ m, $z_{HC} = 68.50$ m, $z_{ob} = 66.78$ m, and $z_{oHC} = 68.28$ m.

With the given flow depth of 2.94 m, the water surface elevation at channel section 2 is 64.77 + 2.94 = 67.71 m. This is higher than the bridge-deck low-chord elevation of $z_{LC} = 67.50$ m at the downstream end of the bridge. Thus, the bridge is submerged at both ends and the hydraulic behavior of the bridge opening will be similar to that of an orifice as long as the water surface at section 3 does not exceed the bridge high-chord elevation. We will use Equation 7.29 to solve this problem.

TABLE 7.6 Trial-and-error results for Example 7.5

y_3 (m)	α_3	V_3 (m/s)	Left-hand side (m)
3.30	2.67	0.60	68.13
3.40	2.61	0.57	68.22
3.50	2.55	0.54	68.32
3.60	2.49	0.51	68.41
3.53	2.53	0.53	68.35

The net bridge opening area at the upstream end of the bridge was found to be 132.16 m^2 in Example 7.4. Similarly, at the downstream end, the net area of the bridge opening is calculated as $A_{BD} = (3.0)(67.50 - 66.77) + \{50.0 - 3(1.0)\}(67.50 - 64.77) + (3.0)(67.50 - 66.77) = 132.69\ m^2$. Then $A_{br} = (132.16 + 132.69)/2 = 132.43\ m^2$. We can now calculate the right-hand side of Equation 7.29 as

$$\frac{1}{2g}\left(\frac{Q}{C_o A_{br}}\right)^2 + z_{b2} + y_2 = \frac{1}{2(9.81)}\left(\frac{375}{(0.8)(132.43)}\right)^2 + 64.77 + 2.94 = 68.35\ \text{m}$$

The correct value of the flow depth at channel Section 3 is the one for which the left-hand side of Equation 7.29 becomes 68.35 m. This will require a trial-and-error procedure. Let us try, for instance, $y_3 = 3.30$ m. Then, referring to Figure 7.7, we have $y_{mc} = 3.30$ m, $A_{mc} = (3.3)(50.0) = 165.0\ m^2$, and $P_{mc} = (50.0) + 2.0 + 2.0 = 54.0$ m. Likewise, for the left overbank, $y_{lob} = 1.3$ m, $A_{lob} = (1.3)(170.0) = 221.0\ m^2$, and $P_{lob} = 1.3 + 170.0 = 171.3$ m. For the right overbank, $y_{rob} = 1.3$ m, $A_{rob} = (1.3)(180.0) = 234.0\ m^2$, and $P_{rob} = 1.3 + 180.0 = 181.3$ m. Assuming that $n_{lob} = n_{rob} = 0.05$ and $n_{mc} = 0.025$, we can use Equation 7.13 to obtain $K_{lob} = 5238.1\ m^3$, $K_{rob} = 5547.8\ m^3$, and $K_{mc} = 13\,897.5\ m^3$. Then, by using Equation 7.12, $S_f = 0.000231$. Next, by using Equation 7.14, we obtain $Q_{lob} = 79.6\ m^3/s$, $Q_{rob} = 84.3\ m^3/s$, and $Q_{mc} = 211.1\ m^3/s$. Then, $V_{lob} = 79.6/221.0 = 0.36$ m/s, $V_{rob} = 84.3/234.0 = 0.36$ m/s, and $V_{mc} = 211.1/165.0 = 1.28$ m/s. The total flow area of the compound channel is $A = 221.0 + 165.0 + 234.0 = 620.0\ m^2$, and the cross-sectional average velocity is $V = 375.0/620.0 = 0.60$ m/s. Also, by using Equation 7.15, we calculate that $\alpha = 2.67$. Substituting these values into the left-hand side of Equation 7.29 and noting that $z_{b3} = 64.78$ m,

$$z_{b3} + y_3 + \frac{\alpha_3 V_3^2}{2g} = 64.78 + 3.30 + \frac{(2.67)0.60^2}{2(9.81)} = 68.13\ \text{m}$$

This is lower than the right-hand side (calculated as 68.35 m), so we need to try higher values of y_3. Table 7.6 summarizes the results for the various flow depths tried. The depth $y_3 = 3.53$ m satisfies Equation 7.29. Also, for this depth the water surface elevation is $64.78 + 3.53 = 68.31$ m. This is above the low-chord elevation and below the high chord, and therefore, the orifice flow formulation applies.

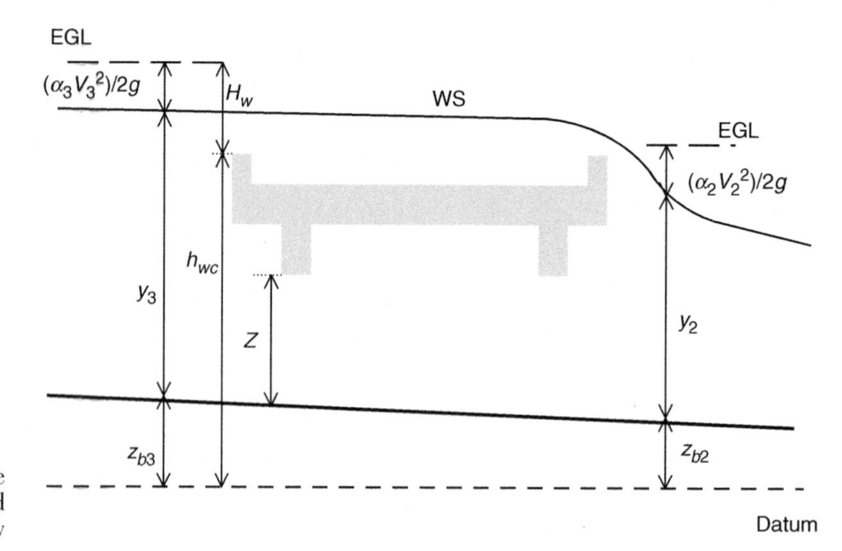

FIGURE 7.9 Bridge with weir and orifice flow

7.1.4.3 Weir-type flow

Weir-type flow occurs when the flow overtops the roadway approaching the bridge, and possibly the bridge itself, as shown schematically in Figure 7.9. In this case, the total flow will be equal to the sum of the flow over the roadway (and possibly the bridge) and the flow through the bridge opening. We can use the weir flow equation to represent the flow over the roadway and the bridge. The flow through the bridge opening can still be calculated by using either the sluice gate equation (Equation 7.27) or the orifice equation (Equation 7.28), depending on the flow depth just downstream of the bridge.

We can write the general weir flow equation as

$$Q = C_w \sqrt{2g} L_w \left(z_{b3} + y_3 + \frac{\alpha_3 V_3^2}{2g} - h_{wc} \right)^{3/2} \tag{7.30}$$

where C_w = weir coefficient, L_w = effective weir crest length, and h_{wc} = elevation of the weir crest.

Usually, we subdivide the weir crest into segments and calculate the discharge over each segment, and then we sum up these discharges to determine the total discharge. For example, three segments are needed if the roadway embankments and the bridge deck are overtopped, as shown schematically in Figure 7.10. The weir discharge of the flow overtopping the roadway to the left of the bridge deck is

$$Q_{wl} = C_{wl} \sqrt{2g} L_{wl} \left(z_{b3} + y_3 + \frac{\alpha_3 V_3^2}{2g} - h_{wlc} \right)^{3/2} \tag{7.31}$$

where C_{wl} = discharge coefficient for roadway to the left of the bridge, L_{wl} = effective crest length for roadway to the left of the bridge, and h_{wlc} = average crest elevation.

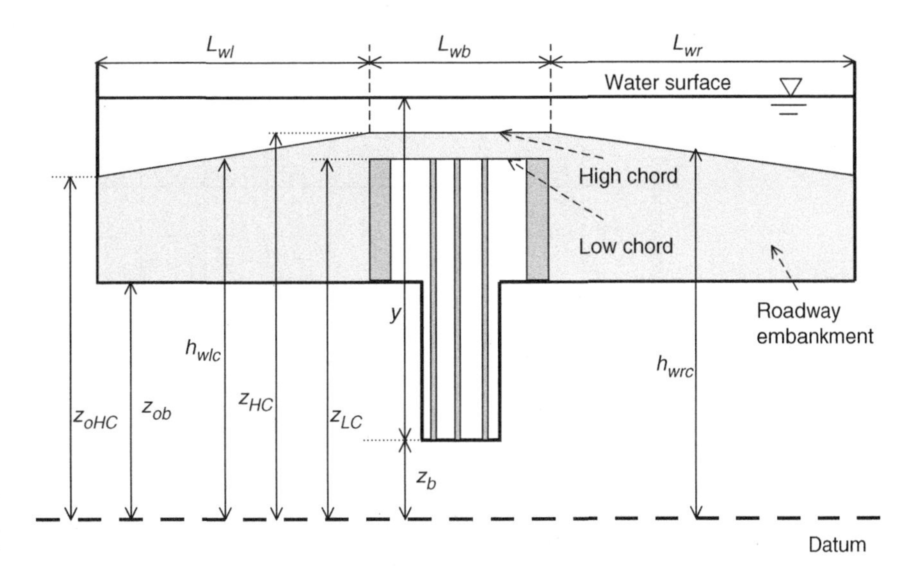

FIGURE 7.10 Cross-sectional view of weir flow

Referring to Figure 7.10, we can use $h_{wlc} = (z_{oHC} + z_{HC})/2$. The weir flow, Q_{wb}, over the bridge can be calculated as

$$Q_{wb} = C_{wb}\sqrt{2g}L_{wb}\left(z_{b3} + y_3 + \frac{\alpha_3 V_3^2}{2g} - z_{HC}\right)^{3/2} \tag{7.32}$$

where C_{wb} = weir coefficient for flow over the bridge, L_{wb} = effective crest length, and z_{HC} = average elevation of the high chord of the bridge deck. The weir discharge of the flow overtopping the roadway to the right of the bridge is

$$Q_{wr} = C_{wr}\sqrt{2g}L_{wr}\left(z_{b3} + y_3 + \frac{\alpha_3 V_3^2}{2g} - h_{wrc}\right)^{3/2} \tag{7.33}$$

where C_{wr} = discharge coefficient for roadway to the right of the bridge, L_{wr} = effective crest length for roadway to the right of the bridge, and $h_{wrc} = (z_{oHC} + z_{HC})/2$ = average crest elevation.

Tables of weir discharge coefficients are available in the literature (e.g. Brater *et al.*, 1996) for broad-crested weirs. However, very few prototype data are available for flow overtopping bridges. The Federal Highway Administration (Bradley, 1978) provides a chart for flow over the roadways in which the weir discharge is about 0.38 for heads over the weir crest larger than 0.2 m (0.6 ft). For smaller heads the coefficient is lower, and it varies between 0.36 and 0.38. In the absence of more reliable data, the US Army Corps of Engineers (2002) suggests weir coefficients of about 0.32 for flow over the bridge deck and 0.37 for flow overtopping the roadways approaching the bridge. These coefficients are suggested when the tailwater (flow depth at Section 2) is low enough not to interfere with the flow overtopping the roadways and the bridge. For higher tailwater elevations, the weir discharge coefficient is reduced by a factor given in Figure 7.11 (Bradley, 1978). In this figure the percentage submergence is defined

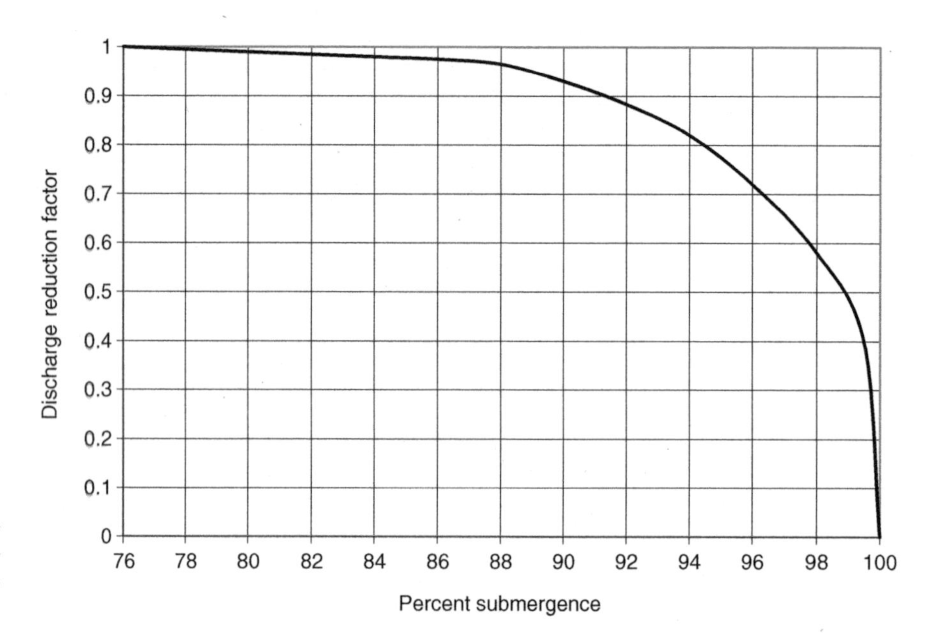

FIGURE 7.11 Weir flow reduction factor for submergence (after Bradley, 1978)

as *100 times depth of water at channel section 2 above minimum weir crest elevation* divided by *energy head at section 3 above minimum weir crest elevation.*

EXAMPLE 7.6 Suppose the discharge in the channel reach considered in Example 7.5 is 500 m^3/s and the corresponding flow depth at section 2 is 3.30 m. Determine the flow depth at section 3.

Recall that Figure 7.6 displays a schematic (not to scale) of the cross-section of the bridge considered in Example 7.5. The horizontal dimensions are the same at the downstream and upstream ends of the bridge, but the elevations are slightly different. At the downstream end, $z_{b2}=64.77$ m, $z_{LC}=67.50$, $z_{HC}=68.50$ m, $z_{ob}=66.77$ m, and $z_{oHC}=68.27$ m. At the upstream end, $z_{b3}=64.78$ m, $z_{LC}=67.50$ m, $z_{HC}=68.50$ m, $z_{ob}=66.78$ m, and $z_{oHC}=68.28$ m.

The water surface elevation at channel section 2 is $64.77+3.30=68.07$ m. This is higher than the bridge-deck low-chord elevation of $z_{LC}=67.50$ m at the downstream end of the bridge. Thus, flow through the bridge opening is of the orifice type. At this point, we do not know whether the roadway embankment and the bridge are overtopped. We will determine this as we proceed with the trial and error solution. If the trial value of y_3 is less than z_{oHC} (see Figure 7.10), weir flow will not occur. If $z_{oHC} < y_3 < z_{HC}$, weir flow will occur over the roadways; if $y_3 > z_{HC}$, the bridge deck will also be overtopped.

Let us try $y_3=3.91$ m. This corresponds to a water surface elevation of $64.78+3.91=68.69$ m, which is higher than $z_{HC}=68.50$ m. Therefore there will be weir flow as well as orifice flow, and the total discharge will be equal to the sum of the discharges obtained by Equations 7.28, 7.31, 7.32, and 7.33. Of course, for the correct value of y_3, the sum of these discharges should be equal to the given discharge, 500 m^3/s.

We first need to evaluate the energy coefficient, α, at section 3 corresponding to $y_3 = 3.91$ ft. Referring to Figure 7.7 and using $y_3 = 3.91$ m, we have $y_{mc} = 3.91$ m, $A_{mc} = (3.91)(50.0) = 195.5\,\text{m}^2$, and $P_{mc} = (50.0) + 2.0 + 2.0 = 54.0$ m. Likewise, for the left overbank, $y_{lob} = 1.91$ m, $A_{lob} = (1.91)(170.0) = 324.7\,\text{m}^2$, and $P_{lob} = 1.91 + 170.0 = 171.91$ m. For the right overbank, $y_{rob} = 1.91$ m, $A_{rob} = (1.91)(180.0) = 343.8\,\text{m}^2$, and $P_{rob} = 1.91 + 180.0 = 181.91$ m. Assuming that $n_{lob} = n_{rob} = 0.05$ and $n_{mc} = 0.025$, we can use Equation 7.13 to obtain $K_{lob} = 9922.7\,\text{m}^3$, $K_{rob} = 10510.8\,\text{m}^3$, and $K_{mc} = 18437.8\,\text{m}^3$. Then, by using Equation 7.12, $S_f = 0.000165$. Next, by using Equation 7.14, we obtain $Q_{lob} = 127.6\,\text{m}^3/\text{s}$, $Q_{rob} = 135.2\,\text{m}^3/\text{s}$, and $Q_{mc} = 237.2\,\text{m}^3/\text{s}$. Then, $V_{lob} = 127.6/324.7 = 0.39$ m/s, $V_{rob} = 135.2/343.8 = 0.39$ m/s, and $V_{mc} = 237.2/195.5 = 1.21$ m/s. The total flow area of the compound channel is $A = 324.7 + 195.5 + 343.8 = 864.0\,\text{m}^2$, and the cross-sectional average velocity is $V = 500.0/864.0 = 0.58$ m/s. Also, by using Equation 7.15, we calculate that $\alpha = 2.29$.

We can now determine the discharge through the bridge opening by using the orifice-flow equation, Equation 7.28. Recalling that $A_{br} = 132.43\,\text{m}^2$ from Example 7.5, and with $C_O = 0.8$,

$$Q_o = C_o A_{br} \sqrt{2g\left(z_{b3} + y_3 + \frac{\alpha_3 V_3^2}{2g} - z_{b2} - y_2\right)}$$

$$= 0.8(132.43)\sqrt{2(9.81)\left[64.78 + 3.91 + \frac{(2.29)0.58^2}{2(9.81)} - 64.77 - 3.30\right]} = 381\,\text{m}^3/\text{s}$$

We will now calculate the weir flow components. First referring to Figures 7.6 and 7.10, we have $L_{wl} = 165$ m, $L_{wb} = 60$ m, $L_{wr} = 175$ m, $h_{wrc} = h_{wlc} = (z_{oHC} + z_{HC})/2 = (68.28 + 68.50)/2 = 68.39$ m, and $z_{HC} = 68.50$ m. The recommended weir discharge coefficients are $C_{wr} = C_{wl} = 0.37$, and $C_{wb} = 0.32$. However, we should determine whether these coefficients need to be reduced due to submergence by tailwater. Recall that percentage submergence is defined as *100 times depth of water at channel section 2 above minimum weir crest elevation* divided by *energy head at section 3 above minimum weir crest elevation.* In this case, the water surface elevation at section 2 is $64.77 + 3.30 = 68.07$ m. This is below the minimum weir crest elevation of both the roadway embankments (68.28 m) and the bridge deck (68.50 m). Therefore, the weir flow is not submerged by tailwater, and there is no need to reduce the weir discharge coefficients.

We are now ready to determine the flow over the roadway embankments and the bridge deck. By using Equations 7.31, 7.32, and 7.33,

$$Q_{wl} = C_{wl}\sqrt{2g}L_{wl}\left(z_{b3} + y_3 + \frac{\alpha_3 V_3^2}{2g} - h_{wlc}\right)^{3/2}$$

$$= (0.37)\sqrt{2(9.81)}(165.0)\left(64.78 + 3.91 + \frac{(2.29)0.58^2}{2(9.81)} - 68.39\right)^{3/2} = 53.4\,\text{m}^3/\text{s}$$

$$Q_{wb} = C_{wb}\sqrt{2g}L_{wb}\left(z_{b3} + y_3 + \frac{\alpha_3 V_3^2}{2g} - z_{HC}\right)^{3/2}$$

$$= (0.32)\sqrt{2(9.81)}(60.0)\left(64.78 + 3.91 + \frac{(2.29)0.58^2}{2(9.81)} - 68.50\right)^{3/2} = 9.3\,\text{m}^3/\text{s}$$

and

$$Q_{wr} = C_{wr}\sqrt{2g}L_{wr}\left(z_{b3} + y_3 + \frac{\alpha_3 V_3^2}{2g} - h_{wrc}\right)^{3/2}$$

$$= (0.37)\sqrt{2(9.81)}(175.0)\left(64.78 + 3.91 + \frac{(2.29)0.58^2}{2(9.81)} - 68.39\right)^{3/2} = 56.7\,\text{m}^3/\text{s}$$

The sum of the orifice flow and the component weir flow discharges become $381.0 + 53.4 + 9.3 + 56.7 = 500.4\,\text{m}^3/\text{s}$, which is very close to the given 500 cfs. Therefore, the guessed depth of $y_3 = 3.91$ m is acceptable.

7.1.4.4 Direct step method for high-flow calculations

This method performs the calculations as though the bridge sections are ordinary channel sections. At the cross-sections inside the bridge, the area obstructed by the piers, the abutments, and the bridge deck are subtracted from the flow area. Also, the wetted perimeter is increased by the amount the water is in contact with the piers, the abutments and the deck. An adjustment to the Manning roughness factor may also be needed. Once these adjustments to the flow area, wetted perimeter, and the roughness factors have been made, the calculations are performed as discussed in Chapter 4. This method should be chosen when the bridge is a small obstruction to the flow, or when the bridge is highly submerged and flow overtopping the road and the bridge deck is not acting like weir flow.

7.2 Evaluating Scour at Bridges

Floods scouring bed material from around bridge foundations are a common cause of bridge failures (Chang, 1973; Brice and Blodgett, 1978; Davis, 1984). It is, therefore, important to estimate the probable scour depths so that the bridge foundations can be designed to support the design structural load safely below the probable scour depth. Richardson and Davis (2001) report the guidelines broadly followed by practicing engineers for designing new bridges to resist scour and evaluating the vulnerability of existing bridges to scour. The discussions and the procedures included in this section are adopted from Richardson and Davis (2001).

Total scour is comprised of long-term *aggradation* and *degradation* of the river bed, *general scour* at the bridge, and *local scour* at the piers and abutments. By *aggradation* we mean the deposition of material eroded from the upstream

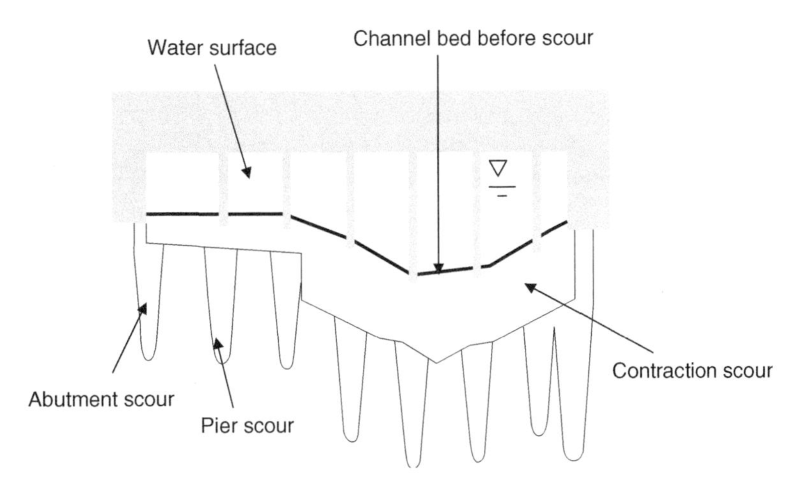

FIGURE 7.12 Various scour types (after US Army Corps of Engineers, 2002)

channel or watershed, while *degradation* refers to lowering of the streambed due to a deficit of sediment supply from upstream. The long-term changes in the streambed elevation can be due to natural or manmade causes. Dams and reservoirs, changes in watershed land use (such as urbanization), or natural lowering of the fluvial system are some of these causes. Procedures for evaluating the long-term changes in the streambed are discussed by Richardson and Davis (2001), and are beyond the scope of this text. *General scour* refers to lowering of the streambed during the passage of a flood wave. At bridge sites, general scour occurs usually owing to contraction of flow, and is called *contraction scour*. Material is removed from the streambed across the channel width by the flow accelerating due to the contraction. The scour depth can be non-uniform. *Local scour* involves removal of material from around piers and abutments as the flow accelerates around the obstructed flow area. Figure 7.12 shows various types of scour.

The bed material is removed by the flow during the scouring process at the bridge section. However, the flow may also be transporting bed material in suspension from the further upstream sections towards the bridge section. Depending on the amount of upstream bed material transported, the contraction and local scour at the bridge section can occur in the form of clear-water or live-bed scour. *Clear-water scour* occurs when there is little or no movement of bed material in the flow upstream of the bridge. *Live-bed scour* occurs when bed material is transported by flow from the upstream reach to the bridge section at a significant rate. In this case, the scour hole that develops during the rising stage of a flood will refill during the falling stage. Clear-water scour occurs mainly in coarse bed-material streams. It reaches its maximum over a long period of time, as shown in Figure 7.13. Indeed, it may take several flood events for the local clear-water scour to reach its maximum. Live-bed scour occurs in coarse bed-material streams only at high discharges of a flood wave. It also occurs in sand-bed streams. Live-bed pier scour in sand-bed streams with a dune-bed configuration fluctuates about the equilibrium scour due to the variability of

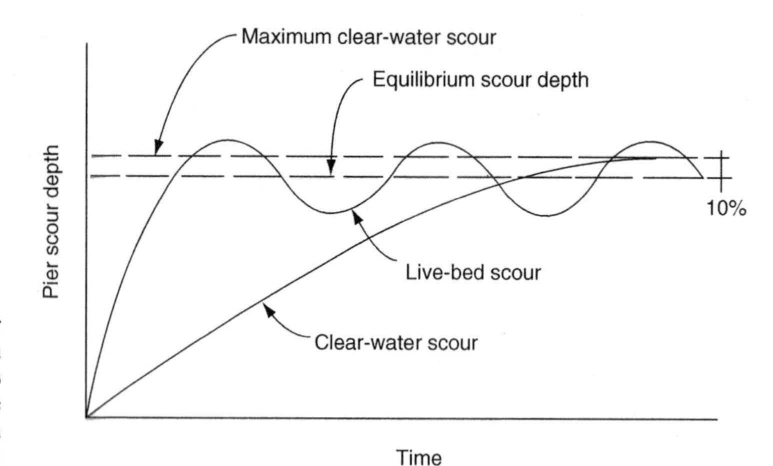

FIGURE 7.13 Pier scour depth in a sand-bed stream as a function of time (after Richardson and Davis, 2001)

the bed-material transport. The maximum pier scour in this case can be 30% higher than the equilibrium scour. However, for general practice, the maximum depth of pier scour in sand-bed streams is about 10% higher than the equilibrium scour.

7.2.1 CONTRACTION SCOUR

Contraction scour occurs where the area of the bridge opening is smaller than the upstream flow area, which may include a main channel and flood plains. A reduction in the flow area causes an increase in the average velocity for the same discharge, as well as an increase in the shear stresses over the streambed. The increased erosive forces will remove more bed material at the contracted section than is being transported from upstream. As a result, the streambed will be lowered across the width of the channel. However, due to the velocity variations within a channel section, the lowering of the streambed may not be uniform across the width of the channel.

Various commonly encountered cases of contraction scour are shown in Figure 7.14. Contraction scour will also occur when a bridge is located over a naturally narrower reach of a river. The case shown in Figure 7.14c can be very complex. If the abutment is set back a small distance from the bank (less than three to five times the average depth of flow through the bridge), there is a danger that the bank will be destroyed under the combined effect of contraction scour and abutment scour. In that event, the bank and bed under the bridge in the overflow area should be protected by rock riprap.

Contraction scour can occur in both the main channel and the overbank areas. Both live-bed scour and clear-water scour are possible, depending on whether the flow upstream of the bridge is transporting bed material (live-bed scour) or is not (clear-water scour). This can be determined by using the concept of a critical velocity above which bed material will be eroded.

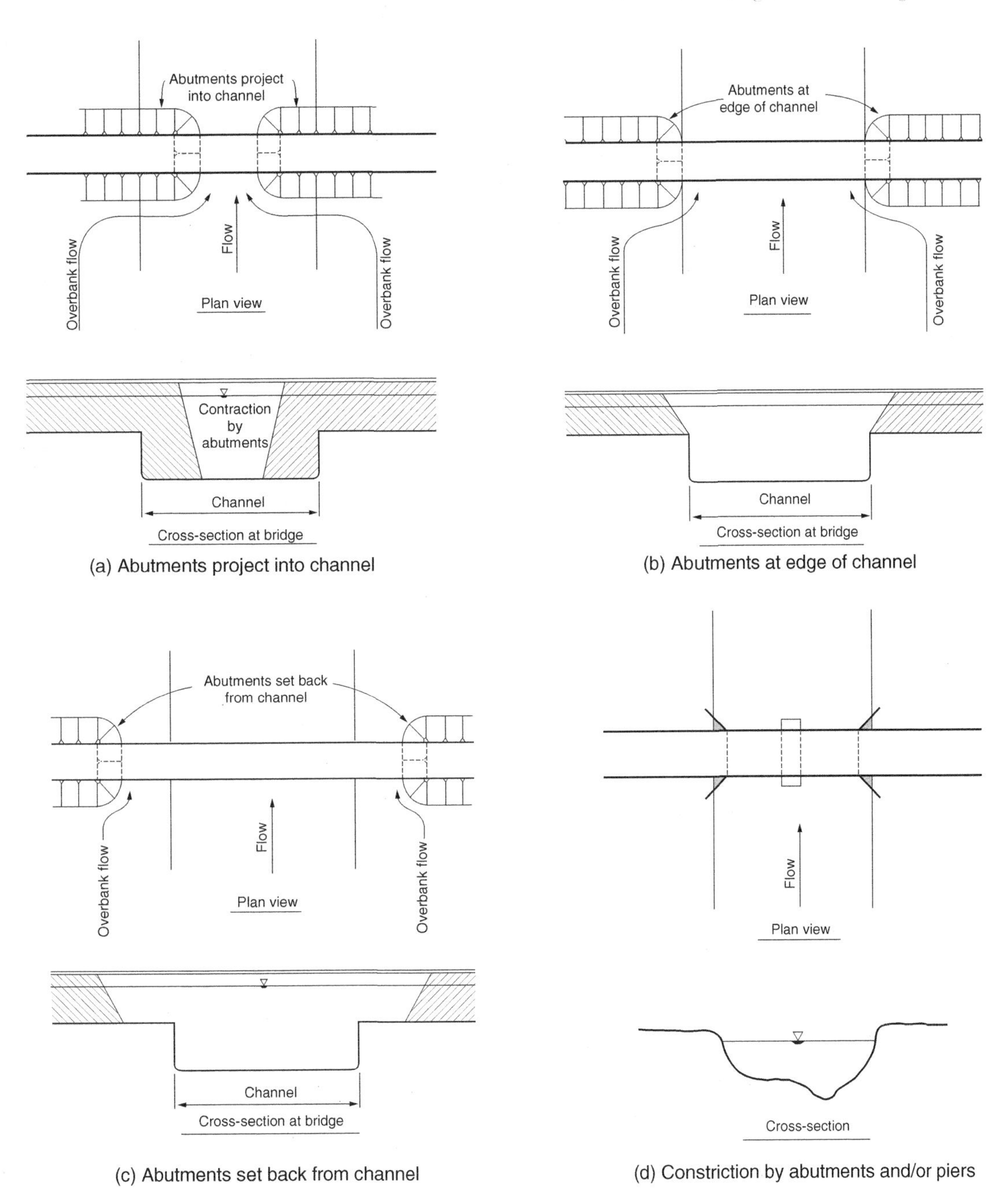

FIGURE 7.14 Common contraction scour cases (after Richardson and Davis, 2001)

7.2.1.1 Critical velocity

The critical velocity here is defined as the velocity above which the bed material of a specified size and smaller will be transported. (This should not to be confused with the velocity corresponding to minimum specific energy discussed in Chapter 2.) Laursen (1963) derived an expression for critical velocity based on the concept of bed shear stress (or tractive force), discussed in Chapter 5. The average bed shear stress on the channel bed is expressed as

$$\tau_0 = \gamma R S_f \tag{7.34}$$

where τ_0 = average shear stress, γ = specific weight of water, R = hydraulic radius, and S_f = friction slope. Using the Manning formula to evaluate the friction slope and approximating the hydraulic radius by the flow depth, y, Equation 7.34 is written as

$$\tau_0 = \gamma R S_f = \frac{\gamma n^2 V^2}{k_n^2 y^{1/3}} \tag{7.35}$$

where n = Manning roughness factor and $k_n = 1.0\,\text{m}^{1/3}/\text{s} = 1.49\,\text{ft}^{1/3}/\text{s}$. For non-cohesive bed material, the critical bed shear stress at incipient motion can be expressed by using the Shield relation,

$$\tau_c = K_s(\gamma_s - \gamma)D_s \tag{7.36}$$

where τ_c = critical shear stress, γ_s = specific weight of sediment particles, K_s = Shield's coefficient, and D_s = particle size. The motion of the indicated particle size is initiated when $\tau_0 = \tau_c$. Therefore, we can determine the critical velocity, V_c, by equating the right-hand sides of Equations 7.34 and 7.36 and solving for $V = V_c$ as

$$V_c = \frac{k_n}{n}\sqrt{K_s(s-1)}y^{1/6}D_s^{1/2} \tag{7.37}$$

where $s = \gamma_s/\gamma$ = specific gravity of particles. Substituting the median diameter, D_{50}, for D_s and using the Strickler equation, $n = 0.034(K_v D_{50})^{1/6}$ with $K_v = 3.28\,\text{m}^{-1} = 1.0\,\text{ft}^{-1}$ to evaluate the Manning roughness factor,

$$V_c = K_u y^{1/6} D_{50}^{1/3} \tag{7.38}$$

where $K_u = k_n\{K_s(s-1)\}^{1/2}/(0.034K_v^{1/6})$, D_{50} = particle size of which 50% are finer, and V_c = critical velocity above which bed material of size D_{50} or smaller will be removed. With typical values of $s = 2.65$ and $K_s = 0.039$, we obtain $K_u = 6.19\,\text{m}^{1/2}/\text{s} = 11.17\,\text{ft}^{1/2}/\text{s}$.

In Equation 7.38, the D_{50} represents the average size of the bed material in the upper 0.3 m (1.0 ft) of the streambed in the reach upstream of the bridge. The critical velocity is calculated at the approach section (channel Section 4 in Figure 7.1) separately for the main channel and the overbank areas. If the main channel average velocity at the approach section is greater than the critical velocity, then live-bed scour condition exists for the main channel. Otherwise, clear-water scour will occur in the main channel. Likewise, if the average

TABLE 7.7 Suggested values for exponents $k1$ and $k2$ (after Richardson and Davis, 2001)

V_*/ω	Mode of bed material transport	$k1$	$k2$
<0.5	Mostly contact bed material discharge	0.59	0.066
0.5–2.0	Some suspended bed material discharge	0.64	0.21
>2.0	Mostly suspended bed material discharge	0.69	0.37

overbank velocity at the approach section is greater than the critical velocity, then the live-bed scour condition exists for the overbank area. Otherwise, clear-water scour will occur.

7.2.1.2 Live-bed contraction scour

Employing various simplifying assumptions, Laursen (1960) derived a live-bed contraction scour equation expressed as

$$y_s = y_4\left(\frac{Q_{BU}}{Q_4}\right)^{6/7}\left(\frac{W_4}{W_{BU}}\right)^{k1}\left(\frac{n_{BU}}{n_4}\right)^{k2} - y_{BU} \qquad (7.39)$$

where y_s = average scour depth in the main channel, y_4 = flow depth at the approach section (section 4 in Figure 7.1), Q_{BU} = discharge in the main channel at the contracted section (section BU in Figure 7.1), Q_4 = discharge in the main channel at channel section 4, W_4 = bottom width of the main channel at channel section 4, W_{BU} = bottom width of the main channel at section BU less piers and abutments if any, n_{BU} = Manning roughness factor in the main channel at section BU, n_4 = Manning roughness factor in the main channel at Section 4, and y_{BU} = existing flow depth in the main channel at section BU before scour. The exponents $k1$ and $k2$ depend on the mode of the bed-material transport as described in Table 7.7.

In Table 7.7, V_* = shear velocity in the approach section (channel section 4) and ω = fall velocity of bed material based on D_{50}. The shear velocity is calculated using

$$V_* = \sqrt{g y_4 S_{f4}} \qquad (7.40)$$

where S_{f4} = friction slope at channel section 4. The fall velocity depends on the temperature, particle size, and specific gravity of sediments. Figure 7.15, depicting the suggested values of the fall velocity, was constructed by using data from Richardson and Davis (2001).

Richardson and Davis (2001) recommend a modified version of Equation 7.39 to determine the live-bed contraction scour. The modified equation is written as

$$y_s = y_4\left(\frac{Q_{BU}}{Q_4}\right)^{6/7}\left(\frac{W_4}{W_{BU}}\right)^{k1} - y_{BU} \qquad (7.41)$$

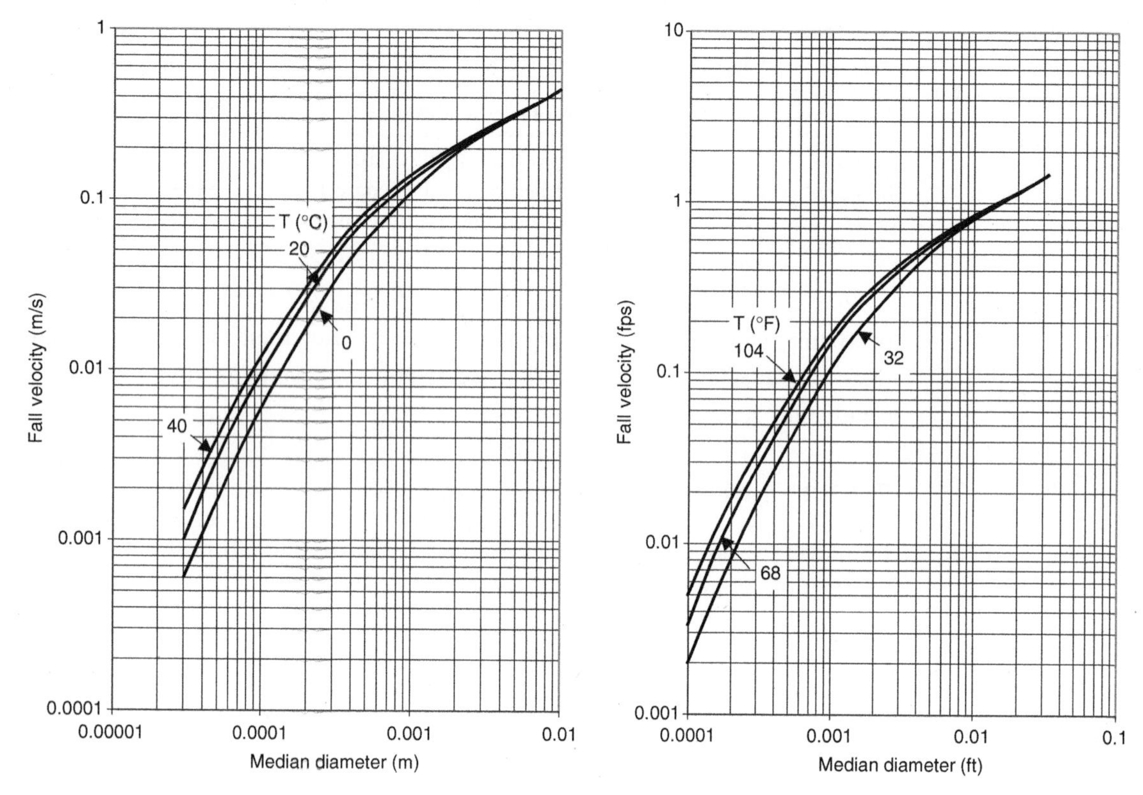

FIGURE 7.15 Particle fall velocity (source: Richardson and Davis, 2001)

Equations 7.39 and 7.41, written for the main channel, can be used for overbank areas as well. However, in this case the overbank flow parameters and variables should be used instead of those of the main channel.

7.2.1.3 Clear-water contraction scour

Clear-water contraction scour occurs if the average velocity at the approach channel section is smaller than the critical velocity discussed in Section 7.2.1.1. The clear-water scour equation, attributed to Laursen, is derived from the bed shear stress concept. The bed in a long contracted section will scour until $\tau_0 = \tau_c$. Thus, equating the right-hand sides of Equations 7.35 and 7.36, and solving for y, we obtain

$$y = \left(\frac{n^2 V^2}{K_s k_n^2 D_m (s-1)} \right)^3 \tag{7.42}$$

Note that, in Equation 7.42, D_s is replaced by $D_m = 1.25 D_{50}$ = effective bed material size, or the size of the smallest non-transportable bed material in the contracted section. Approximating the channel section by a rectangular section of width W, using $V = Q/(yW)$, and approximating the Manning

TABLE 7.8 Summary of data for Example 7.7

Channel section	Flow/section variable	Left overbank	Main channel	Right overbank
4	S_f	0.000308	0.000308	0.000308
	y (m)	0.73	2.73	0.73
	V (m/s)	0.28	1.30	0.28
	Q (m^3/s)	35.15	177.63	37.22
	W (m)	170.0	50.0	180.0
3	y (m)	0.56	2.56	0.56
	V (m/s)	0.37	1.93	0.37
	Q (m^3/s)	1.24	247.52	1.24
	W (m)	6.0	50.0	6.0
BU	y (m)	0.50	2.50	0.50
	V (m/s)	0.41	2.12	0.41
	Q (m^3/s)	0.61	248.78	0.61
	W (m)	3.0	47.0	3.0

roughness factor with the Strickler formula $n = 0.034(K_v D_{50})^{1/6} = 0.034(K_v D_m/1.25)^{1/6}$ where $K_v = 3.28\,\text{m}^{-1} = 1.0\,\text{ft}^{-1}$, Equation 7.42 becomes

$$y = \left(\frac{Q^2}{C_U D_m^{2/3} W^2}\right)^{3/7} \tag{7.43}$$

where W = width of the contracted section and $C_U = k_n{}^2 K_s(s-1)/(0.034K_v{}^{1/6}/1.25^{1/6})^2$. Using $s = 2.65$ and $K_s = 0.039$ as before, we obtain $C_U = 40\,\text{m/s}^2 = 130\,\text{ft/m}^2$. The flow depth calculated by Equation 7.43 is the equilibrium depth, and this is equal to the flow depth at the bridge section before scour plus the scour depth. Therefore, using the subscript BU for the upstream bridge section as in Figure 7.1, the clear-water scour depth becomes

$$y_s = \left(\frac{Q_{BU}^2}{C_U D_m^{2/3} W_{BU}^2}\right)^{3/7} - y_{BU} \tag{7.44}$$

where y_{BU} = flow depth at the bridge section before scour. Equation 7.44 should be applied to the main channel and the overbank areas separately.

EXAMPLE 7.7 Revisit the bridge flow situation discussed in Example 7.1 and calculate the contraction scour for this situation, assuming the median size of the bed material is 0.0005 m and the water temperature is 30°C.

We recall that Figure 7.3 displays the dimensions of the channel sections 4, 3, and BU considered in Example 7.1, and Table 7.2. presents the results of Example 7.1. We will use some of these results to calculate the scour depths in this example. Table 7.8 summarizes the results of Example 7.1 and the channel section dimensions that are relevant to this example.

We will perform the contraction scour calculations for the main channel and the two overbank areas separately. For the main channel, we will first evaluate the critical velocity at Section 4 by using Equation 7.38. Noting that $K_U = 6.19\,\mathrm{m}^{1/2}/\mathrm{s}$ for the metric unit system, $y_4 = 2.73$ m from Table 7.8, and $D_{50} = 0.0005$ m from the problem statement,

$$V_c = K_u y_4^{1/6} D_{50}^{1/3} = (6.19)(2.73)^{1/6}(0.0005)^{1/3} = 0.58\,\mathrm{m/s}$$

As shown in Table 7.8, the flow velocity in the main channel at Section 4 is $V_4 = 1.30$ m/s, and is greater than the critical velocity $V_c = 0.58$ m/s. Hence, live-bed contraction scour will occur, and we will use Equation 7.41 to calculate the scour depth.

Noting that $S_{f4} = 0.000308$ and $y_4 = 2.73$ m for the channel from Table 7.8, the shear velocity in the main channel of channel Section 4 is determined by using Equation 7.40 as

$$V_* = \sqrt{g y_4 S_{f4}} = \sqrt{(9.81)(2.73)(0.000308)} = 0.09\,\mathrm{m/s}$$

Also, we obtain $\omega = 0.08$ m/s from Figure 7.15 for $D_{50} = 0.0005$ m and 30°C. Then, $V_*/\omega = 0.09/0.08 = 1.13$ and, from Table 7.7, $k1 = 0.64$. Noting from Table 7.8 that, for the main channel, $Q_4 = 177.63\,\mathrm{m}^3/\mathrm{s}$, $y_4 = 2.73$ m, $W_4 = 50.0$ m, $Q_{BU} = 248.75\,\mathrm{m}^3/\mathrm{s}$, $y_{BU} = 2.50$ m, and $W_{BU} = 47.0$ m, we can calculate the contraction scour depth by using Equation 7.41 as

$$y_s = y_4 \left(\frac{Q_{BU}}{Q_4}\right)^{6/7} \left(\frac{W_4}{W_{BU}}\right)^{k1} - y_{BU} = 2.73 \left(\frac{248.78}{177.63}\right)^{6/7} \left(\frac{50.0}{47.0}\right)^{0.64} - 2.50 = 1.29\,\mathrm{m}$$

Let us now calculate the contraction scour depth on the left overbank. Using the left overbank flow depth of $y_4 = 0.73$ m (from Table 7.8), the critical velocity is obtained as

$$V_c = K_u y_4^{1/6} D_{50}^{1/3} = (6.19)(0.73)^{1/6}(0.0005)^{1/3} = 0.47\,\mathrm{m/s}$$

This is larger than the left overbank flow velocity of $V_4 = 0.28$ m/s. Then, the contraction scour would be of clear-water type, and Equation 7.44 should be used. From Table 7.8, for the left overbank area, $Q_{BU} = 0.61\,\mathrm{m}^3/\mathrm{s}$, $y_{BU} = 0.50$ m, and $W_{BU} = 3.0$ m. Also, $D_m = 1.25\,D_{50} = 1.25(0.0005) = 0.000625$ m, 0.000625 m, and $C_U = 40\,\mathrm{m/s}^2$. Substituting into Equation 7.44,

$$y_s = \left(\frac{Q_{BU}^2}{C_U D_m^{2/3} W_{BU}^2}\right)^{3/7} - y_{BU} = \left(\frac{0.61^2}{(40)(0.000625)^{2/3}(3.0)^2}\right)^{3/7} - 0.50 = -0.07\,\mathrm{m}$$

Obviously, a negative clear-water scour depth is not possible. Hence, we will conclude that there is no contraction scour over the left bank. In the same manner, we can show that contraction scour will not occur in the right overbank area either.

EXAMPLE 7.8 What would be the contraction scour in the main channel in Example 7.7 if the median grain diameter was $D_{50} = 0.008$ m?

For $D_{50} = 0.008$ m, the critical velocity in the main channel of channel Section 4 is calculated using Equation 7.38 as

$$V_c = K_u y_4^{1/6} D_{50}^{1/3} = (6.19)(2.73)^{1/6}(0.008)^{1/3} = 1.46\,\text{m/s}$$

This is greater than the main channel velocity of $V_4 = 1.30$ m/s, and therefore clear-water contraction scour would occur. From the data given in Table 7.8, for the main channel of Section 4, $Q_{BU} = 248.78\,\text{m}^3/\text{s}$, $y_{BU} = 2.50$ m, and $W_{BU} = 47.0$ m. Also, $D_m = 1.25\,D_{50} = 1.25(0.008) = 0.01$ m, and $C_U = 40\,\text{m/s}^2$. Substituting into Equation 7.34,

$$y_s = \left(\frac{Q_{BU}^2}{C_U D_m^{2/3} W_{BU}^2}\right)^{3/7} - y_{BU} = \left(\frac{248.78^2}{(40)(0.01)^{2/3}(47.0)^2}\right)^{3/7} - 2.50 = 0.70\,\text{m}$$

7.2.2 LOCAL SCOUR AT PIERS

Local scour at piers is caused by *horseshoe vortices* forming at the base of the pier. Obstruction of flow by a pier results in a stagnation line on the front of the pier. As we recall from basic fluid dynamics, the stagnation pressure is larger than the hydrostatic pressure by an amount equal to the dynamic pressure. The dynamic pressure is proportional to the square of the local velocity, and is lower near the bed. Therefore, a downward hydraulic gradient develops in front of the pier that causes downflow directed towards the bed, as shown in Figure 7.16. We also recall from fluid dynamics that flow past a body separates from the body, and *wake vortices* form in the separation zone. The boundary layer separation combined with downflow produces the horseshoe vortex wrapped around the base of the pier. This system of vortices removes bed material from around the base of the pier, producing a local scour hole. The strength of the horseshoe

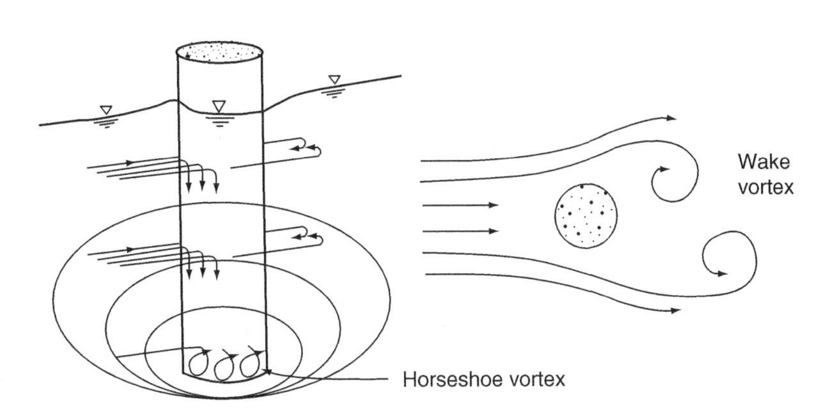

FIGURE 7.16 Horseshoe and wake vortices (after Richardson and Davis, 2001)

vortex is reduced as the depth of the scour increases. For live-bed scour, an equilibrium will be re-established when the amount of the bed material transported from upstream equals the bed material removed by the vortex action. For clear-water scour, equilibrium is reached when the shear stress caused by the vortex action becomes equal to the critical shear stress.

Numerous studies of local scour at piers have been reported in the past, many of which are based on laboratory experiments (e.g. Jain and Fischer, 1979; Laursen, 1980; Melville and Sutherland, 1988; Richardson *et al.*, 2001). Jones (1984) presented comparisons of various pier scour equations. Richardson and Davis (2001) recommend the Colorado State University (CSU) equation (Richardon *et al.*, 2001) for both live-bed and clear-water pier scour. The HEC-RAS model includes the Froechlich (1988) equation as an option in addition to the CSU equation.

7.2.2.1 The CSU equation for pier scour

The CSU equation is used to predict the maximum pier scour depths for both live-bed and clear-water scour conditions. The equation is written as

$$y_s = 2.0K_1K_2K_3K_4a^{0.65}y_3^{0.35}F_{r3}^{0.43} \tag{7.45}$$

where y_s = scour depth, K_1 = correction factor for pier nose shape, K_2 = correction factor for angle of attack of flow, K_3 = correction factor for bed condition, K_4 = correction factor for armoring of bed material, a = pier width, y_3 = flow depth directly upstream of pier (that is at channel Section 3 in Figure 7.1), and F_{r3} = Froude number directly upstream of the pier.

Common pier shapes are shown in Figure 7.17. The correction factors, K_1, corresponding to these shapes are given in Table 7.9. The correction factors, K_2, for angle of attack of the flow is calculated by using

$$K_2 = \left(\text{Cos }\theta_p + \frac{L}{a}\sin\theta_p\right)^{0.65} \tag{7.46}$$

where θ_p = pier angle of attack as shown in Figure 7.17, and L = length of pier. Table 7.10 lists the values of K_2 corresponding to various angles of attack and L/a ratios. The correction factors K_3 for typical bed conditions are given in Table 7.11.

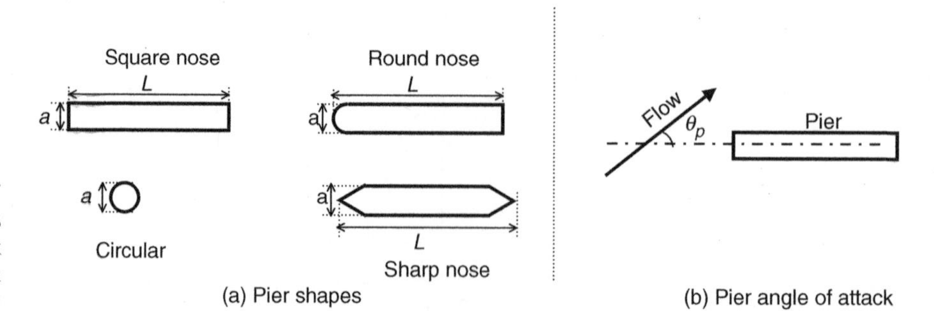

FIGURE 7.17 Common pier shapes and angle of attack (after Richardson and Davis, 2001)

TABLE 7.9 Correction factor K_1 for pier shape (after Richardson and Davis, 2001)

Shape of pier nose	K_1
Square nose	1.1
Round nose	1.0
Circular cylinder	1.0
Sharp nose	0.9

TABLE 7.10 Correction factor K_2 for angle of attack of flow (after Richardson and Davis, 2001)

θ_p (°)	$L/a=4$	$L/a=8$	$L/a=12$
0	1.0	1.0	1.0
15	1.5	2.0	2.5
30	2.0	2.75	3.5
45	2.3	3.3	4.3
90	2.5	3.9	5.0

TABLE 7.11 Correction factor K_3 for bed condition (after Richardson and Davis, 2001)

Bed condition	Dune height	K_3
Clear-water scour	Not applicable	1.1
Plane bed and antidune flow	Not applicable	1.1
Small dunes	0.6–3 m (2–10 ft)	1.1
Medium dunes	3–9 m (10–30 ft)	1.1–1.2
Large dunes	>9 m (>30 ft)	1.3

The correction factor $K_4=1.0$ if $D_{50} < 2\,\text{mm}$ *or* $D_{95} < 20\,\text{mm}$ for the bed material. If $D_{50} > 2\,\text{mm}$ *and* $D_{95} > 20\,\text{mm}$, then K_4 decreases the scour depths for armoring of the bed material. In this case, the correction factor K_4 is calculated as (Mueller and Jones, 1999)

$$K_4 = 0.4(V_R)^{0.15} \tag{7.47}$$

where

$$V_R = \left(\frac{V_3 - V_{i50}}{V_{c50} - V_{i95}}\right) \tag{7.48}$$

$$V_{i50} = (0.645)\left(\frac{D_{50}}{a}\right)^{0.053} V_{c50} \tag{7.49}$$

$$V_{i95} = (0.645)\left(\frac{D_{95}}{a}\right)^{0.053} V_{c95} \tag{7.50}$$

$$V_{c50} = K_u y_3^{1/6} D_{50}^{1/3} \tag{7.51}$$

and

$$V_{c95} = K_u y_3^{1/6} D_{95}^{1/3} \tag{7.52}$$

In Equations 7.47 through 7.52, V_R = velocity ratio, V_3 = average velocity in the main channel or the overbank area (depending on whether the pier is in the main channel or overbank area) at the cross-section just upstream of the bridge (Section 3 in Figure 7.1), V_{i50} = approach velocity required to initiate scour at the pier for grain size D_{50}, V_{i95} = approach velocity required to initiate scour at the pier for grain size D_{95}, V_{c50} = critical velocity for D_{50} bed material size, V_{c95} = critical velocity for D_{95} bed material size, $K_u = 6.19\,\text{m}^{1/2}/\text{s} = 11.17\,\text{ft}^{1/2}/\text{s}$, and y_3 = the depth of water just upstream of the pier (at Section 3).

There are limiting values for certain variables in the application of the CSU equation to predict the pier scour. If L/a is greater than 12, we use $L/a = 12$ as a maximum in Equation 7.46 and Table 7.10 to determine K_2. If θ_p is greater than 5°, K_2 dominates, so K_1 is set equal to 1.0. The minimum value for K_4 is 0.4. For round-nosed piers aligned with flow, the maximum pier scour depth is 2.4 times the pier width if F_{r3} is less than or equal to 0.8, or 3.0 times the pier width otherwise.

The top width of the scour hole from each side of the pier is usually estimated as being twice the scour depth for practical applications.

EXAMPLE 7.9 Suppose the three piers of the bridge considered in Example 7.7 are each 1 m wide and have a round nose, and the angle of attack is $\theta_p = 0°$. Determine the local scour depth for each pier if $D_{50} = 0.0005$ m and $D_{95} = 0.008$ m. Assume that the channel bed is plane.

All the piers are in the main channel, and therefore the calculations will be performed using the main channel flow variables. If there were piers over the banks, a separate set of calculations would be needed for each overbank area.

For the given situation, $K_1 = 1.0$ from Table 7.9 since the piers have a round nose, $K_2 = 1.0$ from Table 7.10 since $\theta_p = 0°$, $K_3 = 1.1$ from Table 7.11 since the river has a plane bed, and $K_4 = 1.0$ since $D_{95} = 0.008\,\text{m} = 8\,\text{mm} < 20\,\text{mm}$. Also, from the data given in Table 7.8 for the main channel of channel Section 3, we have $y_3 = 2.56$ m, and $V_3 = 1.93$ m/s. For a rectangular channel, the hydraulic depth is equal to the flow depth. Therefore,

$$F_{r3} = \frac{V_3}{\sqrt{g y_3}} = \frac{1.93}{\sqrt{(9.81)(2.56)}} = 0.39.$$

Then, noting that the pier width is $a = 1.0$ m, we can calculate the local piers' scour depth by using Equation 7.45 as

$$\begin{aligned} y_s &= 2.0 K_1 K_2 K_3 K_4 a^{0.65} y_3^{0.35} F_{r3}^{0.43} \\ &= (2.0)(1.0)(1.0)(1.1)(1.0)(1.0)^{0.65}(2.56)^{0.35}(0.39)^{0.43} = 2.04\text{m} \end{aligned}$$

7.2.2.2 Froechlich equation for pier scour

The HEC-RAS model includes a pier scour equation developed by Froechlich (1988) as an alternative to the CSU equation. The use of the Froechlich equation is simpler than that of the CSU equation, and it compares well against observed data (Landers and Mueller, 1996). The equation is

$$y_s = 0.32\phi_F(a')^{0.62}y_3^{0.47}F_{r3}^{0.22}D_{50}^{-0.09} + a \tag{7.53}$$

where a = width of the pier, $a' = a(\cos\theta_p)$ = projected pier width with respect to the direction of flow, and $\phi_F = 1.3$ for square-nosed piers, $\phi_F = 1.0$ for round-nosed piers, and $\phi_F = 0.7$ for sharp-nosed piers.

Equation 7.53 is suggested for predicting the maximum pier scour for design purposes, and the term $(+a)$ added to the right-hand side serves as a factor of safety. In the analysis mode (when used for predicting the pier scour of a particular event at a given bridge), the term $(+a)$ can be dropped. Also, the pier scour obtained from this equation is limited to 2.4 times the pier width if F_{r3} is less than or equal to 0.8, and 3.0 times the pier width otherwise.

EXAMPLE 7.10 Determine the local pier scour for the situation considered in Example 7.9, using the Froechlich equation.

We will use Equation 7.53 to determine the scour depth. Because the piers have a rounded nose, $\phi_F = 1.0$. Also, because the flow is aligned with the piers (that is, the angle of attack is zero), $a' = a = 1.0$ m. From Example 7.9, we know that $D_{50} = 0.0005$ m, $y_3 = 2.56$ m and $F_{r3} = 0.39$. Therefore, by using Equation 7.53 with the $(+a)$ term on the right-hand side,

$$\begin{aligned} y_s &= 0.32\phi_F(a')^{0.62}y_3^{0.47}F_{r3}^{0.22}D_{50}^{-0.09} + a \\ &= 0.32(1.0)(1.0)^{0.62}(2.56)^{0.47}(0.39)^{0.22}(0.0005)^{-0.09} + 1.0 = 1.80\,\text{m} \end{aligned}$$

7.2.2.3 Pressure-flow scour

By *pressure flow* we mean the *high-flow* conditions at the bridge site. As we recall from Section 7.1.4, high flow occurs when the water surface elevation at the upstream face of the bridge is equal to or higher than the low chord of the bridge structure. Depending on whether the bridge is submerged at the downstream side, the flow through the bridge opening can be in the form of sluice-gate flow or orifice flow. If the roadway embankment is overtopped, then the sluice-gate or orifice-type flow will be combined with weir flow over the embankments. Of course the bridge itself can also be overtopped, resulting in weir flow over the bridge deck.

The local scour depths at a pier or abutment under pressure-flow conditions can be much higher than scour depths caused by free surface flow (low flow). The increase in the local scour depths under the pressure-flow conditions is due to the vertical contraction of the flow and the increase in the strength of the horseshoe vortex. This effect is somewhat less when the bridge is overtopped, due the increased flow area.

Limited flume studies indicate that local pier scour can increase by 200 to 300% by pressure (Richardson and Davis, 2001). Based on a fairly extensive study of pressure-flow scour under live-bed conditions, Arneson (Richardson and Davis, 2001) suggest that the vertical contraction scour be determined as

$$\frac{y_s}{y_3} = -5.08 + 1.27\left(\frac{y_3}{Z}\right) + 4.44\left(\frac{Z}{y_3}\right) + 0.19\left(\frac{V_B}{V_{cB}}\right) \tag{7.54}$$

where y_s = vertical contraction scour depth, y_3 = flow depth immediately upstream of the bridge, Z = distance from the low chord of the bridge to the average elevation of the stream bed, V_B = average velocity of the flow through the bridge opening before scour occurs, and V_{cB} = critical velocity of the D_{50} of the bed material at the bridge opening (see Equation 7.38). To determine the total pier scour under pressure-flow conditions, the vertical contraction scour obtained from Equation 7.54 should be added to the pier scour calculated by using Equation 7.45 or 7.53.

7.2.3 LOCAL SCOUR AT ABUTMENTS

Local scour at abutments occurs due to the obstruction of flow by the abutments and the approaching road embankment. A vortex similar to the horseshoe vortex discussed in the preceding section forms at the base of the abutment, and a wake vortex forms downstream. The wake vortex may cause the failure of abutments due to the erosion of the fill material. However, most abutment scour studies have focused on the scour hole caused by the horseshoe vortex. As pointed out by Richardson and Davis (2001), Liu *et al.* (1961), Laursen (1980), Melville (1992), and Froechlich (Richardson and Davis, 2001) developed abutment scour equations based on laboratory data.

Richardson and Davis (2001) recommend two equations for the computation of abutment scour for both live-bed and clear-water scour conditions. When the length of embankment projected normal to the flow is greater than 25 times the flow depth upstream of the bridge (channel Section 3 in Figure 7.1), the HIRE equation is recommended. Otherwise, it is suggested that the Froechlich equation be used.

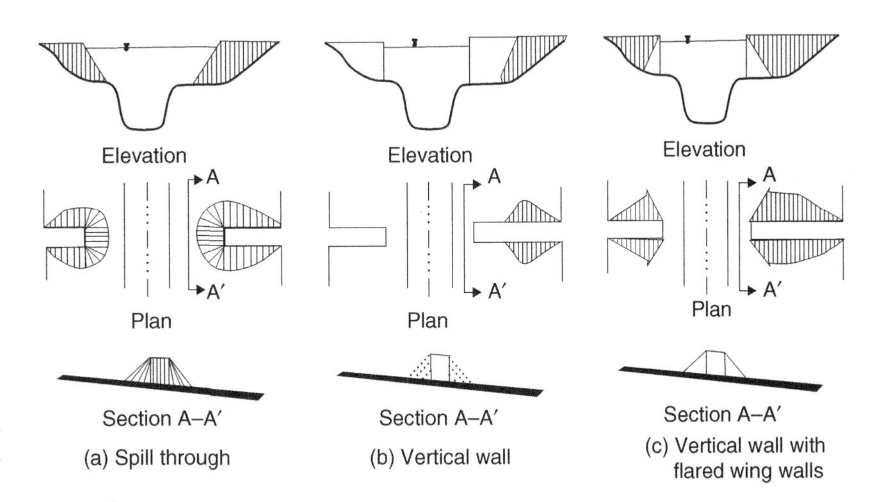

FIGURE 7.18 Abutment shapes (after Richardson and Davis, 2001)

TABLE 7.12 Abutment shape coefficients (after Richardson and Davis, 2001)

Description	K_{a1}
Vertical-wall abutment	1.00
Vertical-wall abutment with wing walls	0.82
Spill-through abutment	0.55

7.2.3.1 The HIRE equation

The HIRE equation (Richardson and Davis, 2001) is

$$y_s = 4y_3\left(\frac{K_{a1}}{0.55}\right)K_{a2}F_{r3}^{0.33} \tag{7.55}$$

where y_s = scour depth, y_3 = main channel or overbank flow depth (depending on whether the abutment is in the main channel or on the overbank) at the toe of the abutment taken at the cross-section just upstream of the bridge (Section 3 in Figure 7.1), K_{a1} = abutment shape coefficient, K_{a2} = correction factor for angle of attack, and F_{r3} = Froude number based on velocity and depth adjacent and just upstream of the abutment.

Figure 7.18 displays various shapes of abutments, and Table 7.12 presents the corresponding K_{a1} values. The correction factor, K_{a2}, is evaluated as $K_{a2} = (90/\theta)^{0.13}$, where θ = angle of attack in degrees as shown in Figure 7.19. Note that $\theta < 90°$ if the embankment points downstream, and $\theta > 90°$ if the embankment points upstream.

EXAMPLE 7.11 Figure 7.3 displays the dimensions of channel sections 4, 3, and BU considered in the Examples 7.1 and 7.7. Table 7.8 summarizes the results of Example 7.1 and the channel section dimensions. Determine the local scour depth for each abutment, using the HIRE equation.

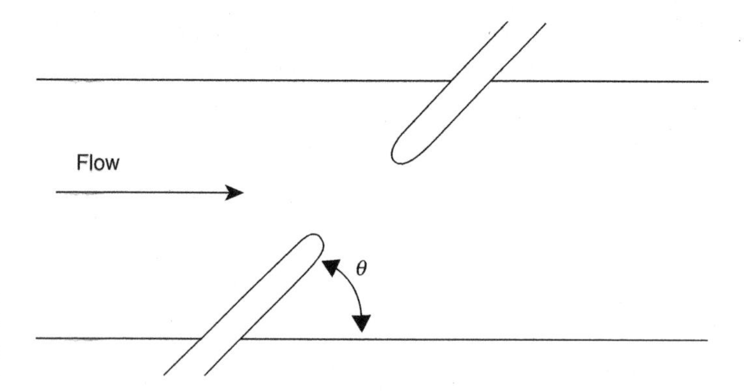

FIGURE 7.19 Embankment angle θ (after Richardson and Davis, 2001)

As we can see from Figure 7.3, both abutments are in the overbank area. Therefore, we will use the overbank flow variables in calculating the scour depth. As obtained from Table 7.8 for the left overbank at Section 3, $y_3 = 0.56$ m and $V_3 = 0.37$ m/s. Because the hydraulic depth is equal to the flow depth in a rectangular channel, we can calculate the Froude number as $F_{r3} = V_3/(gy_3)^{1/2} = (0.37)/\{(9.81)(0.56)\}^{1/2} = 0.16$. Also, $K_{a1} = 1.0$ since the abutments have vertical walls and $K_{a2} = 1.0$ since the flow is in a direction perpendicular to the embankments ($\theta = 90°$). Therefore, for the left abutment, we can evaluate the scour depth by using Equation 7.55 as

$$y_s = 4y_3\left(\frac{K_{a1}}{0.55}\right)K_{a2}F_{r3}^{0.33} = 4(0.56)\left(\frac{1.0}{0.55}\right)(1.0)(0.16)^{0.33} = 2.22\,\text{m}.$$

The scour depth is the same for the right abutment, since the flow depth and the velocity are the same for the left and right overbank areas.

7.2.3.2 The Froechlich equation

This equation is based on a regression analysis of a large number of live-bed scour measurements in a laboratory flume. The Froechlich equation (Richardson and Davis, 2001) for abutment scour is written as

$$y_s = 2.27K_{a1}K_{a2}L^{0.43}y_4^{0.57}F_{r4}^{0.61} + y_4 \tag{7.56}$$

where y_s = scour depth, K_{a1} = abutment shape coefficient (see Table 7.12), $K_{a2} = (\theta/90)^{0.13}$ = correction factor for angle of attack, θ = angle of attack (see Figure 7.19 and note $\theta < 90°$ if embankment points downstream, and $\theta > 90°$ if embankment points upstream), y_4 = average flow depth at the approach section (Section 4 in Figure 7.1), F_r = Froude number at the approach section, and L = length of embankment projected normal to the flow. The terms y_4 and F_r are evaluated for the main channel or an overbank area, depending on whether the abutment is in the main channel or the overbank area.

Equation 7.56 is meant for design purposes. The term ($+y_4$) was added to the equation in order to envelop 98% of the data. If the Froechlich equation is to be used for analysis or scour depth prediction purposes, the term $(+y_4)$ should be left out.

Richardson and Davis (2001) suggest that scour depth is overestimated by Equation 7.56 if L is set equal to the total length of the embankment (and abutment) projected normal to the flow. It is suggested that L be replaced by L' in the equation where L' is defined as the length of the embankment (and abutment) blocking *live* flow. In cases where the flow is distributed non-uniformly over the floodplain with most of the overbank flow occurring near the main channel, the difference between L' and L is significant. Richardson and Davis (2001) describe various procedures to estimate L', which are beyond the scope of this text.

EXAMPLE 7.12 Using the Froechlich equation, determine the abutment scour depths for the bridge considered in Example 7.11.

An inspection of Figure 7.3c reveals that the length of embankment blocking the flow is $L = 167$ m for the left embankment and $L = 177$ m for the right embankment. Also, as we can see from Table 7.8, $y_4 = 0.73$ m and $V_4 = 0.28$ m/s for each of the left and right overbank areas of channel Section 4. Noting that the hydraulic depth of a rectangular channel is the same as the flow depth, we can calculate the Froude number as $F_{r4} = V_4/(gy_4)^{1/2} = (0.28)/\{(9.81)(0.73)\}^{1/2} = 0.105$. Also, $K_{a1} = 1.0$ since the abutments have vertical walls and $K_{a2} = 1.0$ since the flow is in a direction perpendicular to the embankments ($\theta = 90°$). Therefore, for the left abutment we can evaluate the scour depth by using Equation 7.66, with the $(+y_4)$ term on the right-hand side, as

$$y_s = 2.27K_{a1}K_{a2}L^{0.43}y_4^{0.57}F_{r4}^{0.61} + y_4$$

$$= 2.27(1.0)(1.0)(167)^{0.43}(0.73)^{0.57}(0.105)^{0.61} + 0.73 = 5.06\,\text{m}$$

Likewise, for the right abutment

$$y_s = 2.27K_{a1}K_{a2}L^{0.43}y_4^{0.57}F_{r4}^{0.61} + y_4$$

$$= 2.27(1.0)(1.0)(177)^{0.43}(0.73)^{0.57}(0.105)^{0.61} + 0.73 = 5.18\,\text{m}.$$

PROBLEMS

P.7.1 Consider a reach of a river shown schematically in Figure 7.P.1. The distance between sections 1 and 2 and that between 3 and 4 is 100 ft. There is a bridge placed between sections 2 and 3. The bridge is $L_B = 38$ ft long. The distance between sections 2 and BD and that between BU and 3 is 6 ft.

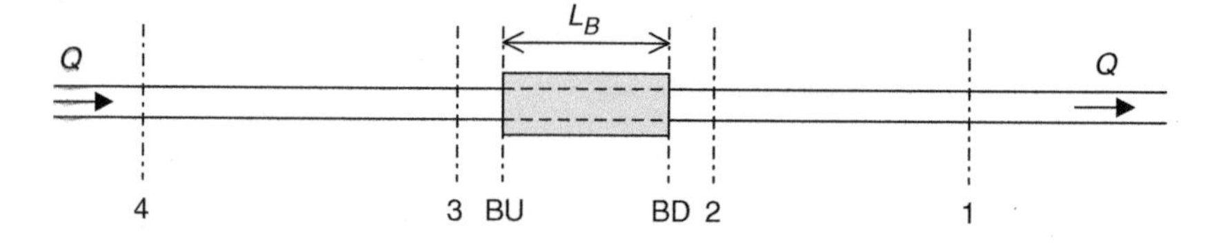

FIGURE 7.P.1 Channel section locations for problems P.7.1 and P.7.2

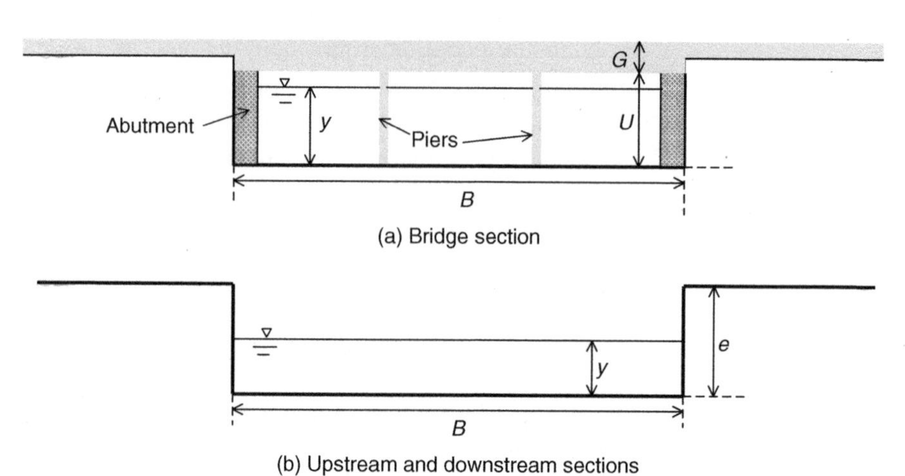

FIGURE 7.P.2 Cross-sections for problems P.7.1 and P.7.2

The cross-section of the river at sections 1, 2, 3, and 4 is approximated by a rectangle shown in Figure 7.P.2b, with $B=120$ ft and $e=15$ ft. The channel is nearly horizontal between sections 4 and 1. The Manning roughness factor is 0.02. The channel geometry at the bridge sections BU and BD is shown schematically in Figure 6.P.2a with $U=13$ ft and $G=3$ ft. The abutments are each 3 ft wide, and the piers are each 3 ft wide. The piers extend the entire length of the bridge, and have semicircular ends. Assume that the Manning roughness factor at bridge sections is also 0.02.

Suppose the flow depth calculated at section 2 for a discharge of 12 000 cfs is 10.29 ft. Using the energy method, calculate the flow depth at sections BD, BU, and 3. Also, verify that choking will not occur.

P.7.2 Consider a reach of a river shown schematically in Figure 7.P.1. The distance between sections 1 and 2 and that between 3 and 4 is 33 m. There is a bridge placed between sections 2 and 3. The bridge is $L_B=13$ m long. The distance between sections 2 and BD and that between sections BU and 3 is 2 m.

The cross-section of the river at sections 1, 2, 3, and 4 is approximated by a rectangle shown in Figure 7.P.2b with $B=37$ m and $e=4.5$ m. The channel is horizontal between sections 4 and 1. The channel geometry at the bridge sections BU and BD is shown schematically in Figure 7.P.2a, with $U=4$ m and $G=1$ m. The abutments are each 2 m wide and the piers are each 1 m wide. The piers extend the entire length of the bridge and have semicircular ends. Assume that the Manning roughness factor is 0.02 at all the sections.

Suppose the flow depth calculated at section 2 for a discharge of 340 m^3/s is 3.10 m. Using the energy method, calculate the flow depth at sections BD, BU, and 3. Also, verify that choking will not occur.

P.7.3 Determine the flow depth at sections BD, BU, and 3 for the situation described in Problem P.7.1, using the momentum approach. Also, verify that choking will not occur.

P.7.4 Determine the flow depth at sections BD, BU, and 3 for the situation described in Problem P.7.2, using the momentum approach. Also, verify that choking will not occur.

P.7.5 Determine the flow depth at section 3 for the situation described in Problem P.7.1, using the Yarnell method.

P.7.6 Determine the flow depth at section 3 for the situation described in Problem P.7.2, using the Yarnell method.

P.7.7 Suppose the bridge described in Problem P.7.1 functions hydraulically like a sluice gate for $Q = 12\,000$ cfs. Determine the flow depth at section 3.

P.7.8 Suppose the bridge described in Problem P.7.2 functions hydraulically like a sluice gate for $Q = 340\ m^3/s$. Determine the flow depth at section 3.

P.7.9 Determine the flow depth at section 3 for the bridge situation described in Problem P.7.1 if the flow depth at section 2 is 13.25 ft for $Q = 12\,000$ cfs.

P.7.10 Determine the flow depth at section 3 for the bridge situation described in Problem P.7.2 if the flow depth at section 2 is 4.15 m for $Q = 340\ m^3/s$.

P.7.11 Suppose that the water surface profile calculations are performed for the bridge problem described in Problem P.7.1, and the results are tabulated below. Determine the contraction scour for this situation, assuming that the median size of the bed material is 0.20 inches and the temperature is 68°F.

Section	S_f	W (ft)	V (fps)	y (ft)
4	0.000813	120	9.26	10.79
3	0.000841	120	9.36	10.68
BU	0.001819	108	11.00	10.10

P.7.12 Suppose the piers are each 3 ft wide in Problems P.7.1 and P.7.11, and have a round nose. The angle of attack is $\theta_p = 0°$. The bed material has $D_{50} = 0.2$ in and $D_{95} = 0.3$ in. Determine the pier scour depth using:

(a) the CSU equation
(b) the Froechlich equation.

P.7.13 The abutments for the bridge situation considered in Problems P.7.1 and P.7.11 are vertical wall abutments with an angle of attack of $\theta = 90°$. Assume the length of embankment on each side is equal to the width of the abutment, and determine the abutment scour by using:

(a) the HIRE equation
(b) the Froechlich equation.

REFERENCES

American Association of State Highway and Transportation Officials (1992). *Standard Specifications for Highway Bridges*, 15th edn. Federal Highway Administration, Washington, DC.

Bradley, J. N. (1978). *Hydraulics of Bridge Waterways*, Hydraulic Design Series No. 1, 2nd edn. Federal Highway Administration, Washington, DC.

Brater, E. F., King, H. W., Lindell, J. E. and Wei, C. Y. (1996). *Handbook of Hydraulics*, 7th edn. McGraw-Hill Book Co., New York, NY.

Brice, J. C. and Blodgett, J. C. (1978). *Countermeasures for Hydraulic Problems at Bridges*, Vols 1 and 2. FHWA/RD-78-162&163, Federal Highway Administration, Washington, DC.

Chang, F. F. M. (1973). *A Statistical Summary of the Cause and Cost of Bridge Failures*. Federal Highway Administration, Washington, DC.

Davis, S. R. (1984). Case histories of scour problems at bridges. In: *Transportation Research Record 950*, Second Bridge Engineering Conference, Vol. 2, pp., 149–155. Transportation Research Board, Washington, DC.

Froechlich, D. C. (1988). Analysis of onsite measurements of scour at piers. In: *Proceedings of the ASCE National Hydraulic Engineering Conference*, Colorado Springs, Colorado, pp. 534–539. American Society of Civil Engineers, New York, NY.

Jain, S. C. and Fischer, R. E. (1979). *Scour Around Bridge Piers at High Froude Numbers*. Report No. FHWA-RD-79-104, Federal Highway Administration, Washington, DC.

Jones, J. S. (1984). Comparison of prediction equations for bridge pier and abutment scour. In: *Transportation Research Record 950*, Second Bridge Engineering Conference, Vol. 2, pp. 202–209. Transportation Research Board, Washington, DC.

Landers, M. N. and Mueller, D. S. (1996). *Channel Scour at Bridges in the United States*. Publication No FHWA-RD-95-184, Federal Highway Administration, Washington, DC.

Laursen, E. M. (1960). Scour at bridge crossings. *Journal of the Hydraulics Division, ASCE*, **86(HY2),** 39–53.

Laursen, E. M. (1963). An analysis of relief bridges. *Journal of the Hydraulics Division, ASCE*, **92(HY3),** 93–118.

Laursen, E. M. (1980). *Predicting Scour at Bridge Piers and Abutments*. General Report Number 3, Arizona Department of Transportation, Phoenix, AZ.

Liu, H. K., Chang, F. M. and Skinner, M. M. (1961). *Effect of Bridge Constriction on Scour and Backwater*. Department of Civil Engineering, Colorado State University, Fort Collins, CO.

Melville, B. W. (1992). Local scour at bridge abutments. *Journal of Hydraulic Engineering, ASCE*, **118(4),** 615–631.

Melville, B.W. and Sutherland, A. J. (1988). Design method for local scour at bridge piers. *Journal of the Hydraulics Division, ASCE*, **114(HY10),** 1210–1226.

Mueller, D. S. and Jones, J. S. (1999). Evaluation of recent field and laboratory research on scour bridge piers in coarse bed materials. In: E. V. Richardson and P. F. Lagasse (eds), *ASCE Compendium, Stream Stability at Highway Bridges*, pp. 298–310. Reston, VA.

Richardson, E. V. and Davis, S. R. (2001). *Evaluating Scour at Bridges*, 4th edn. Hydraulic Engineering Circular No. 18, FHWA NHI 01-001, Federal Highway Administration, Washington, DC.

Richardson, E. V., Simons, D. B., and Lagasse, P. F. (2001). *River Engineering for Highway Encroachments-Highways in the River Environment*. Hydraulic Series No. 6, FHWA NHI 01-004, Federal Highway Administration, Washington, DC.

US Army Corps of Engineers (2002). HEC-RAS river analysis system. *Hydraulic Reference Manual*. Hydrologic Engineering Center, Davis, CA.

8 Introduction to unsteady open-channel flow

In Chapters 2 through 7 we dealt with steady open-channel flow problems in which the discharge, velocity, and flow depth remain constant at a given location. However, generally speaking, the flow in many open channels (such as drainage canals, storm sewers, and natural streams) is unsteady as the flow conditions vary with time. These variations are significant, particularly during and after a storm event. In practice, for flood studies, we sometimes use the steady-flow equations to calculate the maximum flow depths in a channel, assuming the flow is steady at peak discharge. However, this approach is conservative, since it does not account for the attenuation of flood waves due to the storage effect of the channel. Also, we cannot determine the timing of the peak flows in the steady-flow approach, while we are often interested in the timing of the flood elevations at particular locations as well as the flood elevations. Accurate prediction of how a flood wave propagates in a channel is possible only through the use of the unsteady open-channel flow equations. We usually refer to unsteady-flow calculations in open channels as *flood routing* or *channel routing* calculations.

Unsteady-flow equations are complex, and for the most part are not amenable to closed-form analytical solutions. We need to use numerical methods to solve these equations. Since the early 1960 researchers have devoted tremendous efforts to developing efficient solution methods for the unsteady-flow equations, and excellent reviews of these methods are available in the literature (see, for example, Lai, 1986). Most numerical methods can be broadly categorized into the *finite difference* and *finite element* methods. Several basic finite difference schemes are discussed in this chapter.

8.1 Governing Equations

We derived the continuity equation for one-dimensional unsteady open-channel flow in Chapter 1 (Equation 1.27), which is written here as

$$\frac{\partial A}{\partial t} + \frac{\partial Q}{\partial x} = 0 \tag{8.1}$$

where A = flow area, Q = discharge, t = time, and x = displacement in the main flow direction.

We also derived the momentum equation in Chapter 1 (Equation 1.31), rewritten here as

$$\frac{\partial Q}{\partial t}+\frac{\partial}{\partial x}(\beta QV)+gA\frac{\partial y}{\partial x}+gAS_f-gAS_0=0 \qquad (8.2)$$

where β = momentum correction factor, V = cross-sectional average velocity, g = gravitational acceleration, y = flow depth, S_0 = longitudinal channel slope, and S_f = friction slope. Assuming $\beta=1$ for a prismatic channel, and noting that $V=Q/A$, Equation 8.2 becomes

$$\frac{\partial Q}{\partial t}+\frac{\partial}{\partial x}\left(\frac{Q^2}{A}\right)+gA\frac{\partial y}{\partial x}+gAS_f-gAS_0=0 \qquad (8.3)$$

Equations 8.1 and 8.3 describe the one-dimensional unsteady flow in prismatic channels, and are attributed to Saint Venant. As discussed also by Strelkoff (1969), Yen (1973), and Chaudhry (1993), the main assumptions used in the derivation of the Saint Venant equations are that:

1. The pressure distribution is hydrostatic
2. The velocity is uniformly distributed over a channel section
3. The average channel bed slope is small, and therefore the flow depth measured in the vertical is considered equal to that measured perpendicular to the channel bottom
4. The flow is homogeneous and incompressible.

We can also write the momentum equation (Equation 8.3) in terms of the piezometric head or stage, $h=z_b+y$, where h = stage = elevation of the water surface measured from a horizontal datum and z_b = elevation of the channel bottom above the horizontal datum. Substituting $y=h-z_b$ into Equation 8.3 and noting that $S_0=-\partial z_b/\partial x$, we obtain

$$\frac{\partial Q}{\partial t}+\frac{\partial}{\partial x}\left(\frac{Q^2}{A}\right)+gA\frac{\partial h}{\partial x}+gAS_f=0 \qquad (8.4)$$

The Saint Venant equations can also be written in terms of the velocity, as opposed to the discharge. Noting that $Q=AV$, $(\partial A/\partial t)=(\partial A/\partial y)(\partial y/\partial t)=T(\partial y/\partial t)$, and $(\partial A/\partial x)=(\partial A/\partial y)(\partial y/\partial x)=T(\partial y/\partial x)$, where T = top width, Equation 8.1 becomes

$$T\frac{\partial y}{\partial t}+VT\frac{\partial y}{\partial x}+A\frac{\partial V}{\partial x}=0 \qquad (8.5)$$

Likewise, by multiplying both sides of Equation 1.35 by g, we can write the momentum equation as

$$\frac{\partial V}{\partial t}+V\frac{\partial V}{\partial x}+g\frac{\partial y}{\partial x}+gS_f-gS_0=0 \qquad (8.6)$$

Lai (1986) summarized the other common forms of the Saint Venant equations found in the literature. Many researchers, including Boussinesq (Lai, 1986) and Yen (1973), suggested modifications to the Saint Venant equations. However, as pointed out by Lai (1986), the original equations are still considered adequate for many practical problems.

As discussed in Chapter 3, when the Manning formula is used to represent the flow resistance, the friction slope is expressed as

$$S_f = \frac{V^2 n^2}{k_n^2 R^{4/3}} = \frac{n^2 Q^2}{k_n^2 A^2 R^{4/3}} = \frac{P^{4/3} n^2 Q^2}{k_n^2 A^{10/3}} \tag{8.7a}$$

where n = Manning roughness factor, R = hydraulic radius, P = wetted perimeter, and $k_n = 1.0\,\mathrm{m}^{1/3}/\mathrm{s} = 1.49\,\mathrm{ft}^{1/3}/\mathrm{s}$. For unsteady flow we write this expression in a slightly different form, as

$$S_f = \frac{n^2 V|V|}{k_n^2 R^{4/3}} = \frac{n^2 Q|Q|}{k_n^2 A^2 R^{4/3}} = \frac{n^2 P^{4/3} Q|Q|}{k_n^2 A^{10/3}} \tag{8.7}$$

By expressing the friction slope in the form of Equation 8.7, we are able to account for reversed-flow situations in a channel. If the flow occurs in the reversed direction in any part of the channel (from downstream towards upstream), V or Q would have a negative value. Then Equation 8.7 would yield a negative value for the friction slope, in concurrence with the basic principle that the flow always takes place in the direction of decreasing energy head.

The Saint Venant equations are classified as partial differential equations of the hyperbolic type. All the flow variables are functions of both time and distance along the channel. In other words, at a given location, the flow depth, discharge, and the other flow variables vary with time. Likewise, at a fixed time, the flow variables change along the channel. For a given channel of known properties (cross-sectional geometry, roughness factor, longitudinal slope, etc.), the unknowns in Equations 8.1 and 8.3 are the discharge, Q, and the flow depth, y. The other flow variables, such as the area, A, and the friction slope, S_f, can be expressed in terms of Q and y. (The unknowns are V and y if Equations 8.5 and 8.6 are used.) The independent variables are time, t, and distance along the channel, x. An initial condition and two boundary conditions are needed to solve the Saint Venant equations. The initial condition is described by the variation of the unknowns, Q and y (or V and y), along the channel at time zero. For example, the flow in a channel can initially be steady. In this event, a constant discharge everywhere along the channel and the water surface profile corresponding to this discharge describe the initial condition. If the flow is subcritical, one boundary condition at each of the upstream and downstream ends of the channel is required. If the flow is supercritical, both boundary conditions should be given at the upstream end. A boundary condition can be in the form of a specified relationship between the unknowns at one end of the channel. For example, if a channel conveying subcritical flow terminates at a free fall, we can use the critical flow relationship ($F_r = 1.0$) as the downstream boundary condition. Also, one of the unknowns can be prescribed as function of time at a boundary. For example a discharge hydrograph (a relationship between Q and t) could be given as an upstream boundary condition.

8.2 Numerical Solution Methods

There are no closed-form analytical solutions available for the Saint Venant equations, and therefore numerical methods are used to solve them. To apply a numerical method, we discretize a channel into a number of flow reaches, as shown in the upper part of Figure 8.1. We also discretize the time variable, and seek solutions only at discrete time intervals. This allows the use of a *computation grid*, as shown in Figure 8.1, to formulate the finite difference equations. The vertical lines of the computational grid represent different locations along a channel, and the horizontal lines correspond to the discrete times at which we seek a numerical solution. Both the *space increment*, Δx, and the *time increment*, Δt, can vary. However, here they are taken as constant increments for simplicity. The horizontal and the vertical lines intersect at the *nodes* of the computational grid, and the numerical solutions are sought at these nodes. The horizontal line marked 0 represents the initial time, and the flow conditions are known at all the nodes on this line from the initial conditions. The vertical line labeled as 1 represents the upstream end of the channel, and that labeled N represents the downstream end. The boundary conditions apply to

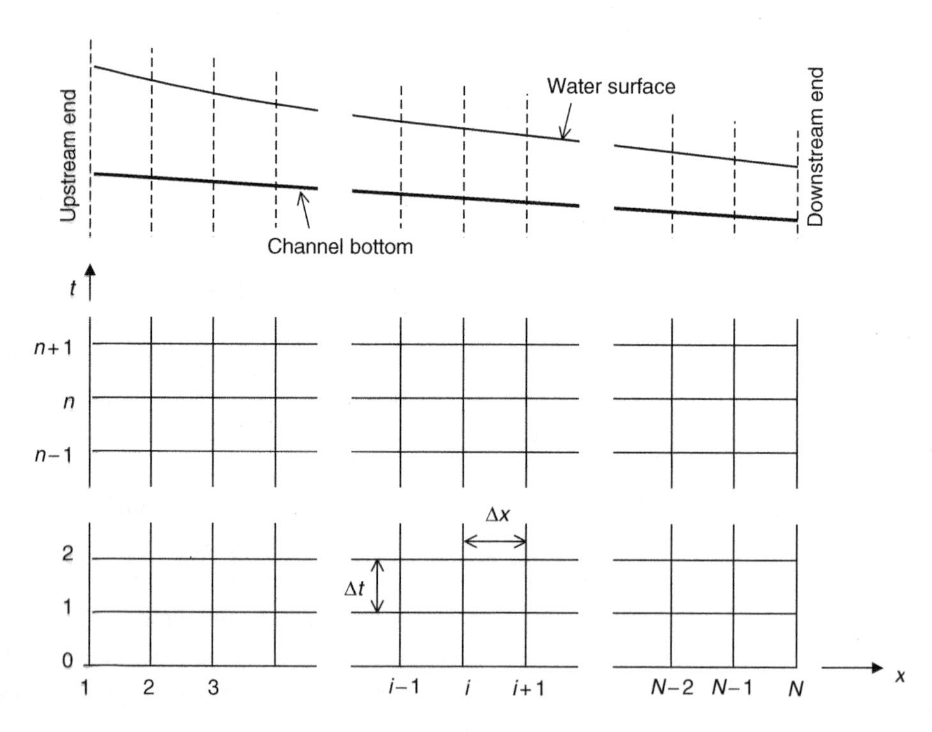

Figure 8.1
Computational grid

the nodes on these lines. Knowing the conditions at time stage 0, that is at $t=0$, the first step of computations determines the flow conditions at all the nodes on the horizontal line labeled 1 (time stage 1), that is at $t=(1)(\Delta t)$. After the conditions at time stage 1 have been determined, the calculations are carried out for time stage 2. The same approach is used for the subsequent time stages. In other words, when we are to calculate the flow conditions at time stage $n+1$, the conditions at n are known either from the given initial conditions or from the previous time step computations. For example, if Equations 8.1 and 8.3 are adopted as the governing equations, for the computations between stages n and $n+1$, the variables Q_i^n and y_i^n are known for $i=1, 2, 3, \ldots, N$. The unknowns are Q_i^{n+1} and y_i^{n+1} for $i=1, 2, 3, \ldots, N$.

The finite difference equations are obtained by using the Taylor series approximations to the partial differential terms of the Saint Venant equations. However, because we truncate the infinite series while replacing the partial differential terms with finite difference quotients, this approximation will cause truncation errors. The truncation errors are related to the size of the space and time increments and the nature of the approximations used to convert the continuous equations to discrete equations. Accumulation of truncation errors can potentially make the results of a numerical method invalid.

A numerical method should be *consistent*, *convergent*, and *stable* to produce acceptable results. As discussed by Franz and Melching (1997), a numerical method is *consistent* if the continuous governing equations are obtained from the finite difference equations as both the space increment, Δx, and the time increment, Δt, approach zero. A numerical method is *convergent* if, as Δx and Δt are reduced, the results approach limiting values that match the true solution of the governing differential equations. Consistency relates to the equations, and convergence relates to the solutions of these equations. A numerical method is computationally *stable* if round-off and truncation errors do not accumulate to cause the solution to diverge. If a numerical method is unstable, small changes in the model input (say, the value of the Manning roughness or the time increment) may lead to very large changes in the results.

8.2.1 EXPLICIT FINITE DIFFERENCE SCHEMES

As discussed in the preceding section, at the time we seek solutions at time stage $n+1$, the conditions at stage n are known either from the previous time step computations or from the initial conditions. If Equations 8.1 and 8.3 are adopted as the governing equations, for the computations between stages n and $n+1$, the variables Q_i^n and y_i^n are known for $i=1, 2, 3, \ldots, N$. The unknowns are Q_i^{n+1} and y_i^{n+1} for $i=1, 2, 3, \ldots, N$. Obviously, A and S_f can be evaluated in terms of Q and y.

In the explicit finite difference schemes, explicit algebraic equations are derived to determine the flow conditions at each node of the computational grid.

However, we can obtain these explicit equations only if we evaluate the spatial derivatives at the previous time stage, n, as well as the friction slopes and the variables with which the partial differential terms are multiplied.

For example, we can use the following approximations for the terms appearing in Equations 8.1 and 8.3:

$$\frac{\partial A}{\partial t} \approx \frac{A_i^{n+1} - ((A_{i+1}^n + A_{i-1}^n)/2)}{\Delta t} \tag{8.8}$$

$$\frac{\partial Q}{\partial t} \approx \frac{Q_i^{n+1} - ((Q_{i+1}^n + Q_{i-1}^n)/2)}{\Delta t} \tag{8.9}$$

$$\frac{\partial Q}{\partial x} \approx \frac{Q_{i+1}^n - Q_{i-1}^n}{2\Delta x} \tag{8.10}$$

$$\frac{\partial (Q^2/A)}{\partial x} \approx \frac{[(Q_{i+1}^n)^2/A_{i+1}^n] - [(Q_{i-1}^n)^2/A_{i-1}^n]}{2\Delta x} \tag{8.11}$$

$$\frac{\partial y}{\partial x} \approx \frac{y_{i+1}^n - y_{i-1}^n}{2\Delta x} \tag{8.12}$$

$$A \approx \frac{(A_{i-1}^n + A_{i+1}^n)}{2} \tag{8.13}$$

$$AS_f = \frac{A_{i-1}^n (S_f)_{i-1}^n + A_{i+1}^n (S_f)_{i+1}^n}{2} \tag{8.14}$$

Substituting Equations 8.8 to 8.14 into Equations 8.1 and 8.3 and simplifying, we obtain

$$A_i^{n+1} = \frac{(A_{i+1}^n + A_{i-1}^n)}{2} - \frac{\Delta t (Q_{i+1}^n - Q_{i-1}^n)}{2\Delta x} \tag{8.15}$$

and

$$\begin{aligned} Q_i^{n+1} = {} & \frac{(Q_{i+1}^n + Q_{i-1}^n)}{2} - \Delta t \frac{[(Q_{i+1}^n)^2/A_{i+1}^n] - [(Q_{i-1}^n)^2/A_{i-1}^n]}{2\Delta x} \\ & - g\Delta t \frac{(A_{i-1}^n + A_{i+1}^n)}{2} \left(\frac{y_{i+1}^n - y_{i-1}^n}{2\Delta x} - S_0 \right) \\ & - g\Delta t \frac{A_{i-1}^n (S_f)_{i-1}^n + A_{i+1}^n (S_f)_{i+1}^n}{2} \end{aligned} \tag{8.16}$$

Obviously, once A_i^{n+1} has been evaluated, y_i^{n+1} can be found from the cross-sectional geometry. This scheme is called the Lax diffusive scheme, and it is stable provided that the Courant condition,

$$\Delta t \leq \frac{\Delta x}{|V \mp c|} \tag{8.17}$$

is satisfied for all the nodes at all the time stages where the celerity c is evaluated in terms of the hydraulic depth, D, as

$$c = \sqrt{gD} \tag{8.18}$$

Liggett and Cunge (1975) showed that the Lax diffusive scheme is not consistent. However, it produces reasonably accurate results if $(\Delta x)^2/\Delta t$ is small enough.

Several other explicit schemes are available in the literature (for example, Chaudhry, 1993; Sturm, 2001). However, the Courant condition must be satisfied for these schemes to be stable. The Courant condition restricts the time increment significantly. For example, suppose a 5-ft wide rectangular drainage channel carrying 60 cfs at a depth of 3 ft has a velocity of 4.0 fps and celerity of 9.8 fps. If we use a space increment of $\Delta x = 100$ ft, the time increment Δt would have to be smaller than $100/(4.0+9.8) = 7.2$ seconds. Also, difficulties may arise in formulating the boundary conditions in the explicit schemes.

8.2.2 IMPLICIT FINITE DIFFERENCE SCHEMES

In the implicit finite difference schemes, the values at time stage $n+1$ as well as stage n are used to approximate the spatial and time derivatives and the dependent variables of the Saint Venant equations. This formulation will not lead to explicit expressions to evaluate the variables at $n+1$; instead, a set of algebraic, non-linear equations will be obtained. These equations will be solved simultaneously to obtain the results at stage $n+1$ all at once.

Suppose the total length of a channel is divided into $N-1$ reaches, as shown in Figure 8.1. Then, for any time step of computations, we will have a total of $2N$ unknowns: Q_i^{n+1} and y_i^{n+1} for $i = 1, 2, 3, \ldots, N$. We will write the continuity and momentum equations in finite difference form for each reach, resulting in $2N-2$ equations. The remaining two equations needed to determine the $2N$ unknowns will come from the boundary conditions. For subcritical flow, one boundary equation for each of the upstream and downstream boundaries is used. For supercritical flow, both boundary equations are for the upstream end.

Various implicit finite difference schemes have been reported in the literature (Chaudhry, 1993; Sturm, 2001). A four-point implicit scheme, sometimes called the Preissman method, is described herein. The method can be applied to any form of the Saint Venant equations, such as Equations 8.1 and 8.3, Equations 8.1 and 8.4, or Equations 8.5 and 8.6. The formulation presented herein is for Equations 8.1 and 8.4. However, similar expressions can be obtained for other forms of the Saint Venant equations.

8.2.2.1 Reach equations

Let us recall that by a reach, we mean a segment of the channel length between two nodes. In the Preissman method, the following approximations are used for the channel reach between nodes i and $i+1$:

$$\frac{\partial A}{\partial t} \approx \frac{(A_{i+1}^{n+1} + A_i^{n+1}) - (A_{i+1}^n + A_i^n)}{2\Delta t} \tag{8.19}$$

$$\frac{\partial Q}{\partial t} \approx \frac{(Q_{i+1}^{n+1} + Q_i^{n+1}) - (Q_{i+1}^n + Q_i^n)}{2\Delta t} \tag{8.20}$$

$$\frac{\partial Q}{\partial x} \approx \frac{\theta(Q_{i+1}^{n+1} - Q_i^{n+1}) + (1-\theta)(Q_{i+1}^n - Q_i^n)}{\Delta x} \tag{8.21}$$

$$\frac{\partial(Q^2/A)}{\partial x} \approx \frac{\theta\{[(Q_{i+1}^{n+1})^2/A_{i+1}^{n+1}] - [(Q_i^{n+1})^2/A_i^{n+1}]\}}{\Delta x} + \frac{(1-\theta)\{[(Q_{i+1}^n)^2/A_{i+1}^n] - [(Q_i^n)^2/A_i^n]\}}{\Delta x} \tag{8.22}$$

$$A\frac{\partial h}{\partial x} \approx \theta\frac{(A_{i+1}^{n+1} + A_i^{n+1})}{2}\frac{(h_{i+1}^{n+1} - h_i^{n+1})}{\Delta x} + (1-\theta)\frac{(A_{i+1}^n + A_i^n)}{2}\frac{(h_{i+1}^n - h_i^n)}{\Delta x} \tag{8.23}$$

$$A \approx \theta\frac{(A_{i+1}^{n+1} + A_i^{n+1})}{2} + (1-\theta)\frac{(A_{i+1}^n + A_i^n)}{2} \tag{8.24}$$

$$AS_f \approx \theta\frac{(A_{i+1}^{n+1} + A_i^{n+1})}{2}\frac{(S_f)_{i+1}^{n+1} + (S_f)_i^{n+1}}{2} + (1-\theta)\frac{(A_{i+1}^n + A_i^n)}{2}\frac{(S_f)_{i+1}^n + (S_f)_i^n}{2} \tag{8.25}$$

where θ = weighting factor between 0 and 1. A weighting factor of $\theta = 1.0$ yields a *fully implicit scheme* (Baltzler and Lai, 1968), and $\theta = 1/2$ produces the *box scheme* (Amein, 1968). This four-point implicit method is unconditionally stable for $0.5 \leq \theta \leq 1$, and the accuracy increases if θ close to 0.5 is chosen (Lai, 1986). Fread (1974) recommended $\theta = 0.55$ for flood waves. Because four nodes (i, n), $(i+1, n)$, $(i, n+1)$, and $(i+1, n+1)$ are used in expressing the spatial and temporal derivatives and the variables in Equation 8.17 to 8.23, this scheme is generally called a four-point implicit scheme.

Substituting Equations 8.19 and 8.21 into Equation 8.1, we obtain the continuity equation in finite difference form as

$$\frac{(A_{i+1}^{n+1} + A_i^{n+1}) - (A_{i+1}^n + A_i^n)}{2\Delta t} + \frac{\theta(Q_{i+1}^{n+1} - Q_i^{n+1}) + (1-\theta)(Q_{i+1}^n - Q_i^n)}{\Delta x} = 0 \tag{8.26}$$

Likewise, substituting Equations 8.20 and 8.22 to 8.25 into Equation 8.4, we obtain the momentum equation in finite difference form as

$$\begin{aligned}
&\frac{(Q_{i+1}^{n+1} + Q_i^{n+1}) - (Q_{i+1}^n + Q_i^n)}{2\Delta t} + \theta\frac{\{[(Q_{i+1}^{n+1})^2/A_{i+1}^{n+1}] - [(Q_i^{n+1})^2/A_i^{n+1}]\}}{\Delta x} \\
&+ (1-\theta)\frac{\{[(Q_{i+1}^n)^2/A_{i+1}^n] - [(Q_i^n)^2/A_i^n]\}}{\Delta x} + g\theta\frac{(A_{i+1}^{n+1} + A_i^{n+1})}{2}\frac{(h_{i+1}^{n+1} - h_i^{n+1})}{\Delta x} \\
&+ g(1-\theta)\frac{(A_{i+1}^n + A_i^n)}{2}\frac{(h_{i+1}^n - h_i^n)}{\Delta x} + g\theta\frac{(A_{i+1}^{n+1} + A_i^{n+1})}{2}\frac{(S_f)_{i+1}^{n+1} + (S_f)_i^{n+1}}{2} \\
&+ g(1-\theta)\frac{(A_{i+1}^n + A_i^n)}{2}\frac{(S_f)_{i+1}^n + (S_f)_i^n}{2} = 0
\end{aligned} \tag{8.27}$$

8.2.2.2 Boundary equations

A variety of boundary conditions are possible. For example, if an upstream hydrograph is given, then the upstream boundary equation is written as

$$Q_1^{n+1} - Q_{up}^{n+1} = 0 \tag{8.28}$$

where Q_{up}^{n+1} is the given upstream inflow rate at time stage $n+1$. Likewise, if a channel conveying subcritical flow terminates at a free overfall, we can assume that the flow will be critical (or $F_r = 1.0$) at the brink. Then the boundary equation becomes

$$\frac{Q_N^{n+1}(T_N^{n+1})^{1/2}}{\sqrt{g}(A_N^{n+1})^{3/2}} - 1 = 0 \tag{8.29}$$

where T = top width. For long channels, we sometimes assume that the flow becomes normal at the downstream end. In this case, the downstream boundary equation is written as

$$(S_f)_N^{n+1} - S_0 = 0 \tag{8.30}$$

where S_0 = longitudinal bottom slope of the channel. Other boundary conditions are also possible. If, for example, a weir is placed at the downstream end, the relationship between Q_N and A_N (or y_N) should satisfy the weir equation at any time stage of $n+1$.

8.2.2.3 Solution procedure

The unknown quantities in implicit finite difference equations formulated in the preceding sections are Q_i^{n+1} and h_i^{n+1} for $i = 1, 2, \ldots, N$. The area, A_i^{n+1}, and the friction slope, $(S_f)_i^{n+1}$, can be expressed in terms of Q_i^{n+1} and h_i^{n+1}. All the other terms are known either from the initial conditions or from the previous time step computations. Hence, Equations 8.26 and 8.27 can be expressed symbolically as

$$C_i\left[Q_i^{n+1}, h_i^{n+1}, Q_{i+1}^{n+1}, h_{i+1}^{n+1}\right] = 0 \tag{8.31}$$

and

$$M_i\left[Q_i^{n+1}, h_i^{n+1}, Q_{i+1}^{n+1}, h_{i+1}^{n+1}\right] = 0 \tag{8.32}$$

where C_i and M_i respectively denote the finite difference form of the equations of continuity and momentum for the flow reach between nodes i and $i+1$. Then, denoting the upstream boundary equation by B_1, the downstream boundary equation by B_N, and writing Equations 8.31 and 8.32 for all the reaches – that is, for $i = 1$ *to* $(N - 1)$, we obtain a total of $2N$ non-linear algebraic equations in $2N$ unknowns as

$$\begin{aligned}
B_1[Q_1^{n+1}, h_1^{n+1}] &= 0 \\
C_1[Q_1^{n+1}, h_1^{n+1}, Q_2^{n+1}, h_2^{n+1}] &= 0 \\
M_1[Q_1^{n+1}, h_1^{n+1}, Q_2^{n+1}, h_2^{n+1}] &= 0 \\
C_2[Q_2^{n+1}, h_2^{n+1}, Q_3^{n+1}, h_3^{n+1}] &= 0 \\
M_2[Q_2^{n+1}, h_2^{n+1}, Q_3^{n+1}, h_3^{n+1}] &= 0 \\
&\cdots\cdots \\
&\cdots\cdots \\
C_i[Q_i^{n+1}, h_i^{n+1}, Q_{i+1}^{n+1}, h_{i+1}^{n+1}] &= 0 \\
M_i[Q_i^{n+1}, h_i^{n+1}, Q_{i+1}^{n+1}, h_{i+1}^{n+1}] &= 0 \\
&\cdots\cdots \\
&\cdots\cdots \\
C_{N-1}[Q_{N-1}^{n+1}, h_{N-1}^{n+1}, Q_N^{n+1}, h_N^{n+1}] &= 0 \\
M_{N-1}[Q_{N-1}^{n+1}, h_{N-1}^{n+1}, Q_N^{n+1}, h_N^{n+1}] &= 0 \\
B_N[Q_N^{n+1}, h_N^{n+1}] &= 0
\end{aligned} \tag{8.33}$$

We can use a generalized Newton iteration method to solve this set of $2N$ simultaneous equations for the $2N$ unknowns Q_i^{n+1} and h_i^{n+1} for $i = 1, 2, \ldots, N$. Computation for the iterative procedure begins by assigning a set of trial values to the unknowns Q_i^{n+1} and h_i^{n+1} for $i = 1, 2, \ldots, N$. Substitution of these trial values in the left-hand side of Equations 8.33 will yield the residuals rB_1, rC_1, rM_1, rC_2, $rM_2, \ldots, rC_i, rM_i, \ldots, rC_{N-1}$, rM_{N-1}, and rB_N. These residuals are likely to be different from zero, since the trial values assigned to the unknowns are probably not the actual solutions. New values for Q_i^{n+1} and h_i^{n+1} for $i = 1, 2, \ldots, N$ for the next iteration are estimated to make the residuals approach zero. We accomplish this by calculating the corrections ΔQ_i and Δh_i to Q_i^{n+1} and h_i^{n+1} for $i = 1, 2, \ldots, N$ such that the total differentials of the functions $B_1, C_1, M_1, C_2, M_2, \ldots, C_i, M_i, \ldots, C_{N-1}, M_{N-1}$, and B_N are equal to the negative of the calculated residuals. In other words,

$$\begin{aligned}
\frac{\partial B_1}{\partial Q_1^{n+1}}\Delta Q_1 + \frac{\partial B_1}{\partial h_1^{n+1}}\Delta h_1 &= -rB_1 \\
\frac{\partial C_1}{\partial Q_1^{n+1}}\Delta Q_1 + \frac{\partial C_1}{\partial h_1^{n+1}}\Delta h_1 + \frac{\partial C_1}{\partial Q_2^{n+1}}\Delta Q_2 + \frac{\partial C_1}{\partial h_2^{n+1}}\Delta h_2 &= -rC_1 \\
\frac{\partial M_1}{\partial Q_1^{n+1}}\Delta Q_1 + \frac{\partial M_1}{\partial h_1^{n+1}}\Delta h_1 + \frac{\partial M_1}{\partial Q_2^{n+1}}\Delta Q_2 + \frac{\partial M_1}{\partial h_2^{n+1}}\Delta h_2 &= -rM_1 \\
\frac{\partial C_2}{\partial Q_2^{n+1}}\Delta Q_2 + \frac{\partial C_2}{\partial h_2^{n+1}}\Delta h_2 + \frac{\partial C_2}{\partial Q_3^{n+1}}\Delta Q_3 + \frac{\partial C_2}{\partial h_3^{n+1}}\Delta h_3 &= -rC_2 \\
\frac{\partial M_2}{\partial Q_2^{n+1}}\Delta Q_2 + \frac{\partial M_2}{\partial h_2^{n+1}}\Delta h_2 + \frac{\partial M_2}{\partial Q_3^{n+1}}\Delta Q_3 + \frac{\partial M_2}{\partial h_3^{n+1}}\Delta h_3 &= -rM_2 \\
&\cdots\cdots \\
&\cdots\cdots
\end{aligned}$$

$$\frac{\partial C_i}{\partial Q_i^{n+1}}\Delta Q_i + \frac{\partial C_i}{\partial h_i^{n+1}}\Delta h_i + \frac{\partial C_i}{\partial Q_{i+1}^{n+1}}\Delta Q_{i+1} + \frac{\partial C_i}{\partial h_{i+1}^{n+1}}\Delta h_{i+1} = -rC_i$$

$$\frac{\partial M_i}{\partial Q_i^{n+1}}\Delta Q_i + \frac{\partial M_i}{\partial h_i^{n+1}}\Delta h_i + \frac{\partial M_i}{\partial Q_{i+1}^{n+1}}\Delta Q_{i+1} + \frac{\partial M_i}{\partial h_{i+1}^{n+1}}\Delta h_{i+1} = -rM_i$$

$$\cdots\cdots\cdots\cdots\cdots\cdots$$

$$\cdots\cdots\cdots\cdots\cdots\cdots \tag{8.34}$$

$$\frac{\partial C_{N-1}}{\partial Q_{N-1}^{n+1}}\Delta Q_{N-1} + \frac{\partial C_{N-1}}{\partial h_{N-1}^{n+1}}\Delta h_{N-1} + \frac{\partial C_{N-1}}{\partial Q_N^{n+1}}\Delta Q_N + \frac{\partial C_{N-1}}{\partial h_N^{n+1}}\Delta h_N = -rC_{N-1}$$

$$\frac{\partial M_{N-1}}{\partial Q_{N-1}^{n+1}}\Delta Q_{N-1} + \frac{\partial M_{N-1}}{\partial h_{N-1}^{n+1}}\Delta h_{N-1} + \frac{\partial M_{N-1}}{\partial Q_N^{n+1}}\Delta Q_N + \frac{\partial M_{N-1}}{\partial h_N^{n+1}}\Delta h_N = -rM_{N-1}$$

$$\frac{\partial B_N}{\partial Q_N^{n+1}}\Delta Q_N + \frac{\partial B_N}{\partial h_N^{n+1}}\Delta h_N = -rB_N$$

These equations form a set of $2N$ linear algebraic equations in $2N$ unknowns ΔQ_i and Δh_i for $i = 1, 2, \ldots, N$. Leaving the superscripts $n+1$ out for brevity, this linear system of equations can be written in matrix form as

$$\begin{bmatrix}
\frac{\partial B_1}{\partial Q_1} & \frac{\partial B_1}{\partial h_1} & 0 & 0 & 0 & 0 & \cdots & 0 & 0 & 0 & 0 \\
\frac{\partial C_1}{\partial Q_1} & \frac{\partial C_1}{\partial h_1} & \frac{\partial C_1}{\partial Q_2} & \frac{\partial C_1}{\partial h_2} & 0 & 0 & \cdots & 0 & 0 & 0 & 0 \\
\frac{\partial M_1}{\partial Q_1} & \frac{\partial M_1}{\partial h_1} & \frac{\partial M_1}{\partial Q_2} & \frac{\partial M_1}{\partial h_2} & 0 & 0 & \cdots & 0 & 0 & 0 & 0 \\
0 & 0 & \frac{\partial C_2}{\partial Q_2} & \frac{\partial C_2}{\partial h_2} & \frac{\partial C_2}{\partial Q_3} & \frac{\partial C_2}{\partial h_3} & \cdots & 0 & 0 & 0 & 0 \\
0 & 0 & \frac{\partial M_2}{\partial Q_2} & \frac{\partial M_2}{\partial h_2} & \frac{\partial M_2}{\partial Q_3} & \frac{\partial M_2}{\partial h_3} & \cdots & 0 & 0 & 0 & 0 \\
\cdot & \cdot & \cdot & \cdot & \cdot & \cdot & \cdots & \cdot & \cdot & \cdot & \cdot \\
\cdot & \cdot & \cdot & \cdot & \cdot & \cdot & \cdots & \cdot & \cdot & \cdot & \cdot \\
\cdot & \cdot & \cdot & \cdot & \cdot & \cdot & \cdots & \cdot & \cdot & \cdot & \cdot \\
0 & 0 & 0 & 0 & 0 & 0 & \cdots & \frac{\partial C_{N-1}}{\partial Q_{N-1}} & \frac{\partial C_{N-1}}{\partial h_{N-1}} & \frac{\partial C_{N-1}}{\partial Q_N} & \frac{\partial C_{N-1}}{\partial h_N} \\
0 & 0 & 0 & 0 & 0 & 0 & \cdots & \frac{\partial M_{N-1}}{\partial Q_{N-1}} & \frac{\partial M_{N-1}}{\partial h_{N-1}} & \frac{\partial M_{N-1}}{\partial Q_N} & \frac{\partial M_{N-1}}{\partial h_N} \\
0 & 0 & 0 & 0 & 0 & 0 & \cdots & 0 & 0 & \frac{\partial B_N}{\partial Q_N} & \frac{\partial B_N}{\partial h_N}
\end{bmatrix}
\begin{bmatrix} \Delta Q_1 \\ \Delta h_1 \\ \Delta Q_2 \\ \Delta h_2 \\ \Delta Q_3 \\ \Delta h_3 \\ \cdot \\ \cdot \\ \cdot \\ \Delta Q_N \\ \Delta h_N \end{bmatrix}
=
\begin{bmatrix} -rB_1 \\ -rC_1 \\ -rM_1 \\ -rC_2 \\ -rM_2 \\ \cdot \\ \cdot \\ \cdot \\ -rC_{N-1} \\ -rM_{N-1} \\ -rB_N \end{bmatrix} \tag{8.35}$$

The solution of this system by any matrix inversion method, like the Gaussian elimination method, provides the corrections to the trial values of Q_i^{n+1} and h_i^{n+1}

for the next iteration. In other words,

$$(Q_i^{n+1})_{k+1} = (Q_i^{n+1})_k + (\Delta Q_i)_k \tag{8.36a}$$

$$(h_i^{n+1})_{k+1} = (h_i^{n+1})_k + (\Delta h_i)_k \tag{8.36b}$$

where k and $(k+1)$ indicate consecutive iteration cycles. This procedure is repeated until the corrections are reduced to tolerable magnitudes. The number of iterations required to achieve the solution of desired accuracy depends on the closeness of the first trial values to the actual results.

8.2.2.4 Elements of the coefficient matrix

We calculate the elements of the coefficient matrix shown in Equation 8.35 by evaluating the partial derivatives of the left-hand sides of the finite difference equations (Equations 8.26 to 8.30) with respect to the unknown variables sought. For example, if the upstream boundary condition is given in the form of Equation 8.28, and omitting the superscript $n+1$ for clarity, we obtain

$$\frac{\partial B_1}{\partial Q_1} = 1.0 \tag{8.37}$$

and

$$\frac{\partial B_1}{\partial h_1} = 0 \tag{8.38}$$

Likewise, suppose the downstream boundary condition is given in the form of Equation 8.30. Recalling the expression for the friction slope, S_f, from Equation 8.7 as

$$S_f = \frac{n^2 P^{4/3} Q|Q|}{k_n^2 A^{10/3}}$$

we can write

$$\frac{\partial B_N}{\partial Q_N} = \frac{\partial S_{fN}}{\partial Q_N} = \frac{2n^2 P_N^{4/3} |Q_N|}{k_n^2 A_N^{10/3}} \tag{8.39}$$

and

$$\frac{\partial B_N}{\partial h_N} = \frac{\partial S_{fN}}{\partial h_N} = \frac{n^2 Q_N |Q_N|}{k_n^2 A_N^{20/3}} \left(\frac{4}{3} P_N^{1/3} \frac{\partial P_N}{\partial h_N} A_N^{10/3} - \frac{10}{3} A_N^{7/3} \frac{\partial A_N}{\partial h_N} P_N^{4/3} \right) \tag{8.40}$$

where $\partial P/\partial h$ and $\partial A/\partial h$ are evaluated based on the cross-sectional geometry.

Equation 8.26 is the continuity equation for the reach between nodes i and $i+1$. Omitting the superscripts $n+1$ for clarity, the partial derivatives of the left-hand side of this equation are:

$$\frac{\partial C_i}{\partial Q_i} = -\frac{\theta}{\Delta x} \tag{8.41}$$

$$\frac{\partial C_i}{\partial h_i} = \frac{1}{2\Delta t} \frac{\partial A_i}{\partial h_i} \tag{8.42}$$

$$\frac{\partial C_i}{\partial Q_{i+1}} = \frac{\theta}{\Delta x} \tag{8.43}$$

$$\frac{\partial C_i}{\partial h_{i+1}} = \frac{1}{2\Delta t}\frac{\partial A_{i+1}}{\partial h_{i+1}} \tag{8.44}$$

Equation 8.27 is the momentum equation for the reach between nodes i and $i+1$. Omitting the superscripts $n+1$ for clarity, the partial derivatives of the left-hand side of Equation 8.27 are:

$$\frac{\partial M_i}{\partial Q_i} = \frac{1}{2\Delta t} - \frac{2\theta Q_i}{A_i \Delta x} + g\theta \frac{(A_{i+1}+A_i)}{4}\frac{\partial (S_f)_i}{\partial Q_i} \tag{8.45}$$

$$\frac{\partial M_i}{\partial h_i} = \frac{\theta}{\Delta x}\frac{(Q_i)^2}{(A_i)^2}\frac{\partial A_i}{\partial h_i} - g\theta\frac{(A_{i+1}+A_i)}{2\Delta x} + g\theta\frac{(h_{i+1}-h_i)}{2\Delta x}\frac{\partial A_i}{\partial h_i}$$
$$+ g\theta\frac{(S_f)_{i+1}+(S_f)_i}{4}\frac{\partial A_i}{\partial h_i} + g\theta\frac{(A_{i+1}+A_i)}{4}\frac{\partial (S_f)_i}{\partial h_i} \tag{8.46}$$

$$\frac{\partial M_i}{\partial Q_{i+1}} = \frac{1}{2\Delta t} + \frac{2\theta Q_{i+1}}{A_{i+1}\Delta x} + g\theta\frac{(A_{i+1}+A_i)}{4}\frac{\partial (S_f)_{i+1}}{\partial Q_{i+1}} \tag{8.47}$$

$$\frac{\partial M_i}{\partial h_{i+1}} = -\frac{\theta}{\Delta x}\frac{(Q_{i+1})^2}{(A_{i+1})^2}\frac{\partial A_{i+1}}{\partial h_{i+1}} + g\theta\frac{(A_{i+1}+A_i)}{2\Delta x} + g\theta\frac{(h_{i+1}-h_i)}{2\Delta x}\frac{\partial A_{i+1}}{\partial h_{i+1}}$$
$$+ g\theta\frac{(S_f)_{i+1}+(S_f)_i}{4}\frac{\partial A_{i+1}}{\partial h_{i+1}} + g\theta\frac{(A_{i+1}+A_i)}{4}\frac{\partial (S_f)_{i+1}}{\partial h_{i+1}} \tag{8.48}$$

With reference to Equation 8.7, we can evaluate the partial derivatives of the friction slope, S_f, as

$$\frac{\partial (S_f)_i}{\partial Q_i} = \frac{2n^2 P_i^{4/3}|Q_i|}{k_n^2 A_i^{10/3}} \tag{8.49}$$

and

$$\frac{\partial (S_f)_i}{\partial h_i} = \frac{n^2 Q_i|Q_i|}{k_n^2 A_i^{20/3}}\left(\frac{4}{3}P_i^{1/3}\frac{\partial P_i}{\partial h_i}A_i^{10/3} - \frac{10}{3}A_i^{7/3}\frac{\partial A_i}{\partial h_i}P_i^{4/3}\right) \tag{8.50}$$

The terms $\partial P/\partial h$ and $\partial A/\partial h$ are evaluated based on the cross-sectional geometry.

EXAMPLE 8.1 An 800-ft long rectangular channel has a bottom width of $b = 10$ ft and a Manning roughness factor of $n = 0.025$. For the purpose of unsteady-flow calculations, the channel is divided into four reaches of equal length, $\Delta x = 200$ ft. Then, with reference to Figure 8.1, in the computational grid, there are five nodes with $N = 5$. Node 1 represents the upstream end, and node 5 represents the downstream end. Suppose a weir structure is placed at the downstream end for which the flow equation is given as

$$Q_5 - 40(h_5 - 1.5)^{3/2} = 0$$

TABLE 8.1 Data for Example 8.1

i	z (ft)	h (ft)	A (ft²)	$\partial A/\partial h$ (ft)	P (ft)	$\partial P/\partial h$	S_f	$\partial S_f/\partial Q$ (s/ft³)	$\partial S_f/\partial h$ (ft⁻¹)
1	0.4	2.27	18.7	10.0	13.74	2.0	0.0002135	0.0000214	−0.0003391
2	0.3	2.23	19.3	10.0	13.86	2.0	0.0001944	0.0000194	−0.0002984
3	0.2	2.19	19.9	10.0	13.98	2.0	0.0001776	0.0000178	−0.0002636
4	0.1	2.16	20.6	10.0	14.12	2.0	0.0001604	0.0000160	−0.0002292
5	0.0	2.13	21.3	10.0	14.26	2.0	0.0001454	0.0000145	−0.0002003

in which Q_5 is the discharge at node 5 in cfs and h_5 is the water surface elevation in ft. The elevations of the channel bottom at nodes 1 to 5 are given in column 2 of Table 8.1. Suppose, initially, the discharge is 20 cfs everywhere in the channel with the water surface elevations given in column (3) of Table 8.1. The discharge at node 1 is raised to 30 cfs over a time increment of $\Delta t = 600$ s. Estimate the discharges and water surface elevations at $t = 600$ s, and formulate the matrix equation (Equation 8.35) for the first iteration cycle using a full implicit scheme with $\theta = 1.0$.

For the first iteration cycle, let us guess that the discharges and the water surface elevations at $t = 600$ s are the same as those at $t = 0$. In other words, $Q^{n+1} = Q^n = 20$ cfs at all the nodes. Likewise $h^{n+1} = h^n$ with the values tabulated in column (3) of Table 8.1 for the five nodes. Omitting the superscript $n+1$ for clarity, let us determine various cross-sectional quantities at node 1. For a rectangular channel section having a width of $b = 10$ ft,

$$A_1 = b(h_1 - z_1) = 10.0(2.27 - 0.4) = 18.7\,\text{ft}^2$$

$$\partial A_1/\partial h_1 = b = 10.0\,\text{ft}$$

$$P_1 = b + 2(h_1 - z_1) = 10.0 + 2(2.27 - 0.4) = 13.74\,\text{ft}$$

$$\partial P_1/\partial h_1 = 2.0$$

Now, by using Equations 8.7, 8.49, and 8.50 respectively,

$$S_{f1} = \frac{n^2 P_1^{4/3} Q_1|Q_1|}{k_n^2 A_1^{10/3}} = \frac{(0.025)^2(13.74)^{4/3}(20)|20|}{(1.49)^2(18.7)^{10/3}} = 0.0002135$$

$$\frac{\partial (S_f)_1}{\partial Q_1} = \frac{2n^2 P_1^{4/3}|Q_1|}{k_n^2 A_1^{10/3}} = \frac{2(0.025)^2(13.74)^{4/3}|20|}{(1.49)^2(18.7)^{10/3}} = 0.0000214\,\text{s/ft}^3$$

and

$$\frac{\partial (S_f)_1}{\partial h_1} = \frac{n^2 Q_1|Q_1|}{k_n^2 A_1^{20/3}}\left(\frac{4}{3}P_1^{1/3}\frac{\partial P_1}{\partial h_1}A_1^{10/3} - \frac{10}{3}A_1^{7/3}\frac{\partial A_1}{\partial h_1}P_1^{4/3}\right)$$

$$= \frac{(0.025)^2(20)|20|}{(1.49)^2(18.7)^{20/3}}\left(\frac{4}{3}(13.74)^{1/3}(2.0)(18.7)^{10/3} - \frac{10}{3}(18.7)^{7/3}(10.0)(13.74)^{4/3}\right)$$

$$= -0.0003391\,\text{ft}^{-1}$$

Similar calculations are performed for all the remaining nodes, and the results are tabulated in columns (4) to (10) of Table 8.1.

We will now substitute the quantities in Table 8.1 into the finite difference equations to calculate the residuals of these equations. The upstream boundary equation is (see Equation 8.28)

$$rB_1 = Q_1 - Q_{up} = 20.0 - 30.0 = -10.0\,\text{cfs}$$

For the continuity equation between nodes 1 and 2 (see Equation 8.26), excluding all the terms multiplied by $(1-\theta)=0$ and omitting the superscripts $n+1$ for clarity, we obtain

$$rC_1 = \frac{(A_2 + A_1) - (A_2^n + A_1^n)}{2\Delta t} + \frac{\theta(Q_2 - Q_1)}{\Delta x}$$
$$= \frac{(19.3 + 18.7) - (19.3 + 18.7)}{2(600)} + \frac{1.0(20.0 - 20.0)}{200} = 0.0\,\text{ft}^2\text{s}$$

Likewise, from the momentum equation (see Equation 8.27) for the reach between nodes 1 and 2,

$$rM_1 = \frac{(Q_2 + Q_1) - (Q_2^n + Q_1^n)}{2\Delta t} + \theta\frac{\{[(Q_2)^2/A_2] - [(Q_1)^2/A_1]\}}{\Delta x}$$
$$+ g\theta\frac{(A_2 + A_1)}{2}\frac{(h_2 - h_1)}{\Delta x} + g\theta\frac{(A_2 + A_1)}{2}\frac{[(S_f)_2 + (S_f)_1]}{2}$$
$$= \frac{(20.0 + 20.0) - (20.0 + 20.0)}{2(600)} + (1.0)\frac{\{[(20.0)^2/19.3] - [(20.0)^2/18.7]\}}{200}$$
$$+ 32.2(1.0)\frac{(19.3 + 18.7)}{2}\frac{(2.23 - 2.27)}{200}$$
$$+ 32.2(1.0)\frac{(19.3 + 18.7)}{2}\frac{0.0001944 + 0.0002135}{2} = -0.00090\,\text{ft}^3/\text{s}^2$$

We can calculate the residuals of the continuity and momentum equations for the remaining reaches in the same manner as being $rC_2 = 0.0$, $rM_2 = -0.01196$, $rC_3 = 0.0$, $rM_3 = 0.00896$, $rC_4 = 0.0$, and $rM_4 = -0.00126$.

The downstream boundary equation was given in the problem statement. We can calculate the residual of this equation as

$$rB_5 = Q_5 - 40(h_5 - 1.5)^{3/2} = 20.0 - 40(2.13 - 1.5)^{3/2} = -0.00188\,\text{cfs}$$

We will now evaluate the partial derivatives of the finite difference equations. For the upstream boundary, using Equations 8.37 and 8.38,

$$\frac{\partial B_1}{\partial Q_1} = 1.0$$
$$\frac{\partial B_1}{\partial h_1} = 0.0\,\text{ft}^2/\text{s}$$

The downstream boundary equation was given in the problem statement as

$$Q_5 - 40(h_5 - 1.5)^{3/2} = 0$$

The partial derivatives of the left-hand side of this equation become

$$\frac{\partial B_5}{\partial Q_5} = 1.0$$

$$\frac{\partial B_5}{\partial h_5} = -\frac{3}{2}(40)(h_5 - 1.5)^{1/2} = -\frac{3}{2}(40)(2.13 - 1.5)^{1/2} = -47.6235\ \text{ft}^2\text{s}$$

We will evaluate the partial derivatives of the continuity equations for all the reaches by using Equations 8.41 to 8.44. For example, for the reach between nodes 1 and 2, we have

$$\frac{\partial C_1}{\partial Q_1} = -\frac{\theta}{\Delta x} = -\frac{1.0}{200} = -0.005\ \text{ft}^{-1}$$

$$\frac{\partial C_1}{\partial h_1} = \frac{1}{2\Delta t}\frac{\partial A_1}{\partial h_1} = \frac{1}{2(600)}(10.0) = 0.0083\ \text{ft}^2/\text{s}$$

$$\frac{\partial C_1}{\partial Q_2} = \frac{\theta}{\Delta x} = \frac{1.0}{200} = 0.005\ \text{ft}^{-1}$$

$$\frac{\partial C_1}{\partial h_2} = \frac{1}{2\Delta t}\frac{\partial A_2}{\partial h_2} = \frac{1}{2(600)}(10.0) = 0.0083\ \text{ft/s}$$

We will use Equations 8.45 to 8.48 to evaluate the partial derivatives of the momentum equations. For the reach between nodes 1 and 2,

$$\frac{\partial M_1}{\partial Q_1} = \frac{1}{2\Delta t} - \frac{2\theta Q_1}{A_1 \Delta x} + g\theta\frac{(A_2 + A_1)}{4}\frac{\partial (S_f)_1}{\partial Q_1}$$

$$= \frac{1}{2(600)} - \frac{2(1.0)20.0}{18.7(200)} + 32.2(1.0)\frac{(19.3 + 18.7)}{4}0.0000214 = -0.0033\ \text{s}^{-1}$$

$$\frac{\partial M_1}{\partial h_1} = \frac{\theta}{\Delta x}\frac{(Q_1)^2}{(A_1)^2}\frac{\partial A_1}{\partial h_1} - g\theta\frac{(A_2 + A_1)}{2\Delta x} + g\theta\frac{(h_2 - h_1)}{2\Delta x}\frac{\partial A_1}{\partial h_1}$$

$$+ g\theta\frac{(S_f)_2 + (S_f)_1}{4}\frac{\partial A_1}{\partial h_1} + g\theta\frac{(A_2 + A_1)}{4}\frac{\partial (S_f)_1}{\partial h_1}$$

$$= \frac{1.0}{200}\frac{(20)^2}{(18.7)^2}10.0 - (32.2)(1.0)\frac{(19.3 + 18.7)}{2(200)} + (32.2)(1.0)\frac{(2.23 - 2.27)}{2(200)}(10)$$

$$+ (32.2)(1.0)\frac{0.0001944 + 0.0002135}{4}(10)$$

$$+ (32.2)(1.0)\frac{(19.3 + 18.7)}{4}(-0.0003391)$$

$$= -3.1049\ \text{ft}^2/\text{s}$$

$$\frac{\partial M_1}{\partial Q_2} = \frac{1}{2\Delta t} + \frac{2\theta Q_2}{A_2 \Delta x} + g\theta\frac{(A_2 + A_1)}{4}\frac{\partial (S_f)_2}{\partial Q_2}$$

$$= \frac{1}{2(600)} + \frac{2(1.0)20.0}{19.3(200)} + 32.2(1.0)\frac{(19.3 + 18.7)}{4}0.0000194 = 0.0171\ \text{s}^{-1}$$

$$\frac{\partial M_1}{\partial h_2} = -\frac{\theta}{\Delta x}\frac{(Q_2)^2}{(A_2)^2}\frac{\partial A_2}{\partial h_2} + g\theta\frac{(A_2 + A_1)}{2\Delta x} + g\theta\frac{(h_2 - h_1)}{2\Delta x}\frac{\partial A_2}{\partial h_2}$$

$$+ g\theta\frac{(S_f)_2 + (S_f)_1}{4}\frac{\partial A_2}{\partial h_2} + g\theta\frac{(A_2 + A_1)}{4}\frac{\partial (S_f)_2}{\partial h_2}$$

$$
\begin{aligned}
&= -\frac{(1.0)\,(20.0)^2}{200\;(19.3)^2}10.0 + 32.2(1.0)\frac{(19.3+18.7)}{2(200)} + 32.2(1.0)\frac{(2.23-2.27)}{2(200)}10.0 \\
&\quad + 32.2(1.0)\frac{0.0001944 + 0.0002135}{4}10.0 \\
&\quad + 32.2(1.0)\frac{(19.3+18.7)}{4}(-0.0002984) \\
&= 2.9147\text{ft}^2/\text{s}^2
\end{aligned}
$$

The derivatives of the continuity and the momentum equations for the other reaches are obtained in the same manner and listed in Table 8.2.

Substituting calculated residuals and the partial derivative terms into Equation 8.35, we obtain

$$
\begin{bmatrix}
1.0 & 0.0 & 0 & 0 & 0 & 0 & 0 & 0 & 0 & 0 \\
-0.0050 & 0.0083 & 0.0050 & 0.0083 & 0 & 0 & 0 & 0 & 0 & 0 \\
-0.0033 & -3.1049 & 0.0171 & 2.9146 & 0 & 0 & 0 & 0 & 0 & 0 \\
0 & 0 & -0.0050 & 0.0083 & 0.0050 & 0.0083 & 0 & 0 & 0 & 0 \\
0 & 0 & -0.0034 & -3.1983 & 0.0165 & 3.0197 & 0 & 0 & 0 & 0 \\
0 & 0 & 0 & 0 & -0.0050 & 0.0083 & 0.0050 & 0.0083 & 0 & 0 \\
0 & 0 & 0 & 0 & -0.0034 & -3.2926 & 0.0158 & 3.1414 & 0 & 0 \\
0 & 0 & 0 & 0 & 0 & 0 & -0.0050 & 0.0083 & 0.0050 & 0.0083 \\
0 & 0 & 0 & 0 & 0 & 0 & -0.0035 & -3.4027 & 0.0152 & 3.2618 \\
0 & 0 & 0 & 0 & 0 & 0 & 0 & 0 & 1.0 & -47.6235
\end{bmatrix}
\begin{bmatrix}
\Delta Q_1 \\ \Delta h_1 \\ \Delta Q_2 \\ \Delta h_2 \\ \Delta Q_3 \\ \Delta h_3 \\ \Delta Q_4 \\ \Delta h_4 \\ \Delta Q_5 \\ \Delta h_5
\end{bmatrix}
=
\begin{bmatrix}
10.0 \\ 0.0 \\ 0.00090 \\ 0.0 \\ 0.01196 \\ 0.0 \\ -0.00896 \\ 0.00 \\ 0.00126 \\ 0.00188
\end{bmatrix}
$$

8.2.2.5 An efficient algorithm to determine corrections

The coefficient matrix shown in Equation 8.35 has a banded structure. There are no more than four non-zero elements in each row, and the non-zero elements are all on the diagonal. This property of the coefficient matrix allows the use of efficient schemes to solve Equation 8.35. One such scheme, developed by Fread (1971), is summarized here.

In this approach we define

$$d_{1,1} = \frac{\partial B_1}{\partial Q_1} \tag{8.51}$$

$$d_{1,2} = \frac{\partial B_1}{\partial h_1} \tag{8.52}$$

TABLE 8.2 Partial derivatives for Example 8.1

i	$\partial C_i/\partial Q_i$	$\partial C_i/\partial h_i$	$\partial C_i/\partial Q_{i+1}$	$\partial C_i/\partial h_{i+1}$	$\partial M_i/\partial Q_i$	$\partial M_i/\partial h_i$	$\partial M_i/\partial Q_{i+1}$	$\partial M_i/\partial h_{i+1}$
1	−0.0050	0.0083	0.0050	0.0083	−0.0033	−3.1049	0.0171	2.9146
2	−0.0050	0.0083	0.0050	0.0083	−0.0034	−3.1983	0.0165	3.0197
3	−0.0050	0.0083	0.0050	0.0083	−0.0034	−3.2926	0.0158	3.1414
4	−0.0050	0.0083	0.0050	0.0083	−0.0035	−3.4027	0.0152	3.2618

$$r_1 = -rB_1 \tag{8.53}$$

$$d_{2N,3} = \frac{\partial B_N}{\partial Q_N} \tag{8.54}$$

$$d_{2N,4} = \frac{\partial B_N}{\partial h_N} \tag{8.55}$$

$$r_{2N} = -rB_N \tag{8.56}$$

Also, for even values of j, that is for $j = 2, 4, 6, \ldots, (2N-2)$,

$$d_{j,1} = \frac{\partial C_{j/2}}{\partial Q_{j/2}} \tag{8.57}$$

$$d_{j,2} = \frac{\partial C_{j/2}}{\partial h_{j/2}} \tag{8.58}$$

$$d_{j,3} = \frac{\partial C_{j/2}}{\partial Q_{(j/2)+1}} \tag{8.59}$$

$$d_{j,4} = \frac{\partial C_{j/2}}{\partial h_{(j/2)+1}} \tag{8.60}$$

$$r_j = -rC_{j/2} \tag{8.61}$$

For odd values of j, that is for $j = 3, 5, 7, \ldots, (2N-1)$,

$$d_{j,1} = \frac{\partial M_{(j-1)/2}}{\partial Q_{(j-1)/2}} \tag{8.62}$$

$$d_{j,2} = \frac{\partial M_{(j-1)/2}}{\partial h_{(j-1)/2}} \tag{8.63}$$

$$d_{j,3} = \frac{\partial M_{(j-1)/2}}{\partial Q_{(j+1)/2}} \tag{8.64}$$

$$d_{j,4} = \frac{\partial M_{(j-1)/2}}{\partial h_{(j+1)/2}} \tag{8.65}$$

$$r_j = -rM_{(j-1)/2} \tag{8.66}$$

Note that Equations 8.51 to 8.66 are used to express Equation 8.35 in a more compact form, and are evaluated using the finite difference equations formulated in the preceding sections. To determine the corrections we will first evaluate the intermediate arrays $p_{j,1}$, $p_{j,2}$, $p_{j,3}$, $p_{j,4}$, and s_j as follows:

$$p_{1,2} = d_{1,2} \tag{8.67}$$

$$s_1 = r_1 \tag{8.68}$$

$$p_{j,1} = d_{j,1} \text{ for } j = 1, 2, 3, \ldots, 2N \tag{8.69}$$

$$p_{j,3} = d_{j,3} \text{ for } j = 2, 4, 6, \ldots, 2N \tag{8.70}$$

$$p_{j,4} = d_{j,4} \text{ for } j = 2, 4, 6, \ldots, (2N-2) \tag{8.71}$$

To calculate the remaining arrays for even values of j (i.e. $j=2, 4, 6, \ldots, 2N$), we will use the recurrent formulae

$$p_{j,2} = -\frac{p_{j,1}}{p_{j-1,m}} p_{j-1,n} + d_{j,2} \tag{8.72}$$

$$s_j = -\frac{p_{j,1}}{p_{j-1,m}} s_{j-1} + r_j \tag{8.73}$$

where $m=1$ and $n=2$ for $j=2$, and $m=3$ and $n=4$ for $j=4, 6, 8, \ldots, 2N$. For odd numbers of j, the recurrent formulae are

$$p_{j,2} = -\frac{p_{j,1}}{p_{j-2,m}} p_{j-2,n} + d_{j,2} \tag{8.74}$$

$$p_{j,3} = -\frac{p_{j,2}}{p_{j-1,2}} p_{j-1,3} + d_{j,3} \tag{8.75}$$

$$p_{j,4} = -\frac{p_{j,2}}{p_{j-1,2}} p_{j-1,4} + d_{j,4} \tag{8.76}$$

$$s_j = -\frac{p_{j,1}}{p_{j-2,m}} s_{j-2} - \frac{p_{j,2}}{p_{j-1,2}} s_{j-1} + r_j \tag{8.77}$$

where $m=1$ and $n=2$ for $j=3$, and $m=3$ and $n=4$ for $j=5, 7, \ldots, (2N-1)$. Note that, in using the recurrent formulae, the computations proceed sequentially for $j=2, 3, 4, \ldots, 2N$.

We can now calculate the components of the solution vector, starting with

$$u_{2N} = \frac{s_{2N}}{p_{2N,4}} \tag{8.78}$$

and proceeding sequentially with $j=(2N-1), (2N-2), \ldots, 4, 3, 2$. For odd values of j

$$u_j = \frac{s_j - p_{j,4} u_{j+1}}{p_{j,3}} \tag{8.79}$$

and for even values of j

$$u_j = \frac{s_j - p_{j,4} u_{j+2} - p_{j,3} u_{j+1}}{p_{j,2}} \tag{8.80}$$

For $j=1$, we have

$$u_1 = \frac{s_1 - p_{1,2} u_2}{p_{1,1}} \tag{8.81}$$

Finally, we can translate these components of the solution vector to the corrections we are seeking by using

$$\Delta Q_i = u_{2i-1} \tag{8.82}$$

and

$$\Delta h_i = u_{2i} \tag{8.83}$$

for $i = 1, 2, 3, \ldots, N$.

We should note that the intermediate parameters p and s are introduced in this formulation for clarity only. When we computer program this scheme we do not need to define and store these intermediate parameters; we can substitute every p in Equations 8.67 to 8.81 with a d, and every s with an r. Obviously, in that event Equations 8.67 to 8.71 will be redundant.

Also, $d_{1,1}$ must have a non-zero value in this formulation. As long as the upstream boundary equation is a function of Q_1, the value $d_{1,1}$ is likely to be different from zero. However, not all upstream boundary equations are functions of Q_1. For example, using a stage hydrograph with a specified stage h_{up} as the upstream boundary condition through the equation

$$h_1^{n+1} - h_{up}^{n+1} = 0 \tag{8.84}$$

will lead to $d_{1,1} = \partial B_1/\partial(Q_1)^{n+1} = 0$. We can avoid this by swapping every Q with an h in Equations 8.31 to 8.35 and Equations 8.51 to 8.83. The rearrangement of the sequence of the unknowns will result in $d_{1,1} = \partial B_1/\partial(h_1)^{n+1} = 1.0$ if Equation 8.84 is used as the upstream boundary equation.

EXAMPLE 8.2 For the channel considered in Example 8.1, determine the corrections to the guessed values of discharge, and the water surface elevation.

We will use the procedure suggested by Fread (1971). First we will calculate the elements $d_{j,1}$, $d_{j,2}$, $d_{j,3}$, $d_{j,4}$, and r_j for $j = 1, 2, \ldots, 2N$. For $j = 1$, using Equations 8.51 to 8.53,

$$d_{1,1} = \frac{\partial B_1}{\partial Q_1} = 1.0$$

$$d_{1,2} = \frac{\partial B_1}{\partial h_1} = 0.0$$

$$r_1 = -rB_1 = 10.0$$

Note that $d_{1,3} = 0.0$ and $d_{1,4} = 0.0$ in this formulation. For $j = 2N = 10$, by using Equations 8.54 to 8.56,

$$d_{10,3} = \frac{\partial B_5}{\partial Q_5} = 1.0$$

$$d_{10,4} = \frac{\partial B_5}{\partial h_5} = -47.6235$$

$$r_{10} = -rB_5 = 0.00188$$

Note that $d_{2N,1} = 0.0$ and $d_{2N,2} = 0.0$ in this formulation. For $j = 2$, 4, 6, and 8, we will use Equations 8.57 to 8.61. For example, for $j = 2$,

$$d_{2,1} = \frac{\partial C_1}{\partial Q_1} = -0.0050$$

$$d_{2,2} = \frac{\partial C_1}{\partial h_1} = 0.0083$$

$$d_{2,3} = \frac{\partial C_1}{\partial Q_2} = 0.0050$$

$$d_{2,4} = \frac{\partial C_1}{\partial h_2} = 0.0083$$

$$r_2 = -rC_1 = 0.0$$

Likewise, for odd values of j, that is for $j = 3$, 5, 7, and 9, we will use Equations 8.62 to 8.66. For example, for $j = 3$,

$$d_{3,1} = \frac{\partial M_1}{\partial Q_1} = -0.0033$$

$$d_{3,2} = \frac{\partial M_1}{\partial h_1} = -3.0149$$

$$d_{3,3} = \frac{\partial M_1}{\partial Q_2} = 0.0171$$

$$d_{3,4} = \frac{\partial M_1}{\partial h_2} = 2.9146$$

$$r_3 = -rM_1 = 0.00090$$

The remaining $d_{j,1}$, $d_{j,2}$, $d_{j,3}$, $d_{j,4}$, and r_j values are determined in the same manner, and listed in columns (2) to (6) of Table 8.3.

We will now calculate the values of $p_{j,1}$, $p_{j,2}$, $p_{j,3}$, $p_{j,4}$, and s_j using Equations 8.67 to 8.77. For example, from Equations 8.67 to 8.71 we have

$$p_{1,2} = d_{1,2} = 0.0$$

$$s_1 = r_1 = 10.0$$

$$p_{1,1} = d_{1,1} = 1.0$$

$$p_{2,3} = d_{2,3} = 0.0050$$

$$p_{2,4} = d_{2,4} = 0.0083$$

Likewise, by using Equations 8.72 and 8.73 for $j = 2$ (and therefore $m = 1$ and $n = 2$),

$$p_{2,2} = -\frac{p_{2,1}}{p_{1,1}} p_{1,2} + d_{2,2} = -\frac{-0.0050}{1.0} 0.0 + 0.0083 = 0.0083$$

$$s_2 = -\frac{p_{2,1}}{p_{1,1}} s_1 + r_2 = -\frac{-0.0050}{1.0} 10 + 0.0 = 0.050$$

TABLE 8.3 Summary calculations for Example 8.2

j	$d_{j,1}$	$d_{j,2}$	$d_{j,3}$	$d_{j,4}$	r	$p_{j,1}$	$p_{j,2}$	$p_{j,3}$	$p_{j,4}$	s_j	u_j
1	1.0000	0.0000	0.0000	0.0000	10.00000	1.0000	0.0000	0.0000	0.0000	10.0000	10.0000
2	−0.0050	0.0083	0.0050	0.0083	0.00000	−0.0050	0.0083	0.0050	0.0083	0.0500	0.2434
3	−0.0033	−3.1049	0.0171	2.9146	0.00090	−0.0033	−3.1049	1.8875	6.0195	18.7381	9.2362
4	−0.0050	0.0083	0.0050	0.0083	0.00000	−0.0050	0.0242	0.0050	0.0083	0.0496	0.2167
5	−0.0034	−3.1983	0.0165	3.0197	0.01196	−0.0034	−3.1875	0.6738	4.1109	6.5713	8.5491
6	−0.0050	0.0083	0.0050	0.0083	0.00000	−0.0050	0.0388	0.0050	0.0083	0.0488	0.1972
7	−0.0034	−3.2926	0.0158	3.1414	−0.00896	−0.0034	−3.2719	0.4374	3.8412	4.1356	7.9343
8	−0.0050	0.0083	0.0050	0.0083	0.00000	−0.0050	0.0522	0.0050	0.0083	0.0473	0.1732
9	−0.0035	−3.4027	0.0152	3.2618	0.00126	−0.0035	−3.3720	0.3381	3.7978	3.0876	7.3893
10	0.0000	0.0000	1.0000	−47.6235	0.00188	0.0000	0.0000	1.0000	−58.8558	−9.1298	0.1551

Also, for $j = 3$ (and therefore $m = 1$ and $n = 2$), by using Equations 8.74 and 8.77,

$$p_{3,2} = -\frac{p_{3,1}}{p_{1,1}} p_{1,2} + d_{3,2} = -\frac{-0.0033}{1.0}(0.0) + -3.1049 = -3.1049$$

$$p_{3,3} = -\frac{p_{3,2}}{p_{2,2}} p_{2,3} + d_{3,3} = -\frac{-3.1049}{0.0083}(0.0050) + 0.0171 = 1.8875$$

$$p_{3,4} = -\frac{p_{3,2}}{p_{2,2}} p_{2,4} + d_{3,4} = -\frac{-3.1049}{0.0083}(0.0083) + 2.9146 = 6.0195$$

$$s_3 = -\frac{p_{3,1}}{p_{1,1}} s_1 - \frac{p_{3,2}}{p_{2,2}} s_2 + r_3 = -\frac{-0.0033}{1.00}(10) - \frac{-3.1049}{0.0083}(0.05) + 0.0009$$
$$= 18.7381$$

The remaining values of $p_{j,1}$, $p_{j,2}$, $p_{j,3}$, $p_{j,4}$, and s_j are calculated in a similar way, and the results are tabulated in columns (7) to (11) of Table 8.3.

We will calculate the elements of the solution vector by using Equations 8.78 to 8.81. For example, by using Equation 8.78:

$$u_{10} = \frac{s_{10}}{p_{10,4}} = \frac{-9.1298}{-58.8558} = 0.1551$$

Proceeding to $j = 9$, from Equation 8.79 we have

$$u_9 = \frac{s_9 - p_{9,4} u_{10}}{p_{9,3}} = \frac{3.0876 - (3.7978)(0.1551)}{0.3381} = 7.3893$$

Next, for $j = 8$, from Equation 8.80 we have

$$u_8 = \frac{s_8 - p_{8,4} u_{10} - p_{8,3} u_9}{p_{8,2}} = \frac{0.0473 - (0.0083)(0.1551) - (0.0050)(7.3893)}{0.0522} = 0.1732$$

The remaining values of u are calculated in the same manner. Column (12) of Table 8.3 lists the u values calculated by use of a spreadsheet program.

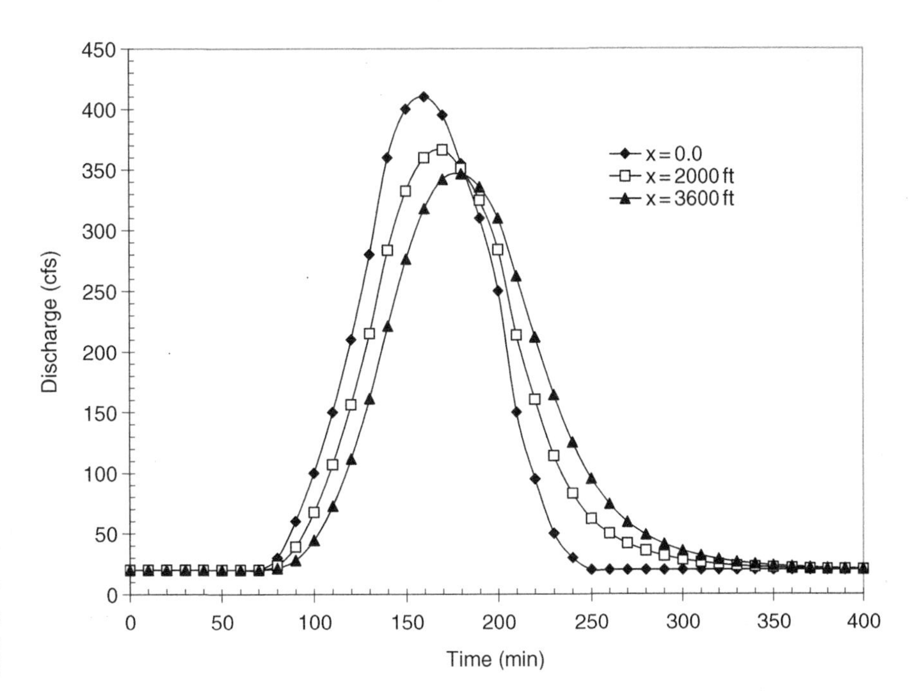

FIGURE 8.2 Flow hydrographs calculated using $\theta = 1.0$

Finally, the corrections to be applied to the guessed values of the discharges and the water surface elevations at the nodes 1 to 5 are determined by using Equations 8.82 and 8.83 for $i = 1$ to 5.

$$\Delta Q_1 = u_1 = 10.0\,\text{cfs}$$
$$\Delta h_1 = u_2 = 0.2434\,\text{ft}$$
$$\Delta Q_2 = u_3 = 9.2362\,\text{cfs}$$
$$\Delta h_2 = u_4 = 0.2167\,\text{ft}$$
$$\Delta Q_3 = u_5 = 8.5491\,\text{cfs}$$
$$\Delta h_3 = u_6 = 0.1972\,\text{ft}$$
$$\Delta Q_4 = u_7 = 7.9343\,\text{cfs}$$
$$\Delta h_4 = u_8 = 0.1732\,\text{ft}$$
$$\Delta Q_5 = u_9 = 7.3893\,\text{cfs}$$
$$\Delta h_5 = u_{10} = 0.1551\,\text{ft}$$

EXAMPLE 8.3 A prismatic, rectangular channel is 10 ft wide and 3600 ft long. The Manning roughness factor is 0.025. The channel has a longitudinal bottom slope of 0.00025 with invert elevations of 0.9 ft and 0.0 ft at the upstream and downstream ends, respectively. Initially, the flow is steady at a rate of 20 cfs and a depth of 1.77 ft. Calculate the unsteady flow in this channel for the upstream hydrograph ($x = 0$) shown in Figure 8.2. Assume that the channel is long, and use a normal flow boundary condition at the downstream end.

A computer program is used to solve this problem. A constant space increment of $\Delta x = 400$ ft is used, resulting in 10 nodes. The time increment is

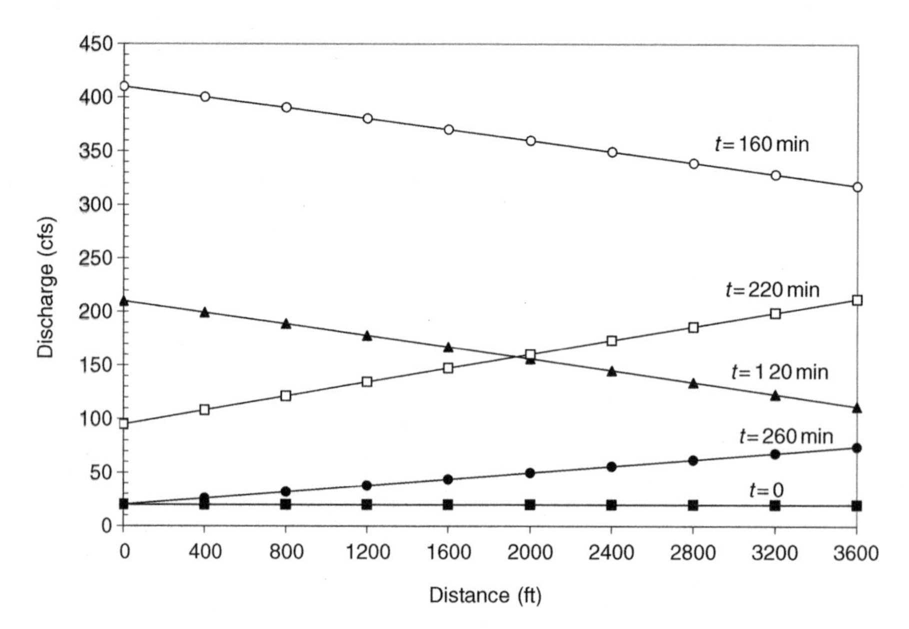

FIGURE 8.3 Variation of discharge along the channel with time

$\Delta t = 10\,\text{min} = 600\,\text{s}$. Equation 8.28 is used as the upstream boundary condition, and the downstream boundary is represented by Equation 8.30. The results shown in Figures 8.2 to 8.4 are obtained by using a fully implicit scheme with $\theta = 1.0$.

Figure 8.2 displays the upstream inflow hydrograph as well as the flow hydrographs calculated at $x = 2000$ ft and $x = 3600$ ft (downstream end). Clearly, the peak discharge is time-shifted and attenuated along the channel. Figure 8.3 displays the variation of the discharge along the channel at different times. At $t = 120$ min, for example, the outflow rate (discharge at $x = 3600$ ft) is lower than the inflow rate (discharge at $x = 0.0$). Therefore the flow is being stored at this time in the channel, causing an overall increase in the water surface elevation. Figure 8.4 displays the variations of the water surface elevation along the channel with time.

Figure 8.5 displays the downstream flow hydrographs (at $x = 3600$ ft) calculated by using the fully implicit ($\theta = 1.0$) and the box ($\theta = 0.5$) schemes. The comparison of the two downstream hydrographs reveals that the results are slightly affected by the weighting factor θ.

8.2.3 SPECIAL CONSIDERATIONS

When unsteady-flow models are applied to drainage channels, numerical difficulties occur if parts of the channel dry up after a rainfall event. To avoid these difficulties, we may superficially maintain a small flow through the channel. This assumed flow should be small enough not to affect the results. Alternatively, to maintain non-zero flow depths, we can assume that there is a rectangular slot attached to the channel bottom (Figure 8.6). The presence of this assumed slot

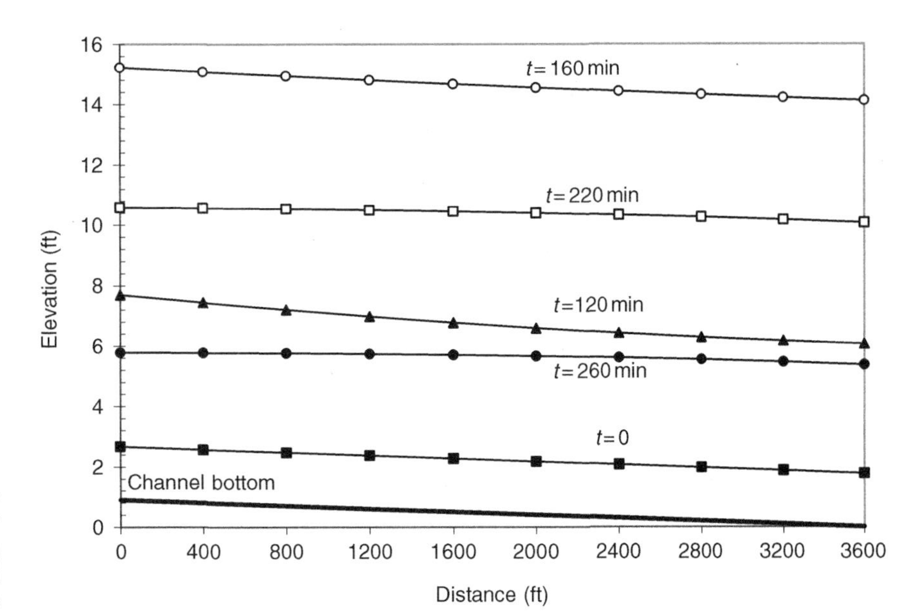

FIGURE 8.4 Variation of water surface elevation along the channel with time

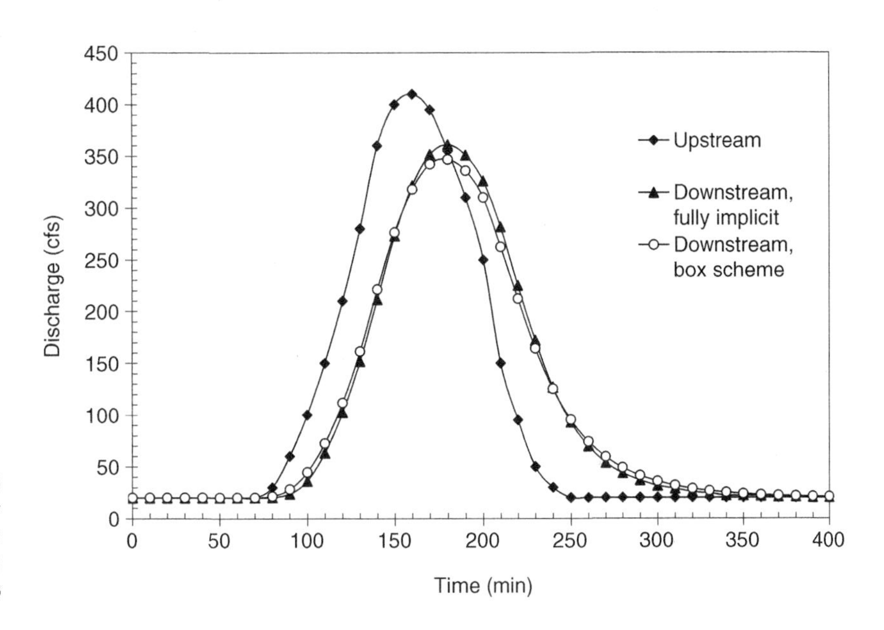

FIGURE 8.5 Comparison of fully implicit ($\theta = 1.0$) and box ($\theta = 0.5$) schemes

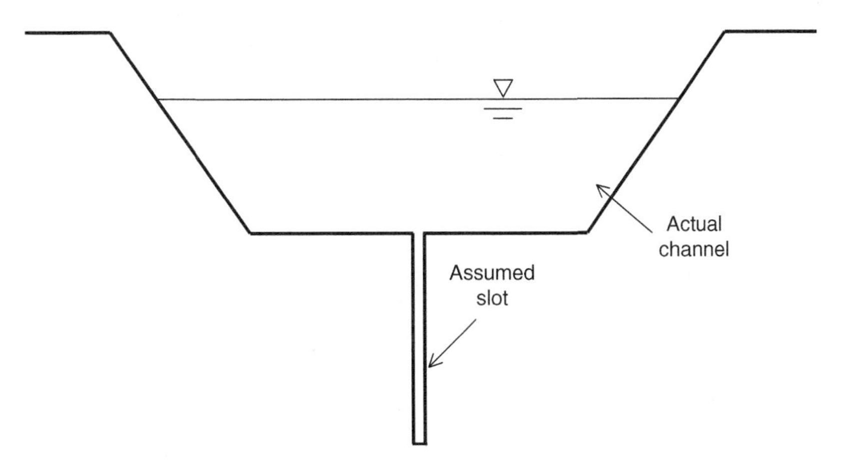

FIGURE 8.6 Assumed bottom slot to model channels drying up

will assure that positive depths will be maintained along the channel when the actual channel dries up. The slot should be narrow enough not to affect the cross-sectional characteristics of the channel section, such as the flow area and the hydraulic radius, so that the results will not be affected. Currey (1998) successfully uses an on/off switch for the bottom slot, activating the slot only when the actual depth in the channel drops below a specified value.

We may encounter another type of problem in unsteady-flow calculations for storm sewers. The Saint Venant equations are applied to flow in a storm sewer when the sewer is partially full. However, due to increased flow rates during flood events, it is possible for a storm sewer to become full (surcharged) at some or all sections. In this event, a different set of equations is needed for the surcharged sections. Although the surcharged flow equations (pressurized pipe flow equations), which are beyond the scope of this text, are available and well understood, switching from the Saint Venant equations to surcharged flow equations is complicated and causes numerical stability problems (Yen, 1986). To facilitate a solution to this problem a hypothetical slot is attached to the sewer crown, as shown in Figure 8.7. With the use of this narrow slot,

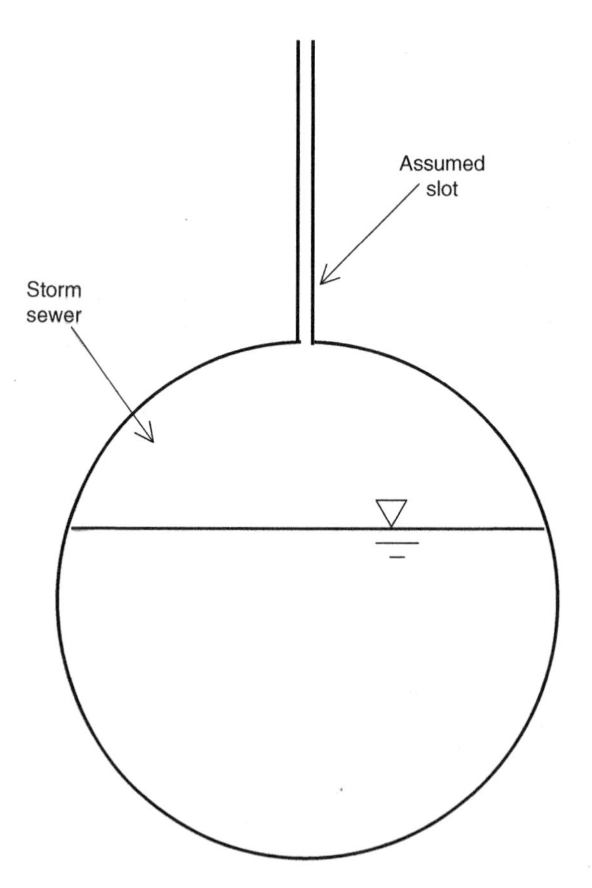

FIGURE 8.7 Preissmann slot

a pressurized flow section is hypothetically transformed to a hypothetical open channel section. When the flow in the actual section is pressurized the water surface will rise in the slot of the hypothetical section, with elevation of the water surface in the slot representing the hydraulic head of the actual pressurized flow. Then, with the introduction of the slot, a free surface (hypothetical at some sections) is maintained along the sewer and the Saint Venant equations can be used without a need to switch to the pressurized pipe flow equations. Yen (1986) provides a brief discussion on how to choose the width of the hypothetical slot.

When the Saint Venant equations are applied to natural channels with irregular cross-sections, it is important to account for the velocity variations within a channel section. This can be achieved by including the momentum correction factor, β, in the momentum equation (see Equation 8.2). With the inclusion of the momentum correction factor, the finite difference formulations presented in the preceding sections for prismatic channels can be applied to irregular channels as well. However, the momentum correction factor, β, will need to be recalculated at every section for each iteration cycle, like the other cross-sectional variables. Moreover, for natural channels, the cross-sectional properties are available only at surveyed sections, and the distance between these sections determines the space increment. Often, different Δx values are needed for different reaches. In this case, the term Δx will be replaced by Δx_i in the finite difference equations.

8.2.4 CHANNEL SYSTEMS

Details of unsteady-flow calculations in channels systems are beyond the scope of this text. However, very briefly, the finite difference equations for individual channels are combined with equations representing the junctions where two or more channels meet. The junction equations given in Section 4.7 for steady flow can be used for unsteady flow as well, if the changes in the volume of water stored in a junction are negligible. Otherwise, a continuity equation is introduced for each junction, balancing the flows entering and leaving the junction with changes in the volume of water stored in the junction. Along with this additional equation, an additional unknown, the junction water surface elevation, is introduced.

The Newton iteration technique can also be used as part of an implicit scheme to solve the system of non-linear algebraic equations, including all the channel and junction equations. However, the coefficient matrices obtained in the iterative procedure may no longer have a banded structure. Still, the coefficient matrix would have relatively few non-zero elements, and one of the many available *sparse matrix* methods can be adopted to obtain efficient solutions.

8.3 Approximate Unsteady-Flow Models

The Saint Venant equations are sometimes referred to as *dynamic-wave* equations, and the models used to solve these equations in complete form are called the *dynamic-wave models*. There are also simplified models reported in the literature in which one or more of the terms in the momentum equation are neglected. These approximate models can be classified with reference to Equation 8.3, rewritten here as

$$\frac{\partial Q}{\partial t}+\frac{\partial}{\partial x}\left(\frac{Q^2}{A}\right)+gA\frac{\partial y}{\partial x}+gAS_f-gAS_0=0 \tag{8.85}$$

Dynamic-wave models employ the equation in its full form. Quasi-steady dynamic wave models simplify the equation by dropping the term $\partial Q/\partial t$. The equation is further simplified in the diffusion-wave models by also dropping the term $\partial/\partial x(Q^2/A)$. In the kinematic-wave models, Equation 8.85 is reduced to $S_f = S_0$.

Numerous studies of the approximate models are available in the literature (see, for example, Gunaratnam and Perkins, 1970; Bettess and Price, 1976; Ponce and Simons, 1977; Katapodes, 1982). A comparison of these models by Sevuk (1973) as applied to a hypothetical storm sewer for various values of the Manning roughness factor and the bed slope indicated that the kinematic wave model was unsatisfactory. Both the diffusion-wave and the quasi-steady dynamic-wave models were satisfactory, but the results of the former were closer to those of the dynamic-wave model. A four-point implicit scheme was used in all the models compared by Sevuk (1973).

Akan and Yen (1977) developed a fully-implicit diffusion-wave model in which the continuity and momentum equations were combined to reduce the total number of algebraic equations by one-half. The model results were compared to those of a fully implicit dynamic-wave model as applied to a hypothetical channel under various downstream boundary conditions. Akan and Yen (1981) extended the diffusion-wave approach to channel systems, and Hromadka and Yen (1986) adopted it for modeling two-dimensional overland flow. Although a stability analysis of the diffusion-wave model has not been reported in the literature, numerical experiments have shown that it produces results very similar to those of a fully implicit dynamic-wave model. A description of the diffusion-wave model follows.

8.3.1 Diffusion-Wave Model for Unsteady Flow

Recalling that $h = z_b + y$, and $S_0 = -\partial z_b/\partial x$, an inspection of Equation 8.85 reveals that the diffusion-wave model approximates the momentum equation as

$$S_f = S_0 - \frac{\partial y}{\partial x} = -\frac{\partial z_b}{\partial x} - \frac{\partial y}{\partial x} = -\frac{\partial h}{\partial x} \tag{8.86}$$

In other words, the friction slope is assumed to be equal to the longitudinal slope of the water surface. Combining Equations 8.86 and 8.7a yields

$$Q = \left(\frac{k_n}{n}\frac{A^{5/3}}{P^{2/3}}\right)\frac{-(\partial h/\partial x)}{(|\partial h/\partial x|)^{1/2}} \tag{8.87}$$

which can account for flows in both the positive and negative x directions. Equations 8.1 and 8.87 are the governing equations for the diffusion-wave formulation presented here.

8.3.2 FINITE DIFFERENCE EQUATIONS

Adopting the computational grid displayed in Figure 8.1 and using a fully-implicit scheme, we can write Equation 8.87 for any node $i = 2, 3, \ldots, (N-1)$ as

$$Q_i = -\left(\frac{k_n}{n}\frac{A_i^{5/3}}{P_i^{2/3}(\Delta x)^{1/2}}\right)\frac{(h_i - h_{i-1})}{|h_i - h_{i-1}|^{1/2}} \tag{8.88}$$

Note that all the variables in Equation 8.88 are at time level $(n+1)$, but the superscript $(n+1)$ is omitted for brevity. The continuity equation (Equation 8.1) is written in finite difference form for a reach between nodes $(i-1)$ and i as

$$\frac{(A_{i+1} + A_i) - (A_{i+1}^n + A_i^n)}{2\Delta t} + \frac{(Q_{i+1} - Q_i)}{\Delta x} = 0 \tag{8.89}$$

Again, the superscript $(n+1)$ is left out for brevity. In Equation 8.89, only the variables with superscript n are at the n-th time level; all the other variables are at time level $(n+1)$. Equation 8.88 will be substituted into Equation 8.89 to obtain a single equation for each reach. However, Equation 8.88 cannot be used to define the discharge at node 1 (upstream end of the channel), since h_{i-1} has no meaning for $i = 1$. Instead, an upstream boundary relationship will be used. This relationship can prescribe either Q_1 or h_1 as a function of time. If the upstream inflow hydrograph is given, for example, then the boundary equation for any time level would be

$$Q_1 = Q_{up} \tag{8.90}$$

Then, substituting Equations 8.88 and 8.90 into Equation 8.89 for $i = 2$, we can obtain the finite difference equation for the reach between nodes 1 and 2 as

$$\frac{(A_2 + A_1) - (A_2^n + A_1^n)}{2\Delta t} + \frac{1}{\Delta x}\left(-\frac{k_n}{n}\frac{A_2^{5/3}}{P_2^{2/3}(\Delta x)^{1/2}}\frac{(h_2 - h_1)}{|h_2 - h_1|^{1/2}} - Q_{up}\right) = 0 \tag{8.91}$$

For the remainder of the reaches between nodes i and $(i+1)$, with $i = 2, 3, 4, \ldots, (N-1)$, substituting Equation 8.87 into Equation 8.88 yields

$$\frac{(A_{i+1}+A_i)-(A_{i+1}^n+A_i^n)}{2\Delta t} + \frac{1}{\Delta x}\left(-\frac{k_n}{n}\frac{A_{i+1}^{5/3}}{P_{i+1}^{2/3}(\Delta x)^{1/2}}\frac{(h_{i+1}-h_i)}{|h_{i+1}-h_i|^{1/2}} + \frac{k_n}{n}\frac{A_i^{5/3}}{P_i^{2/3}(\Delta x)^{1/2}}\frac{(h_i-h_{i-1})}{|h_i-h_{i-1}|^{1/2}}\right) = 0 \tag{8.92}$$

In Equations 8.91 and 8.92, the quantities with superscript n are known either from the initial condition or from the previous time step computations. The quantities A_i and P_i can be expressed in terms of h_i based on the cross-sectional geometry. Therefore, the unknowns are h_i for $i = 1, 2, \ldots, N$. Equation 8.91 for the first reach and Equation 8.92 written for the remaining $(N-2)$ reaches constitute a system of $(N-1)$ algebraic equations. An additional equation is provided by the downstream boundary condition, which may be in the form of a relationship between Q_N and h_N or prescribed values of h_N as a function of time.

8.3.3 SOLUTION OF FINITE DIFFERENCE EQUATIONS

Let G_1 denote Equation 8.91, G_i denote Equation 8.92 written for $i = 2, 3, 4, \ldots, (N-1)$ and G_N denote the downstream boundary condition. This system of N non-linear algebraic equations can be expressed as

$$\begin{aligned}
G_1(h_1, h_2) &= 0\\
G_2(h_1, h_2, h_3) &= 0\\
&\ldots\ldots\ldots\ldots\\
&\ldots\ldots\ldots\ldots\\
G_i(h_{i-1}, h_i, h_{i+1}) &= 0\\
&\ldots\ldots\ldots\ldots\\
&\ldots\ldots\ldots\ldots\\
G_N(h_{N-1}, h_N) &= 0
\end{aligned} \tag{8.93}$$

We can adopt the Newton iteration method to solve these equations. The computations for the iterative procedure begins by assigning a set of trial values to the unknowns h_i for $i = 1, 2, 3, \ldots, N$. Substitution of the trial values into the left-hand sides of Equations 8.93 yield the residuals rG_i for $i = 1, 2, 3, \ldots, N$. New values for the unknowns are estimated for the next iteration to make the residuals approach zero. This is accomplished by calculating the corrections Δh_i such that the total differentials of the functions G_i are equal to the negative of the calculated residuals, i.e.

$$\frac{\partial G_i}{\partial h_{i-1}}\Delta h_{i-1} + \frac{\partial G_i}{\partial h_i}\Delta h_i + \frac{\partial G_i}{\partial h_{i+1}}\Delta h_{i+1} = -rG_i \tag{8.94}$$

Equation 8.94 written for $i = 1, 2, 3, \ldots, N$ forms a set of N linear algebraic equations in N unknowns, Δh_i for $i = 1, 2, 3, \ldots, N$. In matrix notation, this linear system of equations is

$$\begin{bmatrix} \frac{\partial G_1}{\partial h_1} & \frac{\partial G_1}{\partial h_2} & 0 & 0 & \ldots & 0 & 0 & 0 & 0 \\ \frac{\partial G_2}{\partial h_1} & \frac{\partial G_2}{\partial h_2} & \frac{\partial G_2}{\partial h_3} & 0 & \ldots & 0 & 0 & 0 & 0 \\ 0 & \frac{\partial G_3}{\partial h_2} & \frac{\partial G_3}{\partial h_3} & \frac{\partial G_3}{\partial h_4} & \ldots & 0 & 0 & 0 & 0 \\ . & . & . & . & \ldots & . & . & . & . \\ . & . & . & . & \ldots & . & . & . & . \\ . & . & . & . & \ldots & . & . & . & . \\ 0 & 0 & 0 & 0 & \ldots & 0 & \frac{\partial G_{N-1}}{\partial h_{N-2}} & \frac{\partial G_{N-1}}{\partial h_{N-1}} & \frac{\partial G_{N-1}}{\partial h_N} \\ 0 & 0 & 0 & 0 & \ldots & 0 & 0 & \frac{\partial G_N}{\partial h_{N-1}} & \frac{\partial G_N}{\partial h_N} \end{bmatrix} \begin{bmatrix} \Delta h_1 \\ \Delta h_2 \\ \Delta h_3 \\ . \\ . \\ . \\ \Delta h_{N-1} \\ \Delta h_N \end{bmatrix} = \begin{bmatrix} -rG_1 \\ -rG_2 \\ -rG_3 \\ . \\ . \\ . \\ -rG_{N-1} \\ -rG_N \end{bmatrix} \tag{8.95}$$

Note that the coefficient matrix in Equation 8.95 has a banded structure with no more than three non-zero elements, and all the non-zero elements are on the diagonal. We can take advantage of the banded structure of the matrix equation to develop a fast solution by using the recurrence formulae

$$B_i = \frac{\partial G_i / \partial h_{i+1}}{(\partial G_i / \partial h_i) - (\partial G_i / \partial h_{i-1}) B_{i-1}} \tag{8.96}$$

$$H_i = \frac{-rG_i - (\partial G_i / \partial h_{i-1}) H_{i-1}}{(\partial G_i / \partial h_i) - (\partial G_i / \partial h_{i-1}) B_{i-1}} \tag{8.97}$$

$$\Delta h_i = H_i - B_i(\Delta h_{i+1}) \tag{8.98}$$

The quantities B_i and H_i are first calculated from Equations 8.96 and 8.97 for $i = 1, 2, 3, \ldots, N$, noting that $(\partial G_i/\partial h_{i-1}) = 0$ for $i = 1$. Then the corrections Δh_i are calculated from Equation 8.98, starting with $i = N$ and proceeding sequentially with $i = (N-1), (N-2), \ldots, 3, 2, 1$. Note that $\Delta h_{N+1} = 0$. Obviously, when we computer program this procedure we only need to store the non-zero elements of the coefficient matrix in Equation 8.95.

EXAMPLE 8.4 A rectangular channel is 12 ft wide and 4500 ft long, with a Manning roughness factor of 0.020. The longitudinal profile of the bottom of the channel is shown in Figure 8.8. There is a weir placed at the downstream end of the channel, and the weir crest is 1.5 ft above the channel invert as shown in the same figure. Let this weir be represented by the equation $Q_N = 40(h_N - 1.5)^{1.5}$ where Q_N and h_N are, respectively, the discharge in cfs and the water surface elevation in ft at the downstream end of the channel. Initially, the flow is steady at a discharge of 10 cfs with the water surface elevations labeled $t = 0$ in Figure 8.8. Using a diffusion-wave model, $\Delta x = 500$ ft, and $\Delta t = 10$ min $= 600$ s, calculate the water surface in the channel at

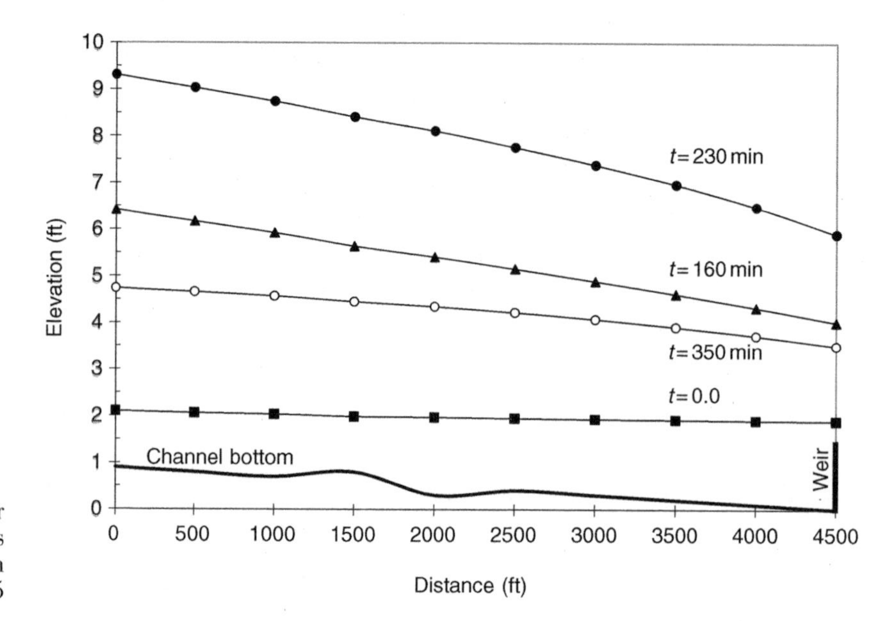

FIGURE 8.8 Water surface profiles calculated in Example 8.5

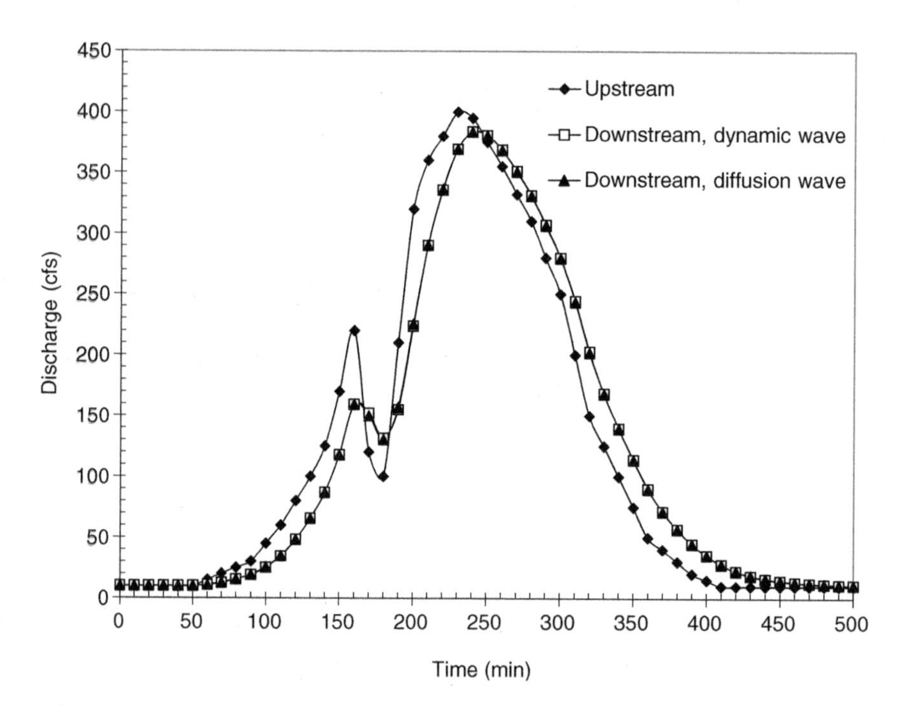

FIGURE 8.9 Downstream hydrographs calculated in Example 8.5

various times and the downstream hydrograph resulting from the upstream hydrograph shown in Figure 8.9. Also calculate the downstream hydrograph using a fully implicit dynamic-wave model, and compare the results of the two models.

Computer programs are used to solve this problem. Figure 8.8 displays the water surface elevations calculated at various times. The water surface rises to its highest position at $t = 230$ min, and falls afterwards. Figure 8.9 shows the downstream hydrographs calculated by using both the diffusion- and dynamic-wave models. The agreement between the two models is excellent.

8.4 Simple Channel-Routing Methods

We often refer to unsteady-flow calculations in open channels as *channel routing*, particularly in conjunction with hydrologic methods. Generally, the hydrologic routing methods are based on the hydrologic storage equation, they are linear, and they cannot account for downstream boundary conditions. Of the two simple channel-routing methods presented here, the Muskingum method is a hydrologic routing method. The Muskingum–Cunge method, however, is based on the kinematic-wave approximation to the Saint Venant equations modified with diffusion contributions of the momentum equation. Neither method can account for the downstream boundary conditions.

8.4.1 The Muskingum Method

The hydrologic storage equation (continuity equation) for a channel reach is

$$\frac{dS}{dt} = I - Q \tag{8.99}$$

where

S = volume of water in storage in the channel reach, I = upstream inflow rate, Q = downstream outflow rate, and t = time. In place of the momentum equation, a linear relationship is assumed between S, I, and Q as

$$S = K[X1 + (1 - X)Q] \tag{8.100}$$

where K = travel time constant, and X = weighting factor between 0 and 1.0. Equations 8.99 and 8.100 form the basis for the Muskingum method. Sometimes, K is vaguely interpreted as the travel time of a flood wave along the channel reach. However, for the most part K and X are treated as calibration parameters into which the channel characteristics are lumped.

8.4.1.1 Routing equation

To solve Equations 8.99 and 8.100 numerically, we discretize the time into finite time increments of Δt. For any time increment, Equation 8.99 is written in finite difference form as

$$\frac{S_2 - S_1}{\Delta t} = \frac{I_1 + I_2}{2} - \frac{Q_1 + Q_2}{2} \tag{8.101}$$

in which subscript 1 refers to the beginning of the time increment and 2 refers to the end. Note that I and Q in Equation 8.101 correspond, respectively, with Q_i and Q_{i+1} in context of the computational grid given in Figure 8.1, and the subscripts 1 and 2 correspond, respectively, with time levels n and $(n+1)$. A different notation is adopted here for simplicity. Rewriting Equation 8.100 in terms of S_1, I_1, Q_1, S_2, I_2, and Q_2, substituting into Equation 8.101 and simplifying, we obtain

$$Q_2 = C_0 I_2 + C_1 I_1 + C_2 Q \tag{8.102}$$

where

$$C_0 = \frac{(\Delta t/K) - 2X}{2(1-X) + (\Delta t/K)} \tag{8.103}$$

$$C_1 = \frac{(\Delta t/K) + 2X}{2(1-X) + (\Delta t/K)} \tag{8.104}$$

$$C_2 = \frac{2(1-X) - (\Delta t/K)}{2(1-X) + (\Delta t/K)} \tag{8.105}$$

Note that $C_0 + C_1 + C_2 = 1.0$. Also, for C_0, C_1, and C_2 to be dimensionless, K and Δt must have the same unit of time.

The only unknown in Equation 8.102 is Q_2 for any time increment, since Q_1 is known either from the initial conditions or from the previous time step computations, and I_1 and I_2 are known from the given inflow hydrograph. The coefficients C_0, C_1, and C_2 are first found from Equations 8.103 to 8.105, and then we determine Q_2 using Equation 8.102.

Various limits are suggested for the parameters used in the Muskingum method. Cunge (1969) suggested that X be non-negative for Equation 8.100 to be physically meaningful, and he also showed that X should be equal to or less than 0.5 for the Muskingum method to be stable. In other words,

$$0 \leq X \leq 0.5 \tag{8.106}$$

The time increment should be equal to or smaller than a fifth of the time to peak of the inflow hydrograph for accurate representation of the rising limb (Sturm, 2001). Meanwhile, C_0 (hence the numerator of Equation 103) should have a non-negative value (Ponce and Theurer, 1982). These two conditions lead to

$$2KX \leq \Delta t \leq \frac{t_p}{5} \tag{8.107}$$

where t_p = time to peak of the inflow hydrograph.

EXAMPLE 8.5 The inflow hydrograph for a channel reach is plotted in Figure 8.10. The flow is initially steady at a rate of $5.0\,m^3/s$. The Muskingum parameters for this channel reach are given as $K = 2\,h$ and $X = 0.10$. Route the inflow hydrograph through this channel reach using $\Delta t = 1\,h$.

The weighting factors are first obtained from Equations 8.103 to 8.107 as being $C_0 = 0.1304$, $C_1 = 0.3043$, $C_2 = 0.5653$. Then, we employ Equation 8.102 to determine the outflow rates. Table 8.4 summarizes the routing calculations.

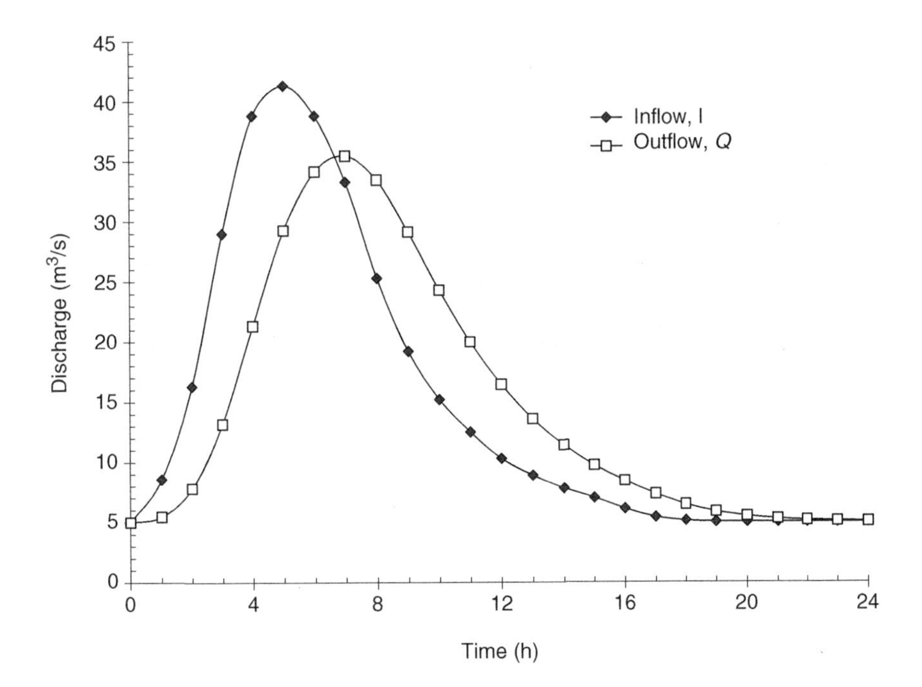

FIGURE 8.10 Muskingum channel routing example

TABLE 8.4 Muskingum channel-routing calculations

Time step	t_1 (h)	t_2 (h)	I_1 (m^3/s)	I_2 (m^3/s)	Q_1 (m^3/s)	Q_2 (m^3/s)
1	0.0	1.0	5.00	8.60	5.00	5.47
2	1.0	2.0	8.60	16.30	5.47	7.83
3	2.0	3.0	16.30	29.00	7.83	13.17
4	3.0	4.0	29.00	38.80	13.17	21.33
5	4.0	5.0	38.80	41.30	21.33	29.25
6	5.0	6.0	41.30	38.80	29.25	34.16
7	6.0	7.0	38.80	33.30	34.16	35.46
8	7.0	8.0	33.30	25.30	35.46	33.48
9	8.0	9.0	25.30	19.20	33.48	29.13
10	9.0	10.0	19.20	15.20	29.13	24.29
11	10.0	11.0	15.20	12.50	24.29	19.99
12	11.0	12.0	12.50	10.30	19.99	16.44
13	12.0	13.0	10.30	8.90	16.44	13.59
14	13.0	14.0	8.90	7.80	13.59	11.41
15	14.0	15.0	7.80	7.00	11.41	9.73
16	15.0	16.0	7.00	6.10	9.73	8.43
17	16.0	17.0	6.10	5.40	8.43	7.32
18	17.0	18.0	5.40	5.10	7.32	6.45
19	18.0	19.0	5.10	5.00	6.45	5.85
20	19.0	20.0	5.00	5.00	5.85	5.48
21	20.0	21.0	5.00	5.00	5.48	5.27
22	21.0	22.0	5.00	5.00	5.27	5.15
23	22.0	23.0	5.00	5.00	5.15	5.09
24	23.0	24.0	5.00	5.00	5.09	5.05

The I_1 and I_2 values in columns (4) and (5) are read off the inflow hydrograph (Figure 8.10) using t_1 and t_2, which are tabulated in columns (2) and (3). For the first time step of computations, $t_1 = 0$ and $t_2 = 1$ hour. Because the flow is initially steady at a constant rate of $5\,\text{m}^3/\text{s}$, we have $Q_1 = 5.0\,\text{m}^3/\text{s}$ in column (6) for the first time step. Next, by using Equation 8.102, we obtain Q_2 as being $5.47\,\text{m}^3/\text{s}$. This is the outflow rate at $t = 1$ hour. For the second time step, $t_1 = 1$ hour and therefore $Q_1 = 5.47\,\text{m}^3/\text{s}$. Now, using Equation 8.102, we obtain $Q_2 = 7.83\,\text{m}^3/\text{s}$. We repeat the same procedure until we obtain the entire outflow hydrograph. The results are displayed in Figure 8.10.

8.4.1.2 Calibration of Muskingum parameters

The Muskingum parameters K and X are treated as calibration parameters, and they can be determined for a channel reach if measured inflow and outflow hydrographs are available for a flood event. Once calibrated, these parameters are assumed to remain constant and are used for other flood events as well.

Aldama (1990) and Sturm (2001) reported a least-square method to calibrate the Muskingum parameters. In this method, K and X are determined by using

$$K = E + F \tag{8.108}$$

and

$$X = \frac{E}{E+F} \tag{8.109}$$

where

$$\begin{aligned} E = \frac{1}{Z}\Big\{&\left[\left(\sum I_jQ_j\right)\left(\sum Q_j\right) - \left(\sum I_j\right)\left(\sum Q_j^2\right)\right]\left(\sum S_j\right) \\ &+ \left[M\left(\sum Q_j^2\right) - \left(\sum Q_j\right)^2\right]\left(\sum I_jS_j\right) \\ &+ \left[\left(\sum I_j\right)\left(\sum Q_j\right) - M\left(\sum I_jQ_j\right)\right]\left(\sum Q_jS_j\right)\Big\} \end{aligned} \tag{8.110}$$

$$\begin{aligned} F = \frac{1}{Z}\Big\{&\left[\left(\sum I_j\right)\left(\sum I_jQ_j\right) - \left(\sum I_j^2\right)\left(\sum Q_j\right)\right]\left(\sum S_j\right) \\ &+ \left[\left(\sum I_j\right)\left(\sum Q_j\right) - M\left(\sum I_jQ_j\right)\right]\left(\sum I_jS_j\right) \\ &+ \left[M\left(\sum I_j^2\right) - \left(\sum I_j\right)^2\right]\left(\sum Q_jS_j\right)\Big\} \end{aligned} \tag{8.111}$$

and

$$\begin{aligned} Z = M&\left[\left(\sum I_j^2\right)\left(\sum Q_j^2\right) - \left(\sum I_jQ_j\right)^2\right] + 2\left(\sum I_j\right)\left(\sum Q_j\right)\left(\sum I_jQ_j\right) \\ &- \left(\sum I_j\right)^2\left(\sum Q_j^2\right) - \left(\sum I_j^2\right)\left(\sum Q_j\right)^2 \end{aligned} \tag{8.112}$$

where I, Q_j, and S_j are, respectively, the observed inflow rate, the observed outflow rate, and the relative storage at the j-th time step. Also, in Equations 8.110

TABLE 8.5 Calibration of Muskingum parameters

j	t (h)	I_j (cfs)	Q_j (cfs)	S_j (ft^3)	I_j^2 (ft^6/s^2)	Q_j^2 (ft^6/s^2)	I_jQ_j (ft^6/s^2)	I_jS_j (ft^6/s)	Q_jS_j (ft^6/s)
1	0	50	50	0	2500	2500	2500	0	0
2	2	86	53	118800	7396	2809	4558	1.0217×10^7	6.2964×10^6
3	4	163	71	568800	26569	5041	11573	9.2714×10^7	4.0385×10^7
4	6	290	114	1533600	84100	12996	33060	4.4474×10^8	1.7483×10^8
5	8	388	187	2890800	150544	34969	72556	1.1216×10^9	5.4058×10^8
6	10	413	263	4154400	170569	69169	108619	1.7158×10^9	1.0926×10^9
7	12	388	317	4950000	150544	100489	122996	1.9206×10^9	1.5692×10^9
8	14	333	339	5184000	110889	114921	112887	1.7263×10^9	1.7574×10^9
9	16	253	331	4881600	64009	109561	83743	1.2350×10^9	1.6158×10^9
10	18	192	298	4219200	36864	88804	57216	8.1009×10^8	1.2573×10^9
11	20	152	255	3466800	23104	65025	38760	5.2695×10^8	8.8403×10^8
12	22	125	215	2772000	15625	46225	26875	3.4650×10^8	5.9598×10^8
13	24	103	180	2170800	10609	32400	18540	2.2359×10^8	3.9074×10^8
14	26	89	151	1670400	7921	22801	13439	1.4867×10^8	2.5223×10^8
15	28	78	127	1270800	6084	16129	9906	9.9122×10^7	1.6139×10^8
16	30	70	108	957600	4900	11664	7560	6.7032×10^7	1.0342×10^8
17	32	61	93	705600	3721	8649	5673	4.3042×10^7	6.5621×10^7
18	34	54	81	493200	2916	6561	4374	2.6633×10^7	3.9949×10^7
19	36	51	71	324000	2601	5041	3621	1.6524×10^7	2.3004×10^7
20	38	50	63	205200	2500	3969	3150	1.0260×10^7	1.2928×10^7
$\Sigma=$		3389	3367	42537600	883965	759723	741606	1.0585×10^{10}	1.0584×10^{10}

to 8.112, the index j runs from 1 to M where $M=$ total number of observed pairs of the inflow and outflow rates. The relative storage is calculated using

$$S_{j+1} = S_j + \frac{\Delta t}{2}(I_j + I_{j+1} - Q_j - Q_{j-1}) \tag{8.113}$$

with $S_1 = 0$.

EXAMPLE 8.6 The observed inflow and outflow rates for a channel reach are tabulated in columns (3) and (4) of Table 8.5 at equal time intervals of $\Delta t = 2\,\text{h} = 7200\,\text{s}$. Determine the Muskingum parameters K and X for this reach.

The calculations are summarized in Table 8.5. The entries in column (5) are obtained by using Equation 8.113. The headings for the remainder of the columns are self-explanatory. Using the sums given in the bottom row of Table 8.5 and with $M=20$, we determine Z, E, and F as being 6.0946×10^{11} ft^{12}/s^4, 2.1655×10^3 s, and 1.5823×10^4 s, respectively, from Equations 8.112, 8.110, and 8.111. Finally, by using Equation 8.108, $K = 17\,990\,\text{s} = 5\,\text{h}$, and from Equation 8.109, $X = 0.12$.

8.4.2 THE MUSKINGUM–CUNGE METHOD

A major drawback of the Muskingum method discussed in the preceding section is that the parameters K and X do not have a physical basis, and we can estimate them only if we have simultaneous inflow and outflow data from a flood event for the channel reach considered. This difficulty is overcome in the Muskingum–Cunge method (Cunge, 1969), where K and X are expressed in terms of the physical characteristics of the channel reach.

The derivation of the Muskingum–Cunge equations is beyond the scope of this text. However, briefly, the method is based on the kinematic-wave approximation to the Saint Venant equations. Theoretically, the peak discharge will be shifted in time but will not be attenuated in a pure kinematic-wave method. In other words, the peak of the downstream outflow hydrograph will occur after the peak of the upstream inflow hydrograph, but the two peaks will be equal in magnitude. In the Muskingum–Cunge method, the kinematic-wave approximation is modified to produce an attenuation based on the diffusion anology (or a linearized diffusion-wave approximation). Therefore, the Muskingum–Cunge method should be categorized as a hydraulic (as opposed to a hydrologic) method. The resulting routing equation, however, is arranged in the same form as that of the Muskingum method.

In the Muskingum–Cunge method, it is assumed that the discharge, Q, and the flow area, A, are related through

$$Q = eA^m \tag{8.114}$$

in which e and m are constant. If the rating curve for a channel section is available (measured values of Q versus A), we can obtain the constants e and m by fitting Equation 8.114 mathematically to the rating curve. In the absence of such data, adopting a normal flow approximation, we can obtain e and m as

$$Q = \left(\frac{k_n S_0^{1/2}}{nP^{2/3}}\right) A^{5/3} \tag{8.115}$$

From Equations 8.114 and 8.115, it is obvious that m = 5/3 and

$$e = \frac{k_n S_0^{1/2}}{nP^{2/3}} \tag{8.116}$$

The wetted perimeter, P, varies with the flow depth, and for most cross-sectional shapes Equation 8.116 does not yield a constant value for e. In this event, we use a representative, constant P in Equation 8.116.

Equations 8.102 to 8.105 given for the Muskingum method are also used in the Muskingum–Cunge method. However, K and X are expressed as

$$K = \frac{L}{mV_0} \tag{8.117}$$

and

$$X = 0.5\left[1 - \frac{Q_0}{mT_0S_0V_0L}\right] \tag{8.118}$$

where Q_0 = a reference discharge, T_0 = top width corresponding to the reference discharge, V_0 = cross-sectional average velocity corresponding to the reference discharge, S_0 = longitudinal slope of the channel, L = length of the channel reach, and m = exponent of the flow area A in Equation 8.114. The base flow rate, the peak of the inflow hydrograph, or the average inflow rate can be used as the reference discharge.

In the Muskingum–Cunge method, X is no longer interpreted as a weighting factor, and it can take negative values. Also, Ponce and Theurer (1982) suggest that Δt be smaller than one-fifth of the time from the beginning to the peak of the inflow hydrograph. Moreover, they recommend that the length of the channel reach for Muskingum–Cunge computations be limited according to

$$L \leq 0.5\left[mV_0\Delta t + \frac{Q_0}{mT_0V_0S_0}\right] \tag{8.119}$$

to obtain accurate results. To satisfy this inequality, we may need to divide a channel reach into shorter segments if it is too long.

The results of the Muskingum–Cunge method depend on the reference discharge employed to calculate the K and X parameters. We can eliminate this dependence by using variable routing coefficients (Ponce and Yevjevich, 1978). In this approach, we update the reference discharge at every time step as

$$Q_0 = \frac{(I_1 + I_2 + Q_1)}{3} \tag{8.120}$$

and recalculate T_0, V_0, X, K, C_0, C_1, and C_2 using the updated reference discharge.

EXAMPLE 8.7 A 3200-ft long trapezoidal channel has a Manning roughness factor of 0.016, a longitudinal bottom slope of 0.0005, a bottom width of 5.0 ft and cross-sectional side slopes of 2H : 1V. Route through this channel the upstream inflow hydrograph shown in Figure 8.11.

Let us pick a time increment of $\Delta t = 0.25\,\text{h} = 900\,\text{s}$, and a reference discharge of $Q_0 = 200$ cfs (half of the peak discharge). Assuming the flow is normal at the reference discharge, we can use the Manning formula to determine the corresponding flow variables as $y_0 = 4.13$ ft, $A_0 = 54.76\,\text{ft}^2$, $T_0 = 21.52$ ft, and $V_0 = 3.65$ fps. Also, note that $m = 5/3$, since the Manning formula is adopted.

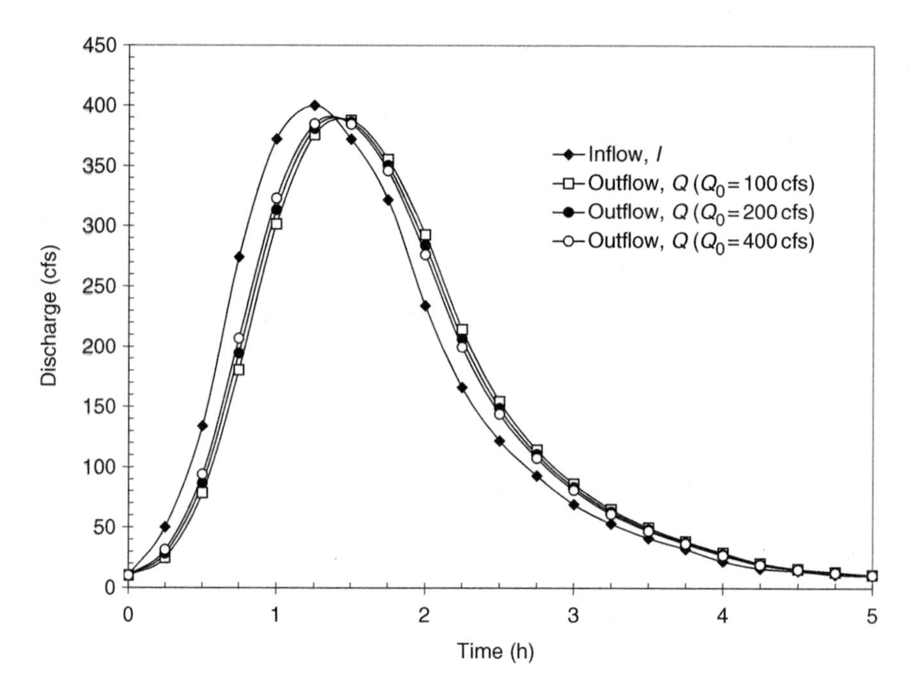

FIGURE 8.11 Muskingum–Cunge hydrographs

Next, we determine the limiting channel length from the right-hand side of Equation 8.119 as

$$0.5\left[mV_0\Delta t+\frac{Q_0}{mT_0V_0S_0}\right]=0.5\left[(5/3)(3.65)(900)+\frac{200}{(5/3)(21.52)(3.65)(0.0005)}\right]$$
$$=4265\text{ ft}$$

The length of the given channel is 3200 ft, and is thus smaller than this limiting value of 4265 ft. Therefore, we can use the whole length of the channel as a single reach. Now, by using Equations 8.117 and 8.118,

$$K=\frac{L}{mV_0}=\frac{3200}{(5/3)(3.65)}=526\,s=0.146\text{ h}$$

$$X=0.5\left[1-\frac{Q_0}{mT_0S_0V_0L}\right]=0.5\left[1-\frac{200}{(5/3)(21.52)(0.0005)(3.65)(3200)}\right]=0.023$$

We next use Equations 8.103 to 8.105 to determine the routing coefficients as being $C_0=0.4545$, $C_1=0.4794$ and $C_2=0.0661$ and perform the routing by using Equation 8.102 as in the Muskingum method. The calculations are summarized in Table 8.6, and the results are plotted in Figure 8.11.

The results of the Muskingum–Cunge method can be somewhat sensitive to the selection of the reference discharge Q_0. Figure 8.11 displays the outflow hydrographs obtained with $Q_0=100$ cfs and 400 cfs, along with that obtained with $Q_0=200$ cfs. It appears that, in this case, the effect of Q_0 on the results is not significant.

TABLE 8.6 Muskingum–Cunge routing

Time step	t_1 (h)	t_2 (h)	I_1 (cfs)	I_2 (cfs)	Q_1 (cfs)	Q_2 (cfs)
1	0.00	0.25	10.00	50.00	10.00	28.18
2	0.25	0.50	50.00	134.00	28.18	86.73
3	0.50	0.75	134.00	274.00	86.73	194.50
4	0.75	1.00	274.00	372.00	194.50	313.28
5	1.00	1.25	372.00	400.00	313.28	380.84
6	1.25	1.50	400.00	372.00	380.84	386.01
7	1.50	1.75	372.00	322.00	386.01	350.20
8	1.75	2.00	322.00	234.00	350.20	283.87
9	2.00	2.25	234.00	166.00	283.87	206.39
10	2.25	2.50	166.00	122.00	206.39	148.67
11	2.50	2.75	122.00	93.00	148.67	110.58
12	2.75	3.00	93.00	69.00	110.58	83.26
13	3.00	3.25	69.00	53.00	83.26	62.67
14	3.25	3.50	53.00	41.00	62.67	48.19
15	3.50	3.75	41.00	32.00	48.19	37.38
16	3.75	4.00	32.00	22.00	37.38	27.81
17	4.00	4.25	22.00	16.00	27.81	19.66
18	4.25	4.50	16.00	14.00	19.66	15.33
19	4.50	4.75	14.00	11.00	15.33	12.72
20	4.75	5.00	11.00	10.00	12.72	10.66
21	5.00	10.00	10.00	10.00	10.66	10.04
22	5.25	5.50	10.00	10.00	10.04	10.00
23	5.50	5.75	10.00	10.00	10.00	10.00
24	5.75	6.00	10.00	10.00	10.00	10.00

EXAMPLE 8.8 Redo Example 8.7 if the length of the channel is 6000 ft.

As we determined in Example 8.7, the reach length should not exceed 4265 ft for the Muskingum–Cunge method to produce reliable results under the given conditions. Therefore, here we divide the length of the channel into two equal reaches of 3000 ft. The inflow hydrograph is first routed through the first reach to obtain the flow hydrograph at $x = 3000$ ft. This becomes the inflow hydrograph for the second reach, and is routed through the second reach to obtain the flow hydrograph at $x = 6000$ ft.

Because the two reaches are equal in length, the routing coefficients are the same for both reaches. As in Example 8.7, $y_0 = 4.13$ ft, $A_0 = 54.76$ ft^2, $T_0 = 21.52$ ft, and $V_0 = 3.65$ fps, corresponding to the reference discharge of $Q_0 = 200$ cfs. From Equations 8.117 and 8.118, respectively, we obtain $K = 0.137$ h and $X = -0.009$. Then, by using Equations 8.103 to 8.105, $C_0 = 0.4797$, $C_1 = 0.4704$, and $C_2 = 0.0499$. As before, we use Equation 8.102 to route the flow hydrographs through each of the two channel reaches. The calculations are summarized in Table 8.7, and the results are shown in Figure 8.12.

TABLE 8.7 Calculations for Example 8.8

					$x = 3000$ ft		$x = 6000$ ft	
Time step	t_1 (h)	t_2 (h)	I_1 (cfs)	I_2 (cfs)	Q_1 (cfs)	Q_2 (cfs)	Q_1 (cfs)	Q_2 (cfs)
1	0.00	0.25	10.00	50.00	10.00	29.19	10.00	19.20
2	0.25	0.50	50.00	134.00	29.19	89.26	19.20	57.50
3	0.50	0.75	134.00	274.00	89.26	198.92	57.50	140.28
4	0.75	1.00	274.00	372.00	198.92	317.26	140.28	252.76
5	1.00	1.25	372.00	400.00	317.26	382.70	252.76	345.43
6	1.25	1.50	400.00	372.00	382.70	385.70	345.43	382.28
7	1.50	1.75	372.00	322.00	385.70	348.70	382.28	367.78
8	1.75	2.00	322.00	234.00	348.70	281.12	367.78	317.23
9	2.00	2.25	234.00	166.00	281.12	203.73	317.23	245.80
10	2.25	2.50	166.00	122.00	203.73	146.78	245.80	178.51
11	2.50	2.75	122.00	93.00	146.78	109.33	178.51	130.40
12	2.75	3.00	93.00	69.00	109.33	82.30	130.40	97.41
13	3.00	3.25	69.00	53.00	82.30	61.99	97.41	73.31
14	3.25	3.50	53.00	41.00	61.99	47.69	73.31	55.70
15	3.50	3.75	41.00	32.00	47.69	37.02	55.70	42.97
16	3.75	4.00	32.00	22.00	37.02	27.45	42.97	32.73
17	4.00	4.25	22.00	16.00	27.45	19.39	32.73	23.85
18	4.25	4.50	16.00	14.00	19.39	15.21	23.85	17.61
19	4.50	4.75	14.00	11.00	15.21	12.62	17.61	14.09
20	4.75	5.00	11.00	10.00	12.62	10.60	14.09	11.73
21	5.00	10.00	10.00	10.00	10.60	10.03	11.73	10.38
22	5.25	5.50	10.00	10.00	10.03	10.00	10.38	10.03
23	5.50	5.75	10.00	10.00	10.00	10.00	10.03	10.00
24	5.75	6.00	10.00	10.00	10.00	10.00	10.00	10.00

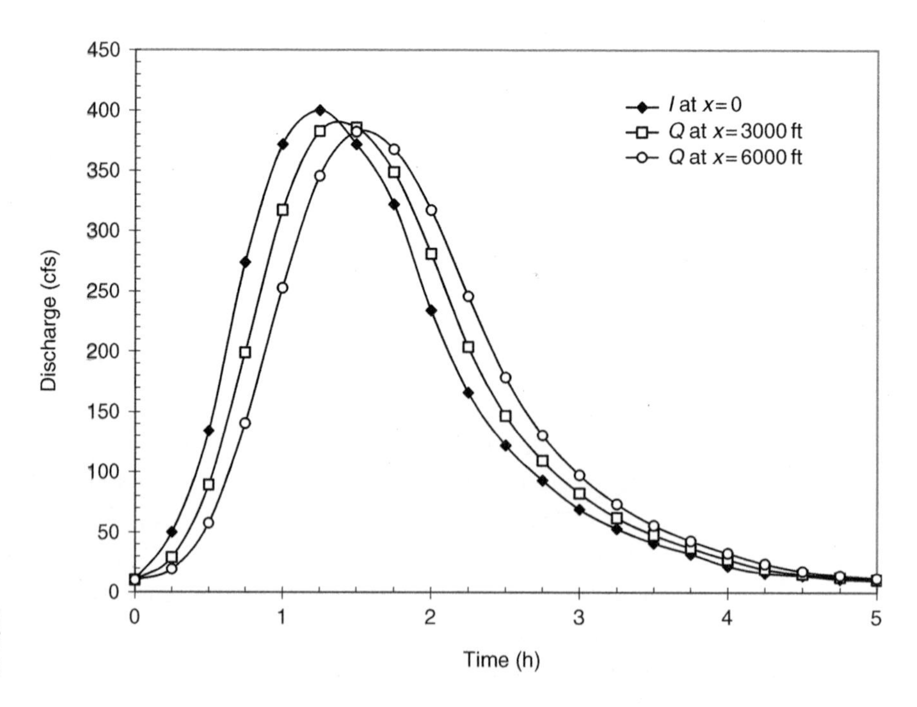

FIGURE 8.12 Results of Example 8.8

PROBLEMS

P.8.1 Can reversed flow occur in a drainage channel if:

(a) the channel terminates at a free fall?
(b) the channel exit is submerged in a tidal river?

P.8.2 Explain why a boundary condition at each end of the channel is needed for subcritical flow while both boundary conditions are at the upstream end for supercritical flow.

P.8.3 Suppose a Lax diffusive scheme with a constant space increment of $\Delta X = 400$ ft is used to perform the unsteady-flow calculations in a rectangular channel having a bottom width of 10 ft. The flow conditions calculated at a certain time stage are tabulated below. Determine the time increment to be used for the next time step calculations.

Node	z_b (ft)	Q (cfs)	h (ft)
1	0.90	150.00	6.11
2	0.80	141.36	5.85
3	0.70	132.76	5.60
4	0.60	124.19	5.37
5	0.50	115.63	5.15

P.8.4 Suppose that in Example 8.1 the channel is trapezoidal with $b = 8$ ft and $m = 2$. Evaluate $\partial A/\partial h$, $\partial S_f/\partial Q$ and $\partial S_f/\partial h$ at the node $i = 1$ if we still have $z_1 = 0.4$ ft, $Q_1 = 20$ cfs, and $h_1 = 2.27$ ft.

P.8.5 Develop a computer program in the computer language of your choice for Fread's algorithm discussed in Section 8.2.2.5. Verify the results of Example 8.2 using your program.

P.8.6 Would the time shift in the peak discharge in Example 8.5 be more pronounced if K were 4 hours? Verify your answer by recalculating the outflow hydrograph using $K = 4$ hours.

P.8.7 Tabulated below are the inflow and outflow hydrographs for a channel reach. Determine the Muskingum parameter K and x for the reach.

t (h)	I (m^3/s)	O (m^3/s)	t (h)	I (m^3/s)	O (m^3/s)
0	10	10.0	9	38.4	59.6
1	17.2	10.6	10	30.4	51.0
2	32.6	14.2	11	25.0	43.0
3	58.0	22.8	12	20.6	36.0
4	77.6	37.4	13	17.8	30.2
5	82.6	52.6	14	15.6	25.4
6	77.6	63.4	15	14.0	21.6
7	66.6	67.8	16	12.2	18.6
8	50.6	66.2	17	10.8	16.2

P.8.8 The channels AC and BC joining at the junction C flow into the channel CD. The Muskingum constants are $K=2$ h and $X=0.10$ for channels AC and CD, and $K=4$ h and $X=0.08$ for channel BC. Initially the flow is steady with a constant discharge of 10 m^3/s in CD. The upstream inflow hydrographs for AC and BC are identical to the inflow hydrograph in Example 8.5. Determine the peak flow rate at the downstream end of channel CD.

P.8.9 Redo Example 8.7 by dividing the channel length into two reaches of equal length. Compare your results to that of Example 8.7. Are the results sensitive to the reach length used in the calculation?

P.8.10 The sides of a triangular channel are sloped at 5H : 1V. The channel has a bottom slope of 0.001 and a Manning roughness factor of 0.05. The length of the channel is 1800 ft. Determine the constants e and m (Equation 8.114) for this channel.

P.8.11 Route the inflow hydrograph tabulated below through the channel discussed in Problem P.8.10. Assume that the flow is initially steady at 10 cfs.

t (h)	I (cfs)
0	10
0.25	16
0.50	31
0.75	50
1.00	58
1.25	60
1.50	54
1.75	42
2.00	32
2.25	25
2.50	20
2.75	17
3.00	15
3.25	13
3.50	12
3.75	11
4.00	10

REFERENCES

Akan, A. O. and Yen, B. C. (1977). A non-linear diffusion-wave model for unsteady open channel flow. In: *Proceedings of the 17th IAHR Congress, Baden-Baden, Germany*, Vol. 2, pp. 181–190. International Association for Hydraulic Research, Delft, The Netherlands.

Akan, A. O. and Yen, B. C. (1981). Diffusion-wave flood routing in channel networks. *Journal of the Hydraulics Division, ASCE*, **107(HY6)**, 719–732.

Aldama, A. A. (1990). Least-squares parameter estimation for Muskingum flood routing. *Journal of Hydraulic Engineering, ASCE*, **116(4)**, 580–586.

Amein, M. (1968). An implicit method for numerical flood routing. *Water Resources Research*, **2**, 123–130.

Baltzler, R. A. and Lai, C. (1968). Computer simulation of unsteady flows in waterways. *Journal of the Hydraulics Division, ASCE*, **94(HY4)**, 1083–1117.

Bettess, R. and Price, R. K. (1976). *Comparison of Numerical Methods for Routing Flow along a Pipe*. Report No. IT162, Hydraulics Research Station, Wallingford.

Chaudhry, M. H. (1993). *Open-Channel Flow*. Prentice Hall, Englewood Cliffs, NJ.

Cunge, J. A. (1969). On the subject of a flood propagation computation method (Muskingum method). *Journal of Hydraulic Research*, **7(2),** 205–230.

Currey, D. L. (1998). A two dimensional distributed hydrological model for infiltrating watersheds with channel networks. MS thesis, Department of Civil and Environmental Engineering, Old Dominion University, Norfolk, VA.

Franz, D. D. and Melching, C. S. (1997). *Full Equations (FEQ) Model for the Solution of the Full, Dynamic Equations of Motion for One-Dimensional Unsteady Flow in Open Channels and Through Control Structures*. Report 96-4240, US Geological Survey, Water-Resources Investigations, Washington, DC.

Fread, D. L. (1971). Discussion of 'Implicit flood routing in natural channels', by Amein and Fang. *Journal of the Hydraulics Division, ASCE*, **97(HY7),** 1157–1159.

Fread, D. L. (1974). *Numerical Properties of Implicit Four-point Finite Difference Equations of Unsteady Flow*. NOAA Technical Memorandum, NWS HYDRO-18, US National Weather Service, Silver Spring, MD.

Gunaratnam, D. and Perkins, F. E. (1970). *Numerical Solution of Unsteady Flows in Open Channels*. Report 127, R. M. Parsons Lab. for Water Resources and Hydrodynamics, MIT, Cambridge, MA.

Hromadka II, T. V. and Yen, C. C. (1986). A diffusion hydrodynamics model. *Advances in Water Resources*, **9**, 118–170.

Katapodes, N. D. (1982). On zero-inertia and kinematic waves. *Journal of the Hydraulics Division, ASCE*, **108(HY11)**, 1380–1387.

Lai, C. (1986). Numerical modeling of unsteady open channel flow. In: B. C. Yen (ed.), *Advances in Hydroscience*, Vol. 14, pp. 162–333. Academic Press, New York, NY.

Liggett, J. A. and Cunge, J. A. (1975). Numerical methods for the solution of the unsteady flow equations. In: Mahmood and Yevjevich (eds), *Unsteady Flow in Open Channels*, pp. 89–172. Water Resources Publications, Littleton, CO.

Ponce, V. M. and Simons, D. B. (1977). Shallow water propagation in open channel flow. *Journal of the Hydraulics Division, ASCE*, **103(HY12)**, 16–28.

Ponce, V. M. and Theurer, F. D. (1982). Accuracy criteria in diffusion routing. *Journal of the Hydraulics Division, ASCE*, **108(HY6)**, 747–757.

Ponce, V. M. and Yevjevich, V. (1978). Muskingum–Cunge method with variable parameters. *Journal of the Hydraulics Division, ASCE*, **104(HY12)**, 1663–1667.

Sevuk, A. S. (1973). Unsteady flow in sewer networks. PhD thesis, Department of Civil Engineering, University of Illinois, Urbana, IL.

Strelkoff, T. (1969). One-dimensional equations of open-channel flow. *Journal of the Hydraulics Division, ASCE*, **95(HY3),** 861–876.

Sturm, T. W. (2001). *Open Channel Hydraulics*. McGraw-Hill, New York, NY.

Yen, B. C. (1973). Open-channel flow equations revisited. *Journal of the Engineering Mechanics Division, ASCE*, **99(EM5),** 979–1009.

Yen, B. C. (1986). Hydraulics of sewers. In: B. C. Yen (ed.), *Advances in Hydroscience*, Vol. 14, pp. 1–122. Academic Press, New York, NY.

Index